REINFORCED AND PRESTRESSED
CONCRETE
THIRD EDITION

The third edition of *Reinforced and Prestressed Concrete* continues to be the most comprehensive text for engineering students, instructors and practising engineers. Theoretical and practical aspects of analysis and design are presented in a clear, easy-to-follow manner and are complemented by numerous illustrative and design examples to aid students' comprehension of complex concepts. The text is divided into two parts: the first addresses the analysis and design of reinforced concrete structures and the second covers topics in prestressed concrete.

This edition has been fully updated to reflect recent amendments and addenda to the Australian Standard for Concrete Structures AS 3600–2009 and allied standards. Two new chapters, covering T-beams, irregular-shaped sections and continuous beams, and strut-and-tie modelling have been added as discrete modules to enhance the progression of topics. Additional information is provided on fire resistance, detailing including cover, long-term deflection and design for torsion. An expanded collection of end-of-chapter tutorial problems consolidate student learning and develop problem-solving skills.

With its thorough coverage of fundamental concepts and abundance of practical examples, *Reinforced and Prestressed Concrete* remains an indispensable resource for students and engineers continuing their professional development.

Yew-Chaye Loo AM, PhD, FIEAust, FICE, FIStructE, CPEng, NPER, CEng was the Foundation Professor of Civil Engineering at Griffith University in Queensland. A former Dean of Engineering and Information Technology and Internationalisation Director for Griffith Sciences Group, he is now a Professor Emeritus at the University.

Sanaul Huq Chowdhury, PhD, MIEAust, MACI, CPEng, NPER is a senior lecturer in Structural Engineering and Mechanics at the School of Engineering and Built Environment, Griffith University. He is a Chartered Professional Engineer and member of both the Civil and Structural Colleges of Engineers Australia.

REINFORCED AND PRESTRESSED
CONCRETE
THIRD EDITION

Yew-Chaye Loo AM
Sanaul Huq Chowdhury

CAMBRIDGE
UNIVERSITY PRESS

CAMBRIDGE
UNIVERSITY PRESS

University Printing House, Cambridge CB2 8BS, United Kingdom

One Liberty Plaza, 20th Floor, New York, NY 10006, USA

477 Williamstown Road, Port Melbourne, VIC 3207, Australia

314–321, 3rd Floor, Plot 3, Splendor Forum, Jasola District Centre, New Delhi – 110025, India

79 Anson Road, #06–04/06, Singapore 079906

Cambridge University Press is part of the University of Cambridge.

It furthers the University's mission by disseminating knowledge in the pursuit of education, learning and research at the highest international levels of excellence.

www.cambridge.org
Information on this title: www.cambridge.org/9781108405645

First edition © Yew-Chaye Loo & Sanaul Huq Chowdhury 2010
Second and third editions © Cambridge University Press 2013, 2018

This publication is copyright. Subject to statutory exception
and to the provisions of relevant collective licensing agreements,
no reproduction of any part may take place without the written
permission of Cambridge University Press.

First published 2010
Second edition 2013
Third edition 2018

Cover designed by Kerry Cooke, eggplant communications
Typeset by SPi Global
Printed in China by C & C Offset Printing Co. Ltd, June 2018

A catalogue record for this publication is available from the British Library

A catalogue record for this book is available from the National Library of Australia

ISBN 978-1-108-40564-5 Paperback

Additional resources for this publication at www.cambridge.edu.au/academic/concrete

Reproduction and communication for educational purposes
The Australian *Copyright Act 1968* (the Act) allows a maximum of
one chapter or 10% of the pages of this work, whichever is the greater,
to be reproduced and/or communicated by any educational institution
for its educational purposes provided that the educational institution
(or the body that administers it) has given a remuneration notice to
Copyright Agency Limited (CAL) under the Act.

For details of the CAL licence for educational institutions contact:

Copyright Agency Limited
Level 15, 233 Castlereagh Street
Sydney NSW 2000
Telephone: (02) 9394 7600
Facsimile: (02) 9394 7601
E-mail: info@copyright.com.au

Cambridge University Press has no responsibility for the persistence or
accuracy of urls for external or third-party internet websites referred to in
this publication and does not guarantee that any content on such websites is,
or will remain, accurate or appropriate.

In memory of our parents
Loo Khai Kee (1900–1989)
Lau Ching (1902–1961)
Shamsul Haque Chowdhury (1920–1999)
Syeda Nurun Nahar Chowdhury (1930–2007)

CONTENTS

Preface to the first edition	page xix
Preface to the second edition	xxi
Preface to the third edition	xxiii
Acknowledgements	xxiv
Notation	xxv
Acronyms and abbreviations	xxxviii

Part 1 Reinforced concrete 1

1 Introduction 3
  1.1 Historical notes 3
  1.2 Design requirements 4
  1.3 Loads and load combinations 5
    1.3.1 Strength design 5
    1.3.2 Serviceability design 7
    1.3.3 Application 7
  1.4 Concrete cover and reinforcement spacing 7
    1.4.1 Cover 7
    1.4.2 Spacing 9
  1.5 Summary 12

2 Design properties of materials 13
  2.1 Introduction 13
  2.2 Concrete 13
    2.2.1 Characteristic strengths 13
    2.2.2 Standard strength grades 14
    2.2.3 Initial modulus and other constants 14
  2.3 Steel 15
  2.4 Unit weight 18
  2.5 Summary 18

3 Analysis and design of rectangular beams for bending 19
  3.1 Introduction 19
  3.2 Definitions 19
    3.2.1 Analysis 19
    3.2.2 Design 19
    3.2.3 Ultimate strength method 19
  3.3 Ultimate strength theory 20
    3.3.1 Basic assumptions 20
    3.3.2 Actual and equivalent stress blocks 20

	3.4	Ultimate strength of a singly reinforced rectangular section	22
		3.4.1 Tension, compression and balanced failure	22
		3.4.2 Balanced steel ratio	24
		3.4.3 Moment equation for tension failure (under-reinforced sections)	25
		3.4.4 Moment equation for compression failure (over-reinforced sections)	26
		3.4.5 Effective moment capacity	27
		3.4.6 Illustrative example for ultimate strength of a singly reinforced rectangular section	28
		3.4.7 Spread of reinforcement	31
	3.5	Design of singly reinforced rectangular sections	35
		3.5.1 Free design	35
		3.5.2 Restricted design	36
		3.5.3 Design example	37
	3.6	Doubly reinforced rectangular sections	40
		3.6.1 Criteria for yielding of A_{sc} at failure	40
		3.6.2 Analysis formulas	41
		3.6.3 Illustrative examples	44
		3.6.4 Other cases	46
		3.6.5 Summary	50
	3.7	Design of doubly reinforced sections	51
		3.7.1 Design procedure	51
		3.7.2 Illustrative example	53
	3.8	Summary	55
	3.9	Problems	56
4	**T-beams and irregular-shaped sections**		**61**
	4.1	Introduction	61
	4.2	T-beams and other flanged sections	61
		4.2.1 Definitions	61
		4.2.2 Effective flange width	61
		4.2.3 Criteria for T-beams	66
		4.2.4 Analysis	67
		4.2.5 Design procedure	68
		4.2.6 Doubly reinforced T-sections	70
		4.2.7 Illustrative examples	71
	4.3	Nonstandard sections	74
		4.3.1 Analysis	74
		4.3.2 Illustrative examples	76

	4.4	Continuous beams	81
	4.5	Detailing and cover	81
	4.6	Summary	81
	4.7	Problems	82

5 Deflection of beams and crack control — 94

	5.1	Introduction	94
	5.2	Deflection formulas, effective span and deflection limits	95
		5.2.1 Formulas	95
		5.2.2 Effective span	96
		5.2.3 Limits	97
	5.3	Short-term (immediate) deflection	98
		5.3.1 Effects of cracking	98
		5.3.2 Branson's effective moment of inertia	99
		5.3.3 Load combinations	100
		5.3.4 Illustrative example	101
		5.3.5 Cantilever and continuous beams	102
	5.4	Long-term deflection	104
		5.4.1 General remarks	104
		5.4.2 The multiplier method	104
		5.4.3 Illustrative example	105
	5.5	Simplified procedure	106
		5.5.1 Minimum effective depth approach	106
		5.5.2 ACI code recommendation	107
	5.6	Total deflection under repeated loading	107
		5.6.1 Formulas	108
		5.6.2 Illustrative example	110
	5.7	Crack control	111
		5.7.1 General remarks	111
		5.7.2 Standard provisions	112
		5.7.3 Crack-width formulas and comparison of performances	114
	5.8	Summary	118
	5.9	Problems	118

6 Ultimate strength design for shear — 123

	6.1	Introduction	123
	6.2	Transverse shear stress and shear failure	123
		6.2.1 Principal stresses	123
		6.2.2 Typical crack patterns and failure modes	125
		6.2.3 Mechanism of shear resistance	127
		6.2.4 Shear reinforcement	127
	6.3	Transverse shear design	128

		6.3.1 Definitions	128
		6.3.2 Design shear force and the capacity reduction factor	128
		6.3.3 Maximum capacity	129
		6.3.4 Shear strength of beams without shear reinforcement	130
		6.3.5 Shear strength checks and minimum reinforcement	131
		6.3.6 Design of shear reinforcement	132
		6.3.7 Detailing	134
		6.3.8 Design example	135
	6.4	Longitudinal shear	138
		6.4.1 Shear planes	138
		6.4.2 Design shear stress	139
		6.4.3 Shear stress capacity	140
		6.4.4 Shear plane reinforcement and detailing	141
		6.4.5 Design example	141
	6.5	Summary	143
	6.6	Problems	144
7	**Ultimate strength design for torsion**		**155**
	7.1	Introduction	155
		7.1.1 Origin and nature of torsion	155
		7.1.2 Torsional reinforcement	155
		7.1.3 Transverse reinforcement area and capacity reduction factor	157
	7.2	Maximum torsion	157
	7.3	Checks for reinforcement requirements	159
	7.4	Design for torsional reinforcement	159
		7.4.1 Design formula	159
		7.4.2 Design procedure	160
		7.4.3 Detailing	161
		7.4.4 Design example	162
	7.5	Summary	167
	7.6	Problems	167
8	**Bond and stress development**		**171**
	8.1	Introduction	171
		8.1.1 General remarks	171
		8.1.2 Anchorage bond and development length	171
		8.1.3 Mechanism of bond resistance	172
		8.1.4 Effects of bar position	173
	8.2	Design formulas for stress development	173
		8.2.1 Basic and refined development lengths for a bar in tension	173
		8.2.2 Standard hooks and cog	176

	8.2.3	Deformed and plain bars in compression	177
	8.2.4	Bundled bars	178
8.3	Splicing of reinforcement		178
	8.3.1	Bars in tension	178
	8.3.2	Bars in compression	179
	8.3.3	Bundled bars	179
	8.3.4	Mesh in tension	179
8.4	Illustrative examples		180
	8.4.1	Example 1	180
	8.4.2	Example 2	181
8.5	Summary		182
8.6	Problems		182

9 Slabs 185

9.1	Introduction		185
	9.1.1	One-way slabs	185
	9.1.2	Two-way slabs	186
	9.1.3	Effects of concentrated load	188
	9.1.4	Moment redistribution	189
9.2	One-way slabs		190
	9.2.1	Simplified method of analysis	190
	9.2.2	Reinforcement requirements	192
	9.2.3	Deflection check	193
	9.2.4	Design example	194
9.3	Two-way slabs supported on four sides		201
	9.3.1	Simplified method of analysis	201
	9.3.2	Reinforcement requirements for bending	206
	9.3.3	Corner reinforcement	208
	9.3.4	Deflection check	209
	9.3.5	Crack control	210
	9.3.6	Design example	210
9.4	Multispan two-way slabs		216
	9.4.1	General remarks	216
	9.4.2	Design strips	216
	9.4.3	Limitations of the simplified method of analysis	218
	9.4.4	Total moment and its distribution	218
	9.4.5	Punching shear	220
	9.4.6	Reinforcement requirements	221
	9.4.7	Shrinkage and temperature steel	222
9.5	The idealised frame approach		222
	9.5.1	The idealised frame	222

		9.5.2 Structural analysis	223
		9.5.3 Distribution of moments	225
	9.6	Punching shear design	226
		9.6.1 Geometry and definitions	226
		9.6.2 Drop panel and shear head	227
		9.6.3 The basic strength	227
		9.6.4 The ultimate strength	228
		9.6.5 Minimum effective slab thickness	228
		9.6.6 Design of torsion strips	229
		9.6.7 Design of spandrel beams	231
		9.6.8 Detailing of reinforcement	232
		9.6.9 Summary	232
		9.6.10 Illustrative example	233
		9.6.11 Semi-empirical approach and layered finite element method	235
	9.7	Slab design for multistorey flat plate structures	236
		9.7.1 Details and idealisation of a three-storey building	236
		9.7.2 Loading details	237
		9.7.3 Load combinations	238
		9.7.4 Material and other specifications	238
		9.7.5 Structural analysis and moment envelopes	239
		9.7.6 Design strips and design moments	242
		9.7.7 Design of column and middle strips	243
		9.7.8 Serviceability check – total deflection	248
		9.7.9 Reinforcement detailing and layout	249
		9.7.10 Comments	249
	9.8	Summary	250
	9.9	Problems	251
10	**Columns**		**255**
	10.1	Introduction	255
	10.2	Centrally loaded columns	257
	10.3	Columns in uniaxial bending	258
		10.3.1 Strength formulas	258
		10.3.2 Tension, compression, decompression and balanced failure	260
		10.3.3 Interaction diagram	262
		10.3.4 Approximate analysis of columns failing in compression	267
		10.3.5 Strengths between decompression and squash points	268

10.4	Analysis of columns with an arbitrary cross-section	269
	10.4.1 Iterative approach	269
	10.4.2 Illustrative example of iterative approach	271
	10.4.3 Semi-graphical method	275
	10.4.4 Illustrative example of semi-graphical method	276
10.5	Capacity reduction factor	279
10.6	Preliminary design procedure	280
	10.6.1 Design steps	280
	10.6.2 Illustrative example	281
10.7	Short column requirements	281
10.8	Moment magnifiers for slender columns	282
	10.8.1 Braced columns	283
	10.8.2 Unbraced columns	284
10.9	Biaxial bending effects	285
10.10	Reinforcement requirements	287
	10.10.1 Limitations and bundled bars	287
	10.10.2 Lateral restraint and core confinement	287
	10.10.3 Recommendations	288
10.11	Comments	288
10.12	Summary	290
10.13	Problems	290

11 Walls — 296

11.1	Introduction	296
11.2	Standard provisions	297
11.3	Walls under vertical loading only	298
	11.3.1 Simplified method	298
	11.3.2 American Concrete Institute code provision	299
	11.3.3 New design formula	300
	11.3.4 Alternative column design method	301
11.4	Walls subjected to in-plane horizontal forces	302
	11.4.1 General requirements	302
	11.4.2 Design strength in shear	302
	11.4.3 American Concrete Institute recommendations	303
11.5	Reinforcement requirements	303
11.6	Illustrative examples	304
	11.6.1 Example 1 – load-bearing wall	304
	11.6.2 Example 2 – tilt-up panel	306
	11.6.3 Example 3 – the new strength formula	307
	11.6.4 Example 4 – design shear strength	308
11.7	Summary	309
11.8	Problems	309

12 Footings, pile caps and retaining walls — 311
  12.1 Introduction — 311
  12.2 Wall footings — 311
   12.2.1 General remarks — 311
   12.2.2 Eccentric loading — 314
   12.2.3 Concentric loading — 317
   12.2.4 Asymmetrical footings — 318
   12.2.5 Design example — 318
  12.3 Column footings — 325
   12.3.1 General remarks — 325
   12.3.2 Centrally loaded square footings — 325
   12.3.3 Eccentric loading — 327
   12.3.4 Multiple columns — 330
   12.3.5 Biaxial bending — 331
   12.3.6 Reinforcement requirements — 332
   12.3.7 Design example — 332
  12.4 Pile caps — 339
   12.4.1 Concentric column loading — 339
   12.4.2 Biaxial bending — 344
  12.5 Retaining walls — 345
   12.5.1 General remarks — 345
   12.5.2 Stability considerations — 348
   12.5.3 Active earth pressure — 353
   12.5.4 Design subsoil pressures — 355
   12.5.5 Design moments and shear forces — 357
   12.5.6 Load combinations — 359
   12.5.7 Illustrative example — 359
  12.6 Summary — 373
  12.7 Problems — 373

13 Strut-and-tie modelling of concrete structures — 377
  13.1 Introduction — 377
  13.2 Fundamentals — 378
  13.3 Struts, ties and nodes — 380
  13.4 Common types of strut-and-tie models — 383
  13.5 Developments — 384
  13.6 Specifications in AS 3600 — 387
   13.6.1 Concrete struts — 388
   13.6.2 Steel ties — 391
   13.6.3 Nodes — 392

		13.6.4 Additional specifications	392
		13.6.5 Illustrative example	392
	13.7	Summary	400

Part 2 Prestressed concrete — 401

14 Introduction to prestressed concrete — 403

14.1	Introduction	403
14.2	Non-engineering examples of prestressing	404
	14.2.1 Wooden barrel	404
	14.2.2 Stack of books	404
14.3	Principle of superposition	405
14.4	Types of prestressing	407
	14.4.1 Pretensioning	408
	14.4.2 Post-tensioning	408
14.5	Partial prestressing	409
14.6	Tensile strength of tendons and cables	411
14.7	Australian Standard precast prestressed concrete bridge girder sections	413
14.8	Summary	413

15 Critical stress state analysis of beams — 414

15.1	Introduction	414
15.2	Notation	414
15.3	Loss of prestress	416
	15.3.1 Standard provisions	416
	15.3.2 Examples of prestress loss due to elastic shortening of concrete	417
	15.3.3 Effective prestress coefficient	420
	15.3.4 Stress equations at transfer and after loss	420
15.4	Permissible stresses c and c_t	421
15.5	Maximum and minimum external moments	422
15.6	Case A and Case B prestressing	425
	15.6.1 Fundamentals	425
	15.6.2 Applying Case A and Case B	426
15.7	Critical stress state (CSS) equations	427
	15.7.1 Case A prestressing	427
	15.7.2 Case B prestressing	429
	15.7.3 Summary of Case A and Case B equations	430
15.8	Application of CSS equations	431
15.9	Summary	433
15.10	Problems	434

16	**Critical stress state design of beams**		**437**
	16.1 Introduction		437
	16.2 Formulas and procedures – Case A		438
		16.2.1 Elastic section moduli	438
		16.2.2 Magnel's plot for Case A	439
		16.2.3 Design steps	440
	16.3 Formulas and procedures – Case B		441
		16.3.1 Elastic section moduli	441
		16.3.2 Magnel's plot for Case B	442
		16.3.3 Design steps	442
	16.4 Design examples		442
		16.4.1 Simply supported beam	442
		16.4.2 Simple beam with overhang	446
		16.4.3 Cantilever beam	451
	16.5 Summary		456
	16.6 Problems		457
17	**Ultimate strength analysis of beams**		**458**
	17.1 Introduction		458
	17.2 Cracking moment (M_{cr})		459
		17.2.1 Formula	459
		17.2.2 Illustrative example	459
	17.3 Ultimate moment (M_u) for partially prestressed sections		460
		17.3.1 General equations	460
		17.3.2 Sections with bonded tendons	461
		17.3.3 Sections with unbonded tendons	462
	17.4 Ductility requirements – reduced ultimate moment equations		463
	17.5 Design procedure		464
		17.5.1 Recommended steps	464
		17.5.2 Illustrative example	464
	17.6 Nonrectangular sections		466
		17.6.1 Ultimate moment equations	466
		17.6.2 Illustrative example	467
	17.7 Summary		469
	17.8 Problems		469
18	**End blocks for prestressing anchorages**		**473**
	18.1 Introduction		473
	18.2 Pretensioned beams		473
	18.3 Post-tensioned beams		475
		18.3.1 Bursting stress	475
		18.3.2 Spalling stress	476

		18.3.3 Bearing stress	477
		18.3.4 End blocks	477
	18.4	End-block design	477
		18.4.1 Geometry	477
		18.4.2 Symmetrical prisms and design bursting forces	477
		18.4.3 Design spalling force	478
		18.4.4 Design for bearing stress	480
	18.5	Reinforcement and distribution	480
	18.6	Crack control	481
	18.7	Summary	482

Appendix A	*Elastic neutral axis*	*483*
Appendix B	*Critical shear perimeter*	*485*
Appendix C	*Australian Standard precast prestressed concrete bridge girder sections*	*487*
References		*489*
Index		*496*

PREFACE TO THE FIRST EDITION

Most of the contents of this book were originally developed in the late 1980s at the University of Wollongong, New South Wales. The contents were targeted towards third-year courses in reinforced and prestressed concrete structures. The book was believed useful for both students learning the subjects and practising engineers wishing to apply with confidence the then newly published Australian Standard AS 3600–1988. In 1995 and following the publication of AS 3600–1994, the contents were updated at Griffith University (Gold Coast campus) and used as the learning and teaching material for the third-year course, 'Concrete structures' (which also covers prestressed concrete). In 2002, further revisions were made to include the technical advances of AS 3600–2001. Some of the book's more advanced topics were used for part of the Griffith University postgraduate course, 'Advanced reinforced concrete'.

In anticipation of the publication of the current version of AS 3600, which was scheduled for 2007, a major rewrite began early that year to expand on the contents and present them in two parts. The effort continued into 2009, introducing in Part 1 'Reinforced concrete', inter alia, the new chapters on walls, as well as on footings, pile caps and retaining walls, plus an appendix on strut-and-tie modelling. In addition, a new Part 2 had been written, which covered five new chapters on prestressed concrete. The entire manuscript was then thoroughly reviewed and revised as appropriate following the publication of AS 3600–2009 in late December 2009.

In line with the original aims, the book contains extensive fundamental materials for learning and teaching purposes. It is also useful for practising engineers, especially those wishing to have a full grasp of the new AS 3600–2009. This is important, as the 2009 contents have been updated and expanded significantly and, for the first time, provisions for concrete compressive strength up to 100 MPa are included. The increase in concrete strength has resulted in major changes to many of the analysis and design equations.

Part 1 contains 11 chapters. An introduction to the design requirements and load combinations is given in Chapter 1, and the properties of and specifications for concrete and reinforcing steel are discussed in Chapter 2. Chapter 3 presents, in detail, the bending analysis and design of rectangular beams, T-beams and other flanged sections. Some significant attention is given to doubly reinforced members. Deflection and crack control are considered in Chapter 4, which also features a section on the effects of repeated loading. Also presented is a unified crack-width formula for reinforced and prestressed beams.

Chapter 5 details transverse and longitudinal shear design, and Chapter 6 presents the design procedure for torsion. Bond and stress development are treated in Chapter 7, and Chapter 8 covers most of the practical aspects of slab analysis and design. It also includes a separate section describing a design exercise that features the complete (multiple-load case) analysis of a three-storey flat plate structure, as well as the detailed design of typical floor panels.

Chapter 9 deals with the analysis and design of columns, including the treatment of arbitrary cross-sections using numerical and semi-graphical methods. The new Chapter 10 examines the use of relevant strength design formulas for walls subjected to vertical axial loads, as well as under combined axial and horizontal in-plane shear forces. This is followed by the new Chapter 11, with an extensive and in-depth coverage of the design of wall and column footings, pile caps and retaining walls.

Part 2 contains five chapters. Prestressed concrete fundamentals, including pre- and post-tensioning processes, are introduced in Chapter 12. Chapters 13 and 14 cover the critical stress state approach to the analysis and design of fully prestressed concrete flexural members, which ensures a crack-free and overstress-free service life for those members. The ultimate strength analysis and design of fully and partially prestressed beams are dealt with in Chapter 15. The final chapter (Chapter 16) presents the design of end blocks for prestressing anchorages.

Appendices A and B present the formulas for computing the elastic neutral axes required in deflection analysis, and those for obtaining various critical punching shear perimeters used in flat plate design, respectively. The development of an integrated personal computer program package for the design of multistorey flat plate systems is described in Appendix C. This may be useful to the reader who has an interest in computer applications. Appendix D highlights the essence of the strut-and-tie modelling approach; it also reviews the advances made in this topic in recent years. Finally, the Australian Standard precast I-girders and super T-girders for prestressed concrete bridge construction are detailed in Appendix E.

In all of the chapters and appendixes, the major symbols used in AS 3600–2009 are adopted. Unless otherwise specified, the term 'Standard' refers to AS 3600–2009 and all the clause numbers referred to in the text are those from AS 3600–2009. For ease of reading, a full notation is provided as well as a subject index.

For the student learning the subject of reinforced and prestressed concrete, sufficient fundamentals and background information are provided in each of the chapters. Most of the analysis and design equations are derived and presented in an explicit form. The practitioner of concrete engineering should find these equations easy to apply in their work. Illustrative and design examples are given throughout to assist the reader with the learning process and with their interpretation of the provisions of the Standard. For the convenience of students and teachers alike, a collection of tutorial problems is included at the end of each relevant chapter. To assist teachers using the book for concrete engineering-related courses, an electronic solution manual is available and posted on a secure website (maintained and continuously updated by the authors).

The book is suitable for use in a university degree course that covers the analysis and design of reinforced and prestressed concrete structures. Selected topics may also be adopted in a postgraduate course in concrete engineering. The practising engineer wanting to apply the Australian Standard with confidence will also find the material helpful. In practice, the book can also serve as a reference manual for and user guide to AS 3600–2009.

Yew-Chaye Loo
Sanaul Huq Chowdhury

PREFACE TO THE SECOND EDITION

The second edition retains all of the features of the original book on the explicit and implicit advice of our peers via the mandatory Cambridge University Press review process. To limit the volume size, the old Appendix C, 'Development of an integrated package for design of reinforced concrete flat plates on personal computer', has been removed, being of diminishing practical importance. To enhance the contents, new and important materials are added, some of which were also on the advice of the reviewers:

- updated tables and figures to reflect the amendments and addenda to AS 3600–2009 promulgated by Standards Australia International since its first publication
- additional information on fire design, detailing and cover, long-term deflection, as well as aspects of partially prestressed concrete design; and
- an expanded Appendix on strut-and-tie modelling, encompassing the latest publications on the topic plus a numerical example.

Just as significant, another 37 tutorial problems have been added to the various chapters of the book. This makes a total of 108.

YCL
SHC

PREFACE TO THE THIRD EDITION

This latest edition of the book is written in response to the domestic and international market needs as well as on the request of the publisher, Cambridge University Press. It retains the features of the original book but to enhance readability, every chapter now begins with an introduction and concludes with a summary, followed if applicable by an enlarged set of tutorial problems. In all 20 new problems are added making a grand total of 128.

The original Chapter 3 on bending analysis and design was judged by some peer-reviewers as being unduly lengthy. It is now split into two consecutive chapters. Based on a previous appendix, a new Chapter 13 on strut-and-tie modelling is added incorporating the latest material on the subject. Since the last edition, many significant amendments have been made and addenda added to the Australian Standard AS 3600–2009. These are incorporated in the current edition in the form of new and updated tables and figures. Also included are the latest information and recommendations on fire resistance, detailing including cover, long-term deflection, as well as torsion design.

YCL
SHC

ACKNOWLEDGEMENTS

Peer review was a mandatory process prior to the publication of this latest edition of the book. It was patiently and professionally handled initially by Ms Emily Thomas and then by Ms Tanya Bastrakova, Development Editor (Academic Publishing) at Cambridge University Press, Australia. Both were under the direction of Ms Lucy Russell, Cambridge's Senior Commissioning Editor, Academic. Under the direction of Ms Jodie Fitzsimmons, Managing Editor, Academic, Cambridge University Press, the final copy editing was done by Ms Karen Jayne, the responsible editor. The authors are grateful for their kind efforts, without which the book would have taken longer to appear. To the 14 unnamed reviewers from academia and the profession, the authors wish to express their sincere thanks and appreciation for their support and constructive suggestions. Their contributions have helped enrich the contents of the book.

The contents of the Australian Standard AS 3600–2009 and allied Standards are extensively quoted in this book. The authors are grateful to SAI Global for the permission to reproduce some of the updated design data in tabulated form and in figures.

NOTATION

A	gross cross-sectional area of a member
A_b	cross-sectional area of a reinforcing bar
A_c	cross-sectional area of concrete only in a reinforced concrete section
A_{cc}	area of additional reinforcement provided for crack control at side faces of beams with $D > 750$ mm
A_g	gross cross-sectional area of a member
A_m	an area enclosed by the median lines of the walls of a single cell
A_p	cross-sectional area of prestressing steel
A_{pt}	cross-sectional area of the tendons in that zone, which will be tensile under ultimate load conditions
$A_{pt.ef}$	effective cross-sectional area of tendons
A_s	cross-sectional area of reinforcement
A_{sc}	cross-sectional area of compression reinforcement
A_{sf}	area of fully anchored shear reinforcement crossing the interface (shear plane)
A_{st}	cross-sectional area of tension reinforcement; the cross-sectional area of reinforcement in the zone that would be in tension under the design loads other than prestressing or axial loads
$A_{st.min}$	minimum cross-sectional area of reinforcement permitted in a beam in tension, or in a critical tensile zone of a beam or slab in flexure
A_{sv}	cross-sectional area of shear reinforcement
$A_{sv.min}$	cross-sectional area of minimum shear reinforcement
A_{sw}	cross-sectional area of the bar forming a closed tie
A_t	area of a polygon with vertices at the centre of longitudinal bars at the corners of the cross-section
A_{tr}	cross-sectional area of a transverse bar along the development length

$A_{tr.min}$	cross-sectional area of the minimum transverse reinforcement along the development length
A_1	a bearing area
A_2	largest area of the supporting surface that is geometrically similar to and concentric with A_1
a	distance; or the maximum nominal size of the aggregate; or depth of equivalent concrete stress block from the extreme compression fibre; or dimension of the critical shear perimeter measured parallel to the direction of M^*_v
a_s	length of a span support
a_v	distance from the section at which shear is being considered to the face of the nearest support
b	width of a cross-section
b^*	overall width of the column head or drop panel, as applicable
b_c	width of the compression strut; or the smaller cross-sectional dimension of a rectangular column
b_{ef}	effective width of a compression face or flange of a member
b_f	width of the shear interface; or width of a footing
b_w	width of the web; or the minimum thickness of the wall of a hollow section
C	force resulting from compressive stresses
c	cover to reinforcing steel or tendons; or the permissible compressive stress
c_d	the smaller of the concrete covers to the deformed bar or half the clear distance to the next parallel
c_t	permissible tensile stress
D	overall depth of a cross-section in the plane of bending
D^*	overall depth of the column head or drop panel, as applicable
D_b	overall depth of a spandrel beam
D_c	diameter of circular column or the smaller dimension of rectangular column
D_f	greater dimension or length of a footing

D_s	overall depth of a slab or drop panel
d	effective depth of a cross-section
d_b	nominal diameter of a bar, wire or tendon
d_c	depth of a compression strut; or the distance from the extreme compressive fibre of the concrete to the centroid of compressive reinforcement
d_o	distance from the extreme compression fibre of the concrete to the centroid of the outermost layer of tensile reinforcement or tendons but for prestressed concrete members not less than $0.8D$
d_{om}	mean value of the shear effective depth (d_o) averaged around the critical shear perimeter
d_p	distance from the extreme compressive fibre of the concrete to the centroid of the tendons in that zone, which will be tensile under ultimate strength conditions
d_{pc}	distance of the plastic centre of a column from the extreme compressive fibre
E_c	mean value of the modulus of elasticity of concrete at 28 days
E_d	design action effect
E_p	modulus of elasticity of tendons
E_s	modulus of elasticity of reinforcement
E_u	action effect due to ultimate earthquake load
e	eccentricity of axial force from a centroidal axis; or the base of Napierian logarithms
e_B	eccentricity of prestressing tendons or cables
e_a	additional eccentricity
F_{BF}	horizontal pressure resultant for a retaining wall due to backfills
F_{SL}	horizontal pressure resultant for a retaining wall due to surcharge load
F^*_c	absolute value of the design force in the compressive zone due to flexure
F_d	uniformly distributed design load, factored for strength or serviceability as appropriate

F_{def}	effective design service load per unit length or area, used in serviceability design
F_r	friction between retaining wall or footing base and soil
f	bending stress
f_b	bearing stress
f_c	concrete strength under working stress condition
f_{cB}	extreme bottom fibre stress
$f_{c.cal}$	calculated compressive strength of concrete in a compression strut
f_{cm}	mean value of cylinder strength
f_{cmi}	mean value of the in situ compressive strength of concrete at the relevant age
f_{cp}	compressive strength of concrete at transfer
f_{cs}	maximum shrinkage-induced tensile stress on the uncracked section at the extreme fibre at which cracking occurs
f_{cT}	extreme top fibre stress
f_{cv}	concrete shear strength
f_{heel}	subsoil pressure at the heel of a retaining wall
f_p	tensile strength of tendons
f_{py}	yield strength of tendons
f_s	maximum tensile stress permitted in the reinforcement immediately after the formation of a crack
f_{sc}	stress in the compression steel
f_{si}	serviceability limit stress in reinforcement
f_{sy}	yield strength of reinforcing steel
$f_{sy.f}$	yield strength of reinforcement used as fitments
f_{toe}	subsoil pressure at the toe of a retaining wall
f'_c	characteristic compressive (cylinder) strength of concrete at 28 days
f'_{cp}	minimum compressive strength of concrete at transfer
f'_{ct}	characteristic principal tensile strength of concrete

$f'_{ct.f}$	characteristic flexural tensile strength of concrete
G	action effect due to dead load
g	dead load, usually per unit length or area
g_p	permanent distributed load normal to the shear interface per unit length (N/mm)
H	height of a retaining wall; or the prestressing force
H_w	overall height of a wall
H_{we}	effective height of a wall
H_{wu}	unsupported height of a wall
I	second moment of area of the uncracked concrete cross-section about the centroidal axis
I_b	second moment of area of a beam
I_c	second moment of area of a column
I_{cr}	second moment of area of a cracked section with the reinforcement transformed to an equivalent area of concrete
I_{ef}	effective second moment of area
$I_{ef.max}$	maximum effective second moment of area
I_f	second moment of area of a flexural member
I_g	second moment of area, of the gross concrete cross-section about the centroidal axis
I_{rep}	equivalent moment of inertia at the Tth loading cycle
J_t	torsional modulus
K	factor that accounts for the position of the bars being anchored with respect to the transverse reinforcement
K_a	active earth pressure coefficient
K_p	passive earth pressure coefficient
k	coefficient, ratio or factor used with and without numerical subscripts
k_A, k_B, k_C	factors for calculating ϕ for backfill materials which are functions of the angularity, grading and density of the backfill particles

xxx NOTATION

k_R	amplification factor
k_{co}	cohesion coefficient
k_{cs}	factor used in serviceability design to take account of the long-term effects of creep and shrinkage
k_r	ratio
k_u	neutral axis parameter, being the ratio, at ultimate strength and under any combination of bending and compression, of the depth to the neutral axis from the extreme compressive fibre, to d
k_{uB}	ratio, at ultimate strength, of the depth to the neutral axis from the extreme compressive fibre, to d, at balanced failure condition
k_{uo}	ratio, at ultimate strength, without axial force of the depth to the neutral axis from the extreme compressive fibre, to d_o
L	centre-to-centre distance between the supports of a flexural member
L_e	effective length of a column
L_{ef}	effective span of a member, taken as the lesser of $(L_n + D)$ and L for a beam or slab; or as $(L_n + D/2)$ for a cantilever
L_l	distance between centres of lateral restraints
L_n	length of clear span in the direction in which moments are being determined, measured face-to-face of supporting beams, columns or walls, or for a cantilever, the clear projection
L_o	distance between the points of zero bending moment in a span
L'_o	length of a span
L_p	development length for pretensioned tendons
L_{pt}	transmission length for pretensioned tendons
L_{st}	development length of a bar for a tensile stress less than the yield stress
$L_{sy.c}$ ($L_{sy.t}$)	development length for compression (tension), being the length of embedment required to develop the yield strength of a bar in compression (tension)
$L_{sy.cb}$ ($L_{sy.tb}$)	basic design development length for compression (tension)

$L_{sy.t.lap}$	tensile lap length for either contact or non-contact splices
L_t	width of a design strip
L_u	unsupported length of a column, taken as the clear distance between the faces of members capable of providing lateral support to the column, where column capitals or haunches are present; L_u is measured to the lowest extremity of the capital or haunch
L_w	overall length of a wall
L_x	shorter effective span of a slab supported on four sides
L_y	longer effective span of a slab supported on four sides
l_b	basic development length
M'	effective or reliable moment capacity of a section (i.e. ϕM_u)
M^*	bending moment at a cross-section calculated using the design load (i.e. the design bending moment)
M^*_v	design bending moment to be transferred from a slab to a support
M^*_x, M^*_y	design bending moment in a column about the major and minor axes, respectively; or the positive design bending moment, at midspan in a slab, in the x and y direction, respectively
M_{cr}	bending moment causing cracking of the section with due consideration to prestress, restrained shrinkage and temperature stresses
M_g	moment due to sustained or dead load
M_o	total static moment in a span; or the decompression moment
M_q	live-load moment
M_s	moment due to service load
M_u	ultimate strength in bending at a cross-section of an eccentrically loaded compression member
$M_{u.dc}$	decompression moment
M_{uB}	particular ultimate strength in bending when $k_{uo} = 0.003/(0.003 + f_{sy}/E_s)$
M_{ud}	reduced ultimate strength in bending without axial force, at a cross-section

M_{uo}	ultimate strength in bending without axial force, at a cross-section
M_{ux}, M_{uy}	ultimate strength in bending about the major and minor axes, respectively, of a column under the design axial force N^*
M_y	yield moment
N^*	axial compressive or tensile force on a cross-section calculated using the design load (i.e. the design axial force)
N_c	buckling load used in column design
N_u	ultimate strength in compression, or tension, at a cross-section of an eccentrically loaded compression or tension member, respectively
N_{uB}	particular ultimate strength in compression of a cross-section when $k_{uo} = 0.003/(0.003 + f_{sy}/E_s)$
N_{uo}	ultimate strength in compression without bending, of an axially loaded cross-section
n	number of bars uniformly spaced around a helix; or the modular ratio (= E_s/E_c)
P	force in the tendons; or the maximum force in the anchorage
P_p	axial load applied to a pile
P_v	vertical component of the prestressing force
p	reinforcement ratio
p_B	balanced steel ratio
p_{all}	maximum allowable steel ratio for beam without special consideration
p_c	compression steel ratio
p_t	tensile steel ratio
p_{th}	tensile steel ratio in horizontal direction
p_{tv}	tensile steel ratio in vertical direction
$p_{t.min}$	minimum steel ratio required for a section
p_w	reinforcement ratio in a wall
Q	action effect due to live load (including impact, if any)

q	live load usually per unit length or area
q_f	allowable soil bearing capacity
R_d	design capacity of a member or structure (equal to ϕR_u)
R_u	ultimate resistance strength
r	radius of gyration of a cross-section
S_u	action effect due to snow load or liquid pressure or earth and/or ground water pressure
S^*	design action effect (E_d)
s	centre-to-centre spacing of shear or torsional reinforcement, measured parallel to the longitudinal axis of a member; or the standard deviation; or the maximum spacing of transverse reinforcement within $L_{sy.c}$; or spacing of stirrups or ties; or spacing of successive turns of a helix – all measured centre to centre, in millimetres
s_b	clear distance between bars of the noncontact lapped splice
T	a temperature; or the force resultant of tensile stresses
T'	maximum permissible torsion
T^*	torsional moment at a cross-section calculated using the design load (i.e. the design torsional moment)
T_b^*	bursting force calculated at the ultimate limit state
$T_{b.s}^*$	bursting force calculated at the serviceability state
$T_{b.cr}$	bursting (or splitting) force across a strut caused at the time of cracking of the strut
T_{uc}	ultimate torsional strength of a beam without torsional reinforcement and in the presence of shear
T_{us}	ultimate torsional strength of a beam with torsional reinforcement
$T_{u.max}$	ultimate torsional strength of a beam limited by web crushing failure
t	thickness of the flange of a flanged section
t_w	thickness of a wall
u	length of the critical shear perimeter for two-way action

u_t	perimeter of the polygon defined for A_t
V_a	inclined shear stress resultant
V_c	concrete shear stress resultant
V_d	dowel force provided by the tension reinforcement
V_u	ultimate shear strength
$V_{u.max}$	ultimate shear strength limited by web crushing failure
$V_{u.min}$	ultimate shear strength of a beam provided with minimum shear reinforcement
V_{uc}	ultimate shear strength excluding shear reinforcement
V_{uo}	ultimate shear strength of a slab with no moment transfer
V_{us}	contribution by shear reinforcement to the ultimate shear strength of a beam or wall
V^*	shear force at a section, calculated using the design load (i.e. the design shear force)
v	percentage by volume of the steel reinforcement in a reinforced or prestressed concrete section; or the shear stress
W_{BF}	weight of the backfill materials over the heel of a retaining wall
W_{FS}	weight of front surcharge materials over the toe of a retaining wall
W_{SL}	weight due to surcharge load over the heel of a retaining wall
W_u	action effect due to ultimate wind load
w_{cr}	average crack width
w_{max}	maximum crack width
x	shorter overall dimension of a rectangular part of a cross-section; or the smaller dimension of a component rectangle of a T-, L- or I-section
y	longer overall dimension of a rectangular part of a cross-section; or the larger dimension of a component rectangle of a T-, L- or I-section
y_t	distance between the neutral axis and the extreme fibres in tension of the uncracked section

y_T, y_B	distance between the neutral axis and the top and bottom extreme fibres of a beam, respectively
Z	section modulus of an uncracked cross-section
Z_T	section modulus of an uncracked cross-section with respect to its top fibre
Z_B	section modulus of an uncracked cross-section with respect to its bottom fibre
z	distance between the resultant compressive and tensile forces in a section
α	coefficient; or the inclination of the initial tangent to the concrete stress–strain curve; or a factor defining the geometry of the actual concrete stress block
α_M	coefficient for the calculation of deflection due to applied moment
α_n	coefficient
α_P	coefficient for the calculation of deflection due to concentrated load
α_v	angle between the inclined shear reinforcement and the longitudinal tensile reinforcement
α_w	coefficient for the calculation of deflection due to uniformly distributed load
α_1, α_2	factors defining the equivalent rectangular stress blocks of concrete under uniform compression and flexure, respectively; or permissible compressive and tensile stress coefficients, respectively, at transfer of prestress
β	coefficient with or without numerical subscripts; or a fixity factor; or a factor defining the geometry of the actual concrete stress block; or slope angle of the backfill for retaining walls
β_n	factor to account for the effect of the anchorage of ties on the effective compressive strength of a nodal zone
β_s	strut efficiency factor
$\beta_{BF}, \beta_{FS}, \beta_{SL}, \beta_W$	load combination factors for calculating different component forces for a retaining wall

β_x, β_y	short and long span bending moment coefficients, respectively, for slabs supported on four sides
γ	ratio, under design bending or combined bending and compression, of the depth of the assumed rectangular compressive stress block to k_{ud}
Δ	deflection
$\Delta_{A.g}$	accumulated sustained or dead-load deflection caused by the dead load and the effects of all the previous live-load repetitions
Δ_I	immediate deflection due to total service load
$\Delta_{I.g}$	immediate deflection caused by the sustained load or in most cases, the dead load
Δ_L	long-term deflection
Δ_T	total deflection
Δ_q	immediate live-load deflection at the Tth cycle
$\delta, \delta_b, \delta_s$	moment magnifiers for slenderness effects
ε	strain
ε_c	strain in concrete in compression
ε_{cu}	ultimate strain of concrete
ε_s	strain in the tensile reinforcement
ε_{sc}	strain in compression steel
ε_{sy}	yield strain in reinforcing steel
η	factor accounting for the difference between the crushing strength of concrete cylinders and the concrete in the beam; or effective prestress coefficient
ν	Poisson's ratio for concrete
θ_v, θ_t	angle between the concrete compression strut and the longitudinal axis of the member
λ_Δ	multiplier for additional deflection due to long-term effects
μ	coefficient of friction
ξ	time-dependent factor for sustained load

ρ	density of concrete, in kilograms per cubic metre (kg/m^3)
ρ_p	transverse compressive pressure (in MPa)
ρ_w	unit weight of reinforced or prestressed concrete
σ_{be}	bearing stress in concrete due to prestressing
σ_{BP}	compressive bottom-fibre stress after loss under prestressing force applied at an eccentricity
$\sigma_{p.ef}$	effective stress in the tendon
σ_{pi}	stress in the tendon immediately after transfer
σ_{pu}	maximum stress that would be reached in a tendon at ultimate strength of a flexural member
σ_{scr}	tensile stress in reinforcement at a cracked section, due to the short-term load combination for the serviceability limit states when direct loads are applied
$\sigma_{scr.1}$	tensile stress in reinforcement at a cracked section, due to the short-term load combination for the serviceability limit states, calculated with $\psi_s = 1.0$, when direct loads are applied
σ_{st}	calculated tensile stress in reinforcement
τ^*	design shear stress acting on the interface
τ_u	unit shear strength
ϕ	capacity reduction factor; or angle of internal friction for soil
ψ_c	combination live-load factor used in assessing the design load for strength
ψ_E	earthquake combination live-load factor used in assessing the design load for strength
ψ_l	long-term live-load factor used in assessing the design load for serviceability
ψ_s	short-term live-load factor used in assessing the design load for serviceability

ACRONYMS AND ABBREVIATIONS

ACI	American Concrete Institute
AS	Australian Standard
BCA	Building Code of Australia
CCAA	Cement Concrete & Aggregates Australia
CE	carbon equivalent
CEB-FIP	Comité Européen du Béton – Fédération Internationale de la Precontrainte
cg	centre of gravity
₡	centre line
CSS	critical stress state
DBM	deep-beam method
ESO	evolutionary structural optimisation
LC	loading case
LDS	linearly distributed stress
NA	neutral axis
pc	plastic centre
SP	shear plane
the Standard	AS 3600–2009 *Concrete Structures*
STM	strut-and-tie model

PART 1

REINFORCED CONCRETE

INTRODUCTION

1.1 HISTORICAL NOTES

Australian Standard (AS) 3600–1988 *Concrete Structures*, the first of the AS 3600 series, was published in March 1988. In line with European practices, it was a unified code covering reinforced and prestressed concrete structures. In effect, AS 3600–1988 *Concrete Structures* was the revised and amalgamated version of AS 1480–1982 *SAA Concrete Structures Code* and AS 1481–1978 *SAA Prestressed Concrete Code*, which it then superseded. Limit state design philosophy was adopted in AS 3600–1988. In practice, especially in strength design, engineers familiar with AS 1480–1982 could make the changeover without too much difficulty. Many of the design equations for shear, torsion, slabs and columns changed, but the strength design procedure was basically the same, that is, to ensure

$$\phi R_u \geq S^* \quad \text{Equation 1.1(1)}$$

where for a given section of any structural member to be designed, S^* was the 'action effect' or axial force, moment, shear or torsion due to the most critical combination of the external service loads, each multiplied by a corresponding load factor; R_u was the computed ultimate resistance (or strength) of the member at that section against the said type of action effect; and ϕ was the capacity reduction factor specified for the type of ultimate strength in question.

Since 1988, AS 3600 has been revised and updated three times and published consecutively at approximately six-year intervals as AS 3600–1994, AS 3600–2001 and the latest AS 3600–2009 (the Standard). However, the limit state design philosophy remains unchanged in the latest version of the Standard in which Clause 2.2.2 states that

$$R_d \geq E_d \quad \text{Equation 1.1(2)}$$

where $R_d = \phi R_u$ is the 'design capacity', and $E_d = S^*$, the design action effect.

Although the strength design procedure is unchanged, the recommended load factors are generally lower than previously specified. However, accompanying these lower load factors are reduced values of ϕ. These changes to ϕ, if seen in isolation, are no doubt retrograde because the implications are that we are less confident now in our design formulas than we were before. A probabilistic-based analytical model was adopted to re-evaluate the reliability of the design procedure. Unfortunately, actual failure statistics were inadequate for the probabilistic analysis to produce a new and more reliable procedure (in terms of load factors and ϕ). Instead, the new procedure was calibrated simply using designs based on the old AS 1480–1982 code. In simplistic terms, the old and the new codes applied in parallel should lead to the same design.

Note also that in AS 3600–2001, which appeared in 2002, N-grade or 500 MPa steel was specified, leading to modifications in serviceability specifications and other consequential changes. In AS 3600–2001, an additional strength grade for concrete was introduced with the characteristic compressive strength f'_c = 65 MPa. Two more grades are provided in AS 3600–2009, i.e. f'_c = 80 MPa and 100 MPa. This has resulted in modification to many of the design equations.

Henceforth, unless otherwise specified, all procedures, clauses, terms, formulas, factors and so forth refer to those given in AS 3600–2009 including relevant subsequent amendments.

1.2 DESIGN REQUIREMENTS

In addition to strength and serviceability, durability and fire resistance are specific design requirements – see Sections 4 and 5 of the Standard, respectively, for details. Since most of the recommended procedures for durability and fire resistance are empirical, they will not be dealt with in as much depth in this book.

Durability requirements mainly affect the choice of concrete strength and the provision of adequate concrete cover for reinforcement (see also Section 1.4). The four exposure conditions originally specified in AS 1480–1982 had been revised. Since AS 3600–1988, there have been six 'classifications': A1, A2, B1, B2, C and U, in order of severity. Minimum concrete strength grades and concrete covers for all except Classification U are detailed in Tables 4.10.3.2 and 4.10.3.3 of AS 3600–2009 for different types of construction method and/or environment. Note that Classification C has been expanded into C1 and C2 in these tables. For Classification U, the designer is responsible for providing their own concrete strength and cover specifications appropriate to the desired design. Additional durability requirements for abrasion, freezing and thawing, and other environmental and chemical actions may be found in Clauses 4.6 to 4.9 of the Standard.

Design for fire resistance is achieved by providing adequate concrete cover. Recommendations are given in the form of tables and charts in Section 5 of the Standard. These are to ensure that reinforced and prestressed concrete members designed for use in buildings meet the fire resistance levels required by the Building Code of Australia (BCA) (BCA 2011).

Drawing on the findings from real fire investigations of concrete buildings, research commissioned in Australia by the Cement Concrete & Aggregates Australia (CCAA) as well as recent overseas studies, CCAA published a Guide (CCAA 2010) for the better understanding of the performance of concrete structures in fire. This provides guidance for structural engineers to assess the structural adequacy of most building elements when subjected to either a 'real' design fire curve or a period of standard fire exposure, should the situations concerned not be covered adequately by the requirements of BCA in conjunction with the provisions of the Standard (CCAA 2010).

1.3 LOADS AND LOAD COMBINATIONS

For the first time, the provisions for loads and load combinations are no longer available in AS 3600–2009. The recommendations adopted herein follow those given in the loading Standard AS/NZS 1170.0–2002 and its amendments.

1.3.1 Strength design

For every member section in question, the design effective resistance (ϕR_u) must be able to sustain the action effects ($\pm S^*$) due to the following load combinations:

- dead load only

$$1.35G \qquad \text{Equation 1.3(1)}$$

- dead load and live load

$$1.2G + 1.5Q \qquad \text{Equation 1.3(2)}$$

- dead load and long-term imposed action

$$1.2G + 1.5\psi_l Q \qquad \text{Equation 1.3(3)}$$

- dead load, wind and live load

$$1.2G + W_u + \psi_c Q \qquad \text{Equation 1.3(4)}$$

- dead load and wind action reversal

$$0.9G + W_u \qquad \text{Equation 1.3(5)}$$

- dead, earthquake and live load

$$G + E_u + \psi_E Q \qquad \text{Equation 1.3(6)}$$

- dead load, liquid pressures and live load

$$1.2G + S_u + \psi_c Q \qquad \text{Equation 1.3(7)}$$

where G is the effect due to permanent action (dead load); Q is due to imposed action (live load); W_u is due to ultimate wind action (wind load); E_u is due to ultimate earthquake action (earthquake load); and S_u is due to snow load or liquid pressure or earth and/or ground water pressure; the combination, earthquake combination and long-term load factors, ψ_c, ψ_E and ψ_l, vary from 0.0 to 1.2, 1.0 and 1.0, respectively, depending on the function of the structure to be designed. For convenience, the values for ψ_c, ψ_E and ψ_l are reproduced from the AS/NZS Standard in Table 1.3(1).

Table 1.3(1) Live-load factors for strength design

	Type of imposed action (live load)	Short-term factor (ψ_s)	Long-term factor (ψ_l)	Combination factor (ψ_c)	Earthquake combination factor (ψ_E)
Distributed imposed actions, Q	**Floors**				
	Residential and domestic	0.7	0.4	0.4	0.3
	Offices	0.7	0.4	0.4	0.3
	Parking	0.7	0.4	0.4	0.3
	Retail	0.7	0.4	0.4	0.3
	Storage	1.0	0.6	0.6	0.6
	Other	1.0	0.6	0.6	0.6
	Roofs				
	Roofs used for floor-type activities	0.7	0.4	0.4	0.3
	All other roofs	0.7	0.0	0.0	0.0
Concentrated imposed actions (including balustrades), Q	Floors	1.0	0.6	0.4	0.3
	Floors of domestic housing	1.0	0.4	0.4	0.3
	Roofs used for floor-type activities	1.0	0.6	0.4	0.3
	All other roofs	1.0	0.0	0.0	0.0
	Balustrades	1.0	0.0	0.0	0.0
	Long-term installed machinery, tare weight	1.0	1.0	1.2	1.0

Source: Standards Australia (2011). *Amendment No. 3 to AS/NZS 1170.0:2002 Structural Design Actions – General Principles*, Standards Australia, Sydney, NSW. © Standards Australia Limited. Copied by Cambridge University Press with the permission of Standards Australia and Standards New Zealand under Licence 1712-c038.

1.3.2 Serviceability design

In addition to other load combinations, the following load combinations must be considered when designing for serviceability:
- short-term effects

$$\begin{aligned} &\bullet\ G \\ &\bullet\ G + W_u \\ &\bullet\ G + \psi_s Q \end{aligned}$$
Equation 1.3(8)

- long-term effects

$$\begin{aligned} &\bullet\ G \\ &\bullet\ G + \psi_l Q \end{aligned}$$
Equation 1.3(9)

Note that, as per Clause 4.2.4 in AS/NZS 1170.0–2002, the following combination is required for fire resistance design

$$G + \text{thermal actions arising from the fire} + \psi_c Q \quad \text{Equation 1.3(10)}$$

For Equation 1.3(8), the values for the short-term load factor (ψ_s) may also be found in Table 1.3(1).

1.3.3 Application

The purpose of load combinations is to obtain the most critical condition for which the structure must be designed. For the same structure, there may be more than one live load (see, for example, Clause 2.4.4 of the Standard) or wind load pattern, each of which, in combination with other loads, may be critical to different members or different sections of the same member. This makes the computation of S^* very tedious indeed. This is particularly true in column design and the design for shear and torsion where the critical condition is not governed by individual actions, but by the interaction of two different effects. For example, in column design, the interaction of axial force and bending moment must be considered.

1.4 CONCRETE COVER AND REINFORCEMENT SPACING

1.4.1 Cover

For durability design of a concrete structure under various environmental conditions, concrete of suitable strength must be specified by the designer and adequate cover for the steel reinforcement must be provided. Note that cover means clear cover, or to the 'outside of the reinforcing steel'. To guide the designer of reinforced and prestressed concrete structures, Table 4.3 of the Standard defines seven classifications of exposure, namely A1, A2, B1, B2, C1, C2 and U, in order of increasing severity. These definitions are re-presented in Table 1.4(1).

Table 1.4(1) Exposure classifications

Surface and exposure environment	Exposure classification
1. Surfaces of members in contact with the ground	
(a) Members protected by a damp-proof membrane	A1
(b) Residential footings in non-aggressive soils	A1
(c) Other members in non-aggressive soils	A2
(d) Members in aggressive soils	
Sulphate bearing (magnesium content $< 1g/L$)[a]	A1, A2, B1, B2 or C2
Sulphate bearing (magnesium content $\geq 1g/L$)	U
Other	U
(e) Salt rich soils and soils in areas affected by salinity[b]	A2, B1 or B2
2. Surfaces of members in interior environments	
(a) Fully enclosed within a building except for a brief period of weather exposure during construction	
Residential	A1
Non-residential	A2
(b) In industrial buildings, the member being subject to repeated wetting and drying	B1
3. Surfaces of members in above-ground exterior environments in areas that are	
(a) Inland (> 50 km from coastline) environment being	
(i) Non-industrial and arid climatic zone	A1
(ii) Non-industrial and temperate climatic zone	A2
(iii) Non-industrial and tropical climatic zone	B1
(iv) Industrial and any climatic zone	B1
(b) Near-coastal (1 km to 50 km from coastline), any climatic zone	B1
(c) Coastal and any climatic zone[c]	B2
4. Surfaces of members in water	
(a) In fresh water	B1
(b) In soft or running water	U

Table 1.4(1) (*cont.*)

Surface and exposure environment	Exposure classification
5. Surfaces of maritime structures in sea water	
(a) Permanently submerged	B2
(b) In spray zone[d]	C1
(c) In tidal/splash zone	C2
6. Surfaces of members in other environments	
Any exposure environment not specified in Items 1 to 5 above[e]	U

a See Table 4.8.1 of the Standard for details.

b See Table 4.8.2 of the Standard for details.

c The coastal zone includes locations within 1 km of the shoreline of large expanses of salt water. Where there are strong prevailing winds or vigorous surf, the distance should be increased beyond 1 km and higher levels of protection should be considered.

d The spray zone is the zone from 1 m above the wave crest level

e Further guidance on measures appropriate to exposure Classification U may be obtained from AS 3735–2001.

Source: Standards Australia (2009). AS 3600–2009 *Concrete Structures*. Standards Australia, Sydney, NSW. © Standards Australia Limited. Copied by Cambridge University Press with the permission of Standards Australia and Standards New Zealand under Licence 1712-c038.

For the first six exposure classifications, combinations of concrete strength and minimum cover are specified for various types of construction process and/or environment. These are contained in two tables in the Standard; for convenience, they are reproduced herein as Tables 1.4(2) and (3). Note that, for crack control, Clause 8.6.1(b) of the Standard states that the maximum cover shall not be greater than (100 mm – $d_b/2$) where d_b is the bar diameter in millimetre (mm).

Adequate concrete cover must also be provided for fire resistance. In-depth and wide-ranging recommendations are provided in Section 5 of the Standard. It is appropriate here to remind the reader that, following durability design for environmental factors, checks must be made against the requirements for fire resistance. Increase the cover if necessary.

1.4.2 Spacing

No minimum spacing for bars in beams or other structural elements is specified in the Standard; it only qualifies that 'the minimum clear distance between parallel bars (including bundled bars), ducts and tendons shall be such that the concrete can be properly placed and compacted ...' (see Clauses 8.1.9, 9.1.5 and 11.7.3 of the Standard).

Table 1.4(2) Required cover (mm), where standard formwork and compaction are used

Exposure classification	Characteristic concrete strength (f'_c)				
	20 MPa	25 MPa	32 MPa	40 MPa	$\geq$ 50 MPa
A1	20	20	20	20	20
A2	(50)	30	25	20	20
B1	–	(60)	40	30	25
B2	–	–	(65)	45	35
C1	–	–	–	(70)	50
C2	–	–	–	–	65

– = not applicable

Notes:

1. Figures in parentheses are the appropriate covers when the concession given in Clause 4.3.2 of the Standard, relating to the strength grade permitted for a particular exposure classification, is applied.
2. Cover should not be less than the greater of the maximum nominal aggregate size and bar diameter.

Source: Standards Australia (2009). AS 3600–2009 *Concrete Structures*. Standards Australia, Sydney, NSW. © Standards Australia Limited. Copied by Cambridge University Press with the permission of Standards Australia and Standards New Zealand under Licence 1712-c038.

Table 1.4(3) Required cover (mm) where repetitive procedures and intense compaction or self-compacting concrete are used in rigid formwork

Exposure classification	Characteristic concrete strength (f'_c)				
	20 MPa	25 MPa	32 MPa	40 MPa	$\geq$ 50 MPa
A1	20	20	20	20	20
A2	(45)	30	20	20	20
B1	–	(45)	30	25	20
B2	–	–	(50)	35	25
C1	–	–	–	(60)	45
C2	–	–	–	–	60

– = not applicable

Notes:

1. Figures in parentheses are the appropriate covers when the concession given in Clause 4.3.2 of the Standard, relating to the strength grade permitted for a particular exposure classification, is applied.
2. Cover should not be less than the greater of the maximum nominal aggregate size and bar diameter.

Source: Standards Australia (2009). AS 3600–2009 *Concrete Structures*. Standards Australia, Sydney, NSW. © Standards Australia Limited. Copied by Cambridge University Press with the permission of Standards Australia and Standards New Zealand under Licence 1712-c038.

For crack control, Clause 8.6.1(b) of the Standard recommends that the centre-to-centre spacing of bars shall not exceed 300 mm near the tension face of the beam. The Standard specifies the maximum spacing of transverse reinforcement or closed ties for shear (Clause 8.2.12.2), torsion (Clause 8.3.8(b)), columns (Clause 10.7.4.3(b)), and for torsion strips and spandrel beams of flat plates (Clause 9.2.6(b)). For crack control of slabs, Clause 9.4.1(b) specifies that the maximum centre-to-centre spacing 'shall not exceed the lesser of 300 mm and $2.0D_s$,' where D_s is the overall depth of the slab. For walls, on the other hand, the minimum clear spacing as per Clause 11.7.3 'shall not be less than $3d_b$'. For general guidance on minimum reinforcement spacing, the designer may refer to AS 1480–1982. Relevant recommendations are reproduced in Table 1.4(4).

Table 1.4(4) Minimum clear spacing of parallel bars

Bars between which clear spacing is measured	Direction in which spacing is measured	Minimum clear spacing (or pitch) shall be the greatest value of ...		
Horizontal bars in beams	Horizontally	25 mm	$1d_b$	$1.5a$
	Vertically	25 mm	$1d_b$	–
Horizontal bars in slabs, walls and footings	Horizontally	50 mm	$3d_b$	$1.5a$
	Vertically	25 mm	$1d_b$	–
Vertical bars	Horizontally	40 mm	$1.5d_b$	$1.5a$
Bars in ribs of hollow-block or concrete-joist slab construction	Horizontally	15 mm	$1d_b$	$1.5a$
Helical reinforcement	Pitch or helix	40 mm	$3d_b$ (pitch)	$1.5a$ (pitch)

where

a = the maximum nominal size of the aggregate

d_b = the diameter of the larger bar, or

= twice the diameter of the larger bar in the bundle, or

= the diameter of a pipe or a conduit, in which case the spacing applied between adjacent parallel pipes or conduits, whichever is the greater or the greatest, as applicable, or

= the diameter of the bar forming the helix.

Source: Standards Association of Australia (1982). SAA 1480–1982 *SAA Concrete Structures Code*, Standards Association of Australia, North Sydney, NSW. © Standards Australia Limited. Copied by Cambridge University Press with the permission of Standards Australia and Standards New Zealand under Licence 1712-c038.

For ease of reference, the various maximum spacing specifications are collated in Table 1.4(5).

Table 1.4(5) Maximum centre-to-centre spacing of parallel bars and closed ties

Bars[1]	Beams	300 mm (Clause 8.6.1(b))
	Slabs	Lesser of D_s and 300 mm (Clause 9.4.1(b))
Ties	Shear	Lesser of $0.5D$ and 300 mm or Lesser[2] of $0.75D$ and 500 mm (Clause 8.2.12.2)
	Torsion	Lesser of $0.12u_t$ and 300 mm (Clause 8.3.8(b))
	Columns	Lesser[3] of D_c and $15d_b$ for single bars or $0.5D_c$ and $7.5d_b$ for bundled bars (Clause 10.7.4.3(b))
	Torsion strips and spandrel beams	Not exceeding the greater of 300 mm and D_b or D_s (Clause 9.2.6(b))

Notes:

1. For crack control purposes.
2. In case of $V^* \leq \phi V_{u.min}$ (see Section 6.3.2 and Equation 6.3(13)).
3. D_c is the diameter of a circular column or the smaller dimension of a rectangular column.

1.5 SUMMARY

In addition to reviewing the historical development of Australia's Concrete Structures Standards, the specifications for strength, serviceability, durability and fire resistance are also collated in this opening chapter. Design loads and load combinations are presented in the form of ready-to-use formulas; the requirements for durability and crack control in terms of concrete covers and bar spacing are tabulated for ease of application.

DESIGN PROPERTIES OF MATERIALS

2.1 INTRODUCTION

Selection of materials and a decision on the associated properties constitute an essential first step in any structural design. In this chapter, the relevant information and data required for the design of reinforced and prestressed concrete structures are covered in some detail.

2.2 CONCRETE

2.2.1 Characteristic strengths

The characteristic compressive strength at 28 days (f'_c) is defined as the strength value of a given concrete that is exceeded by 95% of concrete cylinders tested (made from the same concrete). The test method for determining f'_c is described in AS 1012.9–2014. The value of f'_c is used in all Standard-based design calculations. Thus, in mix design, the strength to aim for (or the 28-day mean in situ compressive strength) is

$$f_{cm} = f'_c + ks \qquad \text{Equation 2.2(1)}$$

where s is the standard deviation of the test data and k varies from 1.25 to 2.5 depending on the grade of concrete and the manufacturing process (see AS 1379–2007 for all aspects of the testing and assessment of concrete or AS 1012.9–2014 for determining the compressive strength of concrete specimens).

As per Clause 3.1.1.2 of AS 3600–2009 (the Standard), f_{cm} at a given age may be taken as 90% of the mean cylinder strength of the same age or f_{cmi}. Refer to Clause 3.1.1.1 for more recommendations. Various design procedures, such as those described by Ryan and Samarin (1992) and by Murdock and Brook (1982), may be used to determine the proportions of the constituent materials (water, cement, fine and course aggregates and admixtures as appropriate).

As a function of f'_c, the characteristic flexural tensile strength as per Clause 3.1.1.3 of the Standard, may be taken as

$$f'_{ct.f} = 0.6\sqrt{f'_c} \quad \text{Equation 2.2(2)}$$

or it may be determined for a given concrete by the laboratory test described in AS 1012.11–2000 (R2014). This value is used mainly in cracking moment and deflection calculations (see Clauses 8.1.6.1 and 8.5.3.1, respectively, in the Standard), and in determining the minimum slab and pad-footing reinforcement (Clause 9.1.1).

A second tensile strength, the characteristic principal (or splitting) tensile strength (f'_{ct}), is required mainly in the design for longitudinal shear (Clause 8.4.3 in the Standard). It may be computed as per Clause 3.1.1.3 as

$$f'_{ct} = 0.36\sqrt{f'_c} \quad \text{Equation 2.2(3)}$$

or determined experimentally according to AS 1012.10–2000 (R2014).

2.2.2 Standard strength grades

Eight standard grades of concrete are specified – concrete with f'_c = 20, 25, 32, 40, 50, 65, 80 and 100 MPa. Among other considerations, a designer's specification of f'_c is governed by the design for durability. For a given exposure classification, there are corresponding minimum f'_c values (see, for example, Table 1.4(2)).

2.2.3 Initial modulus and other constants

The initial modulus (E_c) for concrete, a nonlinear material, is taken to be the modulus of elasticity. This is an accepted international practice. For a given experimental stress–strain curve of a concrete

$$E_c = \tan \alpha \quad \text{Equation 2.2(4)}$$

where α, the inclination of the initial tangent to the stress–strain curve, is defined in Figure 2.2(1).

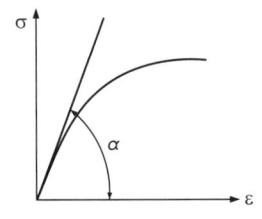

Figure 2.2(1) Concrete stress–strain curve

For design purposes, at a given age of the concrete, the following formulas may be used with ±20% accuracy

$$E_c = \rho^{1.5} 0.043 \sqrt{f_{cmi}} \text{ where } f_{cmi} \leq 40 \text{ MPa} \quad \text{Equation 2.2(5)}$$

or

$$E_c = \rho^{1.5} \left[0.024 \sqrt{f_{cmi}} + 0.12 \right] \text{ where } f_{cmi} > 40 \text{ MPa} \quad \text{Equation 2.2(6)}$$

in which E_c and the mean compressive cylinder strength (f_{cmi}) are given in MPa; for normal weight concrete, the density ρ may be taken as 2400 kg/m³. For the eight Standard strength grades, the recommended values of E_c for a given f_{cmi} or f'_c may be found in Table 3.1.2 of the Standard, which is reproduced herein as Table 2.2(1). Note incidentally that ρ may also be determined by test in accordance with AS 1012.12.1–1998 (R2014) or AS 1012.12.2–1998 (R2014).

Table 2.2(1) Compressive strengths and initial modulus

f'_c (MPa)	20	25	32	40	50	65	80	100
f_{cmi} (MPa)	22	28	35	43	53	68	82	99
E_c (MPa)	24 000	26 700	30 100	32 800	34 800	37 400	39 600	42 200

Source: Standards Australia (2009). AS 3600–2009 *Concrete Structures*, Standards Australia, Sydney, NSW. © Standards Australia Limited. Copied by Cambridge University Press with the permission of Standards Australia and Standards New Zealand under Licence 1712-c038.

Poisson's ratio may be taken as 0.2 or determined experimentally in accordance with AS 1012.17–1997 (R2014). The coefficient of thermal expansion is 10×10^{-6}/°C with ±20% accuracy. Recommendations for shrinkage and creep effects, respectively, are given in Clauses 3.1.7 and 3.1.8 of the Standard. These are normally used in computing long-term deflections.

2.3 STEEL

The AS/NZS 4671–2001 (for steel reinforcing materials) sets out all the requirements for reinforcing steel in Australia and New Zealand. Most steel produced in Australia currently has a specified yield strength of 250, 300 or 500 MPa, respectively, with a ductility designation of low (L), normal (N) or seismic (earthquake) (E) ductility. Normal ductility steel bars, or N bars, are generally used for normal concrete structures. Clause 3.2 of the Standard specifies properties for reinforcement. The designations, bar-group areas and unit mass values for commercially available reinforcing bars are given in Table 2.3(1).

The standard grades of reinforcing steel, characterised by their strength grade and relative ductility class are designated as 250N, 300E, 500L, 500N and 500E.

Letters R, D and I are used to designate plain (round), deformed and deformed indented bars, respectively.

Table 2.3(1) Bar-group areas (mm^2) and unit mass values for commercially available steel reinforcing bars

Number of bars	Cross-sectional area (mm^2) — Bar diameter (mm)								
	10	12	16	20	24	28	32	36	40
1	78	113	201	314	452	616	804	1020	1260
2	156	226	402	628	904	1232	1608	2040	2520
3	234	339	603	942	1356	1848	2412	3060	3780
4	312	452	804	1256	1808	2464	3216	4080	5040
5	390	565	1005	1570	2260	3080	4020	5100	6300
6	468	678	1206	1884	2712	3696	4824	6120	7560
7	546	791	1407	2198	3164	4312	5628	7140	8820
8	624	904	1608	2512	3616	4928	6432	8160	10 080
9	702	1017	1809	2826	4068	5544	7236	9180	11 340
10	780	1130	2010	3140	4520	6160	8040	10 200	12 600
Mass (kg/m)[*]	0.657	0.946	1.682	2.629	3.786	5.153	6.730	8.518	10.516
Min. hole dia. (mm)[#]	12	15	20	25	29	34	39	44	49

[*]Invoice weight – includes rolling margin of 4%

[#]Minimum hole diameter for clearance

Source: <http://www.reinforcing.com/~/media/OneSteel%20Reinforcing/Case%20Study%20PDFs/Technical%20Resources%20PDFs/REODATA43.pdf>

Note that, in practice, the various designated symbols are required to be listed in the order of shape, strength, configuration, ductility, size and spacing. For example, a square mesh consisting of 9 mm diameter deformed ribbed bars at 200 mm centres, of grade 500 MPa low ductility steel, would be designated as 'D500SL92'.

For the convenience of the reader, Table 2.3(1) lists areas of individual (and groups of) bars, as well as the minimum hole diameter for clearance and mass values per unit length. Note that sizes above 40 mm are not generally available and special orders may be required.

For slab design, Table 2.3(2) may be used to obtain the reinforcement areas per metre width at given spacings in millimetres.

Table 2.3(2) Slab reinforcement areas (mm²) per metre width at specific spacings in millimetres

Bar spacing (mm)	Cross-sectional area per unit width (mm²/m) Bar diameter (mm)								
	10	12	16	20	24	28	32	36	40
100	780	1130	2010	3140	4520	6160	8040	10 200	12 600
125	624	904	1608	2512	3616	4928	6432	8160	10 080
150	520	753	1340	2093	3013	4107	5360	6800	8400
175	446	646	1149	1794	2583	3520	4594	5829	7200
200	390	565	1005	1570	2260	3080	4020	5100	6300
225	347	502	893	1396	2009	2738	3573	4533	5600
250	312	452	804	1256	1808	2464	3216	4080	5040
275	284	411	731	1142	1644	2240	2924	3709	4582
300	260	377	670	1047	1507	2053	2680	3400	4200

Source: <http://www.reinforcing.com/~/media/OneSteel%20Reinforcing/Case%20Study%20PDFs/Technical%20Resources%20PDFs/REODATA43.pdf>

Grade 500N bars (or the N bars) are manufactured using the Tempcore process, which produces a reinforcing steel with a carbon equivalent (CE) limit of 0.39–0.44. Like mild steel reinforcement, N bars have excellent bending and rebending characteristics and weldability, with no preheating required. They are also highly ductile having a distinctive plastic zone and as illustrated in Figure 2.3(1), fracture at an ultimate strain of over 20%.

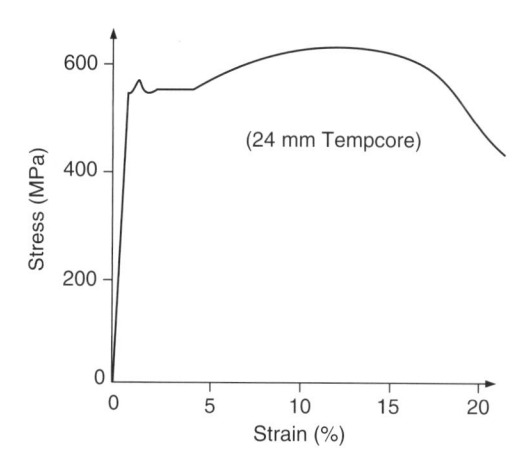

Figure 2.3(1) Steel stress–strain curve

The modulus of elasticity for steel may be taken as $E_s = 200\,000$ MPa and the coefficient of thermal expansion, $12 \times 10^{-6}/°C$.

2.4 UNIT WEIGHT

Clause 3.1.3 of the Standard stipulates that the density of normal-weight concrete (ρ) is to be taken as 2400 kg/m^3 or to be determined by a laboratory test in accordance with either AS 1012.12.1–1998 (R2014) or AS 1012.12.2–1998 (R2014). Table A1 in AS/NZS 1170.1–2002 (R2016) lists the unit weights of materials for use in calculating the self-weight of the structural elements. The unit weight of reinforced or prestressed concrete (ρ_w) is determined as

$$\rho_w = 24.0 + 0.6v \qquad \text{Equation 2.4(1)}$$

where ρ_w is in kN/m^3 and v is the percentage by volume of the steel reinforcement in a reinforced or prestressed concrete section.

2.5 SUMMARY

The relevant mechanical properties of concrete and steel for design are discussed in some detail. For completeness, the standard provisions for these properties and the corresponding laboratory tests specified for their determination are also described. To facilitate selection of the size and number of bars in a given design, relevant information and numerical data are tabulated in a ready-to-use form.

ANALYSIS AND DESIGN OF RECTANGULAR BEAMS FOR BENDING

3.1 INTRODUCTION

The ultimate strength theory underpins the analysis and design of reinforced and prestressed concrete structures, and has been since the promulgation of the Concrete Structures Standard, AS 3600–1988. The fundamentals of the theory are described in this chapter and its application is demonstrated in detail using beams with rectangular cross sections. The discussion covers singly and doubly reinforced beams.

3.2 DEFINITIONS

3.2.1 Analysis

Given the details of a reinforced concrete section, the ultimate resistance (R_u) against a specific action effect or combination of loads may be determined analytically using a method adopted by a code of practice or some other method. The process of determining R_u is referred to as analysis.

3.2.2 Design

Given the action effect (S^*) for moment, shear or torsion, and following a computational process, we can arrive at a reinforced concrete section that would satisfy AS 3600–2009 (the Standard) requirement as given in Equation 1.1(1):

$$\phi R_u \geq S^*$$

The process of obtaining the required section of a member is called design.

3.2.3 Ultimate strength method

If a given action effect at any section of a reinforced concrete member under load is severe enough to cause failure, then the ultimate strength has been reached. The primary responsibility of a designer is to ensure that this will not occur.

The ultimate strength method incorporates the actual stress–strain relations of concrete and steel, up to the point of failure. As a result, such an analysis can predict quite accurately the ultimate strength of a section against a given action effect. Because the design process can be seen as the analysis in reverse, a design should fail when the ultimate load is reached. The capacity reduction factor (ϕ) is used in Equation 1.1(1) to cover unavoidable uncertainties resulting from some theoretical and practical aspects of reinforced concrete, including variation in material qualities and construction tolerances.

3.3 ULTIMATE STRENGTH THEORY

3.3.1 Basic assumptions

The following is a summary of the basic assumptions of ultimate strength theory:
1. Plane sections normal to the beam axis remain plane after bending; that is, the strain distribution over the depth of the section is linear (this requires that bond failure does not occur prematurely – see Chapter 8).
2. The stress–strain curve of steel is bilinear with an ultimate strain approaching infinity (see Figure 3.3(1)a).
3. The stress–strain curve of concrete up to failure is nonlinear (see Figure 3.3(1)b) and may be approximated mathematically.
4. The tensile strength of concrete is negligible.

Assumption 1 makes it easy to establish the compatibility equations, which are important in the derivation of various analysis and design formulas. Assumption 2 leads to the condition that failure of reinforced concrete occurs only when the ultimate strain of concrete is reached (i.e. $\varepsilon_{cu} = 0.003$; Clause 8.1.2(d) of the Standard). Assumption 3 facilitates the use of an idealised stress block to simulate the ultimate (compressive) stress distribution in reinforced concrete sections.

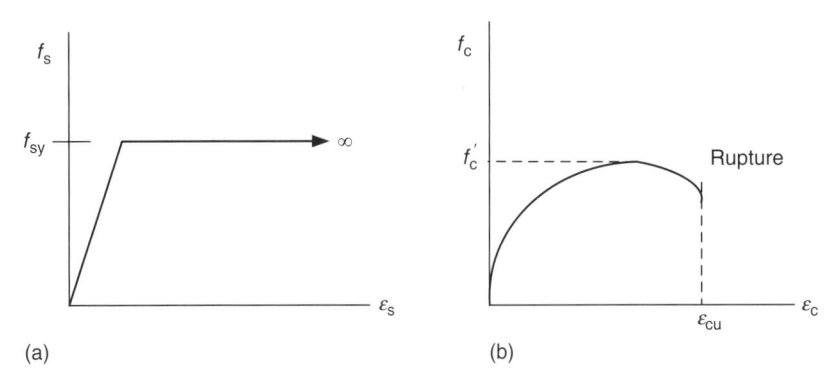

Figure 3.3(1) (a) Idealised stress–strain curve for steel; (b) stress–strain curve for concrete

3.3.2 Actual and equivalent stress blocks

For a beam in bending, the stress distribution over a cross-section at different loading stages can be depicted as in Figure 3.3(2). For an uncracked section and a cracked section at low stress level, linear (compressive) stress distribution may be assumed (see Figure 3.3(2)a, b). However, at the ultimate state, the stress distribution over the depth should be similar to that represented by Figure 3.3(2)c. Note, as can be seen in Figure 3.3(2)c, the resultant C_{ult} is computed as the volume of the curved stress block, which is prismatic across the width of the beam.

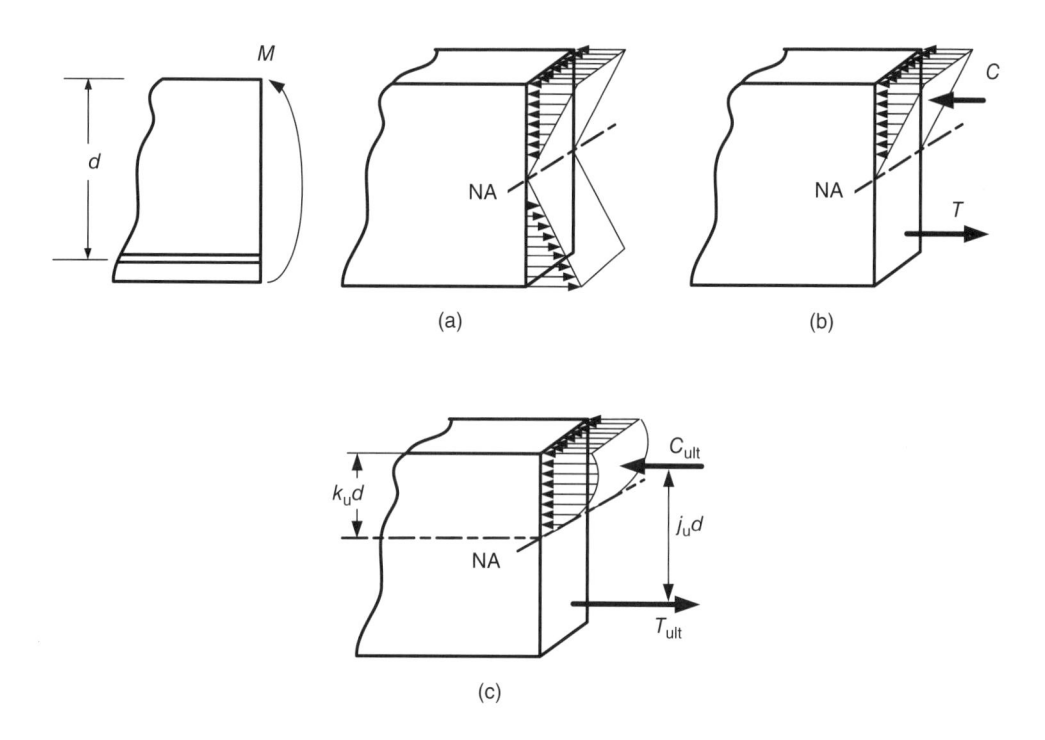

Figure 3.3(2) Stress distributions over a beam cross-section at different loading stages: **(a)** uncracked section; **(b)** transformed section (see Appendix A); and **(c)** at ultimate state

Considering the equilibrium of moments, the ultimate resistance of the section in bending can be expressed as

$$M_u = C_{ult} j_u d = T_{ult} j_u d \quad \text{Equation 3.3(1)}$$

Laboratory studies (e.g. Bridge and Smith 1983) have indicated that the actual shape of the stress–strain curve for concrete (Figure 3.3(1)b) depends on many factors and varies from concrete to concrete. Numerous methods have been proposed to express the curve in terms of only one variable: the characteristic compressive strength of concrete (f'_c). The proposal that has been accepted widely is the 'actual' stress block shown in Figure 3.3(3)a. The factor $\eta < 1$ accounts for the difference between the crushing strength of concrete cylinders and the concrete in the beam; α and β, each being a function of f'_c, define the geometry of the stress block. Empirical but complicated formulas have been given for η, α and β. Although the concept of the curved stress block is acknowledged as an advance, it is considered to be tedious to use. The equivalent (rectangular) stress block, as shown in Figure 3.3(3)b, is so defined that its use gives the same M_u as that computed using the 'actual' stress block. Because of its simplicity and relative accuracy, the use of the rectangular stress block is recommended in many major national concrete structures codes, including AS 3600–2009. According to Clause 8.1.3 of the Standard,

$$\alpha_2 = 1.0 - 0.003 f'_c \quad \text{but} \quad 0.67 \leq \alpha_2 \leq 0.85 \qquad \text{Equation 3.3(2)a}$$

$$\gamma = 1.05 - 0.007 f'_c \quad \text{but} \quad 0.67 \leq \gamma \leq 0.85 \qquad \text{Equation 3.3(2)b}$$

To fully define the stress block, the ultimate concrete strain is taken as $\varepsilon_{cu} = 0.003$ (see Figure 3.3(1)b).

In Equation 3.3(2)a, $\alpha_2 = 0.85$ is for $f'_c \leq 50$ MPa and $\alpha_2 = 0.67$ is for $f'_c \geq 110$ MPa. In Equation 3.3(2)b, $\gamma = 0.85$ is for $f'_c \leq 28.6$ MPa and $\gamma = 0.67$ is for $f'_c \geq 54.2$ MPa.

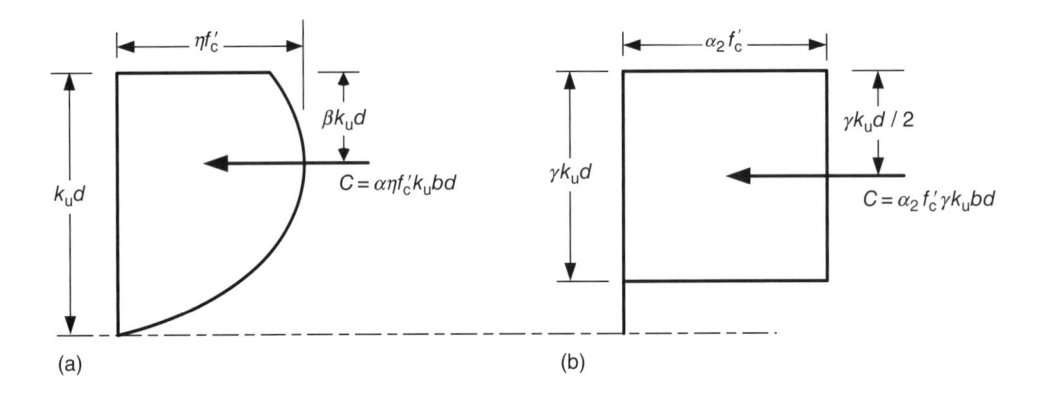

Figure 3.3(3) **(a)** Actual stress block and **(b)** equivalent stress block

3.4 ULTIMATE STRENGTH OF A SINGLY REINFORCED RECTANGULAR SECTION

3.4.1 Tension, compression and balanced failure

Failure of a concrete member occurs invariably when $\varepsilon_{cu} = 0.003$. At this ultimate state, the strain in the tensile reinforcement can be expressed by

$$\varepsilon_s > \varepsilon_{sy} \qquad \text{Equation 3.4(1)}$$

or

$$\varepsilon_s < \varepsilon_{sy} \quad \text{Equation 3.4(2)}$$

If Equation 3.4(1) is true as in Figure 3.4(1)a, the failure is referred to as a tension failure, which is normally accompanied by excessive cracking and deflection giving telltale signs of impending collapse. Such sections are referred to as under-reinforced.

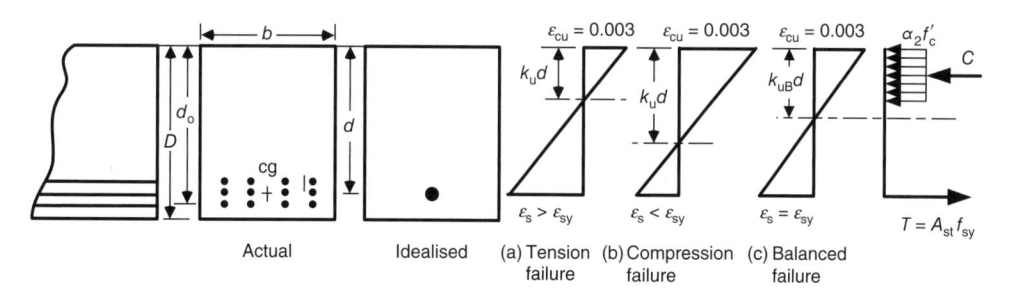

Figure 3.4(1) Strain distributions at the ultimate state

If Equation 3.4(2) is true (see Figure 3.4(1)b), the failure is called a compression failure and is abrupt and without warning. Such sections are referred to as over-reinforced. Tension or ductile failure is the preferred mode.

The ultimate state of a section at which $\varepsilon_s = \varepsilon_{sy}$ when $\varepsilon_c = \varepsilon_{cu}$ is called a balanced failure. The probability of a balanced failure occurring in practice is minute. However, it is an important criterion or threshold for determining the mode of failure in flexural analysis.

By considering Figure 3.4(1)c, we can establish an equation for the neutral axis parameter (k_{uB}). That is

$$\frac{k_{uB}d}{\varepsilon_{cu}} = \frac{d}{\varepsilon_{cu} + \varepsilon_{sy}}$$

from which

$$k_{uB} = \frac{\varepsilon_{cu}}{\varepsilon_{cu} + \varepsilon_{sy}} \quad \text{Equation 3.4(3)}$$

By substituting the known values, we obtain

$$k_{uB} = \frac{0.003}{0.003 + \dfrac{f_{sy}}{200\,000}}$$

or

$$k_{uB} = \frac{600}{600 + f_{sy}} \quad \text{Equation 3.4(4)}$$

For 500 N-grade bars, $k_{uB} = 0.545$.

To ensure that the section is under-reinforced, AS 3600–2001 (Clause 8.1.3) recommended that $k_u \leq 0.4$. Clause 8.1.5 of the current Standard (which supersedes AS 3600–2001), on the other hand, stipulates that $k_{uo} \leq 0.36$, where the neutral axis parameter k_{uo} is, with respect to d_o, the depth of the outermost layer of tensile reinforcement or tendons (see Figure 3.4(1)). This is a more stringent criterion for beams with only a single layer of tensile steel where $d_o = d$. If $k_{uo} > 0.36$, compression reinforcement (A_{sc}) of at least 0.01 times the area of concrete in compression must be provided. This should also be adequately restrained laterally by closed ties. For a definition of A_{sc}, refer to Section 3.6.

The term 'under-reinforced' should not be confused with 'under-designed'.

3.4.2 Balanced steel ratio

The resultants C and T in Figure 3.4(1)c are, respectively,

$$C = \alpha_2 f'_c \gamma k_{uB} bd$$

and

$$T = A_{st} f_{sy} = p_B bd f_{sy}$$

where p_B is the balanced steel ratio.
But $C = T$ or

$$\alpha_2 f'_c \gamma k_{uB} bd = p_B bd f_{sy}$$

from which

$$p_B = \frac{\alpha_2 f'_c \gamma k_{uB}}{f_{sy}} \qquad \text{Equation 3.4(5)}$$

Substituting Equation 3.4(4) into Equation 3.4(5) gives

$$p_B = \alpha_2 \gamma \frac{f'_c}{f_{sy}} \frac{600}{600 + f_{sy}} \qquad \text{Equation 3.4(5)a}$$

If $p_t < p_B$ a beam is under-reinforced; it is over-reinforced if $p_t > p_B$. As per Clause 8.1.5 in the Standard, the maximum allowable $k_{uo} = 0.36$. Being a function of d_o, k_{uo} cannot, at this stage of a design process, be directly related to k_{uB}. As a result, $k_u = 0.4$ is adopted and Equation 3.4(5) becomes

$$p_{all} = 0.4 \alpha_2 \gamma \frac{f'_c}{f_{sy}} \qquad \text{Equation 3.4(6)}$$

CHAPTER 3 ANALYSIS AND DESIGN OF RECTANGULAR BEAMS FOR BENDING   25

This may be taken as the maximum allowable reinforcement ratio. In design, ensure that $k_{uo} \leq 0.36$. Otherwise, provide the necessary A_{sc} as per Clause 8.1.5 of the Standard. Note that Equation 3.4(6) is applicable to all grades of reinforcing bars.

3.4.3 Moment equation for tension failure (under-reinforced sections)

In Figure 3.4(2)b,

$$C = \alpha_2 f'_c \gamma k_u b d$$

and

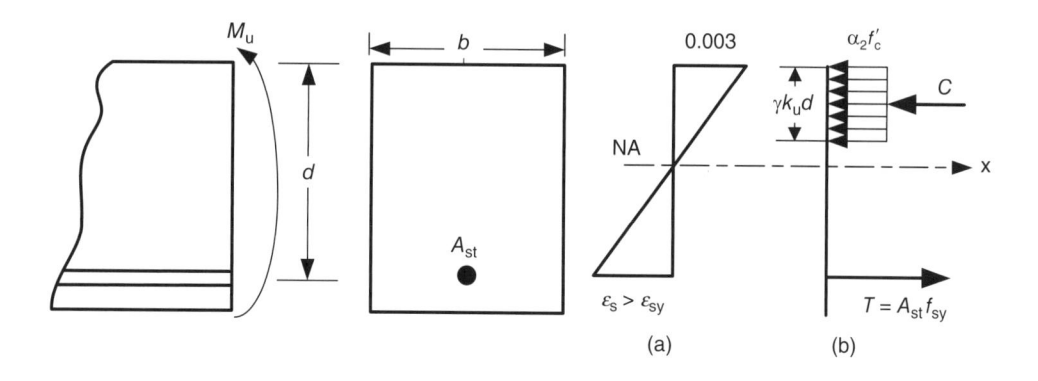

$$T = A_{st} f_{sy}$$

Figure 3.4(2) Stress and strain distributions for tension failure

But $\Sigma F_x = 0$ (i.e. $C = T$) from which

$$k_u = \frac{A_{st} f_{sy}}{\alpha_2 \gamma f'_c b d} \quad \text{Equation 3.4(7)}$$

or if the steel ratio $p_t = A_{st}/bd$, then

$$k_u = \frac{p_t f_{sy}}{\alpha_2 \gamma f'_c} \quad \text{Equation 3.4(8)}$$

Considering $\Sigma M = 0$ at the level of C and following Equation 3.3(1),

$$M_u = A_{st} f_{sy} d \left(1 - \frac{\gamma k_u}{2} \right) \quad \text{Equation 3.4(9)}$$

Substituting Equation 3.4(7) into Equation 3.4(9) leads to

$$M_u = A_{st}f_{sy}d\left(1 - \frac{1}{2\alpha_2}\frac{A_{st}f_{sy}}{bd\,f'_c}\right)$$ Equation 3.4(10)

Equation 3.4(10) supersedes the hitherto widely used formula recommended in AS 1480–1982:

$$M_u = A_{st}f_{sy}d\left(1 - 0.6\frac{A_{st}f_{sy}}{bd\,f'_c}\right)$$ Equation 3.4(11)

which was conservatively based on the equivalent stress block (see Figure 3.3(3)b) assuming a stress intensity of $0.85f'_c$.

3.4.4 Moment equation for compression failure (over-reinforced sections)

For over-reinforced sections, ε_s is less than ε_{sy} at failure and is an unknown. By considering the strain compatibility condition in Figure 3.4(3)a, we obtain

$$\varepsilon_s = 0.003\left(\frac{d - k_u d}{k_u d}\right)$$ Equation 3.4(12)

and

$$f_s = E_s\varepsilon_s = 600\left(\frac{d - k_u d}{k_u d}\right)$$ Equation 3.4(13)

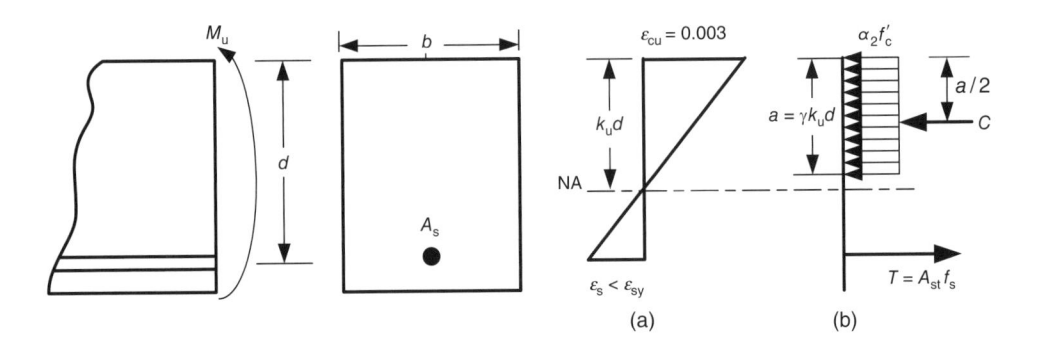

Figure 3.4(3) Stress and strain distribution for compression failure

Since $C = T$, we have

$$\alpha_2 f'_c \gamma k_u b d = A_{st} f_s = 600 A_{st} \left(\frac{d - k_u d}{k_u d} \right) \quad \text{Equation 3.4(14)}$$

Let $a = \gamma k_u d$ and by rearranging terms, Equation 3.4(14) becomes

$$a^2 + \mu a - \mu \gamma d = 0 \quad \text{Equation 3.4(15)}$$

where

$$\mu = \frac{600 p_t d}{\alpha_2 f'_c} \quad \text{Equation 3.4(16)}$$

Solving Equation 3.4(15) leads to

$$a = \frac{\sqrt{\mu^2 + 4\mu\gamma d} - \mu}{2} \quad \text{Equation 3.4(17)}$$

Finally, taking moments about T gives

$$M_u = C\left(d - \frac{a}{2}\right)$$

or

$$M_u = \alpha_2 f'_c ab \left(d - \frac{a}{2}\right) \quad \text{Equation 3.4(18)}$$

3.4.5 Effective moment capacity

In Equations 3.4(10) and 3.4(18), M_u may be seen as the probable ultimate moment capacity of a section. The effective or reliable moment capacity (M') is obtained by

$$M' = \phi M_u \quad \text{Equation 3.4(19)}$$

As per Table 2.2.2 of the Standard, the capacity reduction factor is

$$\phi = 1.19 - 13 k_{uo}/12 \quad \text{Equation 3.4(20)a}$$

but for beams with Class N reinforcement only

$$0.6 \leq \phi \leq 0.8 \quad \text{Equation 3.4(20)b}$$

and for beams with Class L reinforcement

$$0.6 \leq \phi \leq 0.64 \quad \text{Equation 3.4(20)c}$$

In Equation 3.4(20)a, $k_{uo} = \dfrac{k_u d}{d_o}$ in which d_o is the distance between the extreme compression fibre and the centroid of the outermost layer of the tension bars (see Figure 3.4(1)). Note also that, as per Clause 8.1.5, $k_{uo} \leq 0.36$ or M_u shall be reduced, as discussed in Sections 17.4 and 17.5.1.

3.4.6 Illustrative example for ultimate strength of a singly reinforced rectangular section

Problem

For a singly reinforced rectangular section with $b = 250$ mm, $d = 500$ mm, $f'_c = 50$ MPa, and Class N reinforcement only ($f_{sy} = 500$ MPa), determine the reliable moment capacity for the following reinforcement cases:

(a) $A_{st} = 1500$ mm^2
(b) $A_{st} = 9000$ mm^2
(c) a 'balanced' design
(d) with the maximum allowable reinforcement ratio (p_{all})
(e) $A_{st} = 4500$ mm^2.

Then plot M' against p_t.

Solution

Equation 3.3(2)a: $\alpha_2 = 1.0 - 0.003 \times 50 = 0.85$

Equation 3.3(2)b: $\gamma = 1.05 - 0.007 \times 50 = 0.70$

Equation 3.4(4): $k_{uB} = \dfrac{600}{600 + f_{sy}} = 0.545$

Equation 3.4(5): $p_B = \dfrac{0.85 \times 50 \times 0.70 \times 0.545}{500} = 0.0324$

(a) $A_{st} = 1500$ mm^2

$p_t = \dfrac{1500}{250 \times 500} = 0.012 < p_B = 0.0324$, therefore the section is under-reinforced.

Equation 3.4(10): $M_u = 1500 \times 500 \times 500 \times \left(1 - \dfrac{1}{2 \times 0.85} \times \dfrac{1500}{250 \times 500} \times \dfrac{500}{50}\right) \times 10^{-6}$

$= 348.5$ kNm

Equation 3.4(8): $k_u = \dfrac{0.012 \times 500}{0.85 \times 0.70 \times 50} = 0.202$

By assuming the steel reinforcement to be in one layer, we have $d = d_o$ and $k_{uo} = k_u = 0.202$.

Equation 3.4(20)a: $\phi = 1.19 - 13 \times 0.202/12 = 0.971$

For Class N reinforcement

Equation 3.4(20)b: $\phi = 0.8$

Accordingly

Equation 3.4(19): $M' = \phi M_u = 0.8 \times 348.5 = 278.8$ kNm

(b) $A_{st} = 9000$ mm^2

$$p_t = \frac{9000}{250 \times 500} = 0.072 > p_B = 0.0324, \text{ therefore the section is over-reinforced.}$$

Equation 3.4(16): $\mu = \dfrac{600 \times 0.072 \times 500}{0.85 \times 50} = 508.24$

Equation 3.4(17): $a = \dfrac{\sqrt{508.24^2 + 4 \times 508.24 \times 0.70 \times 500} - 508.24}{2}$

$= 238.28$ mm

Equation 3.4(18): $M_u = 0.85 \times 50 \times 238.28 \times 250 \times \left(500 - \dfrac{238.28}{2}\right) \times 10^{-6}$

$= 964.2$ kNm

But $a = \gamma k_u d$ from which $k_u = \dfrac{238.28}{0.70 \times 500} = 0.681$. By assuming one layer of steel, we have $d = d_o$ and $k_{uo} = k_u = 0.681$.

Then

Equation 3.4(20)a: $\phi = 0.452$ but Equation 3.4(20)b stipulates that $\phi \geq 0.6$.

Accordingly

Equation 3.4(19): $M' = \phi M_u = 0.6 \times 964.2 = 578.5$ kNm

This example is for illustrative purposes only. In practice, one layer is not enough to accommodate 9000 mm^2 of bars in the given section.

(c) A 'balanced' design (i.e. $p_B = 0.0324$)

$A_{st} = 0.0324 \times 250 \times 500 = 4050$ mm^2

Equation 3.4(10): $M_u = 4050 \times 500 \times 500$

$$\times \left(1 - \dfrac{1}{2 \times 0.85} \times \dfrac{4050}{250 \times 500} \times \dfrac{500}{50}\right) \times 10^{-6} = 819.5 \text{ kNm}$$

By assuming one layer of steel, we have $d = d_o$ and $k_{uo} = k_u = k_{uB} = 0.545$.
Then

Equation 3.4(20)a with b: $\phi = 0.6$

Accordingly

Equation 3.4(19): $M' = \phi M_u = 0.6 \times 819.5 = 491.7$ kNm

(d) With the maximum allowable reinforcement ratio (p_{all})

Equation 3.4(6): $p_{all} = 0.4 \times 0.85 \times 0.70 \times \dfrac{50}{500} = 0.0238$

$A_{st} = 0.0238 \times 250 \times 500 = 2975$ mm^2

Equation 3.4(10): $M_u = 2975 \times 500 \times 500$
$$\times \left(1 - \dfrac{1}{2 \times 0.85} \times \dfrac{2975}{250 \times 500} \times \dfrac{500}{50}\right) \times 10^{-6} = 639.6 \text{ kNm}$$

By assuming one layer of steel, we have $d = d_o$ and $k_{uo} = k_u = 0.4$.
Then

Equation 3.4(20)a with b: $\phi = 0.757$

Accordingly

Equation 3.4(19): $M' = \phi M_u = 0.757 \times 639.6 = 484.2$ kNm

(e) $A_{st} = 4500$ mm^2

$p_t = \dfrac{4500}{250 \times 500} = 0.036 > p_B = 0.0324$, therefore the section is over-reinforced.

Equation 3.4(16): $\mu = \dfrac{600 \times 0.036 \times 500}{0.85 \times 50} = 254.12$

Equation 3.4(17): $a = \dfrac{\sqrt{254.12^2 + 4 \times 254.12 \times 0.70 \times 500} - 254.12}{2} = 197.11$

Equation 3.4(18): $M_u = 0.85 \times 50 \times 197.11 \times 250 \times \left(500 - \dfrac{197.11}{2}\right) \times 10^{-6}$
$$= 840.7 \text{ kNm}$$

But $a = \gamma k_u d$ from which $k_u = \dfrac{197.11}{0.70 \times 500} = 0.563$

By assuming one layer of steel, we have $d = d_o$ and $k_{uo} = k_u = 0.563$.
Then

Equation 3.4(20)a with b: $\phi = 0.6$

Accordingly

Equation 3.4(19): $M' = \phi M_u = 0.6 \times 840.7 = 504.4$ kNm

The M' versus p_t plot is given in Figure 3.4(4). In the region where $p_t > p_B$ the use of additional A_{st} is no longer as effective. The reason is obvious, since failure is initiated by the rupture of concrete in compression and not by yielding of the steel in tension. Thus, in over-reinforced situations the use of doubly reinforced sections is warranted. This is done by introducing reinforcement in the compressive zone as elaborated in Section 3.6.

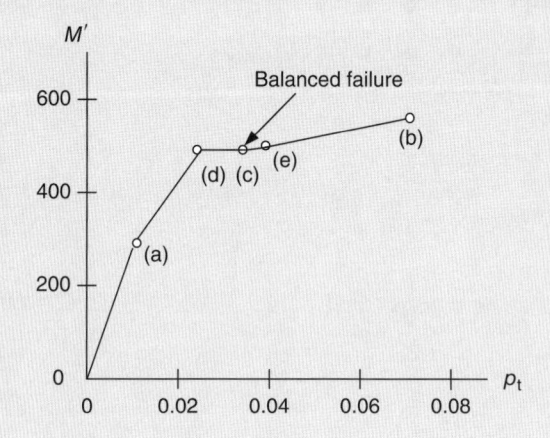

Figure 3.4(4) M' versus p_t for a singly reinforced section

3.4.7 Spread of reinforcement

For computing M_u, Equations 3.4(10) and 3.4(18) are valid only if the reinforcement is reasonably concentrated and can be represented by A_{st} located at the centroid of the bar group. If the spread of reinforcement is extensive over the depth of the beam, some of the bars nearer to the neutral axis may not yield at failure. This leads to inaccuracies. A detailed analysis is necessary to determine the actual M_u. The example below illustrates the general procedure.

Example: Computing M_u from a rigorous analysis

Problem

Compute M_u for the section in Figure 3.4(5), assuming $f'_c = 32$ MPa and $f_{sy} = 500$ MPa.

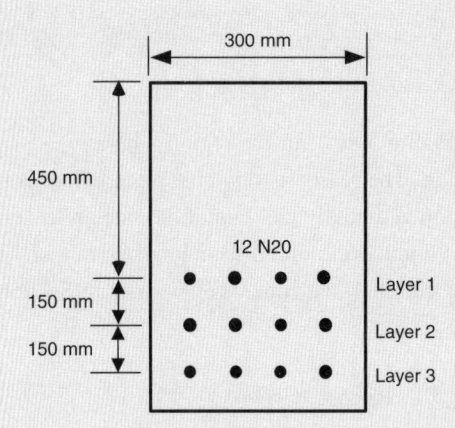

Figure 3.4(5) Cross-sectional details of the example problem

Solution

For f'_c = 32 MPa, Equations 3.3(2)a and b, respectively, give $\alpha_2 = 0.85$ and $\gamma = 0.826$.

And the reinforcement ratios

$$p_t = \frac{3768}{300 \times 600} = 0.0209$$

$$p_B = \frac{0.85 \times 32 \times 0.826}{500} \cdot \frac{600}{600 + 500} = 0.025 > p_t = 0.0209$$

Therefore the section is under-reinforced.

1. Assume all steel yields (i.e. [▦ = ▯]).

    $T = A_{st} f_{sy} = 3768 \times 500 \times 10^{-3} = 1884$ kN
    and
    $C = a \times 300 \times 0.85 \times 32 \times 10^{-3} = T = 1884$ kN
    from which $a = 230.9$ mm
    Thus

    $$k_u d = \frac{a}{\gamma} = \frac{230.9}{\gamma} = 279.5 \text{ mm}$$

    $$\varepsilon_{sy} = \frac{500}{200\,000} = 0.0025$$

From Figure 3.4(6),

$$\varepsilon_{s1} = \frac{170.5}{279.5} \times 0.003 = 0.00183 < \varepsilon_{sy}$$

$$\varepsilon_{s2} > \varepsilon_{sy}$$

Equation 3.4(21)

and

$$\varepsilon_{s3} > \varepsilon_{sy}$$

Therefore, the assumption is invalid.

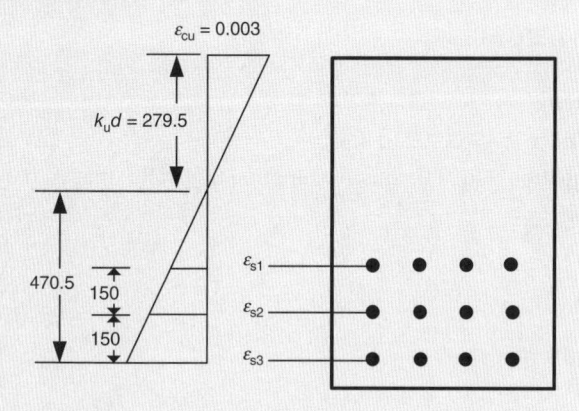

Figure 3.4(6) Strain distribution on the assumption of all steel yielding

2. Assume only the second and third layers yield while the first layer remains elastic (Figure 3.4(7)).

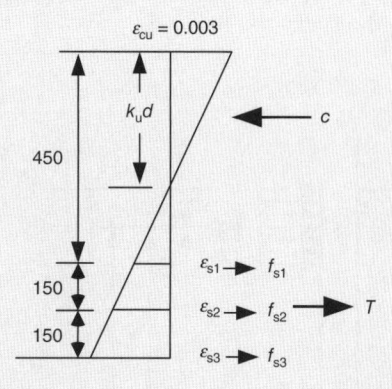

Figure 3.4(7) Strain distribution on the assumption of first steel layer not yielding

From Figure 3.4(7),

$$\frac{\varepsilon_{s1}}{450 - k_u d} = \frac{0.003}{k_u d}$$

Equation 3.4(22)

Therefore

$$f_{s1} = 600 \times \frac{(450 - k_u d)}{k_u d}$$

Since $\sum F_x = 0$, we have $C = T$, that is

$$0.85 \times 32 \times 0.826 k_u d \times 300 = 1256 \times 600 \frac{(450 - k_u d)}{k_u d} + 2512 \times 500$$

or

$$6.74(k_u d)^2 - 502 k_u d - 339120 = 0$$

from which $k_u d = 264.6$ mm
From Equation 3.4(22), we have $\varepsilon_{s1} = 0.00210 < \varepsilon_{sy}$.
Since $\varepsilon_{s3} > \varepsilon_{s2} > \varepsilon_{sy}$ (see Equation 3.4(21)), thus Assumption 2 is valid; that is, only the first steel layer is not yielding.
Hence, from Figure 3.4(8),

$$a = 0.826 \times 264.6 = 218.6 \text{ mm}$$
$$l_1 = 450 - 218.6/2 = 340.7 \text{ mm}$$
$$l_2 = 490.7 \text{ mm and}$$
$$l_3 = 640.7 \text{ mm}$$

Therefore,

$$M_u = (1256 \times 500 \times (490.7 + 640.7)$$
$$+ 1256 \times 200\,000 \times 0.00210 \times 340.7) \times 10^{-6}$$
$$= 890.2 \text{ kNm}$$

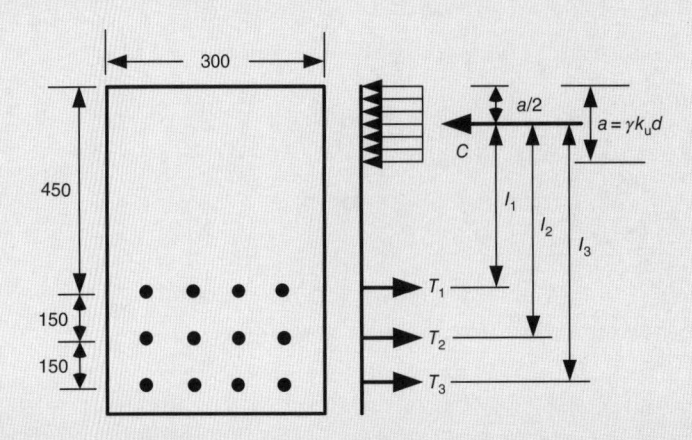

Figure 3.4(8) Lever arms between resultant concrete compressive force and tensile forces at different steel layers

3.5 DESIGN OF SINGLY REINFORCED RECTANGULAR SECTIONS

There are two design approaches for singly reinforced rectangular sections: free design and restricted design.

In a free design, the applied moment is given together with the material properties of the section. The designer selects a value for the steel ratio (p_t), based on which the dimensions b and d can be determined.

In a restricted design, b and d are specified, as well as the material properties. The design process leads to the required steel ratio.

In both approaches, the requirements of the Standard for reinforcement spacing and concrete cover for durability and fire resistance, if applicable, must be complied with. Details of such requirements may be found in Section 1.4.

3.5.1 Free design

In a free design, there is no restriction on the dimensions of the concrete section. Equation 3.4(10) may be rewritten as

$$M_u = p_t f_{sy} bd^2 (1 - \xi_o) \qquad \text{Equation 3.5(1)}$$

where

$$\xi_o = \frac{p_t f_{sy}}{2\alpha_2 f'_c} \qquad \text{Equation 3.5(1)a}$$

Using Equation 1.1(1) gives

$$\phi M_u \geq M^* \qquad \text{Equation 3.5(2)}$$

where M^* is the action effect that results from the most critical load combination as discussed in Section 1.3.3.

Substituting Equation 3.5(1) into 3.5(2) leads to

$$bd^2 = \frac{M^*}{\phi p_t f_{sy}(1 - \xi_o)} \qquad \text{Equation 3.5(3)}$$

Let $R = d/b$, then

$$d = \sqrt[3]{\frac{RM^*}{\phi p_t f_{sy}(1 - \xi_o)}} \qquad \text{Equation 3.5(4)}$$

The upper limit for p_t in design (i.e. p_{all}) is given in Equation 3.4(6) and according to Clause 8.1.6.1 of the Standard, $M_u \geq 1.2 M_{cr}$, which is deemed to be the case if the minimum steel that is provided for rectangular sections is

$$p_{t.min} \geq 0.20(D/d)^2 f'_{ct.f}/f_{sy} \qquad \text{Equation 3.5(5)}$$

For economy, p_t should be about $\frac{2}{3} p_{all}$ (Darval and Brown 1976). A new survey may be conducted to check current practice.

With a value for p_t chosen and the desired R value set, the effective depth of the section can be determined using Equation 3.5(4). Note that the design moment M^* is a function of the self-weight, amongst other variables. This leads to an iterative process in computing d.

As there is no restriction on D, A_{st} can be readily accommodated. Load combination requirements may lead to both the maximum (positive) and minimum (negative) values of M^* to be designed at a given section of the beam. The absolute maximum value should be used to determine d. Then $b = d/R$. These values of b and d must be adopted for a prismatic beam. Thus, for the other sections of the same beam, where M^* (positive or negative) is smaller than the absolute maximum, the design becomes a restricted one. There are other situations where a restricted design is necessary or specified.

3.5.2 Restricted design

In a restricted design, M^*, f_c', f_{sy}, b and D are given, while p_t is to be determined either by solving Equation 3.5(3) or by

$$p_t = \xi - \sqrt{\xi^2 - \frac{2\xi M^*}{\phi b d^2 f_{sy}}} \qquad \text{Equation 3.5(6)}$$

where

$$\xi = \frac{\alpha_2 f_c'}{f_{sy}} \qquad \text{Equation 3.5(7)}$$

The design procedure may be summarised in the following steps:

1. Assume the number of layers of steel bars with adequate cover and spacing. This yields the value for d.
2. With M^*, f_c', f_{sy}, b and d in hand, compute p_t using Equation 3.5(6). This equation will give an imaginary value if b and d are inadequate to resist M^*. If this should occur, a doubly reinforced section will be required (see Section 3.6). At this stage, ϕ as per Equation 3.4(20)a is an unknown, thereby requiring a trial and error process for determining p_t.
3. Ensure that

$$p_{t.min} \leq p_t \leq p_{all} \qquad \text{Equation 3.5(8)}$$

where $p_{t.min}$ and p_{all} are defined in Equations 3.5(5) and 3.4(6), respectively.
4. Compute $A_{st} = p_t bd$ and select the bar group (using Table 2.3(1)), which gives an area greater than but closest to A_{st}.
5. Check that b is adequate for accommodating the number of bars in every layer.

6. Choose a different bar group or arrange the bars in more layers as necessary. Also ensure that $k_{uo} \leq 0.36$ as stipulated in Clause 8.1.5 of the Standard.
7. A final check should be carried out to ensure that

$$\phi M_u \geq M^*$$ Equation 3.5(9)

where M_u is computed using Equation 3.5(1) for the section designed.

3.5.3 Design example

Problem

Using the relevant clauses of AS 3600–2009, design a simply supported beam of 6 m span to carry a live load of 3 kN/m and a superimposed dead load of 2 kN/m plus self-weight. Given that $f'_c = 32$ MPa, $f_{sy} = 500$ MPa for 500N bars, the maximum aggregate size $a = 20$ mm, the stirrups are made up of R10 bars, and exposure classification A2 applies.

Solution

Live-load moment

$$M_q = \frac{wl^2}{8} = \frac{3 \times 6^2}{8} = 13.5 \text{ kNm}$$

Superimposed dead-load moment

$$M_{SG} = \frac{2 \times 6^2}{8} = 9 \text{ kNm}$$

Take $b \times D = 150 \times 300$ mm and assume $p_t = 1.4\%$ (by volume). Then

Equation 2.4(1): $\rho_w = 24 + 0.6 \times 1.4 = 24.84 \text{ kN/m}^3$

Thus, self-weight $= 0.15 \times 0.30 \times 24.84 = 1.118$ kN/m
The moment due to self-weight is

$$M_{SW} = \frac{1.118 \times 6^2}{8} = 5.031 \text{ kNm}$$

and $M_g = M_{SG} + M_{SW} = 9 + 5.031 = 14.031$ kNm.
Then

Equation 1.3(2): $M^* = 1.2 M_g + 1.5 M_q$
or $M^* = 1.2 \times 14.031 + 1.5 \times 13.5 = 37.09$ kNm

Equation 3.3(2)a: $\alpha_2 = 1.0 - 0.003 \times 32 = 0.904$ but $0.67 \leq \alpha_2 \leq 0.85$, therefore $\alpha_2 = 0.85$

Equation 3.3(2)b: $\gamma = 1.05 - 0.007 \times 32 = 0.826$

Equation 2.2(2): $f'_{ct.f} = 0.6 \times \sqrt{32} = 3.394$ MPa

Adopting N20 bars as the main reinforcement in one layer with 25 mm cover gives
$d = D - $ cover $-$ diameter of stirrup $- d_b/2 = 300 - 25 - 10 - 20/2 = 255$ mm

Equation 3.5(5): $p_{t.min} = 0.20 \left(\dfrac{300}{255}\right)^2 \dfrac{3.394}{500} = 0.00191$

Equation 3.4(6): $p_{all} = 0.4 \times 0.85 \times 0.826 \times \dfrac{32}{500} = 0.01797$

Use, say, $p_t = \dfrac{2}{3} p_{all} = 0.01198 > p_{t.min}$; this is acceptable.
Then

Equation 3.5(3): $bd^2 = \dfrac{M^*}{\phi p_t f_{sy} \left(1 - \dfrac{1}{2\alpha_2} \times p_t \times \dfrac{f_{sy}}{f'_c}\right)}$

Equation 3.4(8): $k_u = \dfrac{0.01198 \times 500}{0.85 \times 0.826 \times 32} = 0.267$

For a single layer of bars, we have $d = d_o$. Thus, $k_{uo} = k_u = 0.267$.
Then from Equation 3.4(20)a with b, $\phi = 0.8$. Thus

Equation 3.5(3): $150 d^2 = \dfrac{37.09 \times 10^6}{0.8 \times 0.01198 \times 500 \times \left(1 - \dfrac{1}{2 \times 0.85} \times 0.01198 \times \dfrac{500}{32}\right)}$

from which $d = 240.80$ mm
Finally, $A_{st} = p_t bd = 0.01198 \times 150 \times 240.80 = 432.72$ mm^2
From Table 2.3(1), there are three options:

1. two N20: $A_{st} = 628$ mm^2
2. three N16: $A_{st} = 603$ mm^2
3. four N12: $A_{st} = 452$ mm^2

Taking Option 1, we have two N20 bars (see Figure 3.5(1)) and Table 1.4(2) gives a cover $c = 25$ mm to stirrups at top and bottom.
Hence, the cover to the main bars $= c +$ diameter of stirrup $= 25 + 10 = 35$ mm, and $d = D - $ cover to main bars $- d_b/2 = 300 - 35 - 20/2 = 255$ mm > 240.80 mm; therefore, this is acceptable (see Figure 3.5(1)).

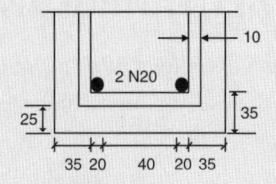

Figure 3.5(1) Checking accommodation for 2 N20 bars

a in Table 1.4(4) specifies a minimum spacing s_{min} of $[25, d_b, 1.5a]_{max}$. Thus $s_{min} = [25, 20, 30]_{max} = 30$ mm.
The available spacing $= 150 - 2 \times 35 - 2 \times 20 = 40$ mm $> s_{min} = 30$ mm; therefore, this is acceptable (see Figure 3.5(1)).
Note also that, since $k_{uo} = 0.267 < 0.36$, the design is acceptable without providing any compression reinforcement (see Section 3.4.1).
Taking Option 2, we have three N16 bars as shown in Figure 3.5(2).
The available spacing $= (150 - 2 \times 35 - 3 \times 16)/2 = 16$ mm $< s_{min} = 30$ mm.
To provide a spacing of 30 mm would require $b > 150$ mm (Figure 3.5(2)); therefore, Option 2 is not acceptable.

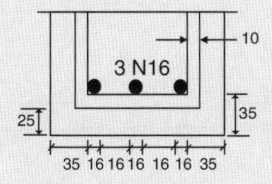

Figure 3.5(2) Checking accommodation for 3 N16 bars

Option 3 is similarly unacceptable. Thus, Option 1 should be adopted, but noting the following qualifications:

(a) Option 1 is slightly over-designed (i.e. $d = 255$ mm is about 5.9% higher than required and $A_{st} = 628$ mm^2 is 45.1% higher than necessary).
(b) The percentage of steel by volume for the section is $[628/(150 \times 300)] \times 100 = 1.396\% = 1.4\%$ as assumed in self-weight calculation; hence this is acceptable.
(c) If the beam is to be used repeatedly or frequently, a closer and more economical design could be obtained by having a second or third trial, assuming different $b \times D$.
(d) If design for fire resistance is specified, ensure that concrete cover of 25 mm is adequate by checking Section 5 of AS 3600–2009.

3.6 DOUBLY REINFORCED RECTANGULAR SECTIONS

As discussed in Section 3.4.6, for singly reinforced sections, if $p_t > p_B$, increasing A_{st} is not effective in increasing the moment capacity, so the use of compression steel (A_{sc}) to reinforce the concrete in the compressive zone is necessary. This situation often arises in a restricted design. Further, A_{sc} may be used to control or reduce creep and shrinkage deflection (see Clause 8.5.3.2 of the Standard).

A section reinforced with both A_{st} and A_{sc} is referred to as being doubly reinforced.

3.6.1 Criteria for yielding of A_{sc} at failure

A typical doubly reinforced section as shown in Figure 3.6(1)a may be for convenience idealised as in Figure 3.6(1)b; Figures 3.6(1)c and 3.6(1)d illustrate the strain and stress (resultant) distributions, respectively. In general, there are two possible cases at the ultimate state: A_{sc} yields and A_{sc} does not yield.

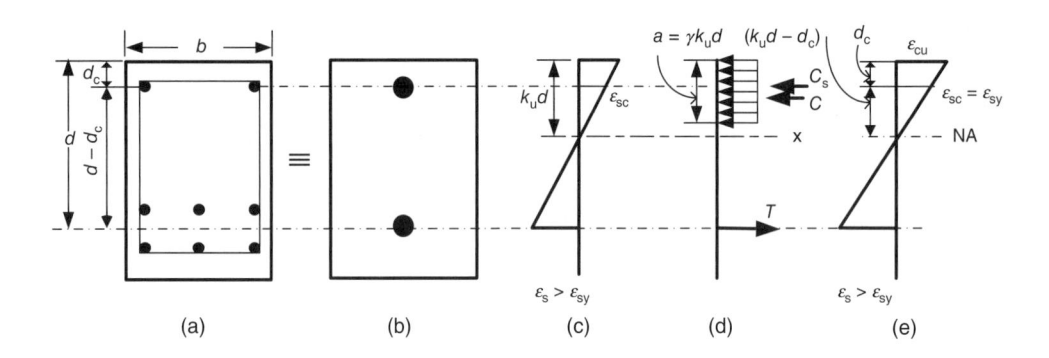

Figure 3.6(1) Stress and strain distributions across a doubly reinforced section

In both cases, A_{st} yields at the ultimate state. However, there are exceptional cases in which A_{st} does not also yield at failure. This section derives formulas that can help identify the first two cases; the two following sections include the analysis formulas and their applications. The more advanced topics for the exceptional cases are presented in Section 3.6.4.

First, we must determine if A_{sc} yields or not at failure (i.e. if $\varepsilon_{sc} >$ or $< \varepsilon_{sy}$). The relevant threshold equation can be derived by considering the compatibility of strains for the limiting case, where $\varepsilon_{sc} = \varepsilon_{sy}$ as shown in Figure 3.6(1)e.

Relating the strains $\varepsilon_{sc} = \varepsilon_{sy}$ to ε_{cu} gives

$$\frac{\varepsilon_{sy}}{k_u d - d_c} = \frac{\varepsilon_{cu}}{k_u d}$$

from which

$$k_u = \frac{\varepsilon_{cu} \frac{d_c}{d}}{\varepsilon_{cu} - \varepsilon_{sy}} = \frac{600 \frac{d_c}{d}}{600 - f_{sy}} \qquad \text{Equation 3.6(1)}$$

Take $\Sigma F_x = 0$ for the limiting case, that is

$$p_c b d f_{sy} + \alpha_2 f'_c \gamma k_u b d = p_t b d f_{sy} \qquad \text{Equation 3.6(2)}$$

Substituting Equation 3.6(1) into Equation 3.6(2) and rearranging the terms gives

$$(p_t - p_c)_{\text{limit}} = \frac{600 \alpha_2 \gamma f'_c \frac{d_c}{d}}{(600 - f_{sy}) f_{sy}} \qquad \text{Equation 3.6(3)}$$

or for 500N bars

$$(p_t - p_c)_{\text{limit}} = \frac{3 \alpha_2 \gamma f'_c d_c}{250 d} \qquad \text{Equation 3.6(3)a}$$

For a given beam section, if $(p_t - p_c)$ is greater than $(p_t - p_c)_{\text{limit}}$ as given in Equation 3.6(3), A_{sc} would yield at failure. Otherwise, it would not.

It is obvious that with a greater amount of A_{st}, the neutral axis at failure will be lowered, leading to a higher value of ε_{sc}. Thus, yielding of A_{sc} would occur. To ensure a tension failure, AS 1480–1982 Clause A1.1.2 states that

$$(p_t - p_c) \leq \frac{3}{4} p_B \qquad \text{Equation 3.6(4)}$$

No recommendation is apparent in the current Standard, but in accordance with the discussion in Section 3.4.2, we can take $k_u \leq 0.4$ (see Equation 3.4(6)), or for Class 500N bars

$$(p_t - p_c) \leq 0.4 \alpha_2 \gamma \frac{f'_c}{f_{sy}} \qquad \text{Equation 3.6(5)}$$

3.6.2 Analysis formulas
Case 1: A_{sc} yields at failure

In Figure 3.6(1)d, imposing $\Sigma F_x = 0$ leads to

$$f_{sy} A_{sc} + \alpha_2 f'_c a b = f_{sy} A_{st}$$

from which

$$a = \frac{(A_{st} - A_{sc})f_{sy}}{\alpha_2 f'_c b}$$ Equation 3.6(6)

Then, taking moments about A_{st} gives

$$M_u = A_{sc}f_{sy}(d - d_c) + \alpha_2 f'_c ab\left(d - \frac{a}{2}\right)$$ Equation 3.6(7)

or taking moments about the resultant C (Figure 3.6(1)d) yields

$$M_u = A_{st}f_{sy}\left(d - \frac{a}{2}\right) + A_{sc}f_{sy}\left(\frac{a}{2} - d_c\right)$$ Equation 3.6(8)

and the effective moment

$$M' = \phi M_u$$ Equation 3.6(9)

Case 2: A_{sc} does not yield at failure

In cases where $(p_t - p_c)$ is less than the right-hand side of Equation 3.6(3), A_{sc} will not yield. Thus, in addition to k_u we also have f_{sc} as an unknown. These two unknowns can be determined using the compatibility equation in conjunction with the equilibrium equation.

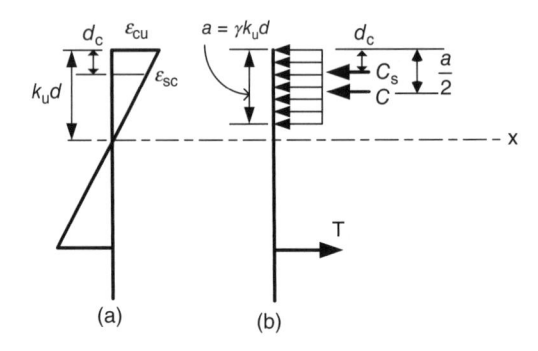

Figure 3.6(2) Stress and strain diagrams across the cross-section

By considering the strain distribution in Figure 3.6(2)a, we establish

$$\frac{\varepsilon_{sc}}{k_u d - d_c} = \frac{\varepsilon_{cu}}{k_u d}$$

or

$$\varepsilon_{sc} = 0.003 \frac{k_u - \frac{d_c}{d}}{k_u}$$ Equation 3.6(10)

then

$$f_{sc} = E_s \varepsilon_{sc} = 600 \frac{k_u - \frac{d_c}{d}}{k_u}$$ Equation 3.6(11)

But $\Sigma F_x = 0$, that is, $T = C + C_s$ in Figure 3.6(2)b, which gives

$$p_t b d f_{sy} = \alpha_2 f'_c \gamma k_u b d + p_c b d f_{sc}$$

or

$$p_t f_{sy} = \alpha_2 f'_c \gamma k_u + 600 p_c \frac{k_u - \frac{d_c}{d}}{k_u}$$

from which

$$k_u = \eta + \sqrt{\eta^2 + v \frac{d_c}{d}}$$ Equation 3.6(12)

where

$$\eta = \frac{p_t f_{sy} - 600 p_c}{2\alpha_2 \gamma f'_c}$$ Equation 3.6(13)

and

$$v = \frac{600 p_c}{\alpha_2 \gamma f'_c}$$ Equation 3.6(14)

Taking moments about T (see Figure 3.6(2)b) and letting $a = \gamma k_u d$ yields

$$M_u = \alpha_2 f'_c a b \left(d - \frac{a}{2} \right) + 600 A_{sc} \left(1 - \frac{d_c}{k_u d} \right) (d - d_c)$$ Equation 3.6(15)a

On the other hand, taking moments about C produces

$$M_u = A_{st} f_{sy} \left(d - \frac{a}{2} \right) + 600 A_{sc} \left(1 - \frac{d_c}{k_u d} \right) \left(\frac{a}{2} - d_c \right)$$ Equation 3.6(15)b

Note that in Equation 3.6(15)b, the contribution by the compression steel automatically becomes negative if C_s is below C (see Figure 3.6(2)b).

3.6.3 Illustrative examples
Example 1

Problem

Given a doubly reinforced section as shown in Figure 3.6(3) with $f'_c = 32$ MPa and $f_{sy} = 500$ MPa. Compute ϕM_u.

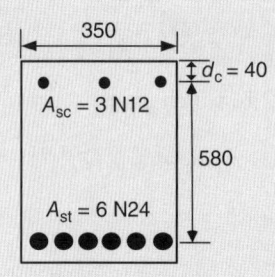

Figure 3.6(3) Cross-sectional details of Example 1

Solution

The reinforcement ratios

$$p_t = \frac{2712}{350 \times 620} = 0.0125$$

and

$$p_c = \frac{339}{350 \times 620} = 0.00156$$

From Section 3.5.3 and for $f'_c = 32$ MPa

$\alpha_2 = 0.85$ and $\gamma = 0.826$

Equation 3.6(3): $(p_t - p_c)_{\text{limit}} = \dfrac{600 \times 0.85 \times 0.826 \times 32 \times \dfrac{40}{620}}{(600 - 500) \times 500} = 0.01739$

But $(p_t - p_c) = 0.01094 < (p_t - p_c)_{\text{limit}} = 0.01739$
Hence A_{sc} does not yield at failure. Then

Equation 3.6(13): $\eta = \dfrac{0.0125 \times 500 - 600 \times 0.00156}{2 \times 0.85 \times 0.826 \times 32} = 0.11826$

Equation 3.6(14): $v = \dfrac{600 \times 0.00156}{0.85 \times 0.826 \times 32} = 0.0417$

Equation 3.6(12): $k_u = 0.11826 + \sqrt{0.11826^2 + 0.0417 \times \dfrac{40}{620}} = 0.247$

Since $a = \gamma k_u d = 0.826 \times 0.247 \times 620 = 126.49$ mm,

Equation 3.6(15)b: $M_u = \left[2712 \times 500 \times \left(620 - \dfrac{126.49}{2}\right) \right.$
$\left. + 600 \times 339 \times \left(1 - \dfrac{40}{0.247 \times 620}\right)\left(\dfrac{126.49}{2} - 40\right)\right] \times 10^{-6}$

That is, $M_u = 758.5$ kNm

With the bars in one layer, we have $d = d_o$, $k_{uo} = k_u = 0.247$, and Equation 3.4(20)a with b gives, for Class N reinforcement $\phi = 0.8$

Finally, $\phi M_u = 0.8 \times 758.5 = 606.8$ kNm

Example 2

Problem

Same as Example 1 (a doubly reinforced section with $f'_c = 32$ MPa and $f_{sy} = 500$ MPa), but $d_c = 35$ mm and A_{st} consists of 6 N28 bars. Compute ϕM_u.

Solution

The reinforcement ratios

$$p_t = \dfrac{3696}{350 \times 620} = 0.01703$$

and

$$p_c = \dfrac{339}{350 \times 620} = 0.00156$$

Equation 3.6(3): $(p_t - p_c)_{\text{limit}} = \dfrac{600 \times 0.85 \times 0.826 \times 32 \times \dfrac{35}{620}}{(600 - 500) \times 500} = 0.01522$

but $(p_t - p_c) = 0.01547 > (p_t - p_c)_{\text{limit}} = 0.01522$

Hence, A_{sc} yields at failure. Then

Equation 3.6(6): $a = \dfrac{(3696 - 339) \times 500}{0.85 \times 32 \times 350} = 176.3$ mm

Equation 3.6(7): $M_u = [339 \times 500 \times (620 - 35) + 0.85 \times 32$
$$\times 176.3 \times 350 \times \left(620 - \dfrac{176.3}{2}\right)] \times 10^{-6}, \text{that is } M_u = 991.8 \text{ kNm}$$

Since $a = \gamma k_u d$ from which $k_u = 0.344$ and for $d = d_o$, we have $k_{uo} = k_u = 0.344$. Thus, $\phi = 0.8$ according to Equations 3.4(20)a and b. Finally, $\phi M_u = 0.8 \times 991.8 = 793.4$ kNm.

It is worth noting here that the American Concrete Institute publication ACI 318–1995 (Commentary 10.3.1(A)(3)) states that A_{sc} may be neglected if $f_{sc} < f_{sy}$ when computing M_u. The AS 1480–1982, on the other hand, recommends an iterative process.

Neither of these approximations is necessary in view of the explicit Equations 3.6(7), 3.6(8), 3.6(15)a or 3.6(15)b developed for computing M_u. Note also that in current versions of ACI 318 and AS 3600, no recommendations are given in this regard.

3.6.4 Other cases

In addition to the two failure conditions discussed in Section 3.6.1, there are other less common cases in which A_{st} does not yield at failure, while A_{sc} may or may not yield.

Again, before we can determine if A_{st} yields or not we need to set up the threshold equation. This is done by considering the balanced failure conditions and by making use of the compatibility and equilibrium conditions.

Balanced failure conditions

Consider strain compatibility in Figure 3.6(4)a for cases with A_{st} yielding at failure.

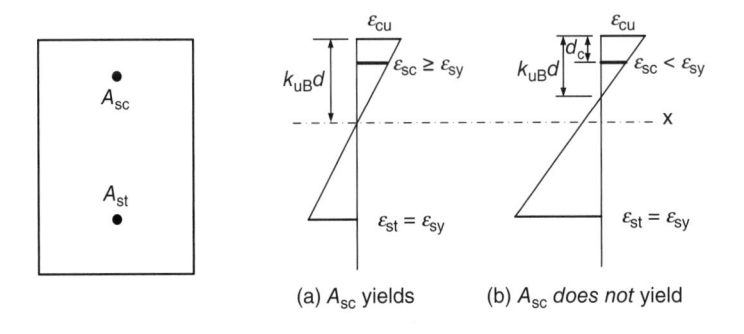

(a) A_{sc} yields   (b) A_{sc} *does not* yield

Figure 3.6(4) Strain profiles for limiting cases

We have

$$\frac{\varepsilon_{cu}}{k_{uB}d} = \frac{\varepsilon_{sy}}{d(1-k_{uB})}$$

from which

$$k_{uB} = \frac{\varepsilon_{cu}}{\varepsilon_{sy} + \varepsilon_{cu}}$$ Equation 3.6(16)

But $\Sigma F_x = 0$ or

$$p_t bd f_{sy} = \alpha_2 f'_c \gamma bd \frac{\varepsilon_{cu}}{\varepsilon_{sy} + \varepsilon_{cu}} + p_c bd f_{sy}$$

That is

$$p_{t.limit} = \frac{\alpha_2 f'_c \gamma}{f_{sy}} \frac{600}{600 + f_{sy}} + p_c$$ Equation 3.6(17)

Thus, if for a given section $p_t \leq p_{t.limit}$, then yielding of A_{st} and A_{sc} will occur. Note that Equation 3.6(17) is similar to Equation 3.4(5)a for singly reinforced sections.

Consider now the strain diagram in Figure 3.6(4)b for cases where A_{sc} does not yield, and we have

$$\frac{\varepsilon_{sc}}{k_{uB}d - d_c} = \frac{\varepsilon_{cu}}{k_{uB}d}$$

That is

$$\varepsilon_{sc} = \varepsilon_{cu} \frac{k_{uB} - \dfrac{d_c}{d}}{k_{uB}}$$ Equation 3.6(18)

The compressive stress in the steel is

$$f_{sc} = 600 \frac{k_{uB} - \dfrac{d_c}{d}}{k_{uB}}$$ Equation 3.6(19)

Take $\Sigma F_x = 0$, and we have

$$p_t bd f_{sy} = \alpha_2 f'_c \gamma bd k_{uB} + p_c bd \left(600 \frac{k_{uB} - \dfrac{d_c}{d}}{k_{uB}}\right)$$

That is

$$p_{t.limit} = \frac{\alpha_2 f'_c \gamma k_{uB}}{f_{sy}} + \frac{600 p_c}{f_{sy}} \left(1 - \frac{d_c}{k_{uB}d}\right)$$ Equation 3.6(20)

where k_{uB} is given in Equation 3.6(16).

For a given section, if $p_t \leq p_{t.limit}$ then yielding of A_{st} will occur, but A_{sc} remains elastic. Note that for 500N bars the limit set in Equation 3.6(20) is lower than that in Equation 3.6(17) whenever

$$d_c > d/11 \qquad \text{Equation 3.6(21)}$$

With Equations 3.6(17) and 3.6(20) in hand, the conditions in which A_{st} yields at failure can be readily determined, with A_{sc} either yielding or not.

Both A_{st} and A_{sc} do not yield

The condition of both A_{st} and A_{sc} not yielding prevails if p_t is greater than $p_{t.limit}$ given in Equation 3.6(20). Figure 3.6(5) illustrates such a condition.

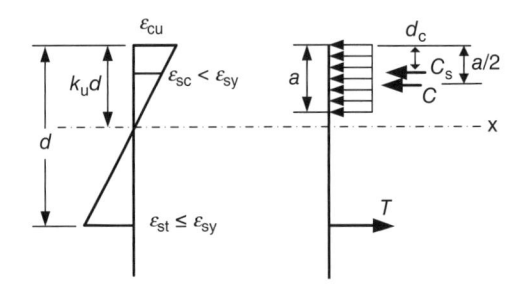

Figure 3.6(5) Stress and strain profiles for both A_{st} and A_{sc} not yielding

Since $\Sigma F_x = 0$, we have

$$T = C + C_s$$

or

$$p_t b d f_s = \alpha_2 f'_c \gamma k_u d b + p_c b d f_{sc} \qquad \text{Equation 3.6(22)}$$

where f_s and f_{sc} are given in Equations 3.4(13) and 3.6(11), respectively. Substituting these equations into Equation 3.6(22) gives

$$600 p_t \left(\frac{1 - k_u}{k_u} \right) = \alpha_2 f'_c \gamma k_u + 600 p_c \left(\frac{k_u - \frac{d_c}{d}}{k_u} \right)$$

or

$$\alpha_2 f'_c \gamma k_u^2 + 600(p_t + p_c) k_u - 600 \left(p_t + p_c \frac{d_c}{d} \right) = 0 \qquad \text{Equation 3.6(23)}$$

the solution of which yields

$$k_u = -\lambda(p_t + p_c) + \sqrt{\lambda^2(p_t + p_c)^2 + 2\lambda\left(p_t + p_c\frac{d_c}{d}\right)}$$ Equation 3.6(24)

where

$$\lambda = \frac{600}{2\alpha_2 f'_c \gamma}$$ Equation 3.6(24)a

For a given section, after determining k_u from Equation 3.6(24), the ultimate moment can be computed by taking moments about T (see Figure 3.6(5)), that is

$$M_u = \alpha_2 f'_c ab\left(d - \frac{a}{2}\right) + 600 A_{sc}\left(1 - \frac{d_c}{k_u d}\right)(d - d_c)$$ Equation 3.6(25)

Alternatively, by taking moments about C, we have

$$M_u = 600 A_{st}\left(\frac{1 - k_u}{k_u}\right)\left(d - \frac{a}{2}\right) + 600 A_{sc}\left(1 - \frac{d_c}{k_u d}\right)\left(\frac{a}{2} - d_c\right)$$ Equation 3.6(26)

A_{st} does not yield but A_{sc} does

A special case of the second case above is where A_{st} does not yield but A_{sc} does, and where Equation 3.6(23) can be written as

$$600 p_t\left(\frac{1 - k_u}{k_u}\right) = \alpha_2 f'_c \gamma k_u + p_c f_{sy}$$ Equation 3.6(27)

or

$$\alpha_2 f'_c \gamma k_u^2 + (600 p_t + p_c f_{sy}) k_u - 600 p_t = 0$$

from which

$$k_u = \zeta + \sqrt{\zeta^2 + \psi}$$ Equation 3.6(28)

where

$$\zeta = \frac{600 p_t + p_c f_{sy}}{2\alpha_2 f'_c \gamma}$$ Equation 3.6(29)

and

$$\psi = \frac{600 p_t}{\alpha_2 f'_c \gamma}$$ Equation 3.6(30)

With Equation 3.6(28) in hand, k_u can be determined readily, after which

$$M_u = \alpha_2 f'_c ab\left(d - \frac{a}{2}\right) + A_{sc}f_{sy}(d - d_c)$$ Equation 3.6(31)

or

$$M_u = 600A_{st}\left(\frac{1 - k_u}{k_u}\right)\left(d - \frac{a}{2}\right) + A_{sc}f_{sy}\left(\frac{a}{2} - d_c\right)$$ Equation 3.6(32)

3.6.5 Summary

Analysis equations for all the possible failure conditions existing for a doubly reinforced section have been presented in detail. So that the reader can have a clearer idea in regard to their applications, a flowchart is given in Figure 3.6(6).

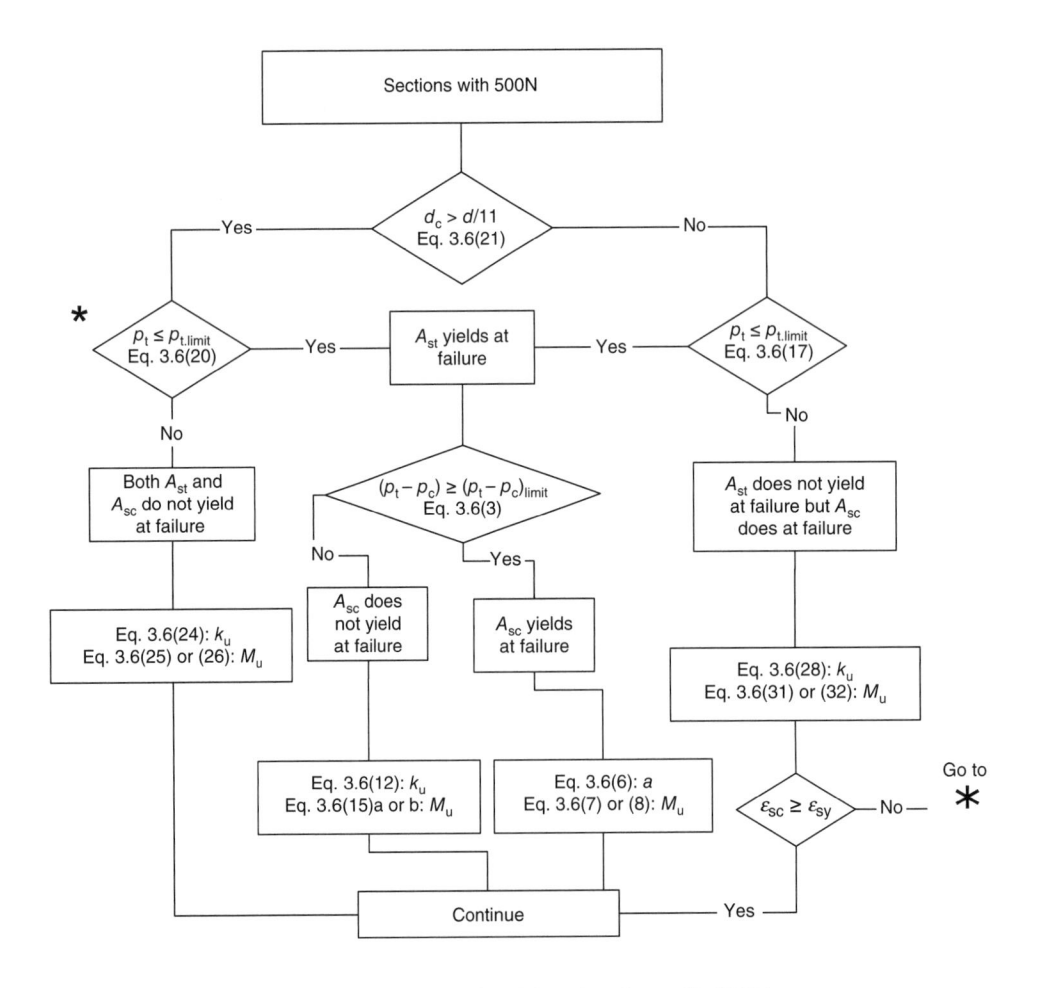

Figure 3.6(6) Summary chart for analysis of doubly reinforced sections with 500N bars

3.7 DESIGN OF DOUBLY REINFORCED SECTIONS

3.7.1 Design procedure

As discussed in Section 3.6, for concrete sections that would otherwise be over-reinforced, the use of A_{sc} will be effective in increasing the moment capacity. A doubly reinforced section may be 'decomposed' into a singly reinforced section and a hypothetical section, with only tension and compression steel reinforcements (but without concrete). These are depicted in Figure 3.7(1). Thus, for the section in Figure 3.7(1)a, we can write

$$M' = \phi M_u = \phi M_{u1} + \phi M_{u2} \qquad \text{Equation 3.7(1)}$$

where M_{u1} and M_{u2}, respectively, are the ultimate moments for the sections in Figure 3.7(1)b and c. Since for the singly reinforced section, p_t must not exceed p_{all}, Equation 3.4(6) yields

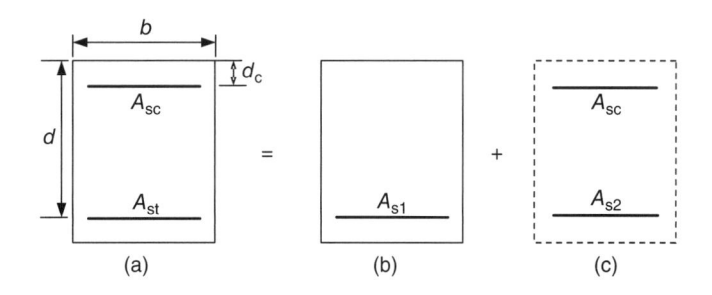

$$A_{s1} = 0.4\alpha_2\gamma \frac{f'_c}{f_{sy}} bd \qquad \text{Equation 3.7(2)}$$

Figure 3.7(1) 'Decomposition' of a doubly reinforced section

With this A_{s1}, ϕM_{u1} can be computed using Equation 3.4(10). As per Equation 1.1(1) and at the lowest safe limit

$$\phi M_{u1} + \phi M_{u2} = M^* \qquad \text{Equation 3.7(3)}$$

where (from Figure 3.7(1)c)

$$M_{u2} = A_{s2} f_{sy}(d - d_c) \qquad \text{Equation 3.7(4)}$$

Substituting Equation 3.7(4) into 3.7(3) leads to

$$A_{s2} = \frac{M^* - \phi M_{u1}}{\phi f_{sy}(d - d_c)} \qquad \text{Equation 3.7(5)}$$

Thus, by superposition

$$A_{st} = A_{s1} + A_{s2} \qquad \text{Equation 3.7(6)}$$

As we know from the discussion in Section 3.6.1, A_{sc} may or may not yield at the ultimate state. It is therefore necessary to set another criterion before we can decide on the amount of compression steel to be used in Figure 3.7(1)c. Following Equation 3.6(3), we have

$$A_{s1.\text{limit}} = \frac{600\alpha_2 f'_c \gamma d_c b}{(600 - f_{sy})f_{sy}} \qquad \text{Equation 3.7(7)}$$

Thus, if A_{s1} from Equation 3.7(2) is greater than $A_{s1.\text{limit}}$, then yielding of A_{sc} would occur, so we have

$$A_{sc} = A_{s2} \qquad \text{Equation 3.7(8)}$$

Otherwise, we need to provide more compression steel than A_{s2}.

Consider the strain diagram given in Figure 3.7(2). The position of the neutral axis is determined by the value of A_{s1} as given in Equation 3.7(2). Thus, we can write

$$\frac{\varepsilon_{sc}}{0.4d - d_c} = \frac{0.003}{0.4d} \qquad \text{Equation 3.7(9)}$$

from which

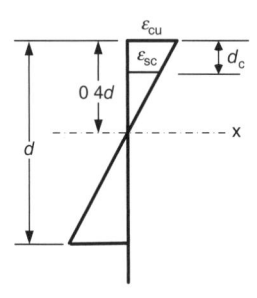

$$\varepsilon_{sc} = 0.003\left(1 - \frac{d_c}{0.4d}\right) \qquad \text{Equation 3.7(10)}$$

Figure 3.7(2) Strain diagram for the doubly reinforced section

Thus, the compression steel that can provide equilibrium of the horizontal forces in Figure 3.7(1)c is

$$A_{sc} = A_{s2}\frac{f_{sy}}{\varepsilon_{sc} E_s} \qquad \text{Equation 3.7(11)}$$

With A_{st} and A_{sc} determined, the most suitable bar groups can be selected; ensure that adequate cover and bar spacing are provided. Revise if necessary.

3.7.2 Illustrative example

Problem

If $b = 230$ mm, $D = 400$ mm, $M^* = 250$ kNm, $f'_c = 25$ MPa and $f_{sy} = 500$ MPa, and where exposure classification A1 applies, determine A_{st} and, as necessary, A_{sc} using only N28 bars. Use R10 ties.

Solution

Assume two layers, say, of N28 bars for A_{st} and one layer for A_{sc} as shown in Figure 3.7(3).

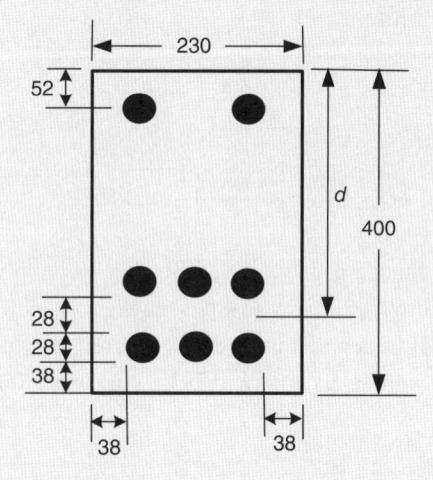

Figure 3.7(3) Section layout for illustrative example
Note: all dimensions are in mm.

Thus

$$d = D - \text{cover up to tie} - \text{tie diameter} - 1.5 \times \text{bar diameter}$$

That is

$$d = 400 - 28 - 10 - 1.5 \times 28 = 320 \text{ mm}.$$

$$d_c = 28 + 10 + 28/2 = 52 \text{ mm}$$

Then

Equation 3.3(2)a: $\alpha_2 = 1.0 - 0.003 \times 25 = 0.925$; but $0.67 \leq \alpha_2 \leq 0.85$; hence
$\alpha_2 = 0.85$, and

Equation 3.3(2)b: $\gamma = 1.05 - 0.007 \times 25 = 0.875$; but $0.67 \leq \gamma \leq 0.85$; hence
$\gamma = 0.85$

Thus

Equation 3.7(2): $A_{s1} = 0.4 \times 0.85 \times 0.85 \times \dfrac{25}{500} \times 230 \times 320 = 1063.52$ mm^2

Equation 3.4(10): $M_{u1} = 1063.52 \times 500 \times 320$

$$\times \left(1 - \dfrac{1}{2 \times 0.85} \times \dfrac{1063.52}{230 \times 320} \times \dfrac{500}{25}\right) \times 10^{-6}$$

That is
$M_{u1} = 141.2$ kNm
and

Equation 3.7(3): $M_2^* = M^* - \phi M_{u1}$

In this case, $k_u = 0.4$

And with $d_o = 400 - 38 - 28/2 = 348$, $k_{uo} = \dfrac{k_u d}{d_o} = \dfrac{0.4 \times 320}{348} = 0.368$

The stipulation of Clause 8.1.5 in the Standard that $k_{uo} \leq 0.36$ may not apply to doubly reinforced sections where $k_u \leq 0.4$, provided that the A_{sc} is not less than the specified minimum, which is true in most cases. If in doubt, double check and revise as necessary. With $k_{uo} = 0.368$, Equation 3.4(20)a with b gives $\phi = 0.791$, with which $M_2^* = 250 - 0.791 \times 141.2 = 138.3$ kNm
And

Equation 3.7(5): $A_{s2} = \dfrac{138.3 \times 10^6}{0.791 \times 500 \times (320 - 52)} = 1304.8$ mm^2

Thus

$A_{st} = A_{s1} + A_{s2} = 1063.52 + 1304.8 = 2368.32$ mm^2

Table 2.3(1) shows that with four N28 bars, $A_{st} = 2464$ mm^2 is acceptable.
Also,

Equation 3.7(7): $A_{s1.\text{limit}} = \dfrac{600 \times 0.85 \times 25 \times 0.85 \times 52 \times 230}{(600 - 500) \times 500}$

$$= 2592.33 \text{ mm}^2$$

Since $A_{s1} < A_{s1.\text{limit}}$, A_{sc} does not yield, and with $k_u = 0.4$

$$\varepsilon_{sc} = 0.003 \times \left(1 - \dfrac{d_c}{0.4d}\right) = 0.003 \times \left(1 - \dfrac{52}{0.4 \times 320}\right) = 0.00178$$

Thus, the compression steel stress

$$f_{sc} = \varepsilon_{sc} \times E_s \leq f_{sy}$$

or

$$f_{sc} = 0.00178 \times 200\,000 = 356 \text{ MPa} < f_{sy} = 500 \text{ MPa}$$

Hence, $f_{sc} = 356$ MPa requires

$$A_{sc} = \frac{A_{s2}f_{sy}}{f_{sc}} = \frac{1304.8 \times 500}{356} = 1832.6 \text{ mm}^2$$

With three N28 bars, $A_{sc} = 1848$ mm^2, which is acceptable.

To check bar accommodations for $b = 230$ mm: $b > (5 \times 28 + 2 \times 10) = 160$ mm is acceptable (use two layers of two bars) or $b > (7 \times 28 + 2 \times 10) = 216$ is acceptable (use three bars in the bottom layer plus one bar above). Details of the two possible reinforcement layouts are shown in Figure 3.7(4).

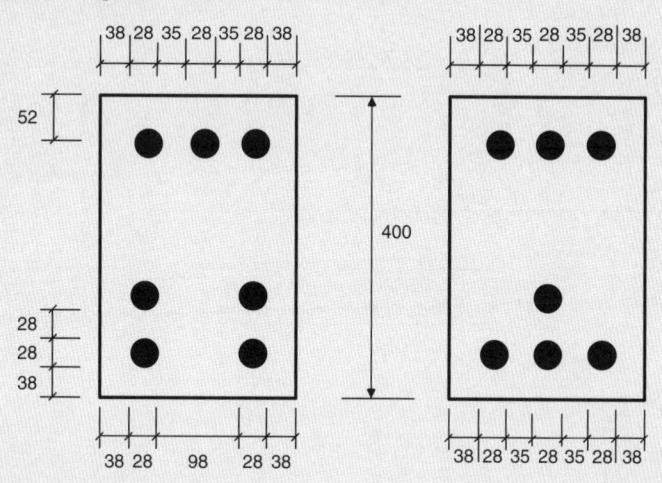

Figure 3.7(4) Section details for illustrative example

Note: all dimensions are in mm.

3.8 SUMMARY

Based on first principles and following on from the recommendations of the Standard, formulas for the analysis and design of reinforced concrete rectangular sections are developed in this chapter. These are carried out for both singly and doubly reinforced rectangular sections under all possible failure conditions. Applications of these formulas and procedures are illustrated with fully worked examples.

3.9 PROBLEMS

1. Repeat the example in Section 3.4.6 with the equivalent stress block, assuming an intensity of $0.85 f'_c$ where $f'_c = 65$ MPa. Draw the M' versus p_t curves for the two values of equivalent stress block intensities on the same diagram (similar to Figure 3.4(4)) and discuss the significance of using α_2 in place of 0.85 in the new Standard.

2. A singly reinforced rectangular section having a cross-section $b = 300$ mm and $d = 600$ mm is reinforced by 10 N32 bars. Assuming $f'_c = 32$ MPa, compute the reliable moment capacity of the section (i.e. ϕM_u).

3. Details of a one-way slab are illustrated in Figure 3.9(1). Based on the load combination formula, ultimate load $= 1.2g + 1.5q$, compute the uniformly distributed live load (q) that may be carried by the slab. Take $f'_c = 25$ MPa and $\rho_w = 24$ kN/m^3. (Hint: take a typical strip 1000 mm wide and analyse it as a simply supported beam.)

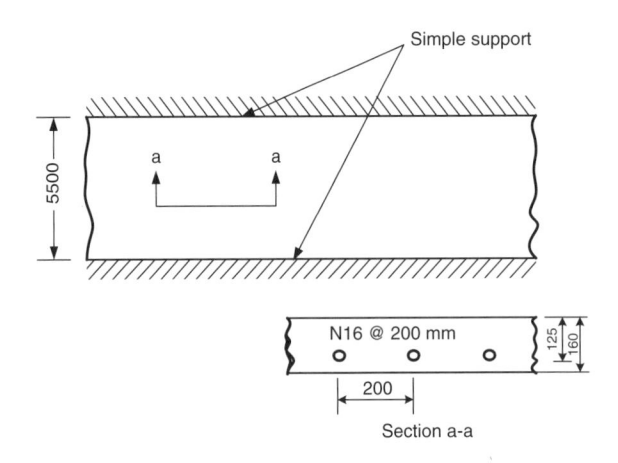

Figure 3.9(1) Details of a slab

Note: all dimensions are in mm.

4. A simply supported beam with a span of 8 m is to carry, in addition to its own weight, a superimposed dead load of 18 kN/m and a live load of 30 kN/m, both over the entire span.

    The beam has a rectangular section, which is to be singly reinforced. Given $p_t = 1.1\%$, design and detail the steel reinforcement for the section where the moment is maximum.

    Take $R = d/b \approx 1.5$, $\rho_w = 24$ kN/m^3, $f'_c = 20$ MPa and use N36 bars only. Exposure classification A1 applies; use R10 bars only for closed ties; maximum aggregate size = 10 mm.

CHAPTER 3 ANALYSIS AND DESIGN OF RECTANGULAR BEAMS FOR BENDING 57

5. Given a beam section $b \times D = 450$ mm $\times$ 950 mm, $M^* = 1500$ kNm, $f'_c = 32$ MPa, $f_{sy} = 500$ MPa and the maximum aggregate size = 20 mm. Design and detail the section. Use N28 bars for the main reinforcement and R10 bars for stirrups. Exposure classification A2 applies.

6. A beam section having $b \times D = 400$ mm $\times$ 800 mm is required to develop an effective ultimate moment (ϕM_u) of 1800 kNm. Design the reinforcement using compression steel if necessary.

    Assume $f'_c = 32$ MPa and $f_{sy} = 500$ MPa. Sketch the cross-section showing the reinforcement details. Use N36 bars only with R10 ties. Exposure classification A2 applies.

7. Evaluate M_u for the section shown in Figure 3.9(2). Assume $f'_c = 20$ MPa.

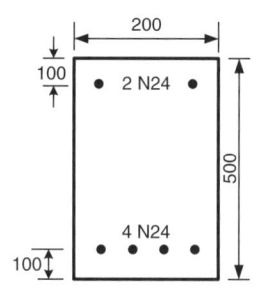

Figure 3.9(2) Cross-sectional details of the example beam section

Note: all dimensions are in mm.

8. Design the reinforcement for the section shown in Figure 3.9(3) so as to resist a design ultimate moment (M^*) of 900 kNm. If multiple bar layers are required, they are to be placed 75 mm centre-to-centre. Use N28 bars only and assume $f'_c = 25$ MPa.

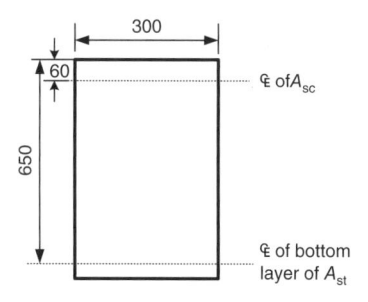

Figure 3.9(3) Cross-sectional details of the example design section

Note: all dimensions are in mm.

9. The details of a reinforced concrete beam are shown in Figure 3.9(4). Taking $f'_c = 32$ MPa, determine its reliable moment capacity.

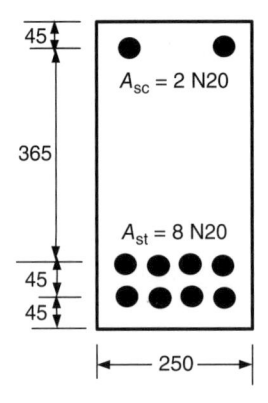

Figure 3.9(4) Cross-sectional details of a reinforced concrete beam

Note: all dimensions are in mm.

10. A doubly reinforced beam section is detailed in Figure 3.9(5).
    (a) Compute M', assuming $f'_c = 40$ MPa.
    (b) If the beam is simply supported over a span of 12 m, what is the maximum superimposed (uniformly distributed) working load permissible?

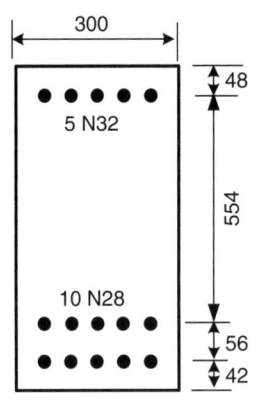

Figure 3.9(5) Cross-sectional details of a doubly reinforced section

Note: all dimensions are in mm.

11. For the doubly reinforced section detailed in Figure 3.9(6), compute the ultimate moment. Take $f'_c = 32$ MPa.

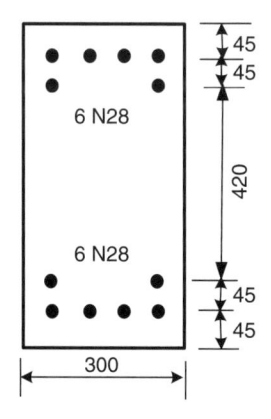

Figure 3.9(6) Cross-sectional details of a doubly reinforced section

Note: all dimensions are in mm.

12. The cantilever beam detailed in Figure 3.9(7) forms part of a loading platform. In addition to self-weight, the beam carries a uniformly distributed dead load of 10 kN/m and a concentrated live load (*q*), positioned as shown in Figure 3.9(7)a. Assuming $f'_c = 32$ MPa, what is the maximum allowable *q*?

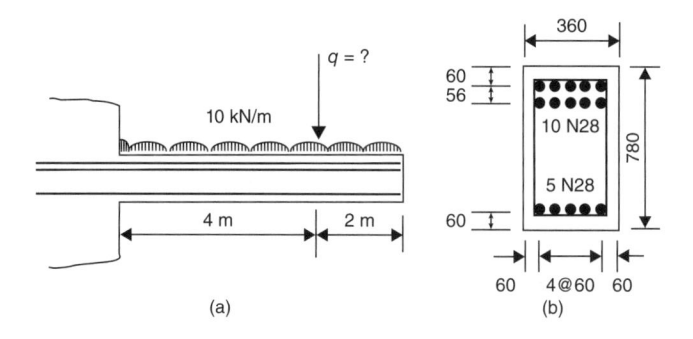

Figure 3.9(7) Details of a cantilever beam

Note: all dimensions are in mm unless otherwise specified.

13. In addition to its own weight (or assuming $\rho_w = 24.6$ kN/m³), the beam shown in Figure 3.9(8)a is to carry a uniformly distributed live load of 10 kN/m plus the three concentrated live loads, as shown. Figure 3.9(8)b shows the cross-sectional details of the beam.

Design for the mid-span section, check if compression steel is required. Take $f_c' = 32$ MPa and for tension steel, use two layers of N40 bars with a cover of 50 mm and a clear spacing of 40 mm between the layers. For compression steel (if required), use one layer of N40 bars with a cover of 50 mm. You may assume $\phi = 0.8$.

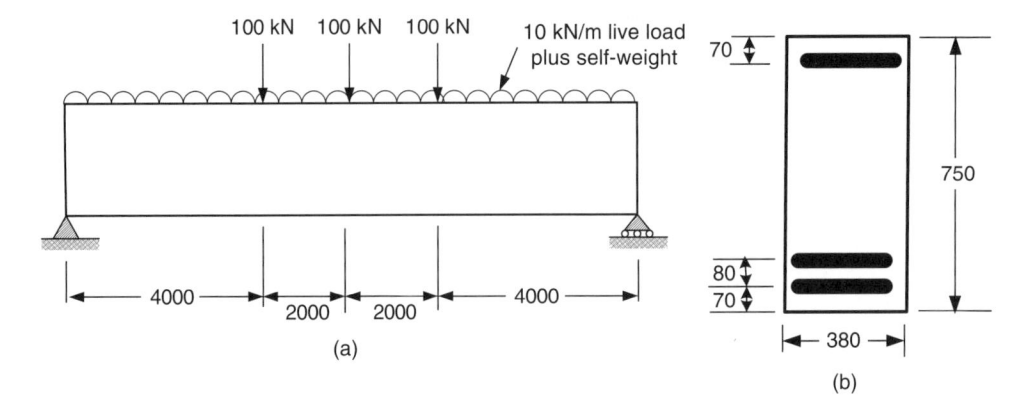

Figure 3.9(8) Loading configuration and cross-sectional details of the design beam

Note: all dimensions are in mm.

14. In addition to its own weight, the beam in Figure 3.9(9)a is to carry a uniformly distributed dead load of 10 kN/m as well as the concentrated live loads of 240 kN each at the tips of the overhangs. Figure 3.9(9)b shows the cross-sectional details of the beam.

    Design for the midspan section between the supports making use of the cover and spacing between bar layers, as shown. Note that A_{st} is to be equally distributed between two layers as shown and use A_{sc} in one layer, as necessary. Use only N24 bars and you can take $\rho_w = 25$ kN/m^3 and $f_c' = 32$ MPa.

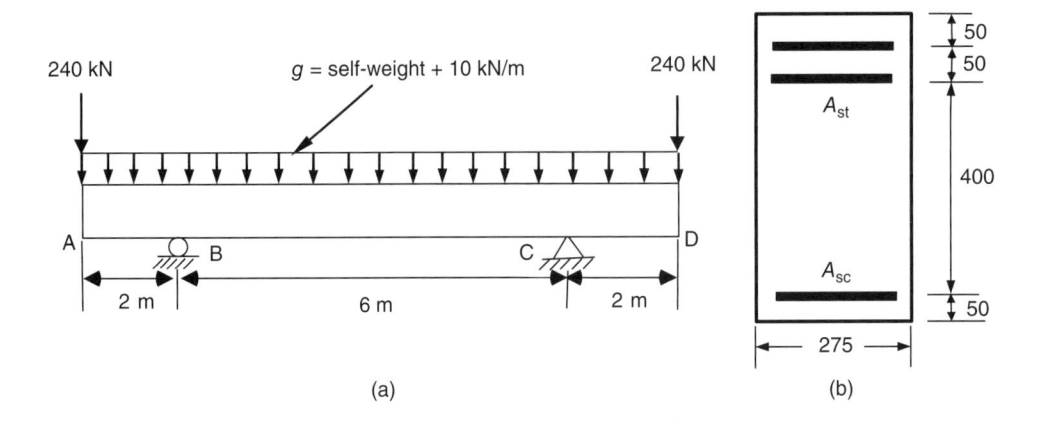

Figure 3.9(9) Loading configuration and cross-sectional details of the design beam

Note: all dimensions are in mm unless otherwise specified.

T-BEAMS AND IRREGULAR-SHAPED SECTIONS 4

4.1 INTRODUCTION

The ultimate strength analysis and design of T-beams and other flanged sections as well as nonstandard or irregular-shaped sections are dealt with in this chapter. The procedures involved together with derivation of relevant formulas and the application of relevant Standard recommendations are presented, with illustrative examples. The treatment of continuous beams and other statically indeterminate structures is briefly described.

4.2 T-BEAMS AND OTHER FLANGED SECTIONS

4.2.1 Definitions

Beams with a cross-sectional shape of T, L, I or a box are referred to collectively as flanged beams. Flanged sections are a popular choice for major structures, such as road and railway bridges, due to their structural efficiency. A beam-and-slab building floor system may be considered as an assembly of T-beams. An example is shown in Figure 4.2(1)a.

The analysis and design of flanged beams may seem to be comparatively more involved than the analysis and design of rectangular beams, due to more complicated geometry. However, using simple assumptions and some approximations, the formulas developed for singly and doubly reinforced rectangular sections can be adapted for use in the analysis and design of flanged beams.

4.2.2 Effective flange width

Figure 4.2(1) shows a typical T-beam. A three-dimensional stress analysis would indicate that, depending on the dimensions of the beam, the bending stress distribution over the width of the flange (b) is not uniform as it would be for a rectangular section. Instead, the stress varies from a maximum (f_1) near the top of the web to a minimum (f_2) at the two edges of the flange, as in Figure 4.2(1)b. However, the stresses f_1 and f_2 are statically indeterminate.

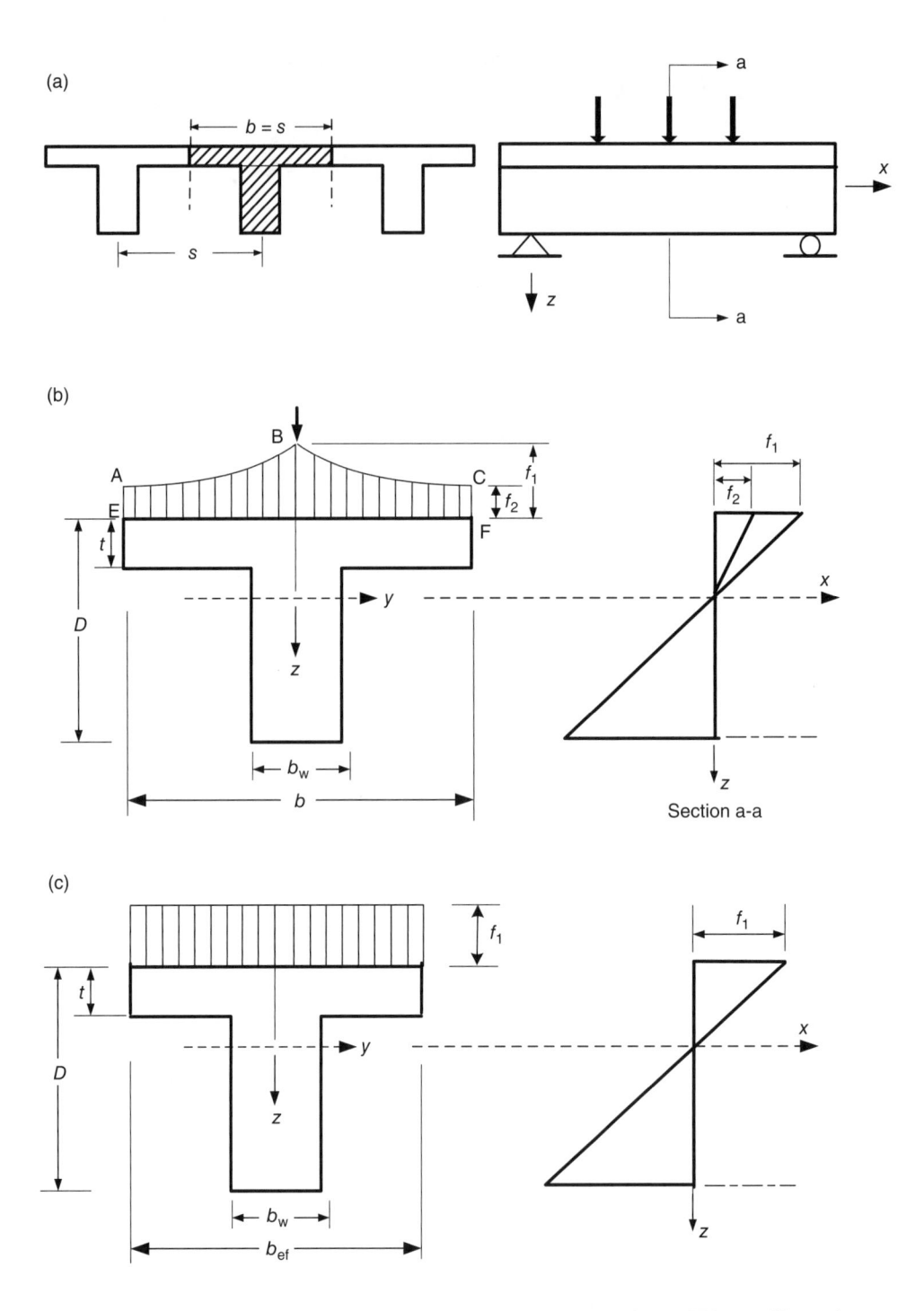

Figure 4.2(1) Stress distributions in T-beams: **(a)** beam-and-slab system and typical T-beam; **(b)** actual stress distribution; and **(c)** effective stress distribution

CHAPTER 4 T-BEAMS AND IRREGULAR-SHAPED SECTIONS

To simplify the problem, the actual section in Figure 4.2(1)b may be replaced by an 'effective' one, as shown in Figure 4.2(1)c, in which the effective flange width is taken as

$$b_{\text{ef}} = \frac{A}{f_1}$$

Equation 4.2(1)

where A is the area of the stress diagram E-A-B-C-F. In the effective section, the maximum stress (f_1) is uniformly distributed over the effective width (b_{ef}) and can be computed in the usual way.

This concept of effective width in the analysis and design of flanged beams has been accepted by the structural engineering profession for many decades. Because of its simplicity, its use is still recommended in all known codes of practice. Table 4.2(1) shows a collection of past and present code recommendations for the values of b_{ef} from 11 countries, plus that of the Comité Européen du Béton – Fédération Internationale de la Precontrainte (CEB-FIP).

Table 4.2(1) Code recommendations for effective width (b_{ef}) for a symmetrical T-beam

Code group	b_{ef}
British Code of Practice 114, Belgium, (Former) Soviet Union	$12t + b_w$ or $\dfrac{L^*}{3}$
American Concrete Institute (ACI) 318–2014	$16t + b_w$ or $\dfrac{L^*}{4}$
Germany; Greece; Spain; Czech and Slovak Republics	$12t + b_w$ or $\dfrac{L^*}{2}$
Italy	$10t + b_w$ or $\dfrac{L^*}{6}$
Netherlands	$16t$ or $\dfrac{L^*}{3}$
Comité Européen du Béton – Fédération Internationale de la Precontrainte CEB-FIP 1978, British Standard BS 8110–1985	$b_w + \dfrac{L}{5}$
Eurocode (EC2) (1992)	For endspan: $b_w + 0.17L$ For interior spans: $b_w + 0.14L$ For cantilever: $b_w + 0.20L$

*Whichever is less

The terms used in these formulas are defined in Figure 4.2(2). The formulas recommended in AS 3600–2009 (Clause 8.8.2) for T-beams (Figure 4.2(2)a) are

$$b_{ef} = b_w + 0.2L_o \leq s \qquad \text{Equation 4.2(2)}$$

and for L-beams (Figure 4.2(2)b) are

$$b_{ef} = b_w + 0.1L_o \leq s' \qquad \text{Equation 4.2(3)}$$

where L_o is the distance between the points of zero bending moment. For simply supported beams

$$L_o = L \qquad \text{Equation 4.2(4)a}$$

For continuous beams

$$L_o = 0.7L \qquad \text{Equation 4.2(4)b}$$

where L is the centre-to-centre span of the beam. Equation 4.2(2) is identical to the one recommended by CEB-FIP and the British Standard BS 8110–1985 (see Table 4.2(1)). Equations 4.2(2) and 4.2(3) have appeared in the Australian Standard since AS 1480–1982.

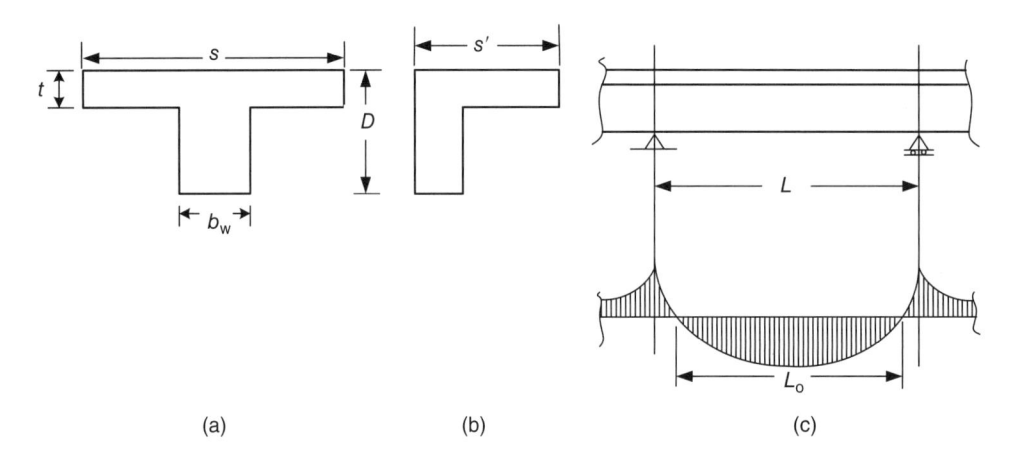

Figure 4.2(2) Definitions of effective width and other terms for T- and L-beams

Following a computer-based study of the stress distributions in 243 T-beams, Loo and Sutandi (1986) recommended the following formulas for b_{ef} under three different loading conditions:
1. for a concentrated load at midspan (load case a)

$$\frac{b_{ef}}{s} = 0.3507 \left(\frac{s}{L}\right)^{-0.4451} \left(\frac{t}{D}\right)^{0.2128} \left(\frac{b_w}{D}\right)^{0.1204} \left(\frac{L}{D}\right)^{0.1451} \qquad \text{Equation 4.2(5)}$$

2. for concentrated loads at third points (load case b)

$$\frac{b_{ef}}{s} = 0.6338 \left(\frac{s}{L}\right)^{-0.3976} \left(\frac{t}{D}\right)^{0.2372} \left(\frac{b_w}{D}\right)^{0.0724} \left(\frac{L}{D}\right)^{0.0014}$$

Equation 4.2(6)

3. for uniformly distributed load (load case c)

$$\frac{b_{ef}}{s} = 0.8651 \left(\frac{s}{L}\right)^{-0.1656} \left(\frac{t}{D}\right)^{0.1370} \left(\frac{b_w}{D}\right)^{-0.0312} \left(\frac{L}{D}\right)^{0.0191}$$

Equation 4.2(7)

The three load cases are defined in Figure 4.2(3); by definition $b_{ef}/s \leq 1$. Unlike the code recommendations, these 'empirical' formulas are expressed in terms of all the dimensional variables describing a given T-beam. A detailed comparative study of Equations 4.2(5), (6) and (7) and the equations given in Table 4.2(1) is given elsewhere (Loo and Sutandi 1986). Equation 4.2(2) compares well with the more rigorous Equations 4.2(5), (6) and (7).

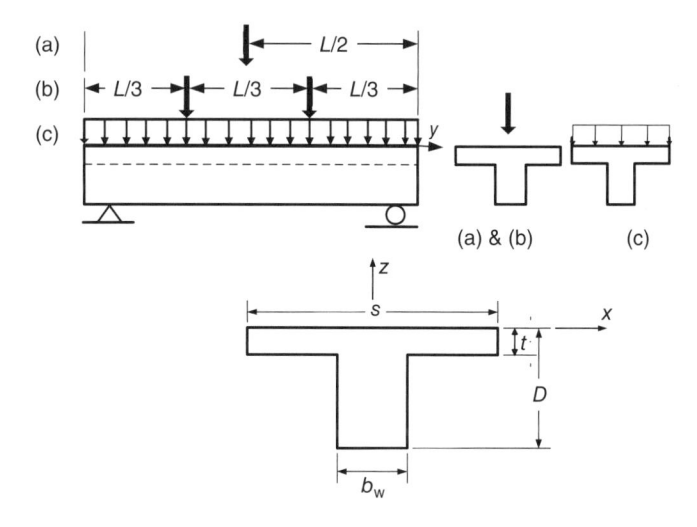

Figure 4.2(3) Load cases: **(a)** concentrated load at midspan; **(b)** concentrated loads at third points; and **(c)** uniformly distributed load

All the formulas for b_{ef} presented in this section are for T-beams. However, a multibox system may be idealised as an assembly of inverted L- and T-beams, as shown in Figure 4.2(4). Note that the bottom flanges are in tension and may be ignored in the analysis and design calculations. The formulas given previously may be safely used for computing the effective width of the box system. Note also that the value of b_{ef} does not significantly affect M_u because of the large compression zone available in the flange area. However, this may not be the case for deflection and other serviceability calculations, in which case Equations 4.2(5), (6) or (7) may be preferable for the respective loading cases.

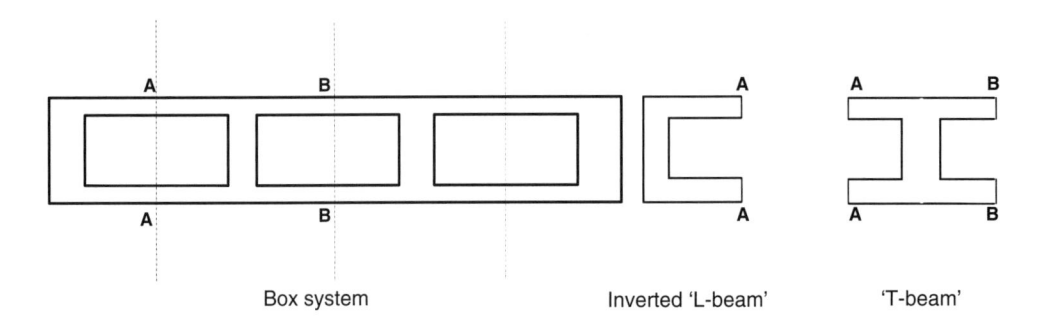

Figure 4.2(4) A multibox system idealised as an assembly of inverted L- and T-beams

4.2.3 Criteria for T-beams

For a typical T-beam, as shown in Figure 4.2(5), if the neutral axis at the ultimate state stays within the thickness (t) of the flange, the T-beam may be analysed or designed as a rectangular beam with the width equal to b_{ef}. This is because the concrete in the tensile zone is ineffective and can be ignored.

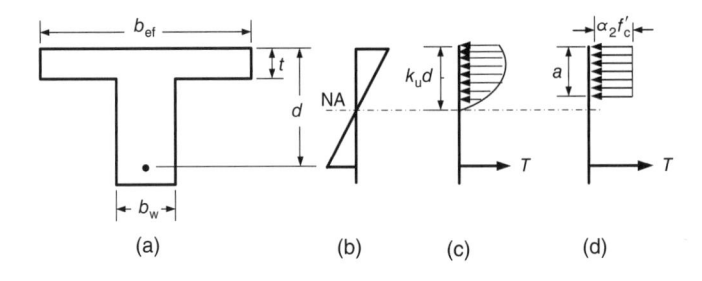

Figure 4.2(5) Stress and strain distributions for a typical T-beam section

For simplicity, and following AS 1480–1982 (Clause A1.2), which recommends that if t is greater than the depth of the rectangular stress block (a) (see Figure 4.2(5)), that is

$$t \geq \frac{A_{\text{st}} f_{\text{sy}}}{\alpha_2 f'_c b_{\text{ef}}}$$

Equation 4.2(8)

the beam may be treated as a rectangular one. Otherwise, if

$$t < \frac{A_{\text{st}} f_{\text{sy}}}{\alpha_2 f'_c b_{\text{ef}}}$$

Equation 4.2(9)

then the beam is to be treated using the procedure developed hereafter for T-sections.

4.2.4 Analysis

Similar to doubly reinforced sections, the T-beam detailed in Figure 4.2(6)a may be 'decomposed' into a web-beam and a flange-beam, respectively, shown in Figures 4.2(6)b and c. Note that a flange-beam should not be confused with a flanged beam.

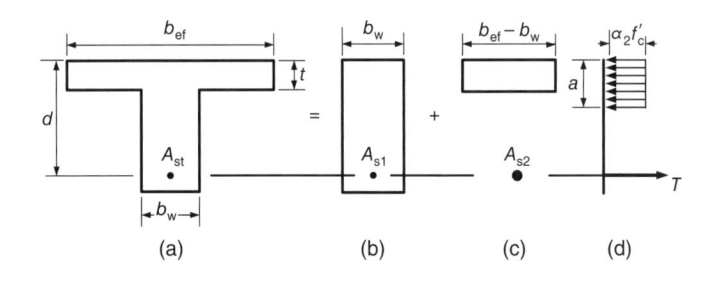

Figure 4.2(6) A T-beam 'decomposed' into a web-beam and a flange-beam: **(a)** typical T-beam; **(b)** web-beam; **(c)** flange-beam; and **(d)** stress diagram

Hence, the ultimate moment for the T-beam

$$M_u = M_{u1} + M_{u2}$$ Equation 4.2(10)

where M_{u1} is the ultimate moment for the web-beam and M_{u2} is the ultimate moment for the flange-beam.

For the flange-beam in Figure 4.2(6)c,

$$M_{u2} = \alpha_2 f'_c t (b_{ef} - b_w)\left(d - \frac{t}{2}\right)$$ Equation 4.2(11)

And by considering $\Sigma F_x = 0$ for the flange-beam, we have

$$A_{s2} f_{sy} = \alpha_2 f'_c t (b_{ef} - b_w)$$

from which

$$A_{s2} = \frac{\alpha_2 f'_c t (b_{ef} - b_w)}{f_{sy}}$$ Equation 4.2(12)

Thus, for the web-beam

$$A_{s1} = A_{st} - A_{s2}$$ Equation 4.2(13)

With A_{s1} computed, the moment M_{u1} may be determined using Equation 3.4(10) if A_{s1} yields at the ultimate state, that is

$$M_{u1} = A_{s1} f_{sy} d \left(1 - \frac{1}{2\alpha_2} \frac{A_{s1} f_{sy}}{b_w d f'_c}\right)$$ Equation 4.2(14)

However, if A_{s1} does not yield at failure, M_{ul} can be calculated using Equation 3.4(18), making use of Equations 3.4(16) and 3.4(17). Note that for the web-beam, the steel ratio in Equation 3.4(16) is taken to be

$$p_t = \frac{A_{s1}}{b_w d} \qquad \text{Equation 4.2(15)}$$

To determine if A_{s1} yields or not at failure, we again need to establish the threshold equation. The neutral axis for the web-beam in Figure 4.2(6)b can be obtained by imposing $\Sigma F_x = 0$, or

$$\alpha_2 f'_c \gamma k_u d b_w = A_{s1} f_{sy} \qquad \text{Equation 4.2(16)}$$

from which

$$k_u = \frac{A_{s1} f_{sy}}{\alpha_2 f'_c \gamma d b_w}$$

For A_{s1} to yield at failure

$$k_u \leq k_{uB}$$

or with Equation 3.4(4)

$$k_u \leq k_{uB} = \frac{600}{600 + f_{sy}} \qquad \text{Equation 4.2(17)}$$

Substituting Equation 4.2(16) into Equation 4.2(17) and rearranging terms gives

$$A_{s1} \leq \alpha_2 \gamma \frac{f'_c}{f_{sy}} b_w d \frac{600}{600 + f_{sy}} \qquad \text{Equation 4.2(18)}$$

If, for a web-beam, Equation 4.2(18) prevails, then A_{s1} yields at ultimate. Otherwise, it does not.

Note that ACI 318–2014 specifies that

$$A_{st} \leq \frac{3}{4}\left[0.85\gamma \frac{f'_c}{f_{sy}} b_w d \frac{600}{600 + f_{sy}} + 0.85 \frac{f'_c}{f_{sy}} t(b_{ef} - b_w)\right] \qquad \text{Equation 4.2(19)}$$

where the two terms within the square brackets are A_{s1} and A_{s2}, respectively, assuming $\alpha_2 = 0.85$.

4.2.5 Design procedure

In practice, the sectional properties of a T-beam are generally known, as they are part of a building floor system or a bridge deck. Given M^*, the design objective is to determine A_{st}. The major steps given below may be followed:

1. Compute the effective flange width using Equation 4.2(2) or Equation 4.2(3), as appropriate. Estimate the value of d.
2. If

$$M^* \leq \phi\alpha_2 f'_c b_{ef} t\left(d - \frac{t}{2}\right)$$ Equation 4.2(20)

the design procedure for singly reinforced rectangular beams may be followed using b_{ef} in place of b.

3. If

$$M^* > \phi\alpha_2 f'_c b_{ef} t\left(d - \frac{t}{2}\right)$$ Equation 4.2(21)

the stress block extends into the web at failure. Thus, the beam should be treated as a T-section, the design of which follows the steps below.

4. Determine the ultimate moment (M_{u2}) for the flange-beam using Equation 4.2(11), and the steel content (A_{s2}) using Equation 4.2(12).
5. Compute

$$M_1^* = M^* - \phi M_{u2}$$ Equation 4.2(22)

6. Design the web-beam using the restricted design procedure for singly reinforced rectangular beams, for which the steel ratio (p_{t1}) may be computed using Equation 3.5(6). Note, however, that the term M^* should be replaced by M^*_1. Then

$$A_{s1} = p_{t1} b_w d$$ Equation 4.2(23)

In accordance with the discussion in Section 3.4.1,

$$p_{t1} \leq p_{all} = 0.4\alpha_2 \gamma \frac{f'_c}{f_{sy}}$$ Equation 4.2(24)

This helps ensure that the T-section is under-reinforced. Alternatively, the American Concrete Institute recommendation may be used, or check that Equation 4.2(19) prevails, where

$$A_{st} = A_{s1} + A_{s2}$$ Equation 4.2(25)

For practical T-beam sections, in view of the large compression zone available in the flange area, the likelihood of either Equation 4.2(24) or Equation 4.2(18) not being satisfied is very small. However, if this is not the case, then the use of doubly reinforced T-sections will be necessary.

As per Clause 8.1.6.1 of the Standard,

$$A_{st} \geq \alpha_b b_w d (D/d)^2 \frac{f'_{ct.f}}{f_{sy}}$$ Equation 4.2(25)a

where

$$\alpha_b = 0.20 + \left(\frac{b_{ef}}{b_w} - 1\right)\left(0.4\frac{D_s}{D} - 0.18\right) \geq 0.20\left(\frac{b_{ef}}{b_w}\right)^{\frac{1}{4}}$$

Equation 4.2(25)b

in which D_s is the overall depth of a slab or drop panel.
Further for inverted T- or L-beams

$$\alpha_b = 0.20 + \left(\frac{b_{ef}}{b_w} - 1\right)\left(0.25\frac{D_s}{D} - 0.08\right) \geq 0.20\left(\frac{b_{ef}}{b_w}\right)^{\frac{2}{3}}$$

Equation 4.2(25)c

7. Check accommodation for A_{st} and ensure that the final d is greater or at least equal to the value assumed in Step (1). Revise as necessary.
8. Do a final check to ensure that $\phi M_u \geq M^*$.

4.2.6 Doubly reinforced T-sections

Occasionally, the need for treating doubly reinforced T-beams arises in practice. An obvious example is a beam-and-slab system in which the bending steel in the slab also acts as the compression steel for the T-section.

For the analysis of a doubly reinforced T-section, as shown in Figure 4.2(7), the first step is to check whether or not

$$t \geq \frac{A_{st}f_{sy} - A_{sc}f_{sy}}{\alpha_2 f'_c b_{ef}}$$

Equation 4.2(26)

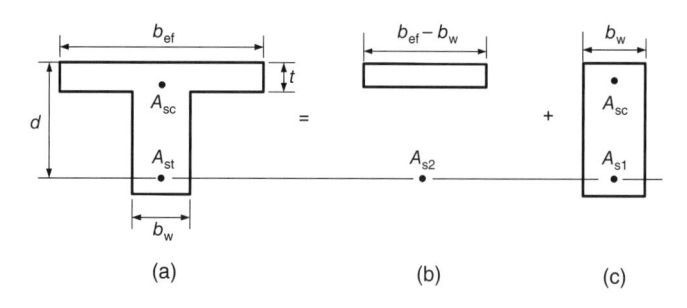

Figure 4.2(7) Decomposition of a doubly reinforced T-section

If Equation 4.2(26) is correct, then the beam may be treated as a rectangular section. Otherwise, a T-beam analysis must be carried out.

The section detailed in Figure 4.2(7)a may be decomposed into a flange-beam and a doubly reinforced web-beam as illustrated in Figures 4.2(7)b and c, respectively. Once again

$$M_u = M_{u1} + M_{u2}$$ **Equation 4.2(27)**

To analyse the flange-beam moment (M_{u2}) and its steel content (A_{s2}), Equations 4.2(11) and 4.2(12), respectively, can be used. Similarly, the tension steel (A_{s1}) is given as

$$A_{s1} = A_{st} - A_{s2}$$ **Equation 4.2(28)**

However, the web-beam is now a doubly reinforced section. As a result, the procedures developed in Sections 3.6.1, 3.6.2 and 3.6.4 must be followed for the various possible ultimate states.

For the design of a doubly reinforced T-beam, the major steps are given in the following list:

1. Compute M_{u2} and A_{s2} using Equations 4.2(11) and 4.2(12), respectively.
2. If M_1^* from Equation 4.2(22) is too large to be taken by a singly reinforced web-beam, double-reinforcement will be necessary.
3. For a doubly reinforced web-beam, the procedure detailed in Section 3.7 may be adopted.

4.2.7 Illustrative examples

Two numerical examples are provided here. Examples 1 and 2 consider the analysis and design of singly reinforced T-sections, respectively.

Example 1: Analysis of singly reinforced T-sections

Problem

Given a T-beam as shown in Figure 4.2(8), reinforced with one layer only of Class N bars. Take $f'_c = 25$ MPa, $f_{sy} = 500$ MPa and compute M'.

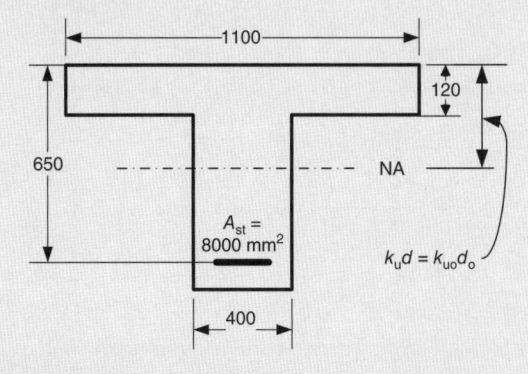

Figure 4.2(8) Cross-sectional details of the example T-beam
Note: all unmarked dimensions are in mm.

Solution

For $f'_c = 25$ MPa, $\alpha_2 = \gamma = 0.85$ and

Equation 4.2(9): $\dfrac{8000 \times 500}{0.85 \times 25 \times 1100} = 171.12 \text{ mm} > t = 120 \text{ mm}$

Thus, the NA at failure is located within the web.
For the flange-beam

Equation 4.2(11): $M_{u2} = 0.85 \times 25 \times 120(1100 - 400) \times \left(650 - \dfrac{120}{2}\right) \times 10^{-6}$

$= 1053 \text{ kNm}$

and

Equation 4.2(12): $A_{s2} = \dfrac{0.85 \times 25 \times 120 \times (1100 - 400)}{500} = 3570 \text{ mm}^2$

Therefore
$A_{s1} = 8000 - 3570 = 4430 \text{ mm}^2$
To check the condition of A_{s1} at failure

Equation 4.2(18): $0.85 \times 0.85 \times \dfrac{25}{500} \times 400 \times 650 \times \dfrac{600}{600 + 500} = 5123 \text{ mm}^2 > A_{s1}$

Therefore
A_{s1} will yield at failure.
For the web-beam

Equation 4.2(14): $M_{u1} = 4430 \times 500 \times 650 \times \left(1 - \dfrac{1}{2 \times 0.85} \times \dfrac{4430 \times 500}{400 \times 650 \times 25}\right) \times 10^{-6}$

$= 1151.2 \text{ kNm}$

and

Equation 3.4(7): $k_u = \dfrac{4430 \times 500}{0.85 \times 0.85 \times 25 \times 400 \times 650} = 0.472$

Assuming that the bars are located in a single layer, we have $d = d_o$, and $k_{uo} = k_u = 0.472$. Thus Equations 3.4(20)a with b: $\phi = 0.679$ and $M' = 0.679(1151.2 + 1053) = 1496.6$ kNm.

Since $k_{uo} > 0.36$, appropriate compression reinforcement must be provided. With reference to Figure 4.2(8), the required $A_{sc} = 0.01 \times [1100 \times 120 + (0.472 \times 650 - 120) \times 400] = 2067.2 \text{ mm}^2$, which may all be placed in the flange at mid-depth.

Example 2: Design of singly reinforced T-sections

Problem

Given a T-beam with the dimensions shown in Figure 4.2(9), f'_c = 32 MPa, f_{sy} = 500 MPa and M^* = 600 kNm. Design the reinforcement for the section.

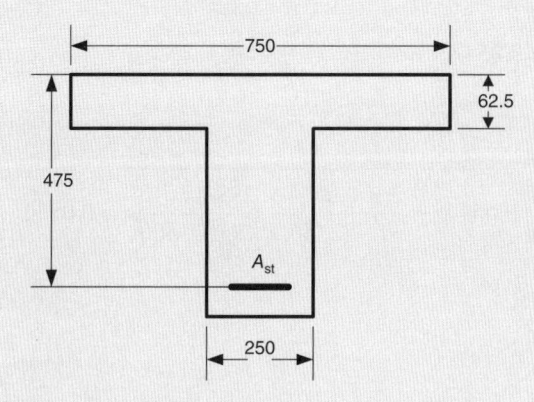

Figure 4.2(9) Cross-sectional details of the design T-beam
Note: all dimensions are in mm.

Solution

For f'_c = 32 MPa, α_2 = 0.85 and γ = 0.826.

To use an alternative method to Section 4.2.3 for criterion checking, assume a rectangular section of $b \times d$ = 750 × 475 and that $a = t$. Then, the effective moment

$$M' = \phi M_u = \phi \alpha_2 f'_c bt\left(d - \frac{t}{2}\right) = 565.8\phi < M^*$$

Therefore, the neutral axis at the ultimate state lies in the web. Or $a > t$.

For the flange-beam

Equation 4.2(11): $M_{u2} = 0.85 \times 32 \times 62.5(750 - 250)\left(475 - \frac{62.5}{2}\right) \times 10^{-6}$

$$= 377.2 \text{ kNm}$$

and

Equation 4.2(12): $A_{s2} = \dfrac{377.2 \times 10^{-6}}{500\left(475 - \dfrac{62.5}{2}\right)} = 1700 \text{ mm}^2$

Since $a = \gamma k_u d = t = 62.5$ mm, we have $k_u = 0.159$. Assuming that the bars are located in a single layer, we have $d = d_o$, and $k_{uo} = k_u = 0.159$. Thus $\phi = 0.8$ as per Equations 3.4(20)a and b.

For the web-beam

Equation 4.2(22): $M_1^* = 600 - 0.8 \times 377.2 = 298.2$ kNm

and

Equation 3.5(7): $\xi = \dfrac{0.85 \times 32}{500} = 0.0544$

with which

Equation 3.5(6): $p_{t1} = 0.0544 - \sqrt{0.0544^2 - \dfrac{2 \times 0.0544 \times 298.2 \times 10^6}{0.8 \times 250 \times 475^2 \times 500}} = 0.0154$

Thus, the web reinforcement

$A_{s1} = 0.0154 \times 250 \times 475 = 1829$ mm^2

As per Equation 4.2(24)

$p_{all} = 0.4 \times 0.85 \times 0.826 \times \dfrac{32}{500} = 0.01797 > p_{t1}$, which is acceptable.

The right-hand side of Equation 4.2(18) gives

$0.85 \times 0.826 \times \dfrac{32}{500} \times 250 \times 475 \times \dfrac{600}{600 + 500} = 2911$ mm$^2 > A_{s1}$

Therefore, A_{s1} yields at the ultimate state.

Finally, $A_{st} = 1829 + 1700 = 3529$ mm^2

Further considerations to complete the design are to:

- select a bar group and check accommodation
- double-check the capacity M_u as necessary.

4.3 NONSTANDARD SECTIONS

4.3.1 Analysis

From time to time, a structural engineer is required to analyse nonstandard sections in bending. For example, the handling of precast concrete piles during construction will lead to bending of the piles (beams), and for aesthetic reasons, the use of nonstandard sections is common. Figure 4.3(1) shows some typical nonstandard sections.

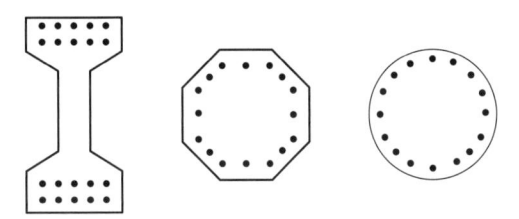

Figure 4.3(1) Typical nonstandard beam sections

For an arbitrary section, it is not possible to develop a closed form solution for analysis or design. Instead, based on the fundamentals, a trial and error approach can be used and the ultimate moment (M_u) obtained numerically. In Figure 4.3(2), the section would fail when $\varepsilon_{cu} = 0.003$ (see Figure 4.3(2)b). With an assumed position of the neutral axis (d_{NA}), all the strain values (ε_{sc1}, ε_{sc2}, ...) for the steel bars can be computed accordingly. Then, in Figure 4.3(2)c, all the steel forces (compressive and tensile) and the concrete compressive force can be determined.

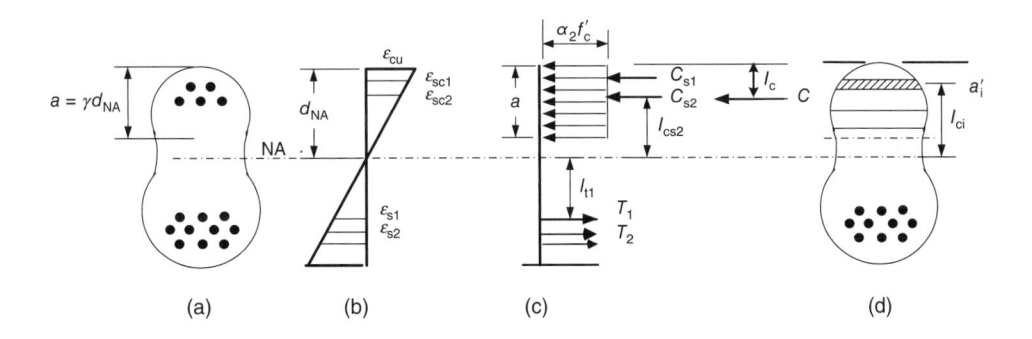

Figure 4.3(2) Details of a nonstandard concrete section

The assumed d_{NA} is the true value only if all the tensile forces (T_1, T_2, ...) are in equilibrium with the compressive forces (C_1, C_{s1}, ...). Otherwise, a second value of d_{NA} should be taken and the process reiterated until the equilibrium criterion is satisfied. Then, M_u can be readily determined by taking moments of all the horizontal forces with respect to a given reference level (e.g. that of the neutral axis). The following steps may be followed:
1. Assume d_{NA}.
2. Compute

$$C = \alpha_2 f'_c A' \qquad \text{Equation 4.3(1)}$$

where A' is the compressive stress block area that may be computed numerically as $\Sigma a'_i$.

3. From Figure 4.3(2)b, compute the tensile and compressive steel strains, $\varepsilon_{sc1}, \varepsilon_{sc2} \ldots, \varepsilon_{s1}, \varepsilon_{s2} \ldots$, etc. (given in terms of d_{NA}).
4. Compute the corresponding steel forces, $C_{s1}, C_{s2}, \ldots, T_1, T_2, \ldots$ using

$$C_{si} = E_{si}\varepsilon_{sci} \qquad \text{Equation 4.3(2)}$$

and

$$T_i = E_{si}\varepsilon_{si} \qquad \text{Equation 4.3(3)}$$

Check equilibrium (i.e. if $\Sigma C = \Sigma T$ or not).

The following flow diagram illustrates this step of the process in detail. Note that the tolerance may be set according to the degree of accuracy required in the analysis.

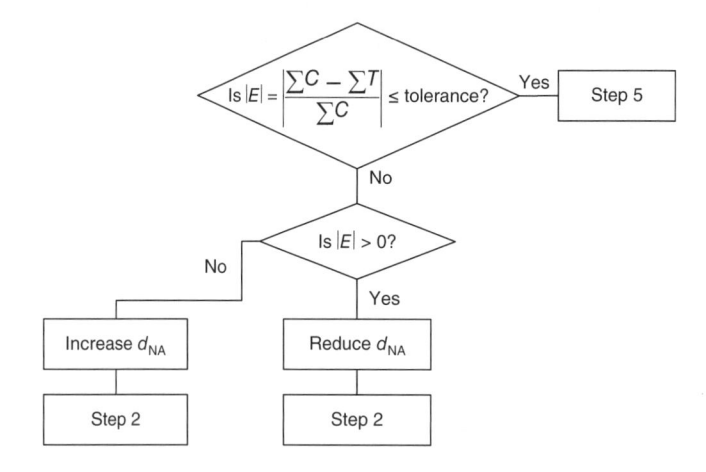

5. Compute the ultimate moment (taken with respect to the neutral axis).

$$M_u = \sum T_i l_{ti} + \sum C_{sj} l_{csj} + \sum \alpha_2 f'_c a'_k l_{ck} \qquad \text{Equation 4.3(4)}$$

Note that if the section is of a mathematically definable shape, it will be possible to set up the equilibrium equation in terms of d_{NA}. Then, no iteration is necessary.

4.3.2 Illustrative examples
Example 1

Problem

For the doubly reinforced section with an irregular shape as shown in Figure 4.3(3), compute the ultimate moment (M_u). Take $f'_c = 25$ MPa.

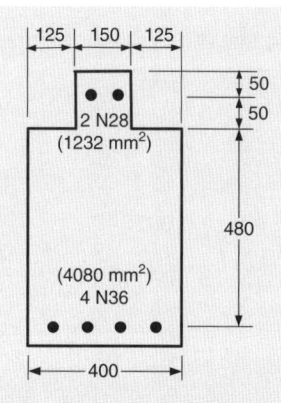

Figure 4.3(3) Cross-sectional details of the irregular shape example problem
Note: all dimensions in mm.

Solution

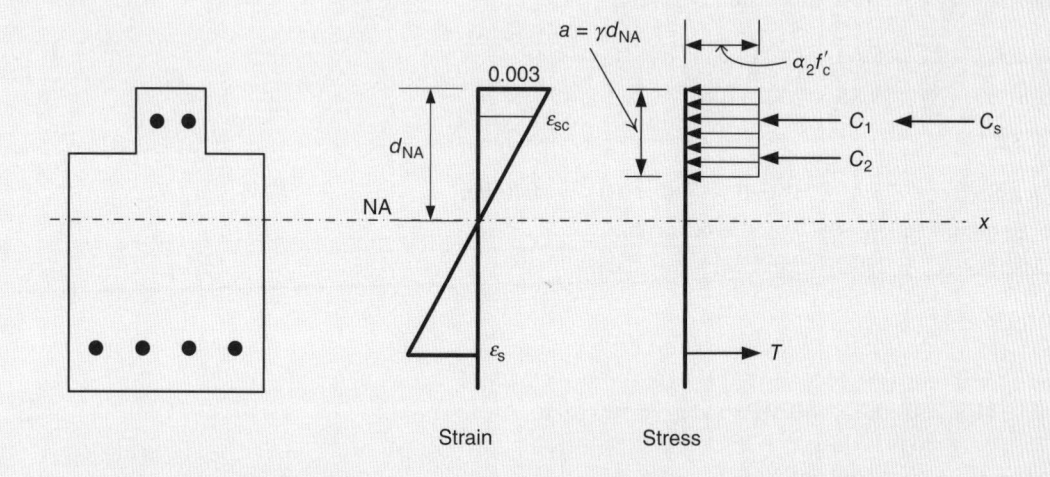

Figure 4.3(4) Stress and strain distribution across the example section

For $f'_c = 25$ MPa, $\alpha_2 = \gamma = 0.85$ as per Equations 3.3(2)a and b, respectively. Based on the strain diagram given in Figure 4.3(4), we obtain

$$\varepsilon_{sc} = \frac{0.003(d_{NA} - 50)}{d_{NA}} \qquad \text{Equation (i)}$$

and

$$\varepsilon_s = \frac{0.003(580 - d_{NA})}{d_{NA}} \qquad \text{Equation (ii)}$$

For the concrete stress over the top area (see the stress diagram in Figure 4.3(4))

$$C_1 = 150 \times 100 \times 0.85 \times f'_c = 318\,750 \text{ N} \qquad \text{Equation (iii)}$$

and for the remaining area

$$C_2 = 400(\gamma d_{NA} - 100) \times 0.85 \times f'_c = 8500(\gamma d_{NA} - 100) \qquad \text{Equation (iv)}$$

The trial and error process below will lead to the required M_u.
Trial 1: Assume $d_{NA} = 200$ mm.

Equation (i): $\varepsilon_{sc} = 0.00225 < \varepsilon_{sy} = \dfrac{500}{200\,000}$
$= 0.0025$; that is, A_{sc} would not yield.

Therefore $f_{sc} = 0.00225 \times 200\,000 = 450$ MPa.
Through **Equation (ii)**, we observe that $\varepsilon_s > \varepsilon_{sy}$. Or $f_s = 500$ MPa.
The total horizontal force in compression is given as

$$\begin{aligned}C &= C_1 + C_2 + C_s = C_1 + C_2 + A_{sc}f_{sc} \\ &= [318\,750 + 8500(0.85 \times 200 - 100) + 1232 \times 450] \times 10^{-3} \\ &= 1468.2 \text{ kN}\end{aligned}$$

The total tensile force $T = A_{st}f_s = 4080 \times 500 \times 10^{-3} = 2040$ kN
Since $T > C$, assume a larger d_{NA} in the next trial.
Trial 2: Assume $d_{NA} = 250$ mm, and we have

$$\varepsilon_{sc} = 0.0024 < \varepsilon_{sy} \text{ and } \varepsilon_s > \varepsilon_{sy}$$

or

$$f_{sc} = 0.0024 \times 200\,000 = 480 \text{ MPa and } f_s = 500 \text{ MPa.}$$

Then $C = C_1 + C_2 + C_s = 1866.4$ kN $< T$.
Try a still larger d_{NA}.
Trial 3: Assuming $d_{NA} = 275$ mm and in a similar process, we obtain

$$C = 2060.4 \text{ kN} \approx T.$$

Accept $d_{NA} = 275$ mm, and by taking moments about the level of T, we have

$$\begin{aligned}M_u &= C_1\left(580 - \frac{100}{2}\right) + C_2\left[580 - 100 - \frac{(\gamma d_{NA} - 100)}{2}\right] + C_s \times 530 \\ &= \left\{318\,750 \times 530 + 1\,136\,875 \times \left[480 - \frac{(0.85 \times 275 - 100)}{2}\right] + 604\,789 \times 530\right\} \times 10^{-6} \\ &= 959.2 \text{ kNm}\end{aligned}$$

Note that for beam sections made up of rectangles and other simple shapes, the exact value of d_{NA} may be determined by equating the total tensile and compressive forces.
In our case

$$C = [318\,750 + 8500(\gamma d_{NA} - 100) + 1232 \times 500] \times 10^{-3}$$

and

$$T = 2040 \text{ kN}$$

But

$$C = T$$

from which

$$d_{NA} = 278.5 \text{ mm}$$

However, in all cases before accepting such an 'exact' d_{NA}, ensure that the resulting stress conditions in A_{st} and A_{sc} (i.e. yielding or otherwise) are as assumed in the first place.

Example 2

Problem

Figure 4.3(5) illustrates the section of a beam in a structure containing prefabricated elements. The total width and total depth are limited to 450 mm and 525 mm, respectively. Tension reinforcement used is 4 N32 bars. Using $f'_c = 32$ MPa, determine the moment capacity M_u of the section.

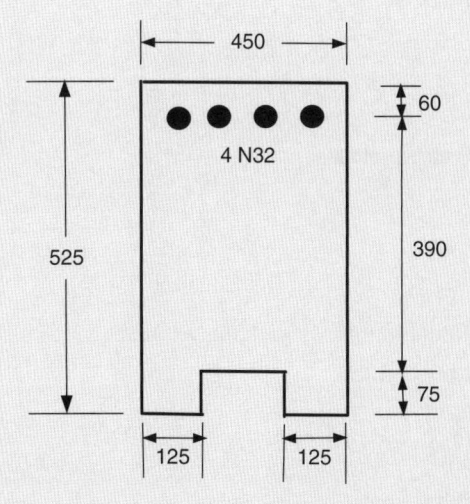

Figure 4.3(5) Cross-sectional details of a prefabricated irregular shape beam section
Note: all dimensions are in mm.

Solution

For $f'_c = 32$ MPa, $\alpha_2 = 0.85$ and $\gamma = 0.826$ as per Equations 3.3(2)a and b, respectively. Assuming that tension reinforcement A_{st} yields at failure, and based on the stress diagram where $d_{NA} = k_u d$ as given in Figure 4.3(6), we obtain

$$C = [0.85 \times 32 \times (\gamma d_{NA} - 75) \times 450 + 2 \times 0.85 \times 32 \times 125 \times 75] \times 10^{-3}$$

and

$$T = 3216 \times 500 \times 10^{-3} = 1608 \text{ kN}$$

But

$$C = T$$

from which

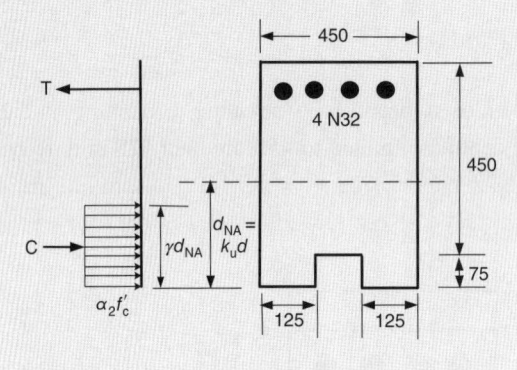

$$d_{NA} = k_u d = 199.4 \text{ mm}$$

Figure 4.3(6) Stress distribution across the example section

Check stress condition for A_{st}:

Now, $k_u = \dfrac{k_u d}{d_o} = \dfrac{199.4}{(525-60)} = 0.429 < k_{uB} = 0.545$ (from Equation 3.4(4))

This confirms the assumption of tension failure or A_{st} yielding at failure.

Finally, taking moment about the level of T gives

$$M_u = 0.85 \times 32 \times (0.826 \times 199.4 - 75) \times 450 \times \left(525 - 60 - 75 - \dfrac{(0.826 \times 199.4 - 75)}{2}\right) \times 10^{-6}$$

$$+ 2 \times 0.85 \times 32 \times 75 \times 125 \times \left(525 - 60 - {}^{75}\!/_2\right) \times 10^{-6}$$

$$= 597.0 \text{ kNm}$$

4.4 CONTINUOUS BEAMS

The analysis and design procedures presented in Chapter 3 and herein are applicable to sections in statically determinate beams, as well as to sections in continuous beams and rigid frame structures. A simplified method for determining the design bending moment (M^*) and shear (V^*) in continuous beams is given in Section 9.2.1. Alternatively, linear–elastic methods of analysis may be used, in which case the Standard allows redistribution of moment at interior supports. Details may be found in Clause 6.2.7 of the Standard.

The design of continuous beams for serviceability requirements is discussed in Sections 5.3.5 and 5.5. The formulas given in Chapters 6, 7 and 8, respectively, for shear, torsion and stress development, are equally applicable to continuous structures.

4.5 DETAILING AND COVER

Reinforcement is generally encased by concrete in reinforced concrete structural members. Once the concrete outline for a structural member is proportioned in accordance with the architectural and structural requirements, the next step in structural design is to determine the quantity of reinforcing steel in each face of the member. It is also important to determine how the steel can be fixed in position using stirrups, tie wires, chairs and spacers until the concrete is placed and has hardened. Detailing of reinforcement is thus the interface between the actual design of the concrete structure and what is to be constructed. Detailing is also important for durability, as poor placement of reinforcement leads to insufficient cover and long-term problems (CIA 2010).

Two primary purposes of cover are for durability and for fire resistance. As a general rule cover is selected from an appropriate table such as Tables 1.4(2) or 1.4(3) for durability and from tables in Section 5 of the Standard for fire resistance. However, there are several cases where additional cross-checks are required. A secondary reason for adequate cover is to ensure that the stresses in steel and concrete can be transferred, one to another, by bond (CIA 2010). These actions, called stress development and anchorage, are dealt with in Chapter 8. The 'cover' required for these purposes is measured not to the 'nearest bar' but to the bar whose stress is being developed. An example is the 'cover' to a longitudinal bar in a beam which is enclosed by a fitment – the latter piece of steel is therefore the 'nearest' surface (CIA 2010).

4.6 SUMMARY

Based on first principles and in accordance with the recommendations of the Standard, formulas for the analysis and design of reinforced concrete T-beams and irregular–shaped sections are derived in this chapter. Reinforcement detailing including concrete cover requirements is explained. Also included are numerical examples to illustrate the applications of the formulas and procedures developed.

4.7 PROBLEMS

1. Figure 4.7(1) details a square beam vertically loaded symmetrically in the diagonal direction. Given $f'_c = 32$ MPa, compute M_u.

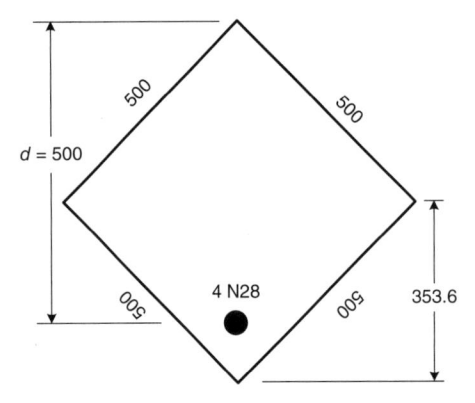

Figure 4.7(1) Cross-sectional details of a square beam

Note: all dimensions are in mm.

2. A symmetrically loaded triangular beam is shown in Figure 4.7(2) with $f'_c = 32$ MPa. Compute M_u.

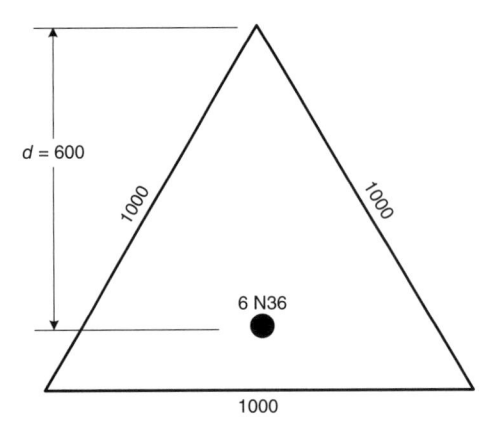

Figure 4.7(2) Cross-sectional details of a triangular beam

Note: all dimensions are in mm.

3. For the section shown in Figure 4.7(3), compute M_u. Take $f'_c = 20$ MPa.

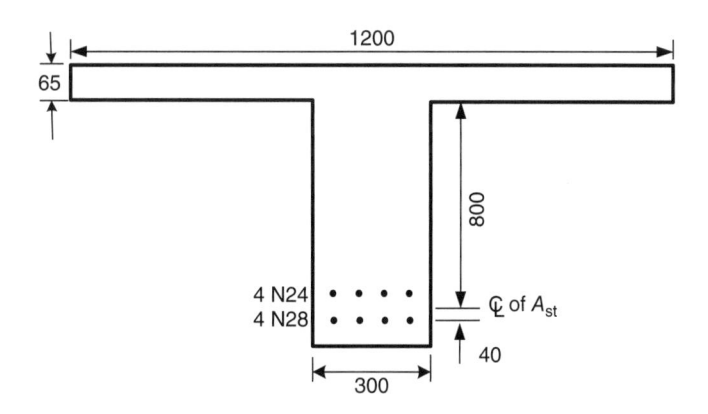

Figure 4.7(3) Cross-sectional details of a T-section

Note: all dimensions are in mm.

4. Figure 4.7(4) details a typical T-beam unit in a beam-and-slab floor with a span of $L = 10$ m. Determine the reliable moment capacity M' for the T-section shown. Take $f'_c = 32$ MPa.

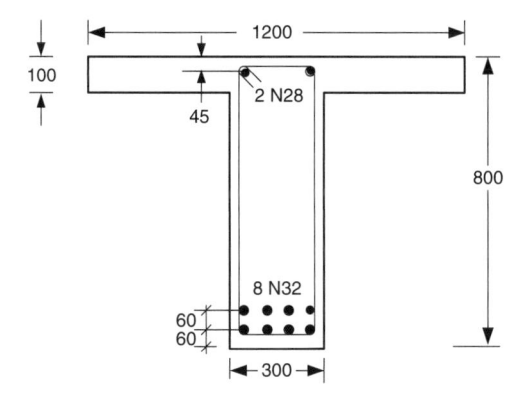

Figure 4.7(4) Cross-sectional details of a typical T-beam unit

Note: all dimensions are in mm.

5. Design and detail the reinforcement for the T-section shown in Figure 4.7(5) for $M^* = 3700$ kNm. Use N32 bars only; centre-to-centre spacing of bar layers is set at 75 mm. Assume $f'_c = 20$ MPa and an A1 exposure classification. A final check must be made on your design for adequacy.

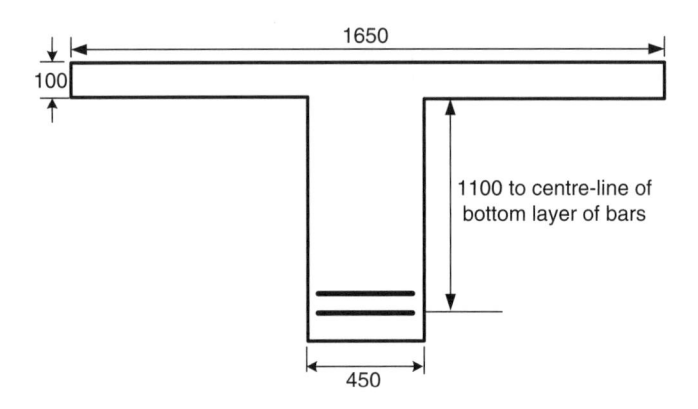

Figure 4.7(5) Cross-sectional details of a T-section

Note: all dimensions are in mm.

6. In addition to its own weight, the beam in Figure 4.7(6)a is to carry a uniformly distributed live load of 12 kN/m. Figure 4.7(6)b shows the cross-sectional details of the beam.

   Design and detail the reinforcement for the section just left of the support at B. Take $f_c' = 32$ MPa and the cover to the centre of the extreme layer of bars as 50 mm. Use only N28 bars, noting that reinforcing bars may be spread over the width of the flange.

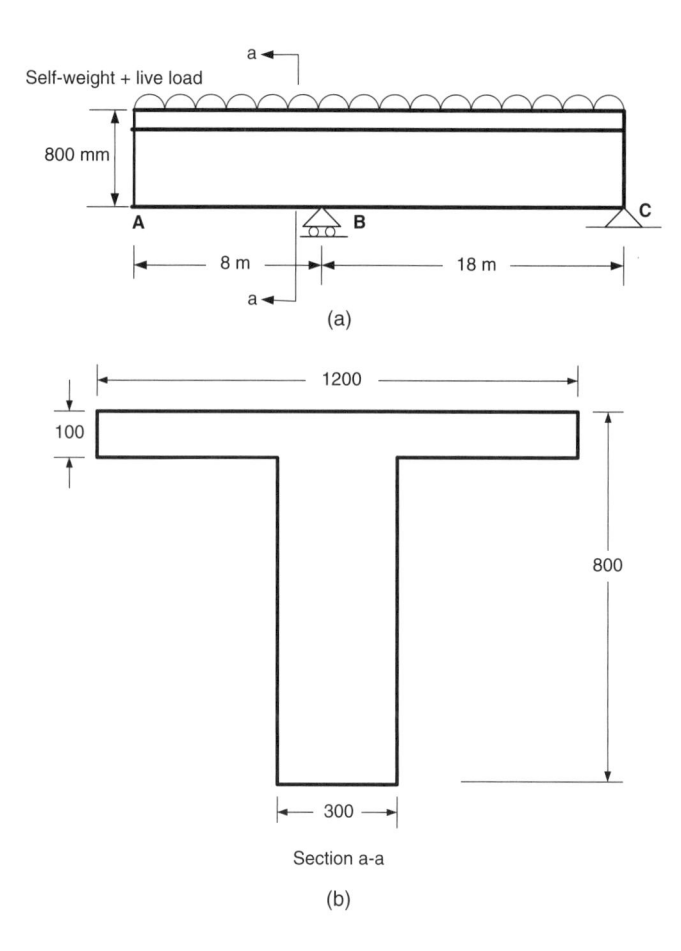

Figure 4.7(6) Loading configuration and cross-sectional details of the design beam

Note: all dimensions are in mm unless otherwise specified.

7. The cross-section of a footbridge structure shown in Figure 4.7(7)a may be idealised as the flanged beam illustrated in Figure 4.7(7)b. For the given loading plus self-weight, design and detail the longitudinal steel reinforcement.

   Take f'_c = 32 MPa and use only N32 bars; exposure classification B1 applies. Note that the full widths of the top and bottom flanges may be used to accommodate the steel reinforcing bars.

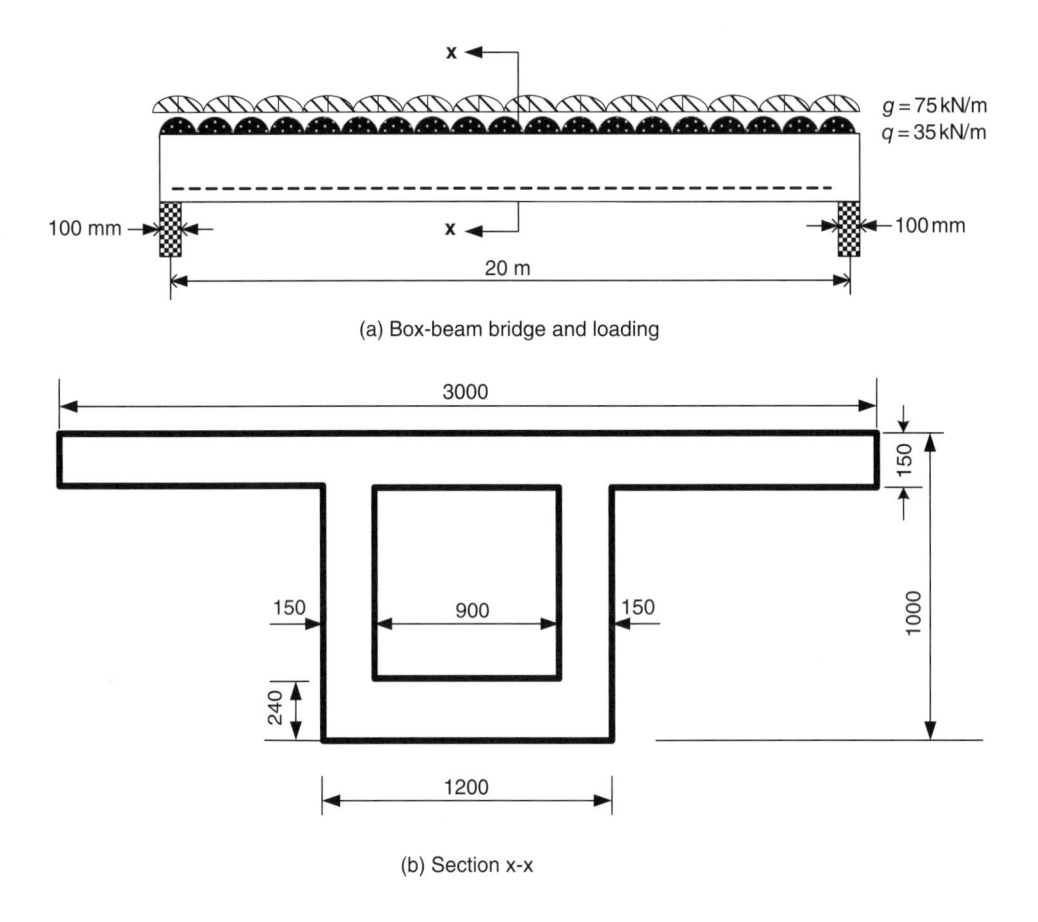

(a) Box-beam bridge and loading

(b) Section x-x

Figure 4.7(7) Loading configuration and sectional details of a footbridge structure

Note: all dimensions are in mm unless otherwise specified.

8. The beam-and-slab floor system detailed in Figure 4.7(8) is of reinforced concrete design. What is the effective moment capacity of a typical T-beam unit, assuming that $d_o = 875$ mm? Assume $f'_c = 25$ MPa.

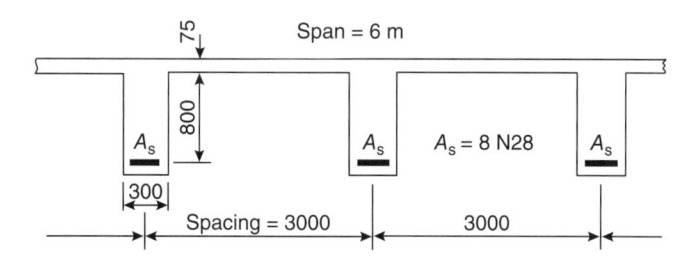

Figure 4.7(8) Details of a beam-and-slab floor system
Note: all dimensions are in mm unless otherwise specified.

9. In addition to its own weight, the beam in Figure 4.7(9)a is to carry a uniformly distributed live load of 18 kN/m and the two concentrated live loads as shown. Figure 4.7(9)b shows the cross-sectional details of the beam.

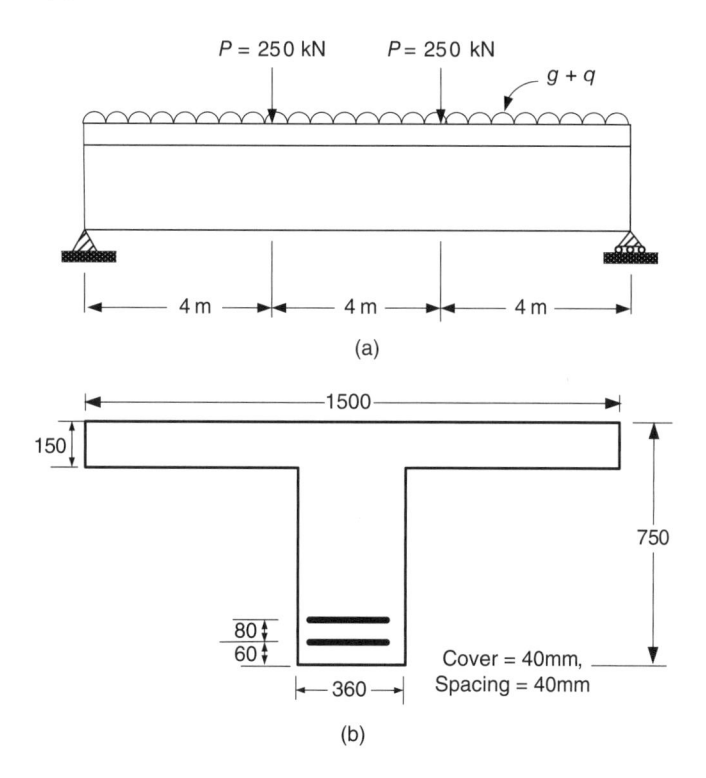

Figure 4.7(9) Loading configuration and cross-sectional details of the design beam
Note: all dimensions are in mm unless otherwise specified.

Design and detail the reinforcement for the midspan section. Take $f_c' = 32$ MPa and use two layers of N40 bars as main reinforcement with a cover of 40 mm and spacing between bars of 40 mm.

10. A reinforced concrete beam section for a precast concrete structural system is as shown in Figure 4.7(10). Assuming $f_c' = 32$ MPa:
    (a) show that the section is under-reinforced
    (b) compute the ultimate moment M_u.
    Hint: make use of the compatibility and equilibrium conditions.

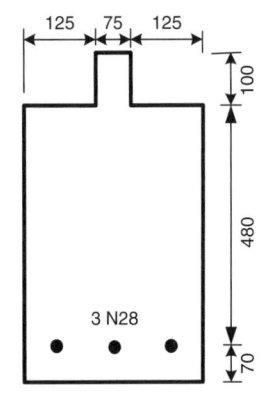

Figure 4.7(10) Cross-sectional details of a precast reinforced concrete beam section

Note: all dimensions are in mm.

11. Figure 4.7(11) represents the section of a beam in a structure containing prefabricated elements. The total width and total depth are limited to 350 and 525 mm, respectively. Tension reinforcement used is 4 N28 bars. Using $f_c' = 32$ MPa, determine the moment capacity M_u of the section.

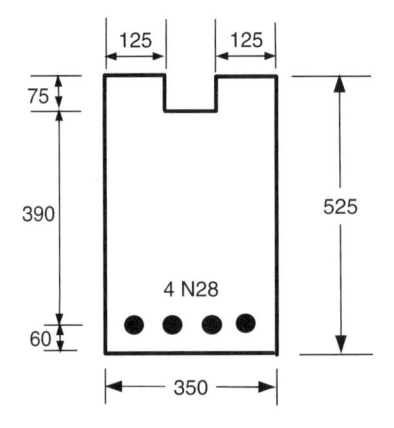

Figure 4.7(11) Cross-sectional details of a prefabricated beam section

Note: all dimensions are in mm.

12. The L-beam illustrated in Figure 4.7(12) is to carry, in addition to its own weight, a dead load $g = 25$ kN/m and a live load $q = 100$ kN/m over a simply supported span of 8 m. Design and detail the tension reinforcement. Use N32 bars and take $f'_c = 50$ MPa; maximum aggregate size is 20 mm, $\rho_w = 24$ kN/m^3 and exposure classification B1 applies.

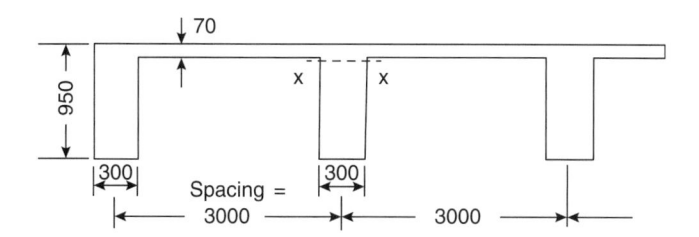

Figure 4.7(12) Details of an L-beam in a beam-and-slab system

Note: all dimensions are in mm.

13. The overhang of a continuous T-beam, illustrated in Figure 4.7(13)a, can be treated as a cantilever beam. In addition to its own weight, the cantilever is to carry a concentrated live load of 75 kN at the tip.

    Figure 4.7(13)b shows the dimensions of the T-section. Design and detail the reinforcement for the support (root) section. Use compression steel if necessary. Take $f'_c = 25$ MPa; the exposure classification is A2 and the maximum aggregate size = 20 mm. Use only N28 bars, noting that reinforcing bars may be spread over the width of the flange.

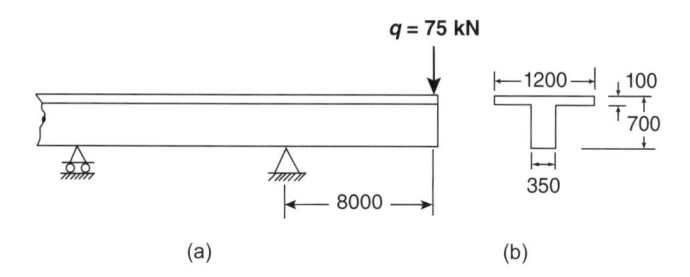

Figure 4.7(13) Details of a continuous T-beam with overhang

Note: all dimensions are in mm.

14. A simply supported beam (AB) with an overhang (BC) is illustrated in Figure 4.7(14)a, with the cross-sectional details given in Figure 4.7(14)b. The dead load comprises self-weight only; $f_c' = 20$ MPa. Compute the maximum allowable concentrated live load q (kN) acting at the tip of the overhang, C.

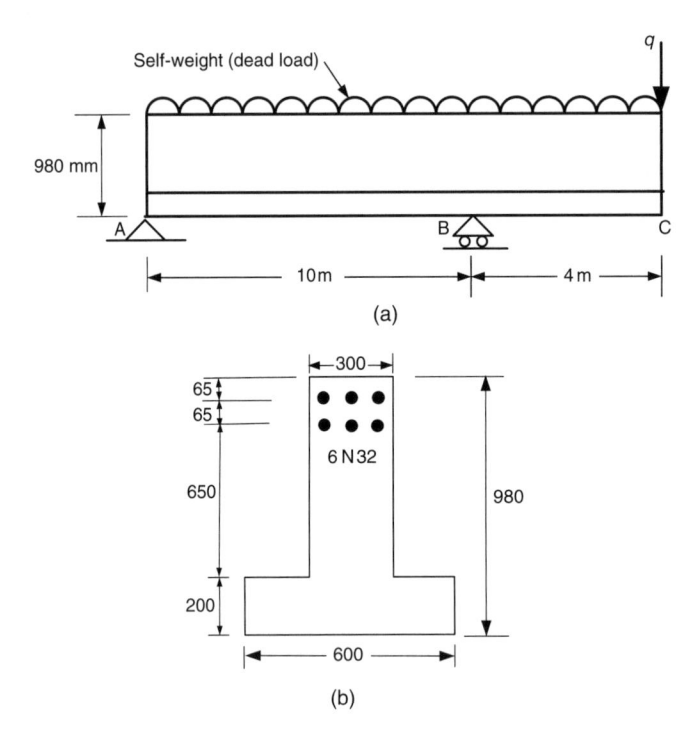

Figure 4.7(14) Details of a simply supported beam with an overhang

Note: all dimensions are in mm unless otherwise specified.

15. In addition to its own weight, the beam in Figure 4.7(15)a is to carry a uniformly distributed dead load of 10 kN/m as well as the concentrated live loads P at the tips of the overhangs. The cross-sectional details of the beam at supports B and C are illustrated in Figure 4.7(15)b.

Considering the support sections only, compute the maximum allowable P. Take $f_c' = 32$ MPa.

(a) Beam configuration

(b) Cross-sectional details at B and C

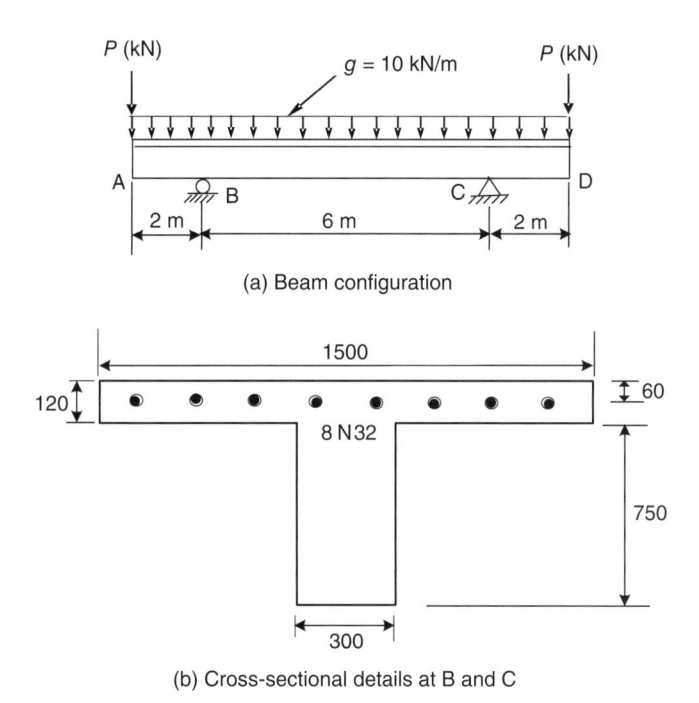

Figure 4.7(15) Details of a beam with overhangs

Note: all dimensions are in mm unless otherwise specified.

16. Figure 4.7(16)a shows a precast reinforced concrete beam subjected to a concentrated live load, P, and a uniformly distributed dead load, g (= 30 kN/m) which includes self-weight of the beam. The cross-sectional details of the beam are shown in Figure 4.7(16)b. Assuming $f_c' = 32$ MPa, compute the maximum allowable value of P that the beam can carry.

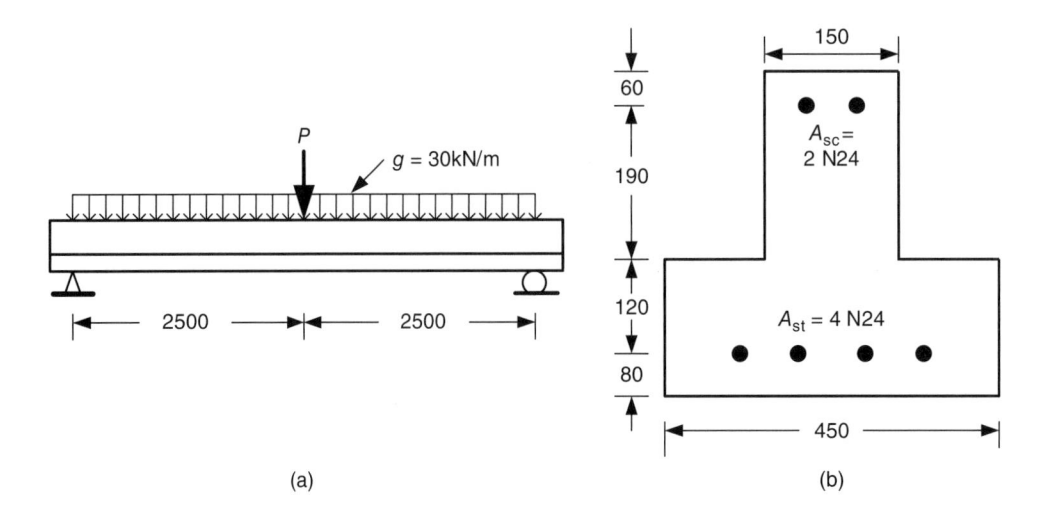

Figure 4.7(16) Details of a precast reinforced concrete beam

Note: all dimensions are in mm.

17. The section shown in Figure 4.7(17) represents a beam in a structure containing prefabricated elements. The total width and total depth are limited to 300 mm and 700 mm, respectively. There is a circular void of 90 mm diameter in the section, as shown. Tension reinforcement used is 3 N32 bars. Taking $f_c' = 32$ MPa, determine the reliable moment capacity of the section.

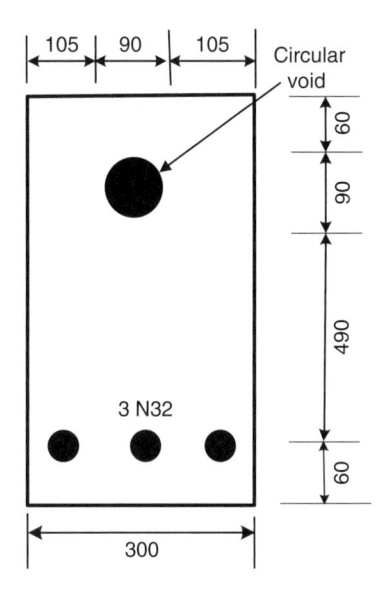

Figure 4.7(17) Cross-sectional details of a prefabricated beam section

Note: all dimensions are in mm.

5 DEFLECTION OF BEAMS AND CRACK CONTROL

5.1 INTRODUCTION

As discussed in Section 1.2, in addition to strength, durability and fire resistance, serviceability is a design requirement specified in the Standard. Practical recommendations are given in Clauses 8.5, 8.6 and 8.9 of the Standard, for the treatments of beam deflection, crack control and slenderness limits for beams, respectively. The Standard also touches very briefly on the vibration of beams, stating qualitatively in Clause 8.7 (and in Clause 9.5 for slabs) that 'vibration of beams shall be considered and appropriate action taken where necessary to ensure that the vibrations induced by machinery, or vehicular or pedestrian traffic, will not adversely affect the serviceability of the structure'.

Although vibration topics are not dealt with in this book, the reader may refer to articles published by Chowdhury and Loo (2001, 2006), and by Salzmann, Fragomeni and Loo (2003) for the damping characteristics of simple and continuous concrete beams.

In this chapter, details are provided on short- and long-term deflection calculations, and on alternative design requirements of maximum span/effective depth ratio, in accordance with the Standard. The analysis of the total deflection of beams under repeated loading is also introduced as an advanced topic. For completeness, crack control of beams is discussed in some detail. All the deflection topics are supplemented by worked out examples. Recommendations for computing effective flange widths have been presented in Section 4.2.2. The reader is referred to the Standard for discussions on slenderness limits and vibrations.

5.2 DEFLECTION FORMULAS, EFFECTIVE SPAN AND DEFLECTION LIMITS

5.2.1 Formulas

The deflection analysis of beams is a topic normally dealt with in courses such as 'mechanics of materials' and 'structural analysis'. Therefore, it is sufficient here to simply summarise the general formulas for computing beam deflections. There are three common types of support conditions for beams:
1. simply supported
2. continuous
3. cantilever.

Loading, on the other hand, includes concentrated or point load, distributed load and applied moment. For a given span (L), the general formulas that encompass all these support and loading conditions may be presented as follows:

- For uniformly distributed loads (w)

$$\Delta = \alpha_w \frac{wL^4}{EI}$$ Equation 5.2(1)

- For concentrated loads (P)

$$\Delta = \alpha_P \frac{PL^3}{EI}$$ Equation 5.2(2)

- For applied moments (M)

$$\Delta = \alpha_M \frac{ML^2}{EI}$$ Equation 5.2(3)

In the above formulas, E and I together define the bending rigidity; α_w, α_P and α_M are coefficients, the values of which are a function of the extent or position of the applied loading and the support conditions. The coefficients also depend on the location at which the deflection is to be calculated. Tables 5.2(1), 5.2(2) and 5.2(3), respectively, present values of α_w, α_P and α_M, for cases that are commonly encountered in the design of reinforced concrete structures. The values of α in these three tables are given for deflections at point C, which is at midspan for the simply supported or statically indeterminate beam, but at the tip for a cantilever beam.

For linear and elastic beams, the principle of superposition holds. Thus, the deflection under a combination of different loading cases can be obtained as the vector sum of the deflections due to individual cases. A more detailed discussion is given in Section 5.3.3.

Table 5.2(1) Values of α_w for distributed loads (in all cases, deflection is positive downwards)

Support and loading conditions	α_w for deflection at C	Support and loading conditions	α_w for deflection at C
Simply supported beam, UDL w, C at midspan, L/2 each side	$\dfrac{5}{384}$	Simply supported beam, UDL w, C at midspan, span L	$\dfrac{5}{192}$
Simply supported beam, triangular load peak at C, span L	$\dfrac{1}{120}$	Fixed-fixed beam, UDL w, C at midspan, span L	$\dfrac{1}{384}$
		Cantilever, UDL w, C at free end, span L	$\dfrac{1}{768}$
Simply supported beam, inverted triangular load, C at midspan, span L	$\dfrac{3}{320}$	Propped cantilever, UDL w, span L	$\dfrac{1}{192}$
Simply supported beam, UDL w over middle portion with a each end, span L	$\dfrac{1}{1920}(5-4a^2)^2$	Cantilever, UDL w, C at free end, span L	$\dfrac{1}{8}$

5.2.2 Effective span

In Equations 5.2(1), 5.2(2) and 5.2(3), the span (L) is taken to be the centre-to-centre distance between the supports for the beam. For concrete beams, the Standard specifies that the effective span (L_{ef}) should be used in computing deflection where, for simple and continuous beams,

$$L_{ef} = \text{the lesser of } (L_n + D) \text{ and } L \qquad \text{Equation 5.2(4)}$$

in which L_n is the clear span, D is the total depth of the beam and L is the centre-to-centre span. For a cantilever beam,

$$L_{ef} = L_n + \frac{D}{2} \qquad \text{Equation 5.2(5)}$$

Table 5.2(2) Values of α_P for concentrated loads (in all cases, deflection is positive downwards)

Support and loading conditions	α_P for deflection at C	Support and loading conditions	α_P for deflection at C
Simply supported, P at L/2	$\dfrac{1}{48}$	Fixed-fixed, P at L/2	$\dfrac{1}{192}$
		Fixed-fixed, P at aL	$(a \leq 1/2)$ $\dfrac{1}{192}(3a - 4a^3)$
Simply supported, P at aL and P at aL (from each end)	$(a \leq 1/2)$ $\dfrac{1}{24}(3a - 4a^3)$	Fixed-fixed, P at aL and P at aL (from each end)	$(a \leq 1/2)$ $\dfrac{1}{96}(3a - 4a^3)$
Simply supported, P at aL	$(a \leq 1/2)$ $\dfrac{1}{48}(3a - 4a^3)$	Cantilever with P at L/2	$\dfrac{7}{768}$
Simply supported, 3 P's at L/4 intervals	$\dfrac{19}{384}$	Cantilever, P at tip	$\dfrac{1}{3}$

Table 5.2(3) Values of α_M for moments

Support and loading conditions	α_M for deflection at C	Support and loading conditions	α_M for deflection at C
Simply supported, M at C	0	Cantilever, M at tip	$\dfrac{1}{16}$
Simply supported, M_1 and M_2 at ends	$\dfrac{1}{16}*$	Cantilever, M at C	$\dfrac{1}{2}**$

$*\Delta_C = \dfrac{L^2}{16EI}(M_1 + M_2)$ $**\Delta_C = \dfrac{ML^2}{2EI}$

5.2.3 Limits

As part of the serviceability design, it is necessary to control the deflection. Clause 2.3.2(a) of the Standard specifies the limits for beams and slabs. An abridged version of Table 2.3.2 in the Standard is given in Table 5.2(4).

Table 5.2(4) Deflection limits

Type of member	Deflection to be considered	Deflection limitation (Δ/L_{ef}) for spans[*]	Deflection limitation (Δ/L_{ef}) for cantilevers[*]
All members	Total deflection	1/250	1/125
Members supporting masonry partitions	The deflection that occurs after the addition or attachment of the partitions[a]	1/500 where provision is made to minimise the effect of movement, otherwise 1/1000	1/250 where provision is made to minimise the effect of movement, otherwise 1/500
Bridge members	The live-load and impact deflection	1/800	1/400

[a]In Equation 5.5(8), use total g and total q.

[*]Note: See Table 2.3.2 in AS 3600–2009 for qualifications.

5.3 SHORT-TERM (IMMEDIATE) DEFLECTION

5.3.1 Effects of cracking

If reinforced concrete beams are uncracked, deflection analysis is a simple matter and can be done readily using Equations 5.2(1), 5.2(2) or 5.2(3), as the case may be. However, reinforced concrete beams crack even under service or working load conditions and cracking occurs at discrete sections along the beam at quite unpredictable positions. Thus, it may be futile to resort to rigorous analytical methods for deflection calculations. Instead, most researchers choose to solve these problems using one semi-empirical method or another. Loo and Wong (1984) studied the relative merits and accuracy of a group of nine such methods. They came to the conclusion that the so-called effective moment of inertia approach is a convenient and accurate one. It is convenient because the traditional deflection formulas (as given in Equations 5.2(1), 5.2(2) and 5.2(3)) are readily applicable with some modifications to the bending rigidity term or EI.

For a cracked reinforced concrete beam, the E term is replaced by E_c as defined in Equations 2.2(5) or 2.2(6), and for I an effective value I_{ef} should be used, where, in general,

$$I_{cr} \leq I_{ef} \leq I_g \qquad \text{Equation 5.3(1)}$$

in which I_g is the gross moment of inertia of the uncracked beam section and I_{cr} is that of a fully cracked beam (see Appendix A).

5.3.2 Branson's effective moment of inertia

The formula for calculating the effective moment of inertia (I_{ef}) adopted in the Standard and several other major codes of practice (including the American Concrete Institute) is originally attributed to Branson (see Loo and Wong 1984). The empirical formula, which takes into consideration the stiffening effects of the concrete in tension between cracks (i.e. tension stiffening), is explicit and all-encompassing. That is

$$I_{ef} = I_{cr} + (I_g - I_{cr}) \left(\frac{M_{cr}}{M_s}\right)^3 \leq I_{ef.\,max} \qquad \text{Equation 5.3(2)}$$

where $I_{ef.max} = I_g$ for $p_t \geq 0.005$ and $I_{ef.max} = 0.6I_g$ for $p_t < 0.005$, which indicates that the Branson formula in its original form underestimates the deflection of very lightly reinforced beams (see Gilbert 2008). The quantities M_{cr}, I_{cr} and M_s in Equation 5.3(2) are discussed in detail below.

The quantity M_s is the maximum bending moment at the section, based on the short-term serviceability load under consideration. For simply supported beams, M_s may be taken as the midspan moment; for cantilever beams, M_s should be taken as the root moment.

The moment of inertia for a fully cracked section (I_{cr}) can be determined in the usual manner, once the position of the neutral axis of the transformed section is known. A brief discussion on this topic may be found in Appendix A. For the specific case of a singly reinforced rectangular section

$$I_{cr} = \frac{bd^3}{12}\left[4k^3 + 12pn(1-k)^2\right] \qquad \text{Equation 5.3(3)}$$

in which the elastic neutral axis parameter

$$k = \sqrt{(pn)^2 + 2pn} - pn \qquad \text{Equation A(5)}$$

Note that Equation 5.3(3) may be used for a conservative evaluation of I_{cr} for a doubly reinforced rectangular section.

Finally, in Equation 5.3(2), the cracking moment

$$M_{cr} = \frac{I_g}{y_t}(f'_{ct.f} - f_{cs}) \qquad \text{Equation 5.3(4)}$$

in which y_t is the distance between the neutral axis and the extreme fibre in tension of the uncracked section; f_{cs} is the maximum shrinkage-induced tensile stress on the uncracked section, which may be computed using a rather tedious process as detailed in the Standard in Clause 8.5.3.1 in conjunction with Clause 3.1.7.

By virtue of Equation 2.2(2)

$$M_{cr} = \frac{I_g}{y_t}\left(0.6\sqrt{f'_c} - f_{cs}\right) \qquad \text{Equation 5.3(5)}$$

This equation is applicable for beam sections of any given shape.

In Clause 3.1.7.2 of the Standard, the formula for shrinkage strains has an accuracy range of ±30%. Relevant provisions in ACI 318–2014 ignore the shrinkage effects.

In view of the cumbersome computational process involved, especially in determining shrinkage-induced stress, Clause 8.5.3.1 of the Standard recommends the following approximate equations for flanged beams:

- For $p = \dfrac{A_{st}}{b_{ef}d} \geq 0.001\dfrac{(f'_c)^{1/3}}{\beta^{2/3}}$ where $\beta = \dfrac{b_{ef}}{b_w} \geq 1$,

$$I_{ef} = \left[(5 - 0.04f'_c)\,p + 0.002\right]b_{ef}d^3 \leq \left(\frac{0.1 b_{ef}d^3}{\beta^{2/3}}\right) \qquad \text{Equation 5.3(6)}$$

- For $p < 0.001\dfrac{(f'_c)^{1/3}}{\beta^{2/3}}$,

$$I_{ef} = \left[\frac{0.055(f'_c)^{1/3}}{\beta^{2/3}} - 50p\right]b_{ef}d^3 \leq \left(\frac{0.06 b_{ef}d^3}{\beta^{2/3}}\right) \qquad \text{Equation 5.3(7)}$$

Note that Equations 5.3(6) and 5.3(7) are valid for flanged beams with the flange in compression. Otherwise, for example, for an inverted T- or L-beam, $b_{ef} = b_w$. Similarly, for a rectangular beam $b_{ef} = b_w = b$.

It is worth mentioning here that some writers opt for a conservative and easy solution and simply recommend $I_{ef} = I_{cr}$ (see, for example, Foster, Kilpatrick and Warner (2010)).

5.3.3 Load combinations

After replacing EI with $E_c I_{ef}$, Equations 5.2(1), 5.2(2) and 5.2(3), which were developed for homogeneous uncracked beams, may be used for cracked reinforced concrete beams. It may be assumed that, under service load, the principle of superposition holds.

The load combination formulas recommended in the Standard for serviceability design have been discussed in Section 1.3.2. Using the relevant formula in Equation 1.3(8), as the case may be, the dead, live and wind loads can be combined accordingly and the bending moment (M_s) can be readily determined. With this value of M_s, the effective moment of inertia (I_{ef}) is computed using Equation 5.3(2). Needless to say, the deflection under combined loads may be obtained as the sum of the individual effects, each of which is calculated using the same I_{ef}.

5.3.4 Illustrative example

Problem

Given a simply supported beam (with L_{ef} = 10 m, b = 350 mm, d = 580 mm, D = 650 mm and p_t = 0.01), compute the midspan deflection under a combination of dead load including self-weight (g = 8 kN/m) and live load (q = 8 kN/m). Take E_c = 26 000 MPa, E_s = 200 000 MPa and f'_c = 32 MPa; assume that the beam forms part of a domestic floor system; ignore the shrinkage effects.

Solution

The gross moment of inertia (see Figure 5.3(1))

$$I_g = 350 \times \frac{650^3}{12} = 8010 \times 10^6 \text{ mm}^4$$

Equation 5.3(5): $M_{cr} = 0.6\sqrt{32}\left(\dfrac{I_g}{325}\right)10^{-6} = 83.65$ kNm

With $n = \dfrac{200\,000}{26\,000} = 7.69$ and $p_t = 0.01$, Equation A(5) (from Appendix A) yields $k = 0.3227$. In turn, Equation 5.3(3) gives $I_{cr} = 3174 \times 10^6$ mm^4.

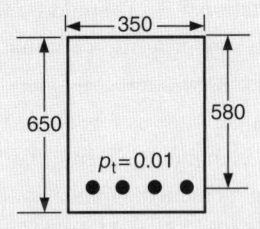

Figure 5.3(1) Cross-sectional details of the example simply supported beam

Note: all dimensions are in mm.

For domestic floor systems, ψ_s and ψ_l are given in Table 1.3(1) as 0.7 and 0.4, respectively. For short-term deflection, the third formula in Equation 1.3(8) governs and we have combined load = $g + 0.7q = 8 + 0.7 \times 8 = 13.6$ kN/m.

The moment at midspan is

$$M_s = 13.6 \times \frac{10^2}{8} = 170 \text{ kNm}$$

Equation 5.3(2) thus gives

$$I_{ef} = 3174 \times 10^6 + (8010 - 3174) \times 10^6 \times \left(\frac{83.65}{170}\right)^3$$

$$= 3750 \times 10^6 \text{ mm}^4 < I_g, \text{ which is acceptable.}$$

Finally, Equation 5.2(1) in conjunction with Table 5.2(1) gives

$$\Delta = \frac{5}{384} \times \frac{13.6 \times 10\,000^4}{26\,000 \times 3750 \times 10^6} = 18.2 \text{ mm}$$

It may be apparent in Table 5.2(4) that the immediate deflection under dead and live loads is not a criterion for serviceability design. Its calculation, however, is essential in the analysis of long-term deflection as discussed in Section 5.4 and the total deflections under normal or repeated loading (Section 5.6).

5.3.5 Cantilever and continuous beams

The discussion so far in this section has centred mainly on simply supported beams. For cantilever beams, the procedure remains basically unchanged, except that for L_{ef} Equation 5.2(5) should be used, and M_s in Equation 5.3(2) should be the bending moment at the root of the cantilever or at the face of the support girder, column or wall, as the case may be.

For continuous beams, the definition of L_{ef} is the same as for simple beams. However, the procedure for computing I_{ef} may require some elaboration. Figure 5.3(2) illustrates typical spans of continuous beams and the corresponding moment diagrams. For the sections i, c and j, the bending moments are M_i, M_c and M_j, respectively, where M_i and M_j are the results from a statically indeterminate analysis of the continuous beam (or output from a computer analysis). For an endspan

$$M_c = M_o - \frac{M_j}{2} \qquad \text{Equation 5.3(8)}$$

and for an interior span

$$M_c = M_o - \frac{(M_i + M_j)}{2} \qquad \text{Equation 5.3(9)}$$

in which

$$M_o = \frac{wL_{ef}^2}{8} \qquad \text{Equation 5.3(10)}$$

is the maximum moment of a simply supported span under uniformly distributed load (w).

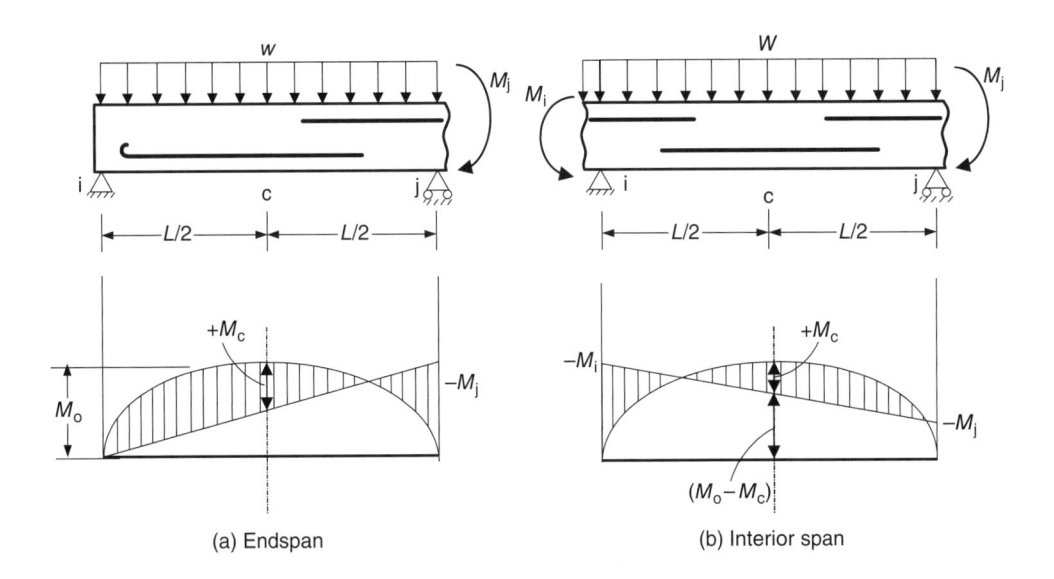

CHAPTER 5 DEFLECTION OF BEAMS AND CRACK CONTROL 103

Figure 5.3(2) Typical spans of continuous beams and the corresponding moment diagrams

With M_s taking the value of M_i, M_c and M_j, respectively, Equation 5.3(2) may be used to compute I_{ef} for the three nominated sections ($I_{ef.i}$, $I_{ef.c}$ and $I_{ef.j}$). Note that AS 3600–2009, like its predecessor AS 3600–2001, has no provisions for the deflection analysis of continuous beams per se. However, Clause 8.5.3 in AS 3600–1994 recommends that, for an endspan

$$I_{ef} = \frac{(I_{ef.c} + I_{ef.j})}{2}$$

Equation 5.3(11)

and for an interior span

$$I_{ef} = \frac{(2I_{ef.c} + I_{ef.i} + I_{ef.j})}{4}$$

Equation 5.3(12)

For computing midspan deflections, these interpolated values of I_{ef} should be used. Following the assumption that the principle of superposition holds for deflection analysis, the midspan deflection of a continuous span, as illustrated in Figure 5.3(2), can be determined by appropriately combining Equations 5.2(1) and/or 5.2(2) with 5.2(3).

5.4 LONG-TERM DEFLECTION

5.4.1 General remarks

Long-term deflection, as the name suggests, is the component of the deflection that occurs after the immediate deflection and that increases with time. Thus, the total deflection

$$\Delta_T = \Delta_I + \Delta_L \qquad \text{Equation 5.4(1)}$$

where Δ_I is the immediate deflection, which can be calculated following the procedure given in Section 5.3, and Δ_L is the long-term deflection. Δ_L consists of two components: deflection due to shrinkage of concrete and deflection due to creep.

Clauses 3.1.7 and 3.1.8 of the Standard provide empirical formulas supplemented by tables and charts for computing the design shrinkage strain and the creep coefficients, respectively. Both are required in a rigorous analysis of the long-term deflection components. The rigorous approach is certainly more tedious and cumbersome, and its reliability depends very much on the accuracy of the formulas for estimating the shrinkage and creep strains. The empirical formulas recommended in the Standard have an accuracy range of approximately ±30% (see Clause 3.1.7.2 in the Standard). This suggests that even the rigorous method is rather inaccurate.

The application of a similarly rigorous method has also been presented by Warner, Rangan, Hall and Faulkes (1998). The reader should refer to this book for more details.

5.4.2 The multiplier method

In the absence of a more accurate analysis, the long-term component of the total deflection in Equation 5.4(1) may be computed using the so-called multiplier method. Equation 5.4(1) may be rewritten as

$$\Delta_T = \Delta_I + k_{cs}\Delta_{I.g} \qquad \text{Equation 5.4(2)}$$

where Δ_I is the immediate deflection due to the combined live (short-term) and sustained (dead or long-term) loads or otherwise known as the total service load; $\Delta_{I.g}$ is the immediate deflection caused by the sustained load or in most cases, the dead load; as per Clause 8.5.3.2 of the Standard, the multiplier

$$k_{cs} = 2 - 1.2\frac{A_{sc}}{A_{st}} \geq 0.8 \qquad \text{Equation 5.4(3)}$$

in which $\dfrac{A_{sc}}{A_{st}}$ is taken at midspan for a simple or continuous beam or at the support for a cantilever beam. It is obvious that the existence of A_{sc} helps reduce the long-term deflection.

5.4.3 Illustrative example

Problem

For the beam analysed in Section 5.3.4, compute the total deflection Δ_T using the multiplier method assuming that the dead load only is the sustained load and $\dfrac{A_{sc}}{bd} = 0.0025$.

Solution

From Section 5.3.4 and with reference to the first formula in Equation 1.3(8), the sustained load moment

$$M_g = 8 \times \frac{10^2}{8} = 100 \text{ kNm}$$

Equation 5.3(2) then gives

$$I_{ef} = 3174 \times 10^6 + (8010 - 3174) \times 10^6 \times \left(\frac{83.65}{100}\right)^3 = 6005 \times 10^6 \text{ mm}^4$$

and, applying Equation 5.2(1),

$$\Delta_{I.g} = \frac{5}{384} \frac{8 \times 10\,000^4}{26\,000 \times 6005 \times 10^6} = 6.67 \text{ mm}$$

Based on Equation 5.4(3), the multiplier

$$k_{cs} = 2 - 1.2 \times \frac{0.0025}{0.01} = 1.7$$

Finally, Equation 5.4(2) yields

$$\Delta_T = 18.2 + 1.7 \times 6.67 = 29.5 \text{ mm}$$

The limit for the total deflection of a beam, as given in Table 5.2(4), is not to be greater than

$$\frac{L_{ef}}{250} = \frac{10\,000}{250} = 40 \text{ mm} > \Delta_T = 29.5 \text{ mm, which is acceptable}$$

Thus, the beam in question is satisfactory as far as total deflection is concerned.

5.5 SIMPLIFIED PROCEDURE

5.5.1 Minimum effective depth approach

It is obvious from the preceding example that even using the multiplier method and ignoring the shrinkage stress (f_{cs}) in computing M_{cr}, the total deflection analysis is tedious and particularly so for continuous beams. In view of this fact, the Standard recommends (in Clause 8.5.4) an alternative check for the effective depth (d) to ensure the adequacy of a beam against excessive deflection.

The recommended process is limited to a beam with uniform cross-section subjected to uniformly distributed load over the entire span, where the intensity of the live load (q) is less than that of the dead load (g). For such a beam, the deflection is deemed to satisfy the limits set in Clause 2.3.2(a) of the Standard (or Table 5.2(4) herein), if the effective depth

$$d \geq \frac{L_{ef}}{\sqrt[3]{\dfrac{k_1(\Delta/L_{ef})b_{ef}E_c}{k_2 F_{def}}}}$$

Equation 5.5(1)

where the limitation (Δ/L_{ef}) is given in Table 5.2(4); L_{ef} is defined in Section 5.2.2; E_c may be computed using Equation 2.2(5) or 2.2(6); b_{ef} is b for rectangular beams and for T- and L-beams it is to be calculated using Equation 4.2(2) or 4.2(3). The geometric factor

$$k_1 = \frac{I_{ef}}{b_{ef}d^3}$$

Equation 5.5(2)

may be determined with the aid of Equation 5.3(6) or 5.3(7).

Further, for a simply supported span, the deflection constant

$$k_2 = \frac{5}{384}$$

Equation 5.5(3)

For a cantilever beam

$$k_2 = 0.125$$

Equation 5.5(4)

On the other hand, for continuous beams, where in adjacent spans the ratio of the longer span to the shorter span does not exceed 1.2 and where no endspan is longer than an interior span

$$k_2 = \frac{2.4}{185} \text{ for an endspan}$$

Equation 5.5(5)

and

$$k_2 = \frac{1.5}{384} \text{ for an interior span}$$

Equation 5.5(6)

Finally, the effective design load per unit length for total deflection consideration is given as

$$F_{\text{def}} = (1.0 + k_{\text{cs}})g + (\psi_s + k_{\text{cs}}\psi_1)q \qquad \text{Equation 5.5(7)}$$

For the deflection that occurs after the addition or attachment of the partition (see Table 5.2(4)), the effective design load per unit length

$$F_{\text{def}} = k_{\text{cs}}g + (\psi_s + k_{\text{cs}}\psi_1)q \qquad \text{Equation 5.5(8)}$$

In Equations 5.5(7) and 5.5(8), k_{cs} is computed using Equation 5.4(3), and ψ_s and ψ_1 are obtained from Table 1.3(1) in Chapter 1.

Although it takes quite an effort to describe this empirical formula for the minimum d, the use of Equation 5.5(1) is rather straightforward. A further advantage is that it is equally applicable to statically determinate and continuous beams. Note that the overhanging span of a simply supported beam cannot be treated as a cantilever beam in this context.

5.5.2 ACI code recommendation

Unless values are obtained by a more comprehensive analysis, ACI code (ACI 318–2014) recommends that additional long-term deflection, Δ_L, resulting from creep and shrinkage of (normal or light-weight concrete) flexural members is to be determined by multiplying the immediate deflection caused by the sustained load concerned, $\Delta_{I.g}$, by a factor λ_Δ. As per Clause 9.5.2.5 of ACI 318–2014, the multiplying factor

$$\lambda_\Delta = \frac{\xi}{1 + 50p_c} \qquad \text{Equation 5.5(9)}$$

where p_c (= A_{sc}/bd) is the value at midspan for simple and continuous spans, and at support for cantilevers; the time-dependent factor for sustained loads, ξ, is assumed as equal to 2.0, 1.4, 1.2 and 1.0 respectively for 5 years or more, 12 months, 6 months and 3 months of loading.

5.6 TOTAL DEFLECTION UNDER REPEATED LOADING

For structural components, such as bridge beams, the loading is often repetitive. A detailed discussion on the deflections of reinforced concrete beams under repeated loading has been presented elsewhere by Loo and Wong (1984). The relevant formulas are reproduced herein followed by an illustrative example.

5.6.1 Formulas

For beams under repeated loading, the total deflection is given as

$$\Delta_T = \Delta_{A.g} + \Delta_q \qquad \text{Equation 5.6(1)}$$

where Δ_q is the immediate live-load deflection at the Tth cycle, and $\Delta_{A.g}$ is the accumulated sustained or dead-load deflection (permanent set) caused by the dead load and the effects of all the previous live-load repetitions.

A typical experimental deflection plot of a reinforced concrete box beam under repeated loading is idealised in Figure 5.6(1). It may be seen in the diagram that the total deflection

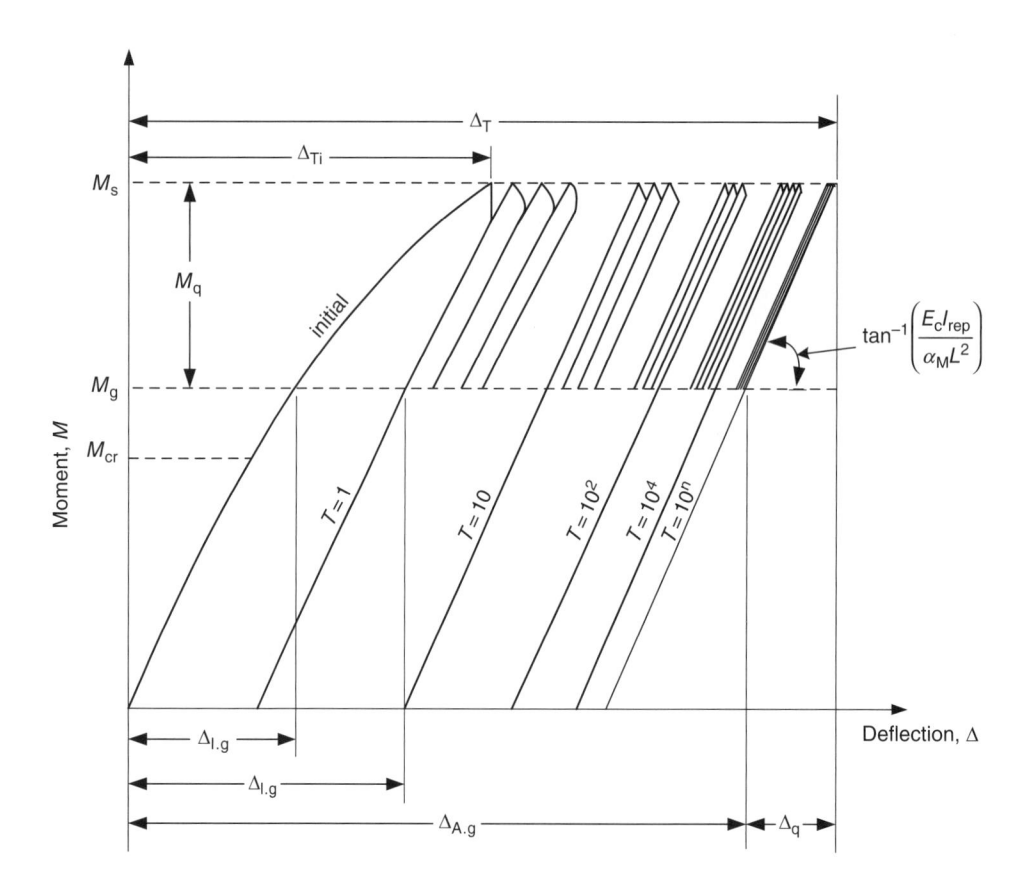

$$\Delta_T = k_R \Delta_{I.g} + \frac{\alpha_M M_q L^2}{E_c I_{rep}} \qquad \text{Equation 5.6(2)}$$

Figure 5.6(1) Deflection of concrete beams under repeated loading (T = number of load repetitions)
$\Delta_{I.g}$ = initial dead-load deflection; Δ_q = live-load deflection; $\Delta_{1.g}$ = 1st cycle dead-load deflection; Δ_{Ti} = initial total deflection; $\Delta_{A.g}$ = accumulated dead-load deflection; Δ_T = total deflection.

where α_M, the deflection multiplication factor, is given in Table 5.2(3), M_q is the live-load moment, L is the centre-to-centre span, E_c may be evaluated using Equation 2.2(5), I_{rep} is the equivalent moment of inertia at the Tth loading cycle and the amplification factor

$$k_R = \frac{\Delta_{A.g}}{\Delta_{I.g}}$$ Equation 5.6(3)

in which $\Delta_{I.g}$ is the initial dead-load deflection. The existence of k_R is a result of the repeated loading.

The increase in dead-load deflection from $\Delta_{I.g}$ to $\Delta_{A.g}$ may be considered analogous to creep of concrete under sustained load. While the creep deflection occurs over a long period, the increase due to repeated loading can occur over a short time period. The amplification factor (k_R) is thus also referred to as the intensive creep factor, and is given as

$$k_R = k_I + R \log_{10} T$$ Equation 5.6(4)

where T is the number of load repetitions

$$k_I = 1.18 + \frac{0.029}{p_t} \frac{M_s - M_g}{M_y - M_{cr}}$$ Equation 5.6(5)

and

$$R = \frac{0.0015}{p_t} \frac{M_s - M_g}{M_y - M_{cr}}$$ Equation 5.6(6)

in which p_t is the tension steel ratio; and as illustrated in Figure 5.6(1), M_s is the total moment, M_g is the moment due to sustained or dead load, M_{cr} is the cracking moment (see Equation 5.3(4)); and M_y is the yield moment. The calculation of M_y is illustrated in the example given in the next section; M_{cr} is determined using Equation 5.3(4), where f_{cs} may be ignored.

The equivalent moment of inertia under repeated loading referred to in Equation 5.6(2) has a value between I_{ef} and I_g. It can be computed as

$$I_{rep} = \left(\frac{M_{cr}}{M_x}\right)^2 I_g + \left[1 - \left(\frac{M_{cr}}{M_x}\right)^2\right] I_{ef}$$ Equation 5.6(7)

where

$$M_x = \frac{(M_s - M_{cr})^2}{(M_y - M_{cr})} + M_{cr}$$ Equation 5.6(8)

With Equations 5.6(2), 5.6(4) and 5.6(7) in hand, the total deflection for a simply supported beam under repeated loading of a given number of repetitions T can readily be computed.

These equations may also be used for cantilever beams. However, note that the various semi-empirical formulas have been derived with the following ranges of experimental data (see Loo and Wong 1984):

- $M_g = 0.3\, M_y$
- $M_s = 0.5\, M_y$ to $0.9\, M_y$
- $T \leq 100\,000$ repetitions.

The method has also been applied to cases with T up to three million repetitions (see Wong and Loo 1985).

The total deflection given in Equation 5.6(1) does not include the time-dependent component. If such exists, the grand total may be taken as

$$\Delta_T = \Delta_{A.g} + \Delta_q + \Delta_L \qquad \text{Equation 5.6(9)}$$

where, following Equation 5.4(2),

$$\Delta_L = k_{cs}\Delta_{I.g} \qquad \text{Equation 5.6(10)}$$

5.6.2 Illustrative example

Problem

Re-analyse the beam in Section 5.4.3, taking into consideration the effects of 31 200 repetitions of full live load (i.e. $q = 8$ kN/m).

Solution

1. I_{ef}

   For $q = 8$ kN/m

   $M_q = 8 \times 10^2/8 = 100$ kNm; $M_s = 200$ kNm

   Note that using $I_{ef.q}$ instead of $I_{ef.g}$ will lead to a conservative I_{rep}.

   Thus, Equation 5.3(2) yields

   $$I_{ef.q} = 3174 \times 10^6 + (8010 - 3174) \times 10^6 \times \left[\frac{83.65}{200}\right]^3 = 3528 \times 10^6\,\text{mm}^4$$

2. Yield moment (M_y)

   The yield moment is the upper limit of the working stress bending moment. For under-reinforced rectangular sections, we have

   $$M_y \approx A_{st}f_{sy}d\left(1 - \frac{k}{3}\right) \qquad \text{Equation 5.6(11)}$$

   where k is obtained using Equation A(5) in Appendix A.
   From the example given in Section 5.3.4, $k = 0.3227$. Thus

   $$M_y = 0.01 \times 350 \times 580 \times 500 \times 580 \times \left(1 - \frac{0.3227}{3}\right) \times 10^{-6} = 525.4\,\text{kNm}$$

3. Intensive creep factor and $\Delta_{A.g}$

    Equation 5.6(5): $k_I = 1.18 + \dfrac{0.029}{0.01}\dfrac{200-100}{525.4-83.65} = 1.84$

    Equation 5.6(6): $R = \dfrac{0.0015}{0.01}\dfrac{200-100}{525.4-83.65} = 0.03396$

    Equation 5.6(4): $k_R = 1.84 + 0.03396 \times \log_{10}(31\,200) = 1.993$

    From Section 5.4.3, $\Delta_{I.g} = 6.67$ mm. Thus
    $\Delta_{A.g} = 1.993 \times 6.67 = 13.29$ mm

4. I_{rep} and Δ_g

    Equation 5.6(8): $M_x = \dfrac{(200-83.65)^2}{(525.4-83.65)} + 83.65 = 114.3$ kNm

    Equation 5.6(7): $I_{\text{rep}} = \left(\dfrac{83.65}{114.3}\right)^2 \times 8010 \times 10^6 + \left[1 - \left(\dfrac{83.65}{114.3}\right)^2\right] \times 3528 \times 10^6$

    $= 5929 \times 10^6$ mm^4

    Comparing Equations 5.6(1) and 5.6(2), we have

    $\Delta_q = \dfrac{5}{48} \times \dfrac{100 \times 10^6 \times 10\,000^2}{26\,000 \times 5929 \times 10^6} = 6.76$ mm

5. Grand total deflection

    From Section 5.4.3 and using Equation 5.6(10),

    $\Delta_L = 1.7 \times 6.67 = 11.34$ mm

    Finally, Equation 5.6(9) gives

    $\Delta_T = 13.29 + 6.76 + 11.34 = 31.39$ mm

5.7 CRACK CONTROL

5.7.1 General remarks

Crack control is gaining a well-deserved place in the design process of concrete structures. The Standard now provides stringent rules for this important aspect of serviceability design. Other major design codes including ACI 318–2014, Eurocode 2–2004 and the British Standard BS 8110–1997 also give crack control prominent treatment. Similar to provisions

112 PART 1 REINFORCED CONCRETE

in ACI 318–2014, the Standard process limits the maximum stress, maximum cover and bar spacing. On the other hand, Eurocode 2–2004 and BS 8110–1997 recommend empirical formulas for calculating the average and maximum crack widths, respectively.

This section discusses in detail provisions in the Standard, and compares the performances of three major codes for reinforced beams and a unified formula (for reinforced and prestressed beams) developed by Chowdhury and Loo (2001).

5.7.2 Standard provisions

AS 3600–2009 states in Clause 2.3.3 that cracking in concrete structures shall be controlled so that:

- the structural performance is not jeopardised
- durability is not compromised
- the appearance is as specified.

Cracking is deemed to be controlled if the beam design complies with the following rules:

- The tension reinforcement must not be less than the minimum as specified in Equation 3.5(5), 4.2(25)a and b, or 4.2(25)a and c, respectively, for rectangular, flanged, or inverted T- and L-beams.
- The dimensional restrictions depicted in Figure 5.7(1) shall be adhered to, noting that:
  - in determining the concrete cover and bar spacing, bars with a diameter less than half that of the largest bar in the section shall be ignored
  - for T- and inverted L-beams, the reinforcement required in the flange shall be distributed across b_{ef}.

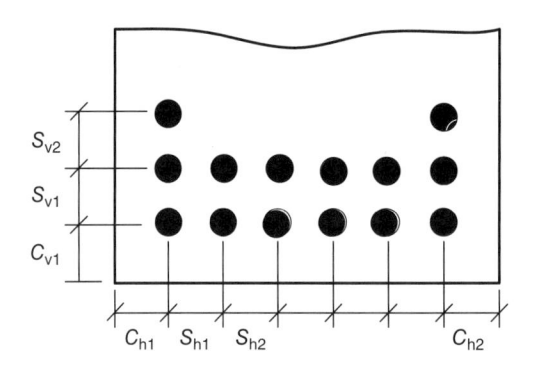

Figure 5.7(1) Dimensional restrictions for crack control in beam design
Requirements: S_{h1}, S_{h2}, etc $\leq$ 300 mm
S_{v1}, S_{v2}, etc $\leq$ 300 mm
C_{v1}, etc $\leq$ 100 mm
C_{h1}, etc $\leq$ 100 mm

- The calculated steel stress (f_{scr}) under service load or M_s and based on the transformed section (see Appendix A) shall not exceed the larger of the maximum steel stresses given in
  - Table 5.7(1) for the largest nominal diameter (d_b) of the bars in the tensile zone
  - Table 5.7(2) for the largest centre-to-centre spacing of adjacent parallel bars in the tensile zone.

Table 5.7(1) Maximum steel stress for tension or flexure in beams

Nominal bar diameter (d_b)	Maximum steel stress (MPa)
10	360
12	330
16	280
20	240
24	210
28	185
32	160
36	140
40	120

Source: Standards Australia (2009). AS 3600–2009 *Concrete Structures*, Table 8.6.1(A). Standards Australia, Sydney, NSW. © Standards Australia Limited. Copied by Cambridge University Press with the permission of Standards Australia and Standards New Zealand under Licence 1712-c038.

Table 5.7(2) Maximum steel stress for flexure in beams

Centre-to-centre spacing	Maximum steel stress (MPa)
50	360
100	330
150	280
200	240
250	200
300	160

Source: Standards Australia (2009). AS 3600–2009 *Concrete Structures*, Table 8.6.1(B). Standards Australia, Sydney, NSW. © Standards Australia Limited. Copied by Cambridge University Press with the permission of Standards Australia and Standards New Zealand under Licence 1712-c038.

Note that Tables 5.7(1) and 5.7(2) are reproduced from Tables 8.6.1(A) and (B) of the Standard, respectively. Under direct loading, the calculated steel stress, $f_{scr.1} \leq 0.8f_{sy}$ where $f_{scr.1}$ is due to M_s with $\psi_s = 1.0$. Again, bars with a diameter less than half the diameter of the largest bar in the section shall be ignored when determining the spacing.

- For tension members or beams mainly subjected to tension, f_{scr} shall not exceed the maximum given in Table 5.7(1) for the largest d_b of the bars in the section, and $f_{scr.1} \leq 0.8f_{sy}$.
- To prevent cracking in the side faces of beams, additional reinforcement (A_{cc}) must be provided as detailed in Figure 5.7(2).

Requirements for A_{cc}:

- $s \leq 200$ mm for 12-mm bars
- $s \leq 300$ mm for 16-mm bars.

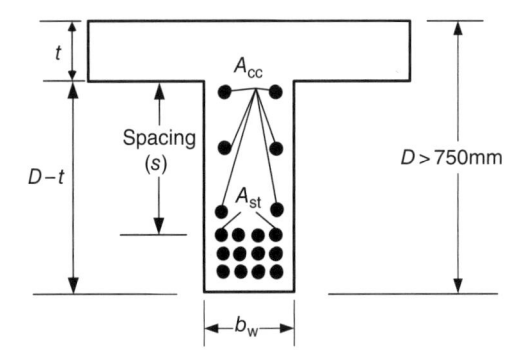

Figure 5.7(2) Additional reinforcement requirement to prevent cracking in the side faces of beams

5.7.3 Crack-width formulas and comparison of performances

Empirical formulas for computing crack widths in reinforced concrete beams may be found in the following national and international codes:

- ACI 318–1995
- Eurocode 2–1992, Eurocode 2–2004
- BS 8110–1985, BS 8110–1997, BS 8007–1987.

Based on the laboratory test results from:

- 10 (full size) simply supported reinforced box beams
- five (half-scale) simply supported reinforced solid beams
- three (full size) two-span continuous reinforced box beams
- 12 (full size) simply supported partially prestressed (pretensioned) box beams,

Chowdhury and Loo (2001) developed two unified crack-width formulas applicable to both reinforced and prestressed concrete beams. In their approach, the average crack width

$$w_{cr} = (f_s/E_s)[0.6(c-s) + 0.1(\phi/\rho)] \quad \text{Equation 5.7(1)}$$

where

f_s = the steel stress in MPa, which, for reinforced beams, may be determined using the transformed section approach given in Appendix A

E_s = Young's modulus of the reinforcement (MPa)

c = clear concrete cover (mm)

s = average clear spacing between bars (mm)

ϕ = average bar diameter (mm)

ρ = tension reinforcement ratio.

The maximum crack width may simply be computed as

$$w_{max} = 1.5 w_{cr} \qquad \text{Equation 5.7(2)}$$

Equation 5.7(1) is compared with the test results from 30 reinforced and prestressed beams in Figure 5.7(3). With an error range of ±30%, the linear regression process used to derive the average crack-width formula is considered satisfactory.

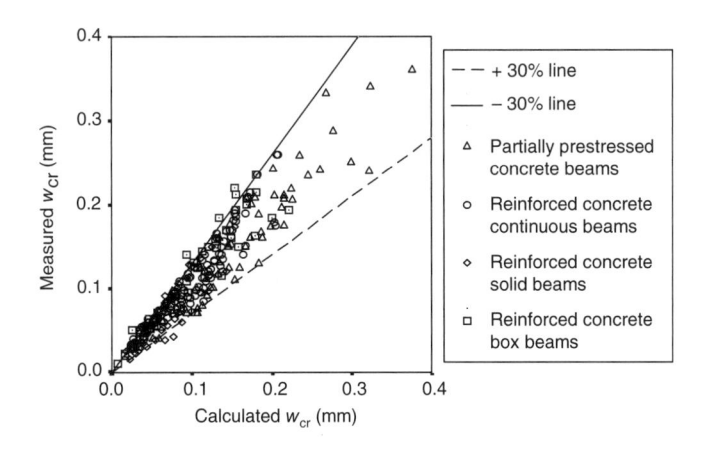

Figure 5.7(3) Comparison of Equation 5.7(1) with test results from 30 reinforced and prestressed beams

In a comparative study on the available literature, Chowdhury and Loo (2001, 2002) collated the following simply supported beam test data:
- w_{cr} from 26 reinforced concrete beams by Clark (1956)
- w_{cr} from 16 reinforced concrete beams by Chi and Kirstein (1958)
- w_{cr} from 8 reinforced concrete beams by Hognestad (1962)
- w_{cr} from 9 reinforced concrete beams by Kaar and Mattock (1963)
- w_{max} from 34 prestressed pre- and post-tensioned beams by Nawy (1984).

Figure 5.7(4) compares these laboratory test results and the predicted values using Equations 5.7(1) and 5.7(2). The two formulas for average and maximum crack widths perform satisfactorily in that almost all the correlation points lie between the ±30%

lines. Note that points located below the diagonal unbroken thick line represent conservative or safe predictions of crack widths. Satisfactory comparison with the test data published by these independent researchers confirms the reliability of the two empirical formulas.

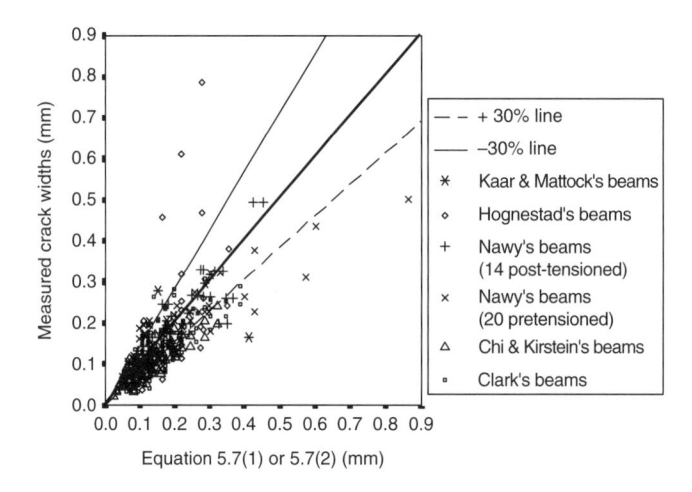

Figure 5.7(4) Comparison of empirical formulas with all available reinforced and prestressed concrete beam test results

Based on the published test data for reinforced concrete beams (only), Chowdhury and Loo (2001) also compared the performance of Equations 5.7(1) or 5.7(2) with each of the recommendations made in Eurocode 2–1992, ACI 318–1995 and BS 8110–1985 – the results are presented in Figures 5.7(5)a, b and c, respectively. For a better viewing of the comparison graphs, the reader can refer to the original paper.

The comparisons found that:
- Equation 5.7(1) for average crack widths compares favourably with, or gives more conservative results than, the prediction procedure recommended in Eurocode 2–1992.
- Equation 5.7(2) compares reasonably well with the maximum crack-width formula recommended in ACI 318–1995.
- the maximum crack-width formula given in the British Standards, BS 8110–1985 and BS 8007–1987, tends to underestimate maximum crack widths considerably and consistently.

Equations 5.7(1) and 5.7(2) are considered superior to all of the code procedures, because they are also applicable to partially prestressed beams – the code formulas are restricted to reinforced concrete beams only. In addition, the recommendations in Eurocode 2–2004 and BS 8110–1997 remain the same as in their respective predecessors.

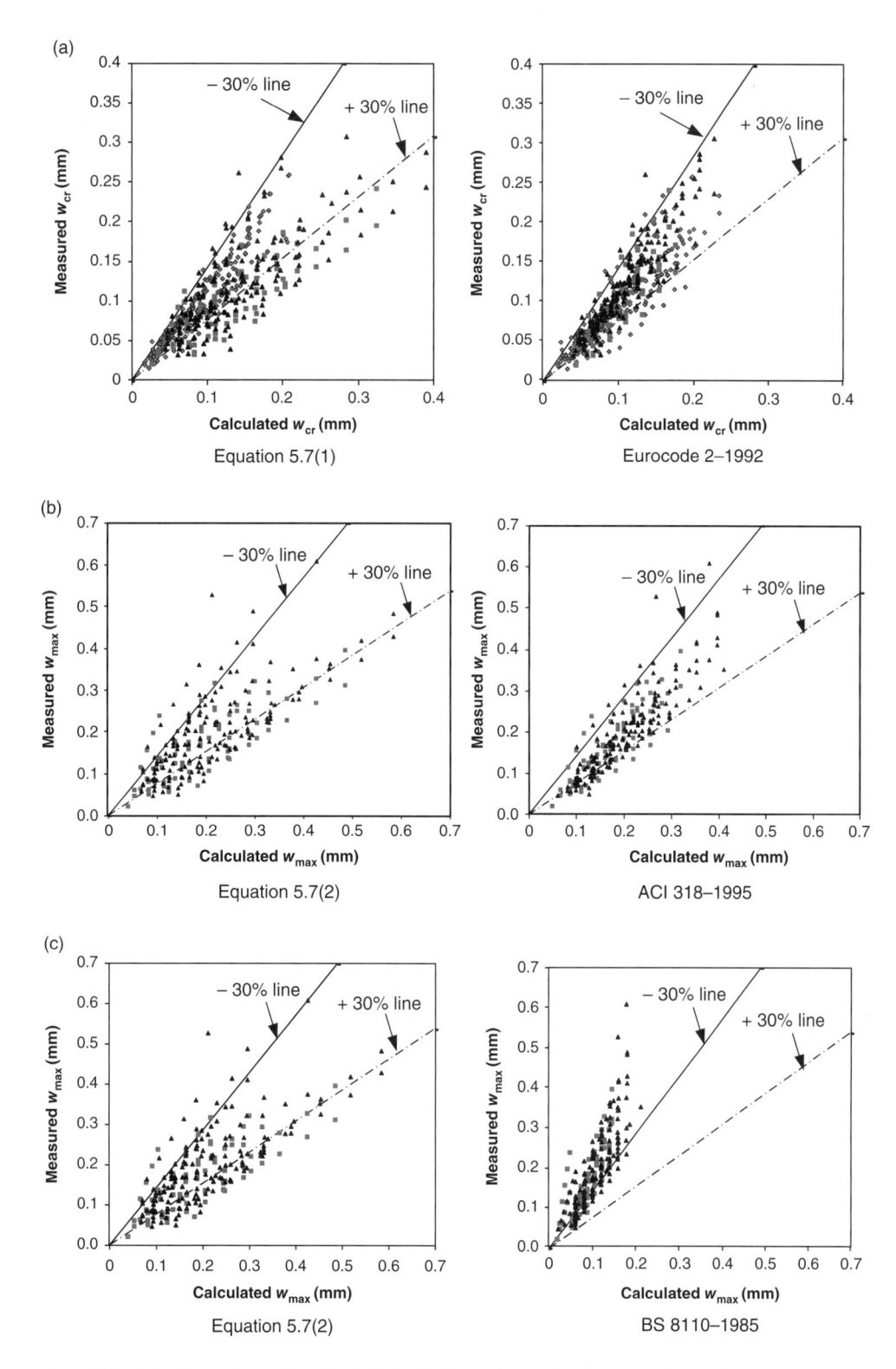

Figure 5.7(5) **(a)** Comparison of Equation 5.7(1) with Eurocode 2–1992; **(b)** comparison of Equation 5.7(2) with ACI 318-1995; **(c)** comparison of Equation 5.7(2) with BS 8110–1985

5.8 SUMMARY

Serviceability design of reinforced concrete beams in terms of deflection and crack control are dealt with in this chapter. The procedures involved are presented together with derivations of useful formulas and the application of relevant Standard recommendations. Illustrative examples are included. For completeness, the general formulas for computing deflections of beams with common support conditions and under different loading types are summarised in a ready-to-use form. As an advanced topic, the analysis of the total deflection of beams under repeated loading is also introduced.

5.9 PROBLEMS

1. For a fully cracked reinforced concrete section as shown in Figure 5.9(1), the neutral axis coefficient is given in Equation A(5) (see Appendix A) as

$$k = \sqrt{(pn)^2 + 2pn} - pn$$

   Show that this expression can also be obtained by making use of the equilibrium and compatibility conditions (respectively, $C = T$ and ε_c is proportional to ε_s).
   (Hint: equate $\frac{f_c}{f_s}$ from both conditions.)

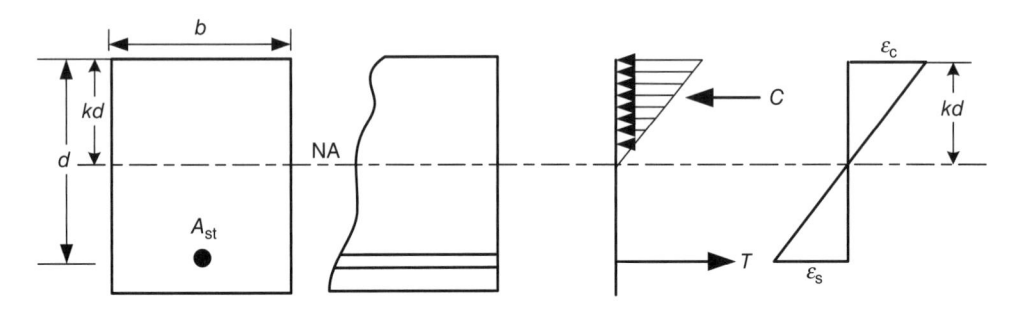

Figure 5.9(1) Neutral axis location for a fully cracked reinforced concrete section

2. For each of the reinforced concrete sections detailed in Figure 5.9(2), compute the value of kd. Take $E_s = 200\,000$ MPa $f'_c = 25$ MPa and 32 MPa for the sections in Figure 5.9(2)a and b, respectively.

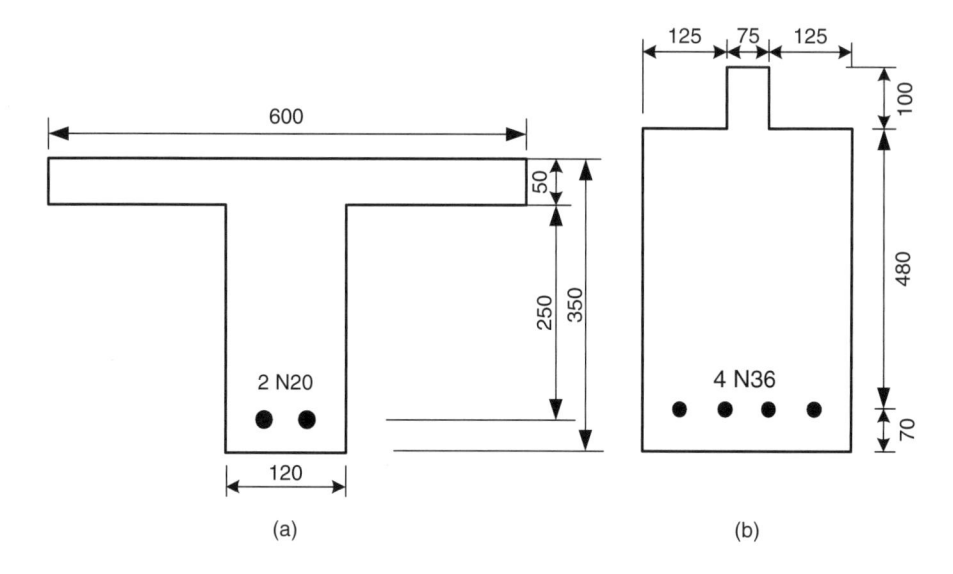

Figure 5.9(2) Reinforced concrete sections with **(a)** $f'_c = 25$ MPa and **(b)** $f'_c = 32$ MPa

Note: all dimensions are in mm.

3. A simply supported rectangular beam, 340 mm wide and 630 mm deep (with $d = 568$ mm), is part of a floor system that supports a storage area. The steel ratios $p_t = 0.008$ and $p_c = 0.0025$; $E_s = 200\,000$ MPa and $f'_c = 32$ MPa. For $g = 8$ kN/m (including self-weight) and $q = 8$ kN/m, what is the maximum effective span (L_{ef}) beyond which the beam is not considered by AS 3600–2009 as complying with the serviceability requirement for total deflection?

4. A simply supported beam of 12 m effective span having a cross-section as shown in Figure 5.9(2)b, is subjected, in addition to self-weight, to a dead load of 3 kN/m and a live load of 3 kN/m. Take $f'_c = 32$ MPa, ignore the shrinkage effects and consider the beam to be part of a domestic floor.
   (a) What is the short-term maximum deflection?
   (b) Does the beam satisfy the Australian Standard's minimum effective depth requirement for total deflection?

5. Figure 5.9(3) details a cantilever beam that forms part of a domestic floor.
   (a) Under the given loading, compute the total maximum deflection.
   (b) Does the design satisfy the requirement of span/depth ratio?

   You may ignore the shrinkage effects and assume that self-weight is the only sustained load, $f'_c = 32$ MPa, $E_c = 27\,000$ MPa and $n = 7.4$.

   Note: the deflections at the tip of a cantilever are $\delta = \dfrac{wL^4}{8EI}$ and $\delta = \dfrac{PL^3}{3EI}$, where w is the uniformly distributed load, P is the concentrated load and L is the span.

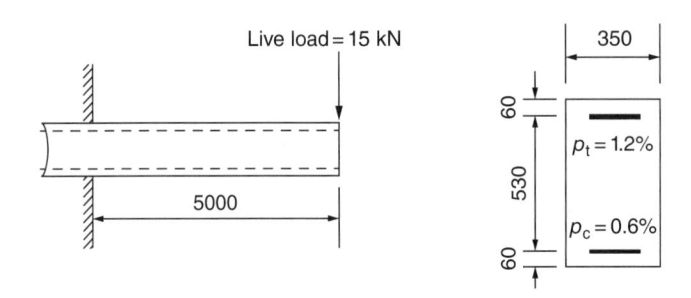

Figure 5.9(3) Details of a cantilever beam

Note: all dimensions are in mm.

6. For the flanged beam considered in Problem 7 (Figure 4.7(7)) in Section 4.7, assume that all other details remain unchanged, but $p_t = 0.015$ and $p_c = 0.005$. What is the minimum acceptable d for the section in order to satisfy the alternative serviceability requirement of the Australian Standard? You may assume the beam to be part of a parking area.

7. For crack control of the beam in Problem 12 in Section 4.7, what necessary measures must be taken?

8. The cantilever beam shown in Figure 5.9(4) is to support a dead load (including self-weight) $g = 10$ kN/m and a live load $q = 30$ kN/m. Take $f'_c = 32$ MPa, $E_s = 200\,000$ MPa and consider the beam forming part of an office floor. You may ignore the shrinkage effects.
   (a) Compute the instantaneous deflection.
   (b) Does the design satisfy the alternative Australian Standard requirement for total deflection?

   Note:
   1. L_{ef} = clear span + $D/2$
   2. For a cantilever beam under uniformly distributed load of w over a span L,

$$\Delta_{max} = \frac{wL^4}{8EI}$$

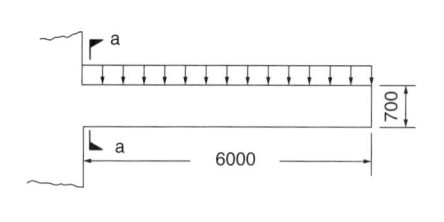

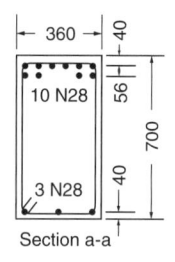

Figure 5.9(4) Load configuration and sectional details of a cantilever beam

Note: all dimensions are in mm.

9. Taking $E_s = 200\,000$ MPa and $f'_c = 32$ MPa and for the T-section shown in Figure 5.9(5):
   (a) determine the location of the neutral axis of the fully cracked section (i.e. kd)
   (b) compute the fully cracked moment of inertia, I_{cr}.

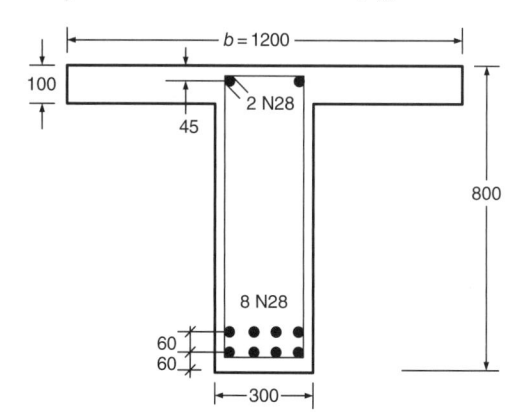

Figure 5.9(5) Details of a T-section

Note: all dimensions are in mm.

10. A reinforced concrete beam, which forms part of a domestic floor system, has an overall depth of 450 mm, a width of 300 mm, an effective span of 4.5 m and a concrete strength of $f'_c = 20$ MPa. Tension reinforcement consists of 2464 mm^2 of N bars (i.e. $E_s = 200\,000$ MPa) with an effective depth of 400 mm. The dead- and live-load moments are 95 kNm and 103 kNm, respectively. All dead load is in place before the formwork is removed. If 30% of the superimposed live load may be considered sustained, determine whether the deflection of the beam may be considered satisfactory as per AS 3600 requirements. You may ignore the shrinkage effects.

11. The reinforced concrete T-beam shown in Figure 5.9(6) is reinforced with N bars at the positions indicated. Design and detail the additional reinforcement required to prevent cracking in the side faces and check that the spacing of the tension reinforcement complies with crack control requirements of AS 3600.

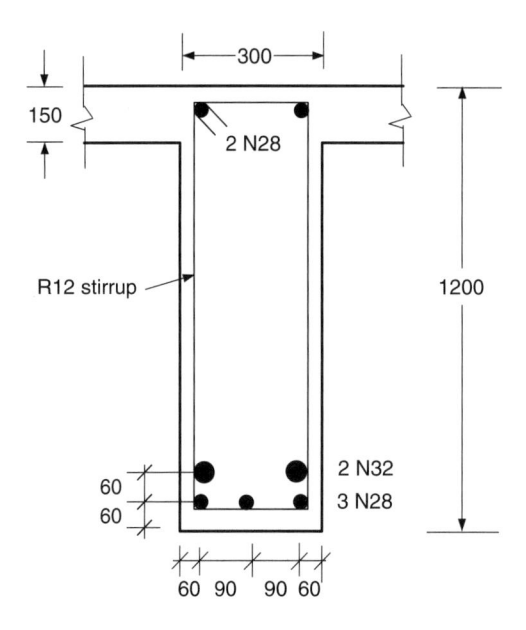

Figure 5.9(6) Details of a reinforced concrete T-beam

Note: all dimensions are in mm.

ULTIMATE STRENGTH DESIGN FOR SHEAR

6.1 INTRODUCTION

This chapter deals with the ultimate strength design of reinforced concrete beams for shear. The derivation of relevant formulas and application of the Standard recommendations are presented in some detail. The design examples cover both the transverse and longitudinal shear.

6.2 TRANSVERSE SHEAR STRESS AND SHEAR FAILURE

6.2.1 Principal stresses

For a beam in bending as depicted in Figure 6.2(1)a, the bending stress is given as

$$f = \frac{My}{I}$$

Equation 6.2(1)

and the shear stress as

$$v = \frac{VQ'}{Ib'}$$

Equation 6.2(2)

where M and V are the moment and shear force at the beam section; y is the distance from the neutral axis (positive downwards); Q' is the first moment of the area beyond the level where shear stress (v) is being computed, with respect to the neutral axis; b' is the width of the section at the level in question; and I is the moment of inertia.

In Figure 6.2(1)a, Element 1 is located at the neutral axis level and Element 4 is at the extremity of the tensile region, whereas Elements 2 and 3, respectively, are somewhere within the tensile and compressive zones of the beam. All of the elements are, at a given beam section, a distance x from the left support. The bending and shear stresses acting on each of the elements are readily computed using Equations 6.2(1) and 6.2(2), respectively. Figure 6.2(1)b shows the free-body diagrams of these elements of the beam. For each of the free-bodies, a Mohr circle can be constructed (see Figure 6.2(1)c), which yields the magnitudes and

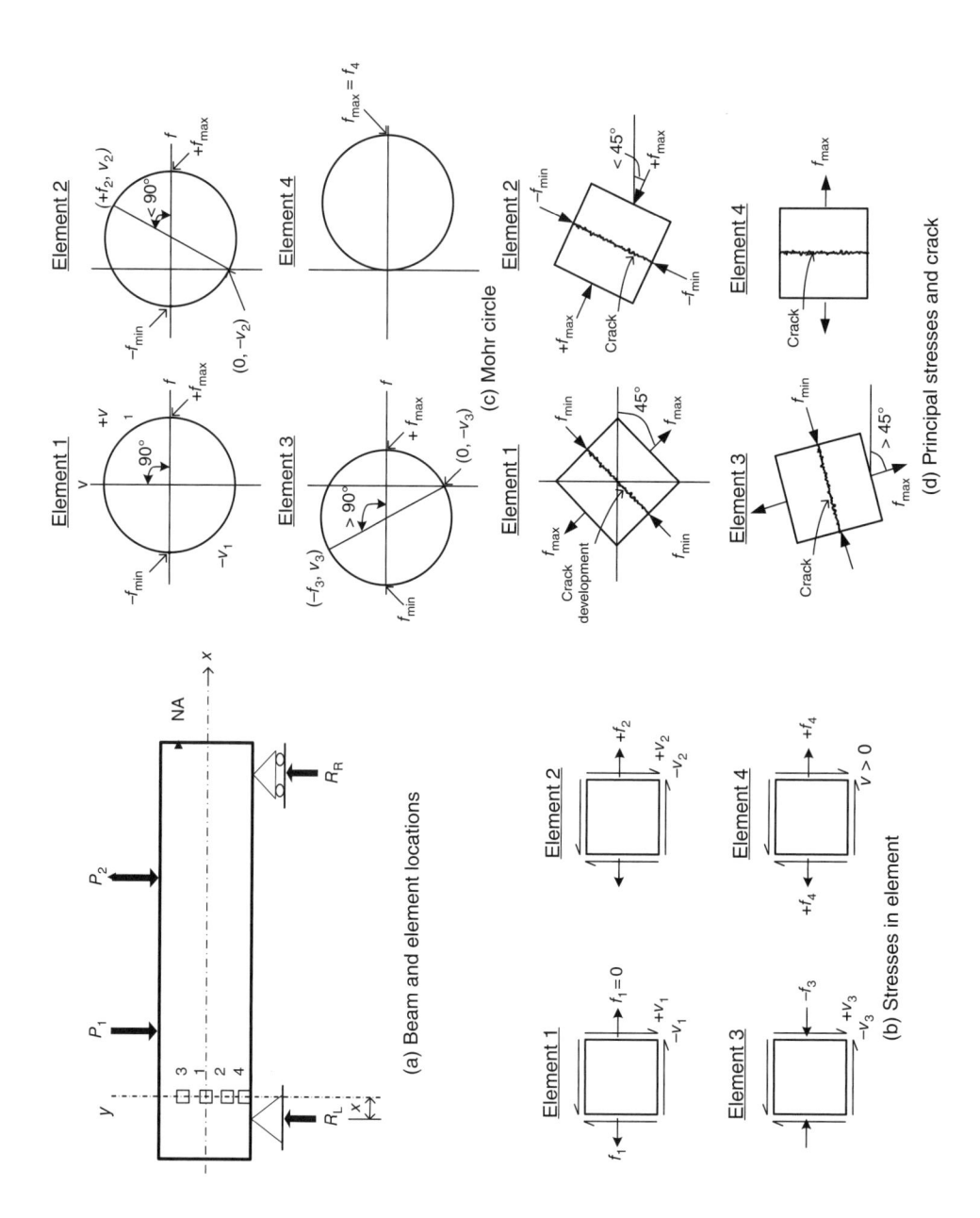

Figure 6.2(1) A beam in bending – principal and element stresses

orientations of the principal stresses (f_{max}) and (f_{min}). Figure 6.2(1)d shows the four free-body diagrams acted upon by the principal stresses.

It is obvious in Figure 6.2(1)d that, as concrete is weak in tension, the principal tensile stress (f_{max}) will cause cracking in each element in a direction 90° to itself. In Element 1, the direction of the crack is 45° anticlockwise from the direction of the neutral axis (plane) as indicated by the letters NA. This crack is caused by shear stress alone. On the other hand, only bending stress exists in Element 4 (i.e. $f_{max} = f_1$). The crack is vertical (i.e. 90°) from the neutral axis. Element 2 is influenced by both shear stress and tensile bending stress; the crack direction is greater than 45°, but less than 90° anticlockwise from the neutral axis. The exact inclination depends on the position of the element or the value of y. Finally, shear stress and compressive bending stress act on Element 3 and the crack direction is greater than 0°, but less than 45° from the neutral axis.

The 45° crack at the neutral axis level is referred to as the diagonal crack – it is often catastrophic (i.e. the beam abruptly fails as a result).

6.2.2 Typical crack patterns and failure modes

A typical reinforced concrete beam subjected to uniform and point loads can be divided into three regions that are dominated by shear, shear and bending, and bending. Figure 6.2(2) shows the typical crack patterns of an overloaded beam. A crack in a given section of a region will take the directions as depicted in Figure 6.2(2) due to the action of the principal stresses explained in Section 6.2.1.

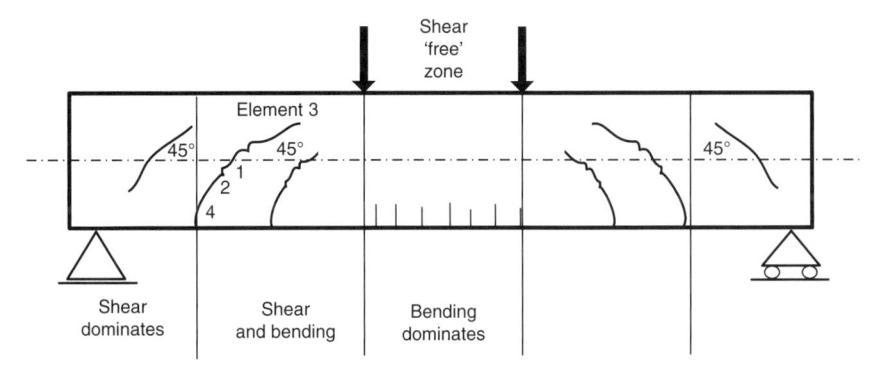

Figure 6.2(2) Typical crack patterns in an overloaded beam

For a beam without shear reinforcement, failure can take one of four modes:
1. shear failure
2. shear-compression failure
3. shear-tension failure
4. bending failure.

Figure 6.2(3) shows the failure patterns of a total of 10 beam models with spans ranging from 0.9 m to 5.8 m, which were tested by Leonhardt and Walther (see Park and Paulay 1975).

126 PART 1 REINFORCED CONCRETE

For Models 1, 2 and 3 (Figure 6.2(3)), the mode of failure is shear mode, which is characterised by excessive 45° (or diagonal) cracks. Failure is normally due to the separation of the portions above and below the major diagonal crack.

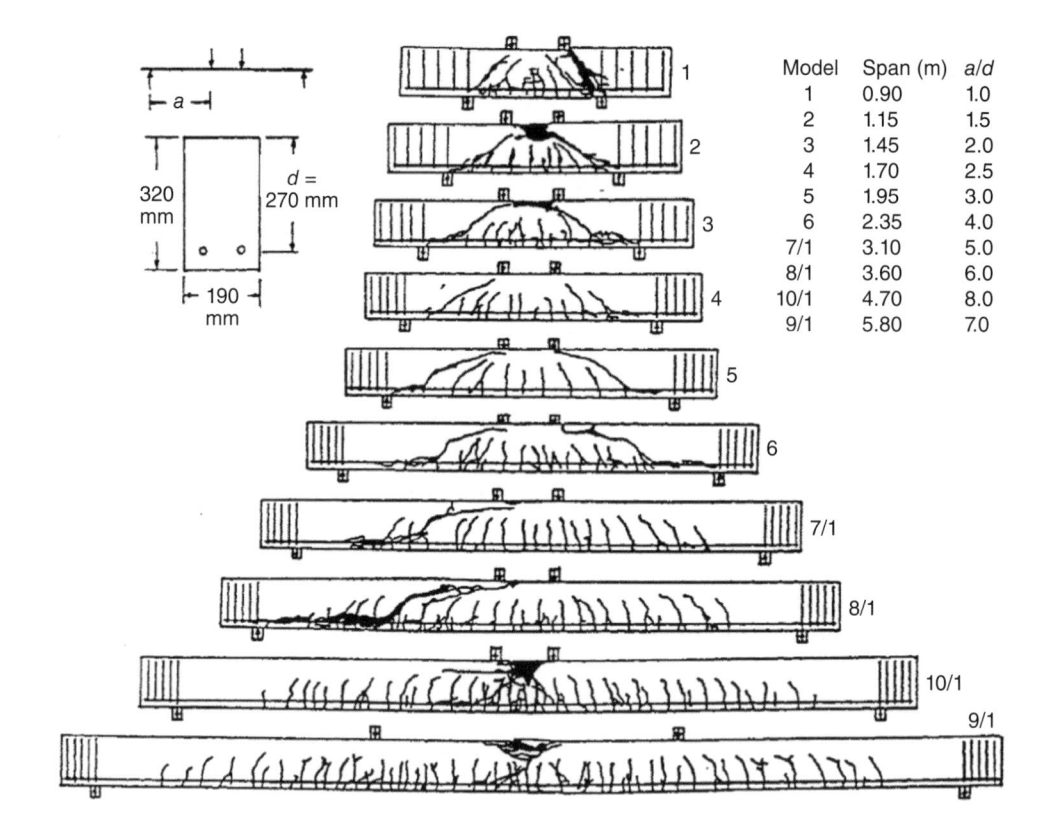

Model	Span (m)	a/d
1	0.90	1.0
2	1.15	1.5
3	1.45	2.0
4	1.70	2.5
5	1.95	3.0
6	2.35	4.0
7/1	3.10	5.0
8/1	3.60	6.0
10/1	4.70	8.0
9/1	5.80	7.0

Figure 6.2(3) Crack pattern in beam models tested by Leonhardt and Walther (see Park and Paulay 1975)

Shear-compression failure is evident in Models 4 and 5 in Figure 6.2(3). Because of the extension of the shear-compression cracks well into the compressive zone of the beam, failure occurs when there is inadequate concrete area to take a combination of the shear stress (V_c) (see Section 6.2.3) and the rapidly increasing compressive bending stress.

Models 6, 7 and 8 (Figure 6.2(3)) exhibit a typical shear-tension mode of failure. The continuation of the shear-tension, shear and shear-compression cracks is obvious, and, like the shear failure mode, failure is due to separation of the beam portions above and below the major crack.

Finally, bending failure is clearly demonstrated by Models 9 and 10 (Figure 6.2(3)), where vertical cracks dominate the scene. Failure occurs by crushing of concrete at the compressive zone (i.e. when the compressive strain reaches the ε_{cu} value). Note that all the three shear-related failures are 'brittle' or they occur abruptly.

6.2.3 Mechanism of shear resistance

Figure 6.2(4) illustrates a typical continuous crack in the bending-shear region. Along the section of the crack, shear force (V) is resisted by a combination of the following forces:
1. concrete shear stress (V_c) over the uncracked zone
2. the vertical component of the inclined shear stress (V_a) due to the interlocking of aggregate particles
3. a dowel force (V_d) provided by the tension reinforcement (A_{st}).

It has been estimated that Type 1 constitutes between 20% and 40% of the total shear resistance, whereas Types 2 and 3, respectively, provide 25% to 50% and 15% to 25% of the resistance.

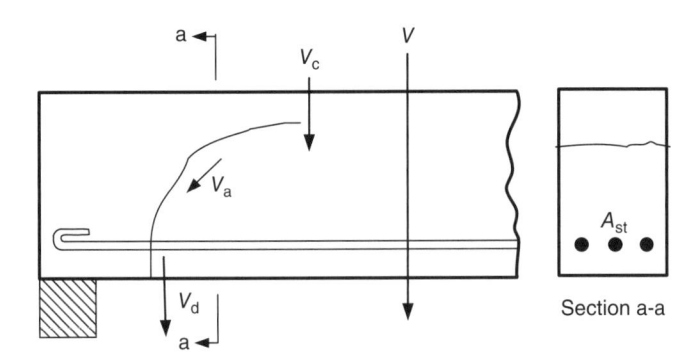

Figure 6.2(4) A typical crack in the bending-shear region

Note that the dowel action of the steel bars tends to split the concrete cover horizontally from the rest of the beam. The reasons are obvious.

6.2.4 Shear reinforcement

In a beam, the longitudinal bars (main reinforcement) resist shear by dowel action, which is not very effective. As shear cracks are mainly 45° from the neutral axis, shear reinforcement should either be vertical or at a right angle to the potential crack direction, which is even more effective. These are illustrated in Figure 6.2(5)a.

Shear reinforcement can take the form of (open) stirrups or closed ties, which may be of single or multiple formation (for wider beams). Note that in AS 3600–2009 (the Standard), stirrups and ties are also referred to as 'fitments' and 'closed fitments', respectively. Figure 6.2(5)b shows some typical arrangements. Stirrups and ties can be either vertical or inclined, but bent bars (i.e. bending up of main bars in a high-shear and low-moment region to provide inclined shear reinforcement) are no longer specified since AS 3600–1988. Although not discussed here, welded wire mesh may also be used as shear reinforcement (Clause 8.2.12.1(b) in the Standard).

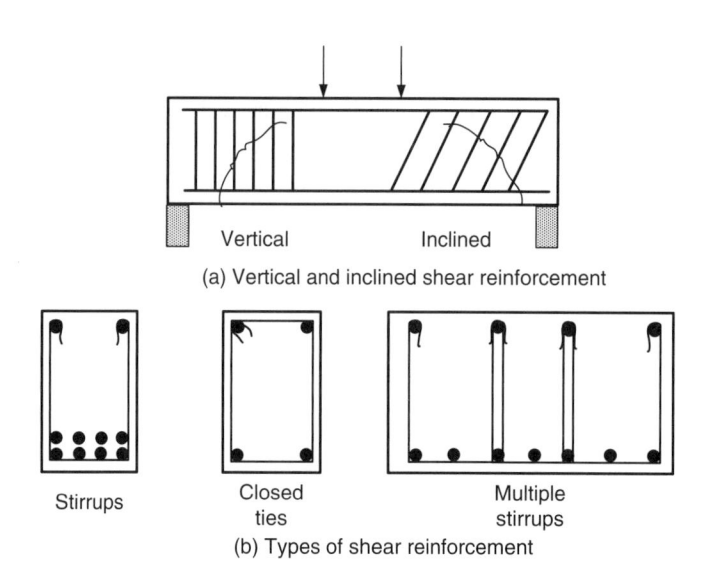

Figure 6.2(5) Details of shear reinforcement in concrete beams

6.3 TRANSVERSE SHEAR DESIGN

6.3.1 Definitions

Most of the symbols used in the shear formulas are the same as those used in bending. The existing and the new symbols are defined in Figure 6.3(1). It should be noted that:

- $d_o \geq 0.8D$ is the effective beam depth measured to the reinforcement layer farthest from the extreme compressive fibres
- A_{sv} is the total cross-sectional area of the shear reinforcement for a given beam section (i.e. two times the bar area for single stirrups or ties – similarly for multiples)
- A_g is the gross area of the beam section
- $f_{sy.f}$ is the yield stress of the shear reinforcement (i.e. stirrups or ties)
- the maximum centre-to-centre width of a stirrup or tie shall not be greater than the lesser of D and 600 mm.

If the stipulation in the last dot point cannot be complied with, multiple stirrups or ties must be used (see Figure 6.2(5)b).

6.3.2 Design shear force and the capacity reduction factor

It will be recalled in Equation 1.1(1) that V^* is the design shear force, which is taken to be caused by the most severe load combination amongst those computed using Equations 1.3(1) to 1.3(7). As each of the seven equations includes load factors, V^* is also referred to as the factored shear force.

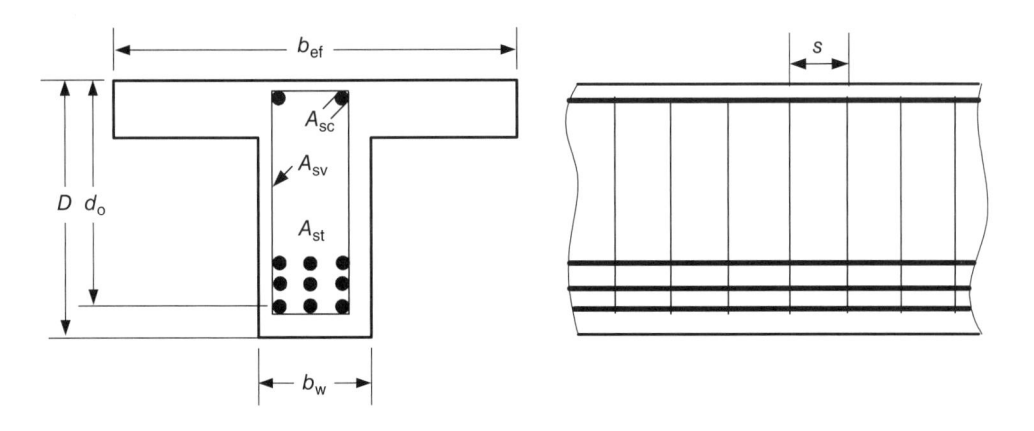

Figure 6.3(1) Definition of symbols used in shear formulas

The maximum shear normally occurs near a support. Clause 8.2.4 of the Standard stipulates that the design shear shall be taken as that at the face of the support or a distance (d_o) from the face of the support, provided that
- diagonal cracking cannot take place at the support or extend into it
- there are no concentrated loads closer than $2d_o$ from the face of the support
- $\beta_3 = 1$ in Equation 6.3(4)
- the shear reinforcement so determined at d_o from the support is continued unchanged to the face of the support.

In both options, the A_{st} required at d_o from the face of the support shall be continued onto the support and anchored past the face.

The underlying reasons for these stipulations may be found in the failure crack patterns of the first five beams shown in Figure 6.2(3) and in the discussion in Section 6.2.3.

Table 2.2.2 of the Standard specifies that the capacity reduction factor for shear considerations (ϕ) is = 0.7.

6.3.3 Maximum capacity

The nominal maximum shear force that can be carried by a beam is limited by the crushing strength of the web. According to Clause 8.2.6 of the Standard

$$V_{u.max} = 0.2 f'_c b_w d_o \quad \text{Equation 6.3(1)}$$

Thus, if the design shear force

$$V^* > \phi V_{u.max} \quad \text{Equation 6.3(2)}$$

then either b or D (or both) must be increased such that

$$V^* \leq \phi V_{u.\max} \qquad \text{Equation 6.3(3)}$$

No amount of shear reinforcement is considered effective if Equation 6.3(3) is not satisfied.

6.3.4 Shear strength of beams without shear reinforcement

The ultimate shear strength not including shear reinforcement (V_{uc}) is given in Clause 8.2.7.1 of the Standard as a force in newtons (N), or

$$V_{uc} = \beta_1 \beta_2 \beta_3 b_w d_o f_{cv} \sqrt[3]{\frac{A_{st}}{b_w d_o}} \qquad \text{Equation 6.3(4)}$$

where

$$f_{cv} = \sqrt[3]{f'_c} \leq 4 \text{ MPa} \qquad \text{Equation 6.3(4)a}$$

$$\beta_1 = 1.1\left(1.6 - \frac{d_o}{1000}\right) \geq 1.1 \qquad \text{Equation 6.3(5)a}$$

if the shear reinforcement is greater than the minimum specified in Equation 6.3(12). Otherwise

$$\beta_1 = 1.1\left(1.6 - \frac{d_o}{1000}\right) \geq 0.8 \qquad \text{Equation 6.3(5)b}$$

$\beta_2 = 1$ for beams with $N^* = 0$

$$\beta_2 = 1 - \frac{N^*}{3.5 A_g} \geq 0 \qquad \text{Equation 6.3(6)}$$

for beams subjected to axial tension, and

$$\beta_2 = 1 + \frac{N^*}{14 A_g} \qquad \text{Equation 6.3(7)}$$

for beams subjected to axial compression.

In the above equations, N^* is the absolute value of the design axial force in newtons and A_g is the gross cross-sectional area in mm². Finally, $\beta_3 = 1$, or for a case where 'arching' action is provided by the applied load and the support

$$\beta_3 = \frac{2d_o}{a_v} \leq 2 \qquad \text{Equation 6.3(8)}$$

in which a_v is the distance between the section (where shear is being considered) and the nearest support. A typical case where Equation 6.3(8) may be applicable is shown in Figure 6.3(2), noting that in any case β_3 is not less than 1.

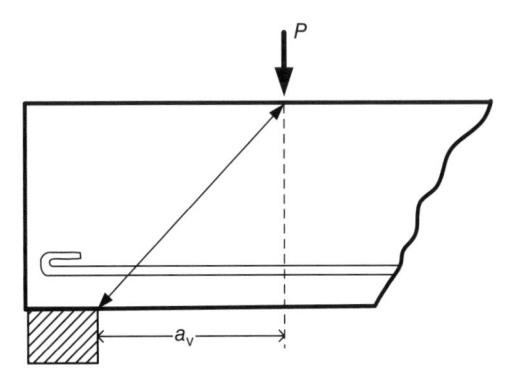

Figure 6.3(2) Definition of a_v

Note that Equation 6.3(4) may not apply in cases where:
- secondary stresses due to creep, shrinkage and differential temperatures are significant (see Clause 8.2.7.3 of the Standard)
- reversal of loads and/or torsion cause cracking in a zone usually in compression, in which case V_{uc} may be taken as zero (see Clause 8.2.7.4 of the Standard and also refer to Item (3) in the solution (b) for the design example in Section 7.4.4).

6.3.5 Shear strength checks and minimum reinforcement

After completing the bending and serviceability design, the resulting beam section is ready for shear considerations. The following checks (Clause 8.2.5 of the Standard) must be carried out to ensure the adequacy of the beam in resisting shear.

Case A
If

$$V^* \leq \phi V_{uc} \qquad \text{Equation 6.3(9)}$$

where V_{uc} is computed using Equation 6.3(4), and if $D <$ (the greater of 250 mm and $b_w/2$), the section is safe or no shear reinforcement is needed.

Case B
If

$$V^* \leq 0.5\phi V_{uc} \qquad \text{Equation 6.3(10)}$$

and $D \leq 750$ mm, the section is also adequate against shear.

Case C

If

$$V^* \leq 0.5\phi V_{uc} \quad \text{Equation 6.3(11)}$$

and $D > 750$ mm, the section requires 'minimum' shear reinforcement, that is

$$A_{sv.min} = \frac{0.06\sqrt{f'_c}b_w s}{f_{sy.f}} \geq \frac{0.35 b_w s}{f_{sy.f}} \quad \text{Equation 6.3(12)}$$

at a centre-to-centre spacing (see Figure 6.3(1)) $s \leq$ (the lesser of $0.75D$ and 500 mm).

Case D

If

$$0.5\phi V_{uc} < V^* \leq \phi V_{u.min} \quad \text{Equation 6.3(13)}$$

then

$$A_{sv} = A_{sv.min} \quad \text{Equation 6.3(14)}$$

In Equation 6.3(13), $V_{u.min}$ is the shear strength of a beam with the minimum reinforcement ($A_{sv.min}$). Or

$$V_{u.min} = V_{uc} + 0.1\sqrt{f'_c}b_w d_0 \geq V_{uc} + 0.6 b_w d_0 \quad \text{Equation 6.3(15)}$$

at a centre-to-centre spacing $s \leq$ (the lesser of $0.75D$ and 500 mm).

6.3.6 Design of shear reinforcement

The discussions in the preceding sections are for beams subjected to low shear force. For cases of higher shear force, the quantity of shear reinforcement must be determined using a more rigorous procedure.

If

$$V^* > \phi V_{u.min} \quad \text{Equation 6.3(16)}$$

the shear to be carried by shear reinforcement is given as

$$V_{us} = \left(\frac{V^* - \phi V_{uc}}{\phi}\right) \quad \text{Equation 6.3(17)}$$

in which case, vertical or inclined shear reinforcement is required to resist this shear.

Vertical stirrups or ties

If the spacing s is specified, the total cross-sectional area of the shear reinforcement is

$$A_{sv} = \frac{V_{us}s}{f_{sy.f}d_o \cot\theta_v} \qquad \text{Equation 6.3(18)}$$

where

$$\theta_v = 45° \qquad \text{Equation 6.3(19)}$$

or chosen within the range of 30° to 60° with the minimum (conservative) value taken as

$$\theta_v = 30° + 15° \times \frac{V^* - \phi V_{u.min}}{\phi V_{u.max} - \phi V_{u.min}} \qquad \text{Equation 6.3(20)}$$

in which $V_{u.max}$ and $V_{u.min}$ are given in Equations 6.3(1) and 6.3(15), respectively.

On the other hand, if the shear reinforcement area (A_{sv}) is specified, which is frequently the case, then the spacing

$$s = \frac{A_{sv}f_{sy.f}d_o \cot\theta_v}{V_{us}} \qquad \text{Equation 6.3(21)}$$

but

$$s < \text{the lesser of } D/2 \text{ and } 300 \text{ mm} \qquad \text{Equation 6.3(22)}$$

Inclined stirrups or ties

If properly arranged, inclined shear reinforcement is more effective than vertical shear reinforcement. Figure 6.3(3) illustrates the correct orientation of inclined reinforcement in relation to the direction of the design shear (V^*).

It is obvious in Figure 6.3(3) that inclination to the left or right depends on the direction of V^*. This is because the reinforcement must cross the shear cracks (see Figure 6.2(1)d), but not be parallel to them. This consideration is unnecessary in the case of vertical shear reinforcement, as either left or right inclined cracks are crossed by the vertical arrangement. Thus, the designer must be able to anticipate the direction of V^* before adopting inclined reinforcement, especially in the design of frame members where wind load effects are significant.

The design equations for inclined reinforcement are similar to those for a vertical arrangement. If spacing (s) is specified, the reinforcement area

$$A_{sv} = \frac{V_{us}s}{f_{sy.f}d_o(\sin\alpha_v \cot\theta_v + \cos\alpha_v)} \qquad \text{Equation 6.3(23)}$$

in which α_v is the inclination of the reinforcement where $45° \leq \alpha_v \leq 90°$ and θ_v is determined using Equations 6.3(19) and 6.3(20).

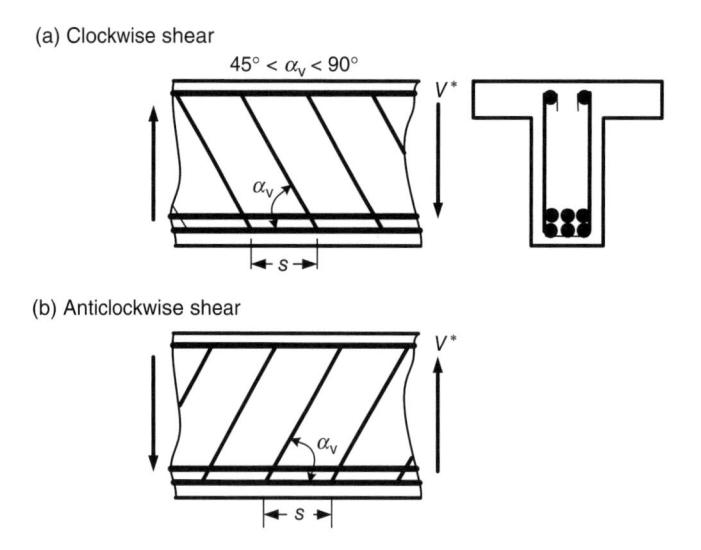

Figure 6.3(3) Examples of inclined shear reinforcement: **(a)** clockwise shear and **(b)** anticlockwise shear

On the other hand, if A_{sv} is given, the spacing

$$s = \frac{A_{sv} f_{sy.f} d_o (\sin \alpha_v \cot \theta_v + \cos \alpha_v)}{V_{us}}$$ Equation 6.3(24)

6.3.7 Detailing

In general, the maximum spacing s is governed by Equation 6.3(22), that is, not greater than the lesser of $D/2$ and 300 mm. For light shear regions, these rules are somewhat relaxed, the details of which may be found in Cases C and D in Section 6.3.5. Note also that Clause 8.2.12.2 in the Standard stipulates that the maximum transverse spacing over the width of wider beams (see Figure 6.2(5)b) is not to exceed the lesser of D and 600 mm.

In locating the stirrups or ties, it is a good practice to start from inside a support moving towards the centre of the span. Note that the computed amount of reinforcement for a beam section must be extended from the section over a distance D in the direction of decreasing shear, even if theoretically the shear force beyond this section drops abruptly (as in the case of a concentrated load). This requirement is specified in Clause 8.2.12.3 in the Standard, which also states that the first stirrup or tie at each end of a span must be positioned 'not more than 50 mm from the face of the adjacent support'.

The minimum cover requirements discussed in Section 1.4.1 for the main reinforcement also apply to shear reinforcement. The reader should consult Tables 1.4(1), 1.4(2) and 1.4(3) for details.

Although shear reinforcement design requires consideration of shear stress, the dimensional details of individual stirrups or ties are a stress development problem. For a detailed discussion, the reader is referred to Section 8.4. In practice, stirrups and ties of nominated dimensions can be ordered or bought in bulk.

6.3.8 Design example

Problem

A T-beam with a simply supported span of 6 m is subjected to a concentrated live load $P = 700$ kN, as shown in Figure 6.3(4)a; the cross-sectional details are given in Figure 6.3(4)b. Design the beam for shear, assuming $f_c' = 20$ MPa.

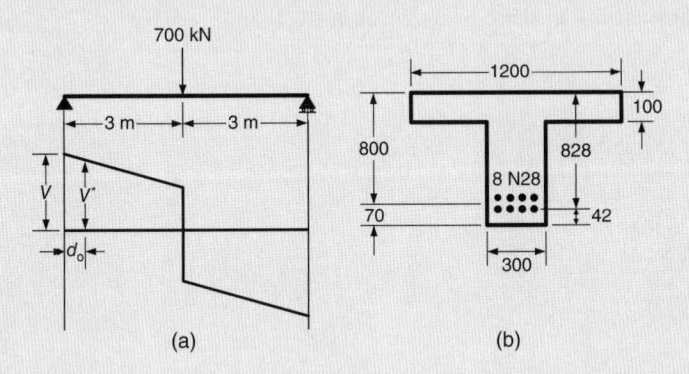

Figure 6.3(4) Details of the example T-beam: **(a)** loading configuration and shear force diagram and **(b)** cross-sectional details

Note: cross-sectional dimensions are in mm unless specified otherwise.

Solution

Design shear

From Table 2.3(1) we obtain $A_{st} = 4928$ mm^2 and the gross sectional area
$$A_g = 1200 \times 100 + 770 \times 300 = 351\,000 \text{ mm}^2$$
based on which the steel percentage by volume

$$v = \frac{4928}{351\,000} \times 100 = 1.4\%$$

Then Equation 2.4(1): $\rho_w = 24 + 0.6 \times 1.4 = 24.84$ kN/m^3 and the self-weight

$$g = (0.1 \times 1.2 + 0.77 \times 0.3) \times 24.84 = 8.72 \text{ kN/m}$$

The maximum shear can be determined via Equation 1.3(2) as

$$\begin{aligned}
V &= 1.2 V_g + 1.5 V_q \\
&= 1.2 \times 8.72 \times 3 + (1.5 \times 700)/2 \\
&= 31.4 + 525 \\
&= 556.4 \text{ kN}
\end{aligned}$$

The design shear may be taken to be the shear at a distance d_o (= 828 mm) from the support. Or from Figure 6.3(4)a

$$V^* = 525 + \frac{3-d_o}{3} \times 31.4 = 525 + \frac{3-0.828}{3} \times 31.4 = 547.7 \text{ kN}$$

Section adequacy

Check the maximum section capacity in shear using Equation 6.3(1) and we have

$$V_{u.max} = 0.2 \times 20 \times 300 \times 828 \times 10^{-3}$$
$$= 993.6 \text{ kN} > \frac{V^*}{\phi} = 782.4 \text{ kN}$$

Thus D = 870 mm is acceptable.

Concrete shear capacity

Next, compute V_{uc} – Equation 6.3(5)a gives

$$\beta_1 = \left(1.6 - \frac{828}{1000}\right) \times 1.1 = 0.849 \geq 1.1$$

Therefore

$$\beta_1 = 1.1$$

Equation 6.3(6): $\beta_2 = 1$

Equation 6.3(8): $\beta_3 = 2 \times \frac{0.828}{3} = 0.552$

Therefore, $\beta_3 = 1$, according to Equation 6.3(8).

Equation 6.3(4)a gives $f_{cv} = \sqrt[3]{f'_c} = \sqrt[3]{20} = 2.71$ MPa < 4 MPa, which is acceptable. Then Equation 6.3(4):

$$V_{uc} = 1.1 \times 1 \times 1 \times 300 \times 828 \times 2.71 \times \sqrt[3]{\frac{4928}{300 \times 828}} \times 10^{-3} = 200.5 \text{ kN}$$

and

$$\phi V_{uc} = 0.7 V_{uc} = 140.4 \text{ kN}.$$

The minimum shear according to Equation 6.3(15) is

$$V_{u.min} = V_{uc} + 0.6 \times 300 \times 828 \times 10^{-3} = 349.5 \text{ kN}$$

and

$$\phi V_{u.min} = 0.7 V_{u.min} = 244.7 \text{ kN} < V^*$$

Thus, shear reinforcement is required.

Shear reinforcement

If, say, vertical ties made of N12 bars are used, we have

$$A_{sv} = 2 \times 113 = 226 \text{ mm}^2$$

Equation 6.3(20): $\theta_v = 30° + 15° \times \dfrac{547.7 - 244.7}{0.7 \times 993.6 - 244.7} = 40.1°$

Equation 6.3(17): $V_{us} = \dfrac{(547.7 - 140.4)}{0.7} = 581.9 \text{ kN}$

That is,

$$s = \frac{226 \times 500 \times 828 \cot 40.1°}{581.9 \times 10^3}$$

$= 191.0$ mm (say 180 mm), which also satisfies Equation 6.3(22).

Reinforcement arrangement

The layout is as shown in Figure 6.3(5).

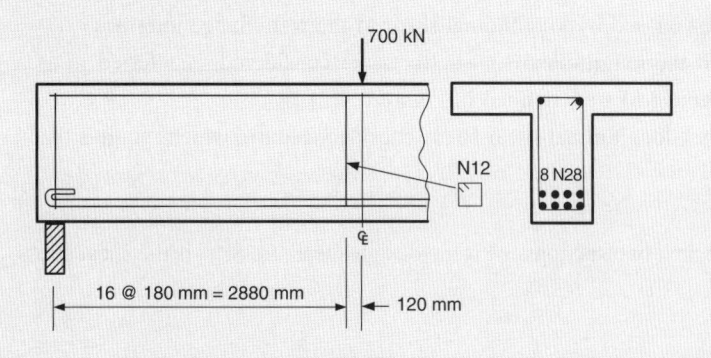

Figure 6.3(5) The layout of shear reinforcement for the example in Section 6.3.8

For the full length of the beam, we require a total of (17 + 17 + 1) = 35 ties, with the spacings next to the load reduced to 120 mm for obvious reasons. As the shear force is due mainly to the 700 kN of concentrated load and the distribution is almost uniform, it is acceptable that the chosen tie spacing (*s*) is used throughout. In cases where the shear distribution varies greatly, *s* may vary accordingly along the span to suit, but the detailing requirements specified in Clause 8.2.12.3 of the Standard must be met (see Section 6.3.7).

6.4 LONGITUDINAL SHEAR

6.4.1 Shear planes

In accordance with the principle of complementary shear, longitudinal shear stress acting in the spanwise direction exists as a result of the vertical shear. This phenomenon is illustrated in Figure 6.4(1).

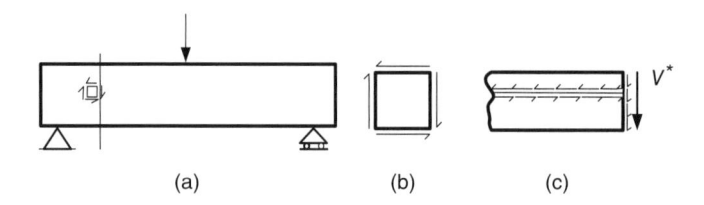

Figure 6.4(1) Illustration of longitudinal shear: **(a)** beam elevation; **(b)** complementary shear; **(c)** longitudinal shear

Longitudinal shear is another mandatory design consideration for reinforced concrete structures. In practice, even supposedly monolithic concrete is normally cast in layers. For example, for a beam-and-slab structure, the beams may be cast first on a given date, then the slab on a later date after the beams have hardened. This is commonly the case for composite construction with precast beams and 'cast in situ' flanges or other toppings. If not properly designed, damage could be caused by longitudinal shear to the web–flange interface.

An interface, at which the longitudinal shear effects are considered, is referred to as a shear plane (SP). Some typical SPs are defined in Figure 6.4(2).

The Standard specifies a longitudinal shear stress check procedure, which requires that

$$\phi\tau_u \geq \tau^*$$ Equation 6.4(1)

where, as discussed in the next two sections, τ^* is the design shear stress, τ_u is the shear stress capacity and $\phi = 0.7$.

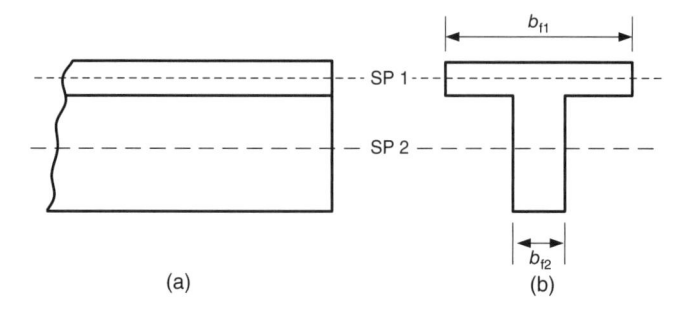

Figure 6.4(2) Definition of shear planes: **(a)** elevation and **(b)** cross-section

6.4.2 Design shear stress

The design longitudinal shear stress is given in Clause 8.4.2 of the Standard as

$$\tau^* = \beta V^*/(zb_f)$$ Equation 6.4(2)

where b_f = width of the shear plane; the internal moment lever arm

z = distance between the resultant compressive and tensile forces in the section Equation 6.4(3)

and

$$\beta = {A_1}/{A_2}$$ Equation 6.4(4)

in which A_1 and A_2 are defined in Figure 6.4(3), noting that there are two cases in terms of the shear plane location:

1. in a compression zone. A_1 is the concrete area above the shear plane (Figure 6.4(3)a) and A_2 is the total area covered by the rectangular stress block or that above the line, a distance $\gamma k_u d$ from the extreme compressive fibre.
2. in a tension zone. A_1 is the sum of the tension reinforcement area above the shear plane (Figure 6.4(3)b) and A_2 is the total tension reinforcement area.

In cases where the shear plane is located in the flange, closed ties are preferable to stirrups as shear reinforcement, because they can better contain the concrete.

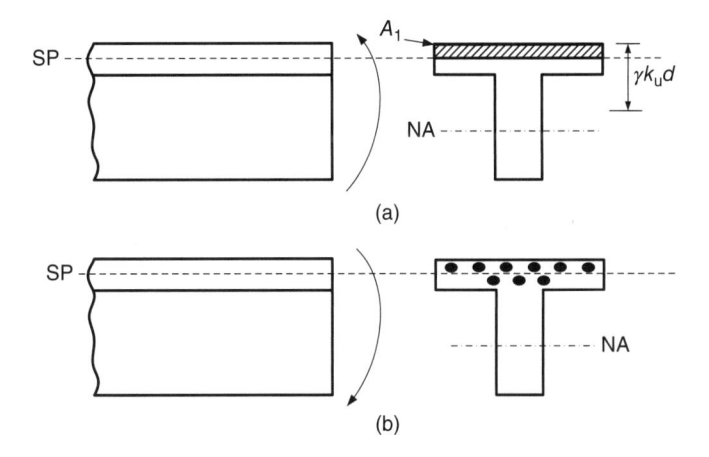

Figure 6.4(3) Shear plane locations: **(a)** in a compression zone and **(b)** in a tension zone

6.4.3 Shear stress capacity

In Equation 6.4(1), the longitudinal shear stress capacity is given in Clause 8.4.3 of the Standard as

$$\tau_u = \mu\left(\frac{A_{sf}f_{sy.f}}{sb_f} + \frac{g_p}{b_f}\right) + k_{co}f'_{ct} \leq \text{the lesser of } \begin{cases} 0.2f'_c \\ 10 \text{ MPa} \end{cases} \quad \text{Equation 6.4(5)}$$

where

A_{sf} = area of the fully anchored shear reinforcement crossing the shear plane (in mm^2)
$f_{sy.f}$ = yield strength of the shear reinforcement but $\leq$ 500 MPa
s = spacing of the shear reinforcement
b_f = width of the shear plane (in mm)
g_p = permanent distributed load normal to the shear plane (in N/mm)
f'_{ct} = direct tensile strength of concrete (see Equation 2.2(3))
μ = friction coefficient
k_{co} = cohesion coefficient.

Note that the recommended values of μ and k_{co} may be found in Table 8.4.3 of the Standard, which is reproduced here as Table 6.4(1).

Table 6.4(1) Shear plane surface coefficients

Surface condition of the shear plane	Coefficients μ	k_{co}
A smooth surface, as obtained by casting against a form, or finished to a similar standard.	0.6	0.1
A surface trowelled or tamped, so that the fines have been brought to the top, but where some small ridges, indentations or undulations have been left; slip-formed and vibro-beam screeded, or produced by some form of extrusion technique.	0.6	0.2
A surface deliberately roughened: (a) by texturing the concrete to give a pronounced profile (b) by compacting but leaving a rough surface with coarse aggregate protruding but firmly fixed in the matrix (c) by spraying when wet, to expose the coarse aggregate without disturbing it, or (d) by providing mechanical shear keys.	0.9	0.4
Monolithic construction	0.9	0.5

Source: Standards Australia (2009). AS 3600–2009 *Concrete Structures*. Standards Australia, Sydney, NSW. © Standards Australia Limited. Copied by Cambridge University Press with the permission of Standards Australia and Standards New Zealand under Licence 1712-c038.

6.4.4 Shear plane reinforcement and detailing

The design for longitudinal shear can readily be carried out using Equations 6.4(2) and 6.4(5). Note that both transverse shear reinforcement and torsional reinforcement (see Section 7.4) resist the corresponding design actions by tension in the legs of the stirrups or closed ties. Thus, stirrups or closed ties are effective in providing resistance to longitudinal shear, which is by dowel action of the vertical legs. In design, follow these steps:

1. Check that Equation 6.4(1) prevails.
2. If not, the required longitudinal reinforcement may be obtained from Equation 6.4(5) as

$$\frac{A_{sf}}{s} = \frac{b_f}{f_{sy.f}} \left[\left(\tau^*/\phi - k_{co}f'_{ct} \right)/\mu - \left(g_p/b_f \right) \right] \qquad \text{Equation 6.4(6)}$$

3. Ensure that the existing transverse shear and/or torsional reinforcement equals or exceeds that given in Equation 6.4(6). Otherwise, provide the balance of the shear reinforcement as required in Equation 6.4(6).

    Note that Clause 8.4.4 of the Standard limits the spacing to

$$s \leq 3.5 t_f \qquad \text{Equation 6.4(7)}$$

where t_f = thickness of the topping or flange anchored by the shear reinforcement.

Further, Clause 8.4.5 of the Standard stipulates that the average thickness of the structural components subject to longitudinal shear shall be not less than 50 mm, with no local thickness less than 30 mm.

6.4.5 Design example

Problem

For the T-beam in the preceding design example (Section 6.3.8), check and design for longitudinal shear, if applicable. Take $f'_c = 20$ MPa and use N12 or N16 bars for ties as necessary. Assume monolithic construction.

Solution

From Section 6.3.8, self-weight = 8.72 kN/m and V^* = 547.7 kN.

The critical shear plane, although not specified, may be taken as at the rectangular stress block depth level located (in this case) in the web – thus b_f = 300 mm.

Equation 6.4(4): $\beta = 1$

With reference to Figure 6.3(4), imposing $\Sigma F_x = 0$ gives

$$4928 \times 500 = 0.85 \times 20 \times [1200 \times 100 + 300(\gamma k_u d - 100)]$$

from which

$$\gamma k_u d = 183.1 \text{ mm}$$

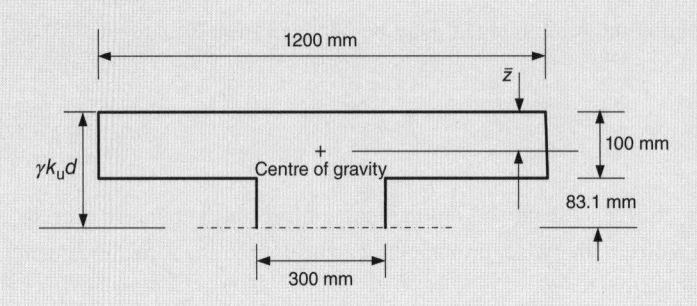

Figure 6.4(4) Determination of z for the example T-beam

Then, in Figure 6.4(4), the distance between the centre of gravity of the compressive resultant and the extreme compressive fibre

$$\bar{z} = \frac{1200 \times 100 \times 100/2 + (83.1 \times 300) \times (100 + 83.1/2)}{1200 \times 100 + 83.1 \times 300} = 65.7 \text{ mm}$$

Equation 6.4(3): $z = d - \bar{z} = 800 - 65.7 = 734.3$ mm

Equation 6.4(2): $\tau^* = 547.7 \times 10^3/(734.3 \times 300) = 2.49$ MPa

The quantities required for Equation 6.4(5) are:

- A_{sf}/s for the existing transverse reinforcement or N12 ties @ 180 mm = 226/180 = 1.256 mm²/mm
- $g_p = (100 \times 1200 + 83.1 \times 300) \times 10^{-6} \times 24.84 \times 10^3 = 3600.1$ N/m = 3.6 N/mm
- From Equation 2.2(3), $f'_{ct} = 0.36\sqrt{20} = 1.61$ MPa
- From Table 6.4(1), $\mu = 0.9$; $k_{co} = 0.5$

Thus Equation 6.4(5):

$$\tau_u = 0.9(1.256 \times 500/300 + 3.6/300) + 0.5 \times 1.61 = 2.70 \text{ MPa}$$

Check Equation 6.4(1):

$$\phi\tau_u = 0.7 \times 2.70 = 1.89 < \tau^* = 2.49 \text{ MPa, which is inadequate.}$$

Thus, additional shear reinforcement is required.

Equation 6.4(6): $\dfrac{A_{sf}}{s} = \dfrac{300}{500} \times \left[\left(\dfrac{2.49}{0.7} - 0.5 \times 1.61\right)/0.9 - (3.6/300)\right]$

$$= 1.828 \text{ mm}^2/\text{mm}$$

Try N12 ties: $s = \dfrac{2 \times 113}{1.828} = 123.6$ mm, which is rather close

Thus, try N16 ties and

$$s = \dfrac{2 \times 201}{1.828} = 219.9 \text{ mm}$$

Say, use N16 ties @ 215 mm, which is less than $3.5t_f = 3.5 \times 100 = 350$ mm as required in Equation 6.4(7).

In summary, the final reinforcement arrangement for both transverse and longitudinal shear is N16 closed ties @ 215 mm, and the total number required over the 6-m span: $(28 + 1) = 29$.

6.5 SUMMARY

The fundamentals of shear stress and shear failure as well as the mechanism of shear resistance are discussed in detail. Considering the relevant Standard recommendations, formulas are derived for the transverse and longitudinal shear design of reinforced concrete beams. Applications of these design formulas are illustrated using fully worked examples.

6.6 PROBLEMS

1. Figure 6.6(1) shows a simple beam with overhang with its loading configuration and cross-sectional details. The concrete strength is $f'_c = 25$ MPa.

   Assuming that the concentrated load of 90 kN is a live load, design and detail the transverse shear reinforcement as necessary for the critical shear section of the beam. Use R10 ties. The maximum shear reinforcement requirement may be adopted throughout the beam.

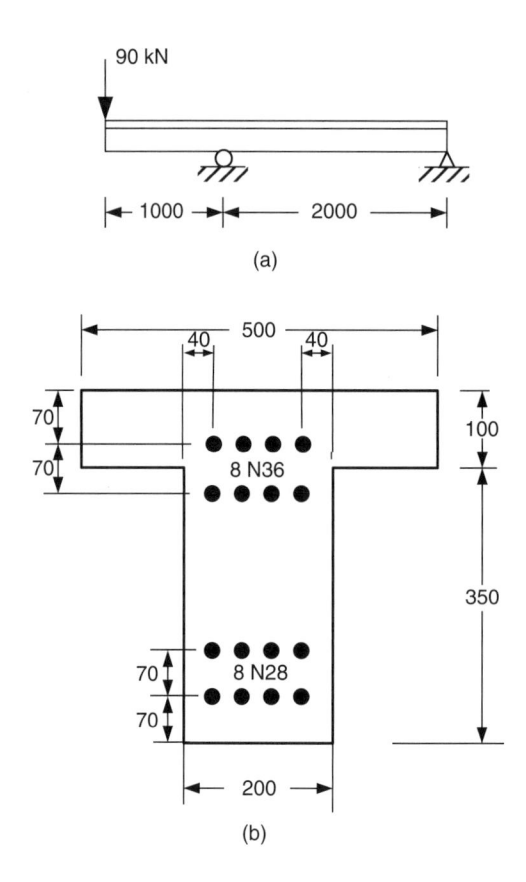

Figure 6.6(1) Details of a simple beam with overhang: **(a)** loading configuration and **(b)** cross-sectional details

Note: all dimensions are in mm.

2. A simple beam with a cantilever overhang and its loading configuration and cross-sectional details are shown in Figure 6.6(2). The concrete strength $f'_c = 25$ MPa. Note that the load combination formula $1.2g + 1.5q$ applies, in which g and q are dead and live loads, respectively.

   Design and detail the transverse shear reinforcement as necessary for the critical shear section of the beam, which may occur at the left or right of the support at B. Use R10 ties and the maximum shear reinforcement requirement may be adopted throughout the beam.

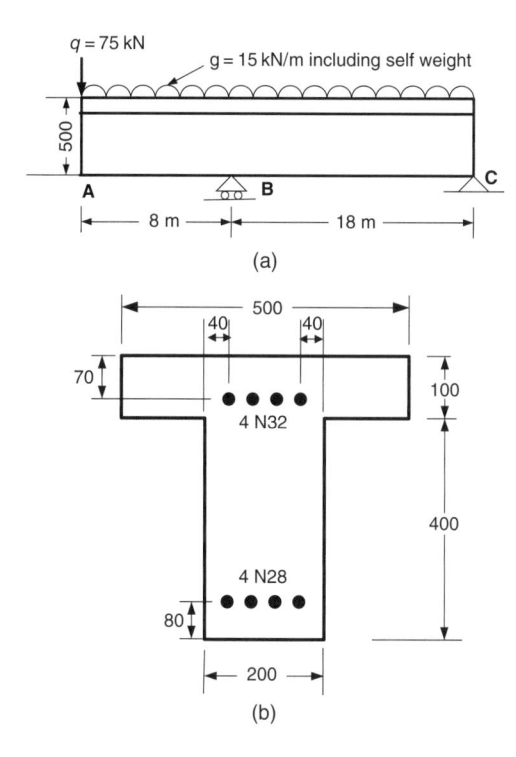

Figure 6.6(2) Details of a simple beam with a cantilever overhang: **(a)** loading configuration and **(b)** cross-sectional details

Note: all dimensions are in mm unless otherwise specified.

3. For the beam shown in Figure 6.6(3), design the transverse and longitudinal shear reinforcement. Take f'_c = 25 MPa and use N16 bars for closed ties. Assume that the construction is monolithic and the 700 kN is a concentrated live load.

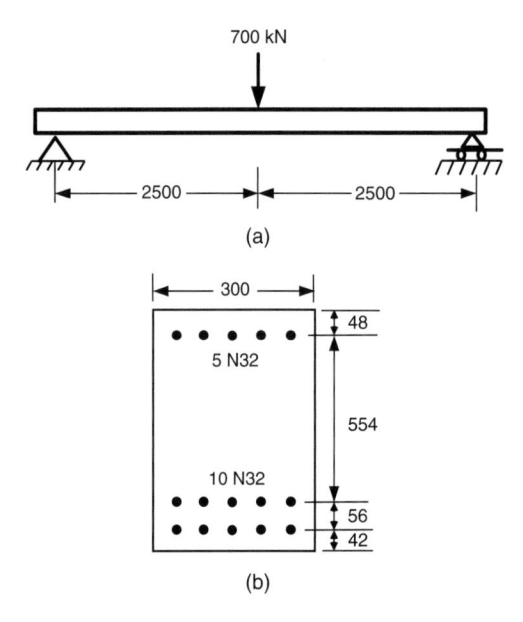

Figure 6.6(3) Details of a simple beam: **(a)** loading configuration and **(b)** cross-sectional details

Note: all dimensions are in mm.

4. For the cantilever beam shown in Figure 3.9(7) (for Problem 12 in Section 3.9), if q = 150 kN and there is no change to any other condition as given in Problem 12, design and detail, as necessary, the transverse shear reinforcement. Use N12 closed ties only.

5. Details of a simply supported beam are given in Figures 6.6(4)a and b. Note that the dead load, g, includes the self-weight.

   Design the transverse shear reinforcement in terms of N10 ties @ s, the required spacing at the critical section, which in this case may be taken at the appropriate support (not d_o from it). Use the same spacing throughout. Take f'_c = 32 MPa and you may assume $\theta_v = 45°$.

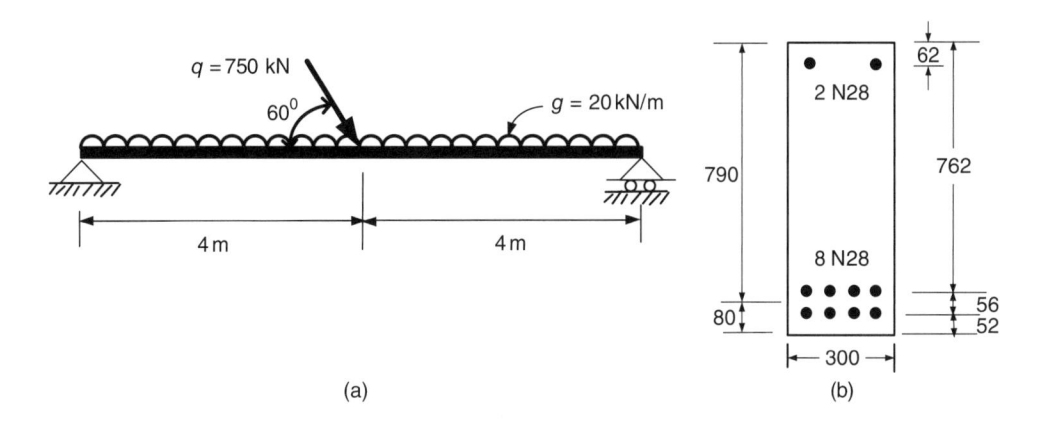

Figure 6.6(4) Details of a simply supported beam: **(a)** loading configuration and **(b)** cross-sectional details

Note: all dimensions are in mm unless otherwise specified.

6. A simply supported beam is shown in Figure 6.6(5)a with details of the longitudinal reinforcement given in Figure 6.6(5)b. Taking f'_c = 30 MPa, and that both the P_u are concentrated live loads, design the shear reinforcement (using vertical N10 stirrups only). Sketch the final design.

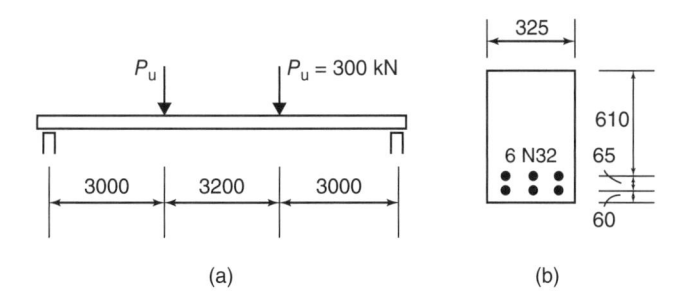

Figure 6.6(5) Details of a simple beam under two concentrated loads

Note: all dimensions are in mm.

7. A cantilever beam and its loading configuration are detailed in Figure 6.6(6). Considering transverse shear only, design and detail the shear reinforcement as necessary. Use R10 ties, assume the 80 kN to be a concentrated live load and adopt the maximum shear reinforcement requirement for the entire span. Take $f'_c = 32$ MPa.

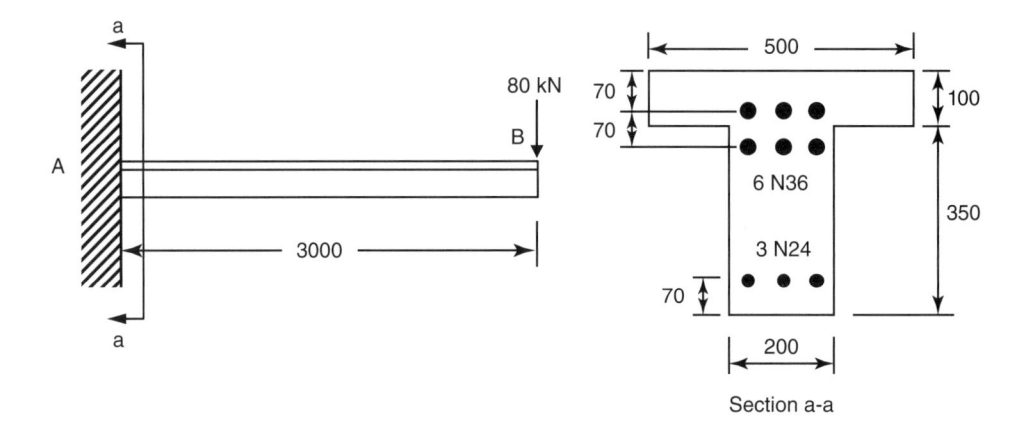

Figure 6.6(6) Details of a cantilever beam

Note: all dimensions are in mm.

8. The beam-and-slab system in Figure 4.7(12) (for Problem 12 in Section 4.7) is to be cast in two steps: first the webs up to level x-x, then the slab.

    In addition to a dead load (including self-weight) $g = 35$ kN/m and a live load $q = 80$ kN/m, each of the T-beams has to carry a concentrated live load of 150 kN at the middle of a simply supported span of 8 m.

    As the structural engineer, you are required to design the slab–web interface against longitudinal shear failure. You may assume $\mu = 0.9$ and $k_{co} = 0.4$; use N12 bars for ties if necessary. Take $f'_c = 20$ MPa, $\rho_w = 24$ kN/m³, and the effective depth may be taken as $d = 870$ mm for this design.

9. Figures 6.6(7)a and b, respectively, give details of a simply supported bridge beam and its (idealised) cross-section. In addition to the dead load $g = 26$ kN/m (including self-weight), the beam is to carry the two concentrated live loads as shown. The characteristic strength of the concrete is $f'_c = 32$ MPa.
   (a) Design the transverse reinforcement to resist the vertical shear. Use (vertical) N12 closed ties only; show the design details in a sketch.
   (b) Assuming monolithic construction, carry out a check for the longitudinal shear. Is the design adequate?

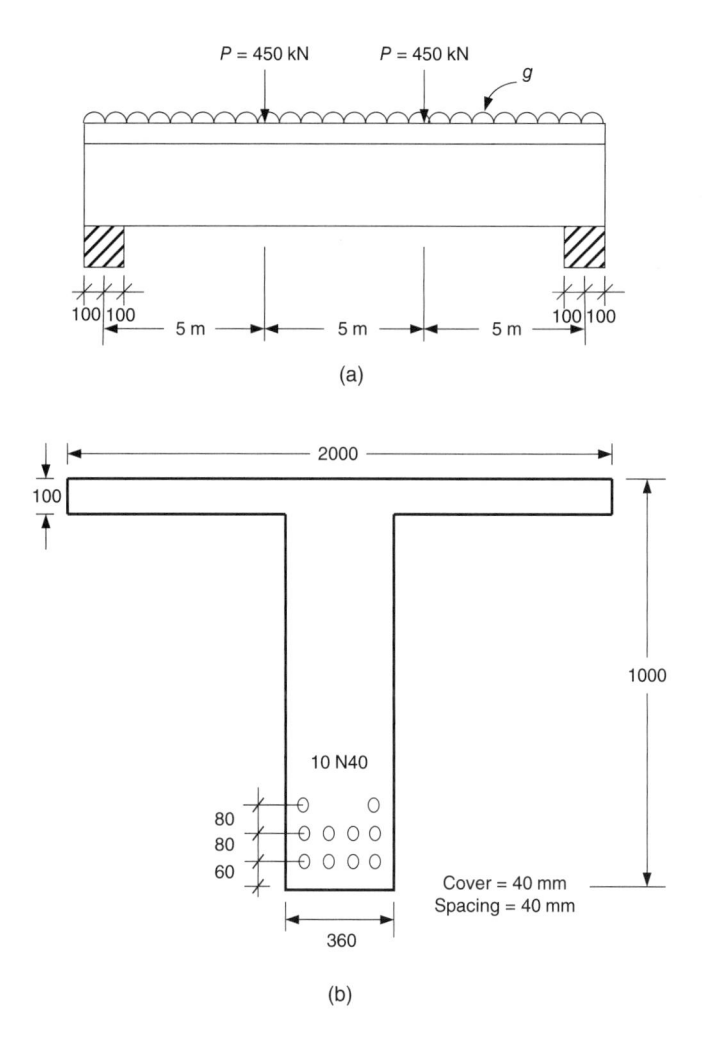

Figure 6.6(7) A simply supported bridge beam and its (idealised) cross-section

Note: all dimensions are in mm unless otherwise specified.

10. A simply supported beam and its loading configuration and cross-sectional details are shown in Figure 6.6(8). The concrete strength is $f'_c = 25$ MPa.

    Design and detail the transverse shear reinforcement as necessary for the critical shear section of the beam. Use R10 ties and adopt the maximum shear reinforcement requirement throughout the beam.

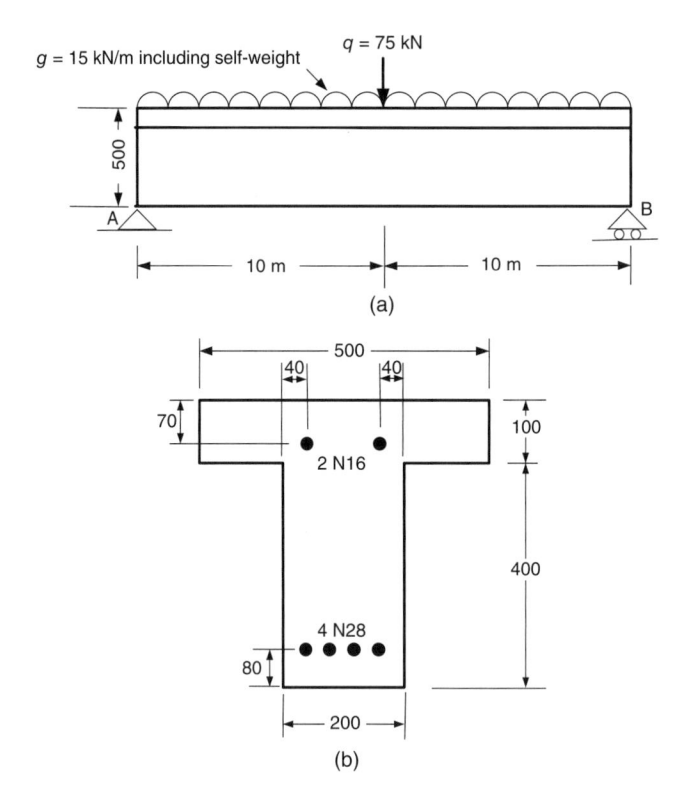

Figure 6.6(8) Details of a simply supported beam: **(a)** loading configuration and **(b)** cross-sectional details

Note: all dimensions are in mm unless otherwise specified.

11. Details of a cantilever beam are given in Figures 6.6(9)a and b. Note that the dead load, g, includes the self-weight.

Design the transverse shear reinforcement at the critical section, which in this case may be taken at the root or support of the cantilever (not d_o from it). The answer should be given in terms of N12 ties @ s, the required spacing. Use the same spacing throughout the span. Take $f'_c = 40$ MPa.

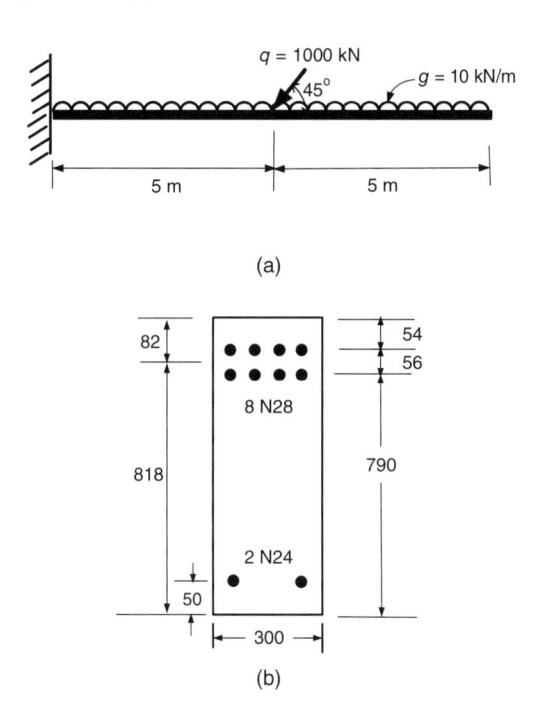

Figure 6.6(9) Details of a cantilever beam: **(a)** loading configuration and **(b)** cross-sectional details

Note: all dimensions are in mm unless otherwise specified.

152 PART 1 REINFORCED CONCRETE

12. A simply supported T-beam under loads and its cross-section are detailed respectively in Figures 6.6(10)a and b. Note that the dead load, g, includes the self-weight. Design the transverse shear reinforcement at the critical section. The answer should be given in terms of N12 ties @ s, the required spacing. Use the same spacing throughout the span. Take $f'_c = 40$ MPa and you may assume $\theta_v = 45°$.

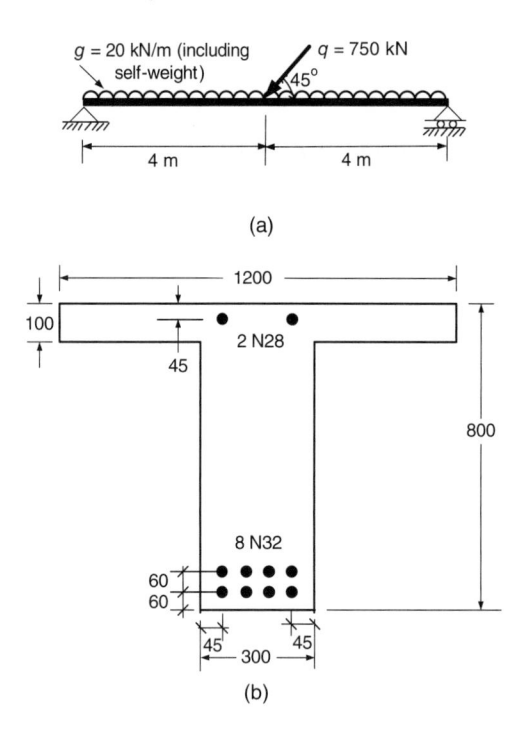

(figure: loading configuration — $g = 20$ kN/m (including self-weight), $q = 750$ kN at 45°, spans 4 m + 4 m)

(a)

(figure: T-beam cross-section — flange 1200 × 100, web 300 wide, overall depth 800; 2 N28 in flange at 45 mm from top; 8 N32 in web at 60/60 from bottom and 45/45 from sides)

(b)

Figure 6.6(10) Details of a simply supported T-beam: **(a)** loading configuration and **(b)** cross-sectional details

Note: all dimensions are in mm unless otherwise specified.

13. Figures 6.6(11)a and b illustrate a simply supported beam under loads and its rectangular cross-section, respectively. Note that the dead load, g, includes the self-weight. Design the transverse shear reinforcement at the critical section, which in this case may be taken as at the appropriate support (not d_o from it). The answer should be given in terms of N12 ties @ s, the required spacing. Use the same spacing for the entire span. Take $f_c' = 32$ MPa.

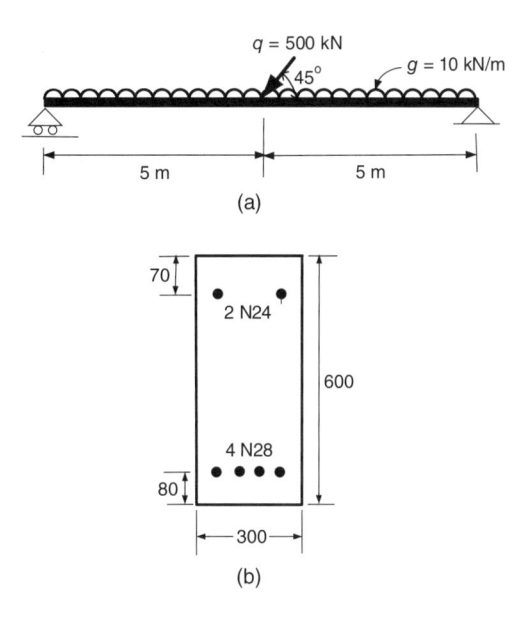

Figure 6.6(11) Details of a simply supported beam: **(a)** loading configuration and **(b)** cross-sectional details

Note: all dimensions are in mm unless otherwise specified.

14. A simply supported beam with an overhang, its loading configuration and the cross-sectional details are shown in Figure 6.6(12). The concrete strength $f'_c = 25$ MPa. Under the inclined concentrated live load $P = 100$ kN plus the self-weight, design if necessary and then detail the transverse shear reinforcement for the critical shear section of the beam. Use R10 ties and the maximum shear reinforcement requirement may be adopted throughout the span.

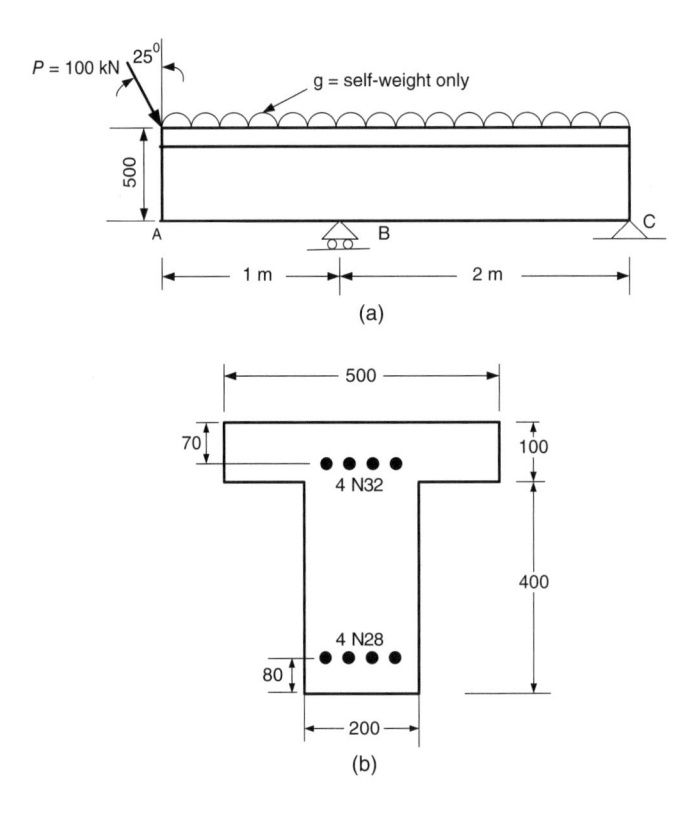

Figure 6.6(12) Details of a simply supported beam with an overhang: **(a)** loading configuration and **(b)** cross-sectional details

Note: all dimensions are in mm unless otherwise specified.

ULTIMATE STRENGTH DESIGN FOR TORSION

7.1 INTRODUCTION

7.1.1 Origin and nature of torsion

Torsion is a three-dimensional action; it is the moment about the longitudinal axis of the structural member. Occasionally, torsional moment is also referred to as twisting moment or torque.

In a three-dimensional structure, there are numerous situations in which torsion occurs. Figure 7.1(1) shows two typical cases.

For the case of the cantilever bent beam or bow girder in Figure 7.1(1)a, the torsional moment (T) is produced by the transverse load (P) acting eccentrically with respect to the axis of the beam. As it is a statically determinate structure, adequate design for torsion is vitally important – collapse of the system will result if failure in torsion occurs.

The grillage system shown in Figure 7.1(1)b is often used for beam-and-slab floor structures. The system is statically indeterminate. Torsion of the girder is a result of the unbalanced end moments at C of the two cross beams spanning in the z-direction. Note that, for convenience, torsional moments may be indicated in the x–y plane by double-headed arrows following the right-hand screw rule. As the system is statically indeterminate, failure of beam AB in torsion would not automatically mean collapse of the grillage. However, serious serviceability problems of the beam (torsional cracking) can be expected as well as the redistribution of bending moments in the two cross beams (DC and CE).

7.1.2 Torsional reinforcement

Figure 7.1(2) depicts a rectangular beam in pure torsion. On each face of the beam, shear stresses are induced by the torsional moment. At the instance when the principal tensile stress exceeds the tensile strength of concrete, cracking occurs at 90° to the principal stress direction.

Typical failure patterns of beams in pure torsion are shown in Figure 7.1(3). These are typical results of the tests carried out by Lei (1983). It is obvious that, on each face of the beam, the torsional shear cracks are inclined with respect to the axis of the beam. Thus, the use of

156 PART 1 REINFORCED CONCRETE

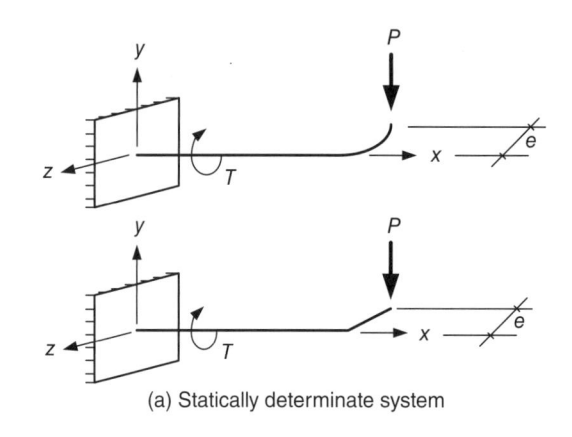

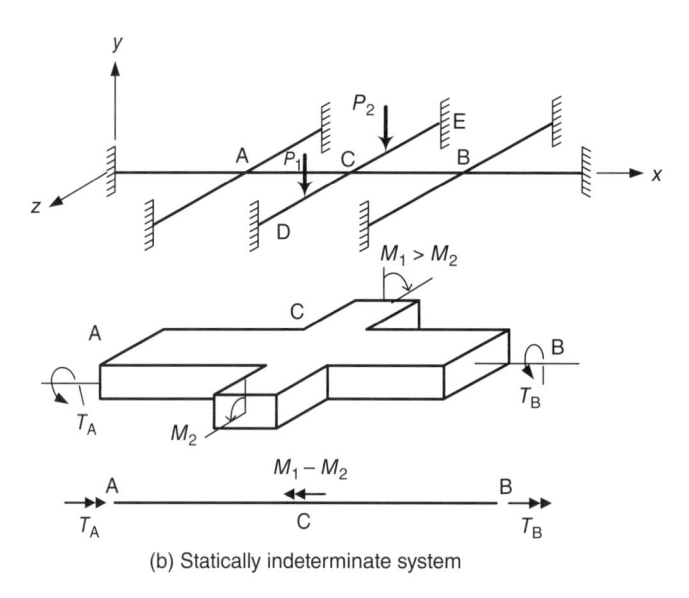

Figure 7.1(1) Typical cases of torsion in three-dimensional structures

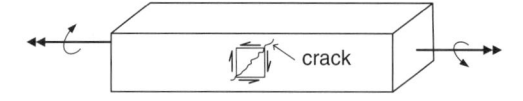

Figure 7.1(2) A rectangular beam in pure torsion

closed ties only is effective in resisting torsional moment. Longitudinal steel placed behind the four faces of the beam, through dowel and other actions, is also helpful in resisting torsion.

It should be noted that (open) stirrups are not used in torsion design as torsional cracks occur on all faces of the beam. Further, torsional shear is additive to the transverse

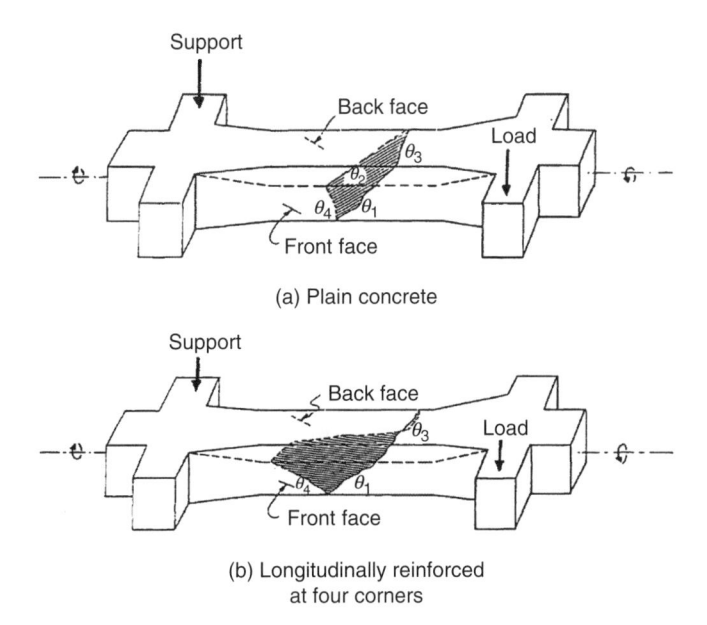

Figure 7.1(3) Typical failure patterns of beams in pure torsion

bending shear. Therefore, closed ties designed for the bending shear of a beam may not be relied on to resist torsion.

For a beam subject to torsion and bending, the torsional moment interacts with the transverse shear force. Design and analysis equations presented here reflect this interaction.

7.1.3 Transverse reinforcement area and capacity reduction factor

The torsion in a beam is resisted by the combined action of the concrete, the transverse steel (closed ties), including steel used to resist longitudinal shear and the longitudinal steel in the tensile and compressive zones. A_{sw} is the cross-sectional area of the transverse reinforcement and is the area of the reinforcing bar that makes up the tie. (For transverse shear, A_{sv} for a single stirrup or tie is twice the bar area.) This is obvious, as the tie (or A_{sw}) is required to resist torsional shear on all faces of the beam.

Similar to bending shear design, the capacity reduction factor for torsion (see Clause 2.2.2(ii) of AS 3600–2009 [the Standard]) is $\phi = 0.7$.

7.2 MAXIMUM TORSION

The maximum capacity of a beam in torsion, from Clause 8.3.3 of the Standard, is

$$T_{u.\max} = 0.2 f'_c J_t \qquad \text{Equation 7.2(1)}$$

where J_t is the torsional modulus and is given as

$$J_t = 0.33x^2y \qquad \text{Equation 7.2(2)}$$

$$J_t = 0.33\sum(x_i^2 y_i) \qquad \text{Equation 7.2(3)}$$

and

$$J_t = 2A_m b_w \qquad \text{Equation 7.2(4)}$$

respectively, for solid beams, flanged beams and box sections. The terms used in these equations are defined in Figure 7.2(1). Note that b_w is the smallest wall thickness of the box.

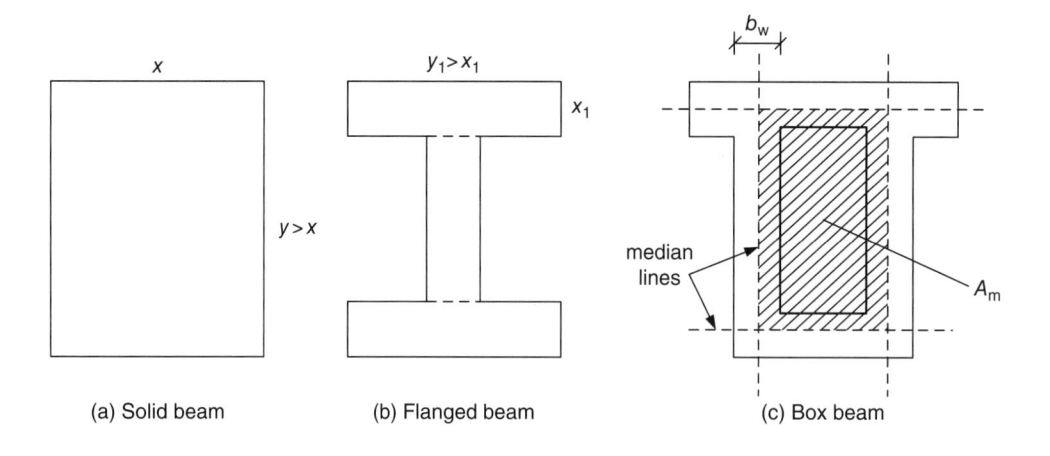

Figure 7.2(1) Definition of terms to calculate the torsional modulus (J_t) of a beam section

For a section that is also subjected to a transverse shear (V^*), the maximum torsion permissible (T') is limited by the crushing strength of the web, where

$$T' = \phi T_{u.max}\left(1 - \frac{V^*}{\phi V_{u.max}}\right) \qquad \text{Equation 7.2(5)}$$

in which $T_{u.max}$ is given in Equation 7.2(1) and $V_{u.max}$ is computed using Equation 6.3(1).

If the design or factored torsional moment $T^* > T'$ then f'_c and/or the sectional dimensions must be increased. If $T^* \leq T'$, then checks should be conducted to see if torsional reinforcement is required. If it is, it needs to be designed – details for this are given in the next two sections.

7.3 CHECKS FOR REINFORCEMENT REQUIREMENTS

Torsional reinforcement is required in Amendment No. 2, Clause 8.3.4 of the Standard, if one of the following conditions is satisfied:

1. For pure torsion

$$T^* > 0.25\phi T_{uc} \qquad \text{Equation 7.3(1)}$$

where, as given in Clause 8.3.5(a) of the Standard, the plain-concrete beam strength in pure torsion is

$$T_{uc} = J_t \left(0.3\sqrt{f'_c}\right) \qquad \text{Equation 7.3(2)}$$

in which J_t may be determined using Equation 7.2(2), 7.2(3) or 7.2(4) as the case may be.

2. For beams also subjected to a transverse shear (V^*),

$$T^* > \phi T_{uc}\left(0.5 - \frac{V^*}{\phi V_{uc}}\right) \qquad \text{Equation 7.3(3)}$$

where V_{uc}, the shear strength of the beam excluding shear reinforcement, is given in Equation 6.3(4).

Or, for beams with $D \leq b_w/2$ and $D \leq 250$ mm

$$T^* > \phi T_{uc}\left(1 - \frac{V^*}{\phi V_{uc}}\right) \qquad \text{Equation 7.3(4)}$$

7.4 DESIGN FOR TORSIONAL REINFORCEMENT

7.4.1 Design formula

Up until AS 3600–2001, torsion was considered in conjunction with transverse shear, but this has been simplified in the current Standard. For beams described in Section 7.3, Amendment No. 2, Clause 8.3.4(b) of the Standard requires that

$$T^* \leq \phi T_{us} \qquad \text{Equation 7.4(1)}$$

where, as specified in Clause 8.3.5(b),

$$T_{us} = f_{sy.f}\left(\frac{A_{sw}}{s}\right)2A_t \cot\theta_v \qquad \text{Equation 7.4(2)}$$

in which θ_v is as given in Equation 6.3(20) for a conservative estimate, noting that $\theta_v \geq 30°$.

In Equation 7.4(2), $A_t = \bar{x}.\bar{y}$, with $\bar{x}$ and $\bar{y}$ as defined in Figure 7.4(1); in the equation, A_{sw} is the area of the steel bar making up the closed ties and s is the spacing of the ties.

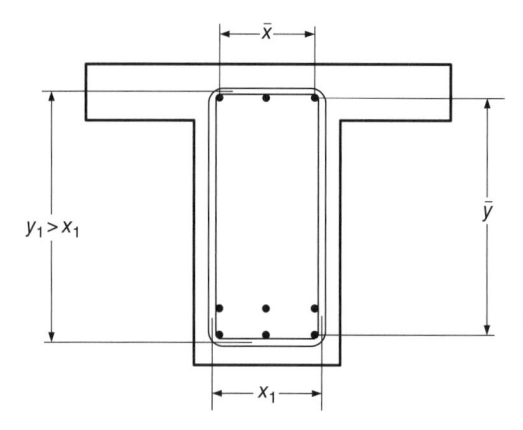

Figure 7.4(1) Definition of $\bar{x}$ and $\bar{y}$ for a typical beam section

Note that the Standard does not specify inclined transverse reinforcement for torsion design. This is because the effectiveness of this reinforcement depends entirely upon the correct orientation of the ties. If the direction of the torsional moment was reversed due to load reversal, the 'correct' orientation for transverse shear resistance would turn out to be parallel to the torsional crack! A similar discussion on inclined shear reinforcement may be found in Section 6.3.6.

7.4.2 Design procedure

In practice, torsion design normally takes place after shear design. Shear reinforcement may take the form of stirrups or ties, neither of which can be relied upon to resist torsion.

Substituting Equation 7.4(2) into the design formula (Equation 7.4(1)) and rearranging terms leads to

$$A_{sw}/s_w = \frac{T^*}{\phi f_{sy.f} 2 A_t \cot \theta_v} \quad \text{Equation 7.4(3)}$$

As per Clause 8.3.7 of the Standard, the minimum torsional reinforcement shall be the greater of

$$A_{sw}/s_w = A_{sv.min}/s_v \quad \text{Equation 7.4(4)}$$

where $A_{sv.min}$ for closed ties only (i.e. excluding stirrups) is computed using Equation 6.3(12), and the A_{sw}/s_w, which ensures that $T_{us} \geq 0.25 T_{uc}$, that is,

$$A_{sw}/s_w = 0.25 T_{uc} / (2 f_{sy.f} A_t \cot \theta_v) \quad \text{Equation 7.4(5)}$$

Note that Equation 7.4(5) is obtained by substituting $T_{us} = 0.25T_{uc}$ into Equation 7.4(2).

For the transverse reinforcement (ties) to be fully effective, longitudinal bars are needed. Thus, longitudinal torsional steel in addition to the main reinforcement for bending must be provided in the bending tensile and compressive zones as follows.

In the tensile zone

$$\Delta A_{st} = \left[0.5 f_{sy.f}(A_{sw}/S_w)u_t \cot^2\theta_v\right]/f_{sy} \qquad \text{Equation 7.4(6)}$$

where

$$u_t = 2(\bar{x} + \bar{y}) \qquad \text{Equation 7.4(7)}$$

in which $\bar{x}$ and $\bar{y}$ are defined in Figure 7.4(1), but

$$\Delta A_{st} \geq 0.2 y_1 u_t / f_{sy} \qquad \text{Equation 7.4(8)}$$

where y_1 is defined in Figure 7.4(1).

In the compressive zone

$$\Delta A_{sc} = \left[0.5 f_{sy.f}\left(\frac{A_{sw}}{S_w}\right)u_t \cot^2\theta_v - F_c^*\right]/f_{sy} \geq 0 \qquad \text{Equation 7.4(9)}$$

where F_c^* is the absolute value of the design force in the compressive zone due to bending, and

$$F_c^* \leq \alpha_2 f'_c \gamma k_u b d + A_{sc} f_{sy} \qquad \text{Equation 7.4(10)}$$

where α_2 and γ, respectively, are given in Equations 3.3(2)a and b, and k_u is described in Sections 3.4.3, 3.4.4 and 3.6.2.

In general, a conservative estimate of F_c^* can be obtained using the working load or transformed section approach (see Appendix A). However, the F_c^* calculated in this way is only a nominal value, which may not have practical significance. The procedure for calculating F_c^* is given in Section 7.4.4.

7.4.3 Detailing

The spacing for transverse reinforcement for torsional reinforcement (i.e. closed ties) is

$$s_w \leq \text{the lesser of } 0.12 u_t \text{ and } 300 \text{ mm} \qquad \text{Equation 7.4(11)}$$

where u_t is given in Equation 7.4(7).

Additional longitudinal reinforcement obtained from Equations 7.4(6) and 7.4(9) should be placed as close to the corners of the respective stress zones as possible. In all cases, at least one bar must be provided at each corner of the closed ties.

7.4.4 Design example

Problem

A cantilever bent beam subjected to torsion, shear and bending is detailed in Figure 7.4(2). Transverse reinforcement (i.e. N10 ties at 300 mm) is provided to resist transverse and longitudinal shear only.
(a) Is the beam section adequate to resist the torsion?
(b) If not, what should the A_{sw}/s_w be and the corresponding longitudinal steel? Take $f'_c = 20$ MPa.

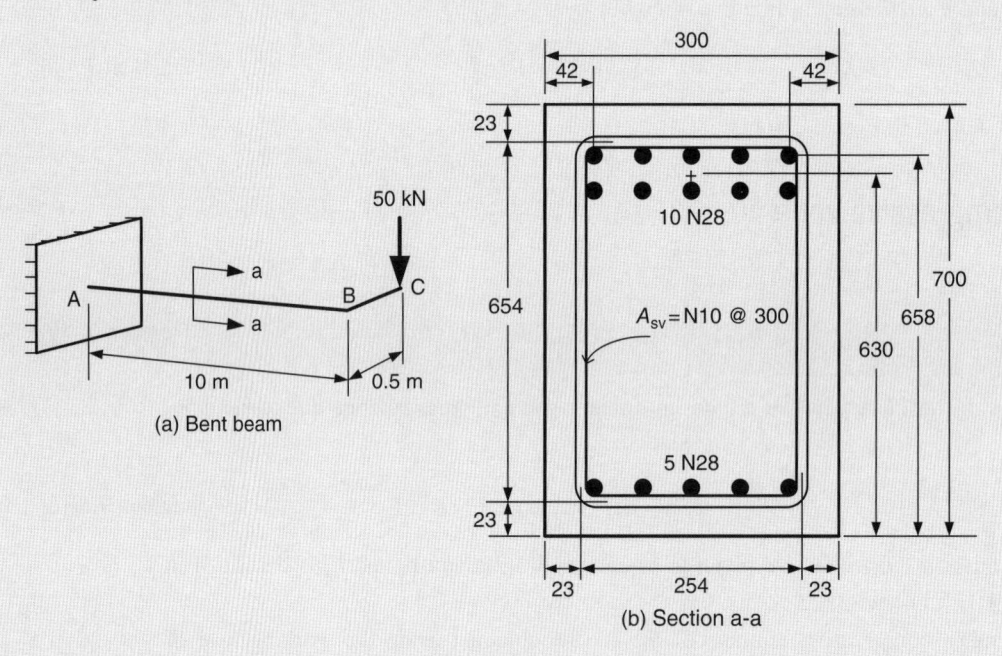

Figure 7.4(2) Details of the example beam for torsion design

Solution

For (a)
1. Compute design shear, torsion and moment.
   From Table 2.3(1), we obtain the total reinforcement area for the section
   $A_{st} + A_{sc} = 6160 + 3080 = 9240$ mm^2
   Thus, the percentage of steel by volume $= \dfrac{9240}{300 \times 700} \times 100 = 4.4\%$

According to Equation 2.4(1), $\rho_w = 24.0 + (0.6 \times 4.4)$
$$= 26.64 \text{ kN/m}^3$$
Then the self-weight $= 0.3 \times 0.7 \times 26.64 = 5.6$ kN/m.
At support A:
$V^* = 1.5 \times 50 + 1.2 \times 5.6 \times 10.5 = 145.56$ kN (which is a conservative value)
$T^* = 1.5 \times 50 \times 0.5 + 1.2 \times 5.6 \times 0.5 \times \dfrac{0.5}{2} = 38.34$ kNm
$M^* = 1.5 \times 50 \times 10 + 1.2\, (5.6 \times 0.5 \times 10 + 5.6 \times \dfrac{10^2}{2}) = 750 + 369.6$
$ = 1119.6$ kNm

2. Check acceptability of the concrete section.

Equation 6.3(1): $\phi V_{u.\max} = 0.7 \times 0.2 \times 20 \times 300 \times 658 \times \times 10^{-3} = 552.7$ kN
Equation 7.2(2): $J_t = 0.33 \times 300^2 \times 700 = 20.79 \times 10^6$ mm^3
Equation 7.2(1): $T_{u.\max} = 0.2 \times 20 \times 20.79 = 83.16$ kNm
Equation 7.2(5): $T' = 0.7 \times 83.16 \times \left(1 - \dfrac{145.56}{552.7}\right) = 42.88$ kNm $> T^*$

Therefore, the section is reinforceable.

3. Check torsional reinforcement requirement.

Equation 6.3(4): $\phi V_{uc} = 0.7 \times 1.1 \times 1 \times 1 \times 300 \times 658 \times \sqrt[3]{20} \times \sqrt[3]{\dfrac{6160}{300 \times 658}} \times 10^{-3}$
$\phantom{\text{Equation 6.3(4): } \phi V_{uc}} = 129.9$ kN

Equation 7.3(2): $T_{uc} = 20.79 \times 10^6 \times 0.3 \times \sqrt{20} \times 10^{-6} = 27.89$ kNm

Equation 7.3(3): right-hand side $= 0.7 \times 27.89 \times \left(0.5 - \dfrac{145.56}{129.9}\right)$, which is negative

and hence $< T^*$

Thus, the beam section is inadequate and torsional reinforcement is required.

For (b)
1. Compute θ_v.

Equation 6.3(15): $\phi V_{u.\min} = 129.9 + 0.7 \times 0.6 \times 300 \times 658 \times 10^{-3} = 212.8$ kN
Equation 6.3(20): $\theta_v = 30° + 15° \times \dfrac{145.56 - 212.8}{552.7 - 212.8} < 30° = 30°$

2. Compute A_{sw}/s_w.

Equation 7.4(3): $A_{sw}/s_w = \dfrac{38.34 \times 10^6}{0.7 \times 500 \times 2 \times 216 \times 616 \times \cot 30°}$

That is, $A_{sw}/s_w = 0.2377$ mm^2/mm.

3. Minimum reinforcement requirement checks

   Equation 6.3(12): $A_{sv.min}/s_v = \dfrac{0.35 \times 300}{500} = 0.21$ mm^2/mm

   Equation 7.4(4): right-hand side $= 0.21 < 0.2377$, which is acceptable.

   (Check 1)

   Equation 7.4(5): $A_{sw}/s_w = \dfrac{0.25 \times 27.89 \times 10^6}{2 \times 500 \times (216 \times 616) \times \cot 30°} = 0.0303 < 0.2377$,

   which is acceptable.

   (Check 2)

   Thus, the required transverse torsional reinforcement

   $A_{sw}/s_w = 0.2377$ mm^2/mm.

   Note that the stipulation in Clause 8.2.7.4 of the Standard regarding the torsional effects on shear strength (as mentioned in the last paragraph of Section 6.3.4) is naturally satisfied because the beam is reinforced against shear (as well as torsion).

4. Longitudinal reinforcement

   The modular ratio, $n = \dfrac{E_s}{E_c} = \dfrac{200\,000}{24\,000} = 8.33$

   where for reinforcing steel $E_s = 200\,000$ MPa and from Table 2.2(1), $E_c = 24\,000$ MPa for $f'_c = 20$ MPa

   $p = \dfrac{6160}{300 \times 630} = 0.0326$ and using these values,

   Equation A(5): $k = \sqrt{(0.0326 \times 8.33)^2 + 2 \times 0.0326 \times 8.33} - 0.0328 \times 8.33$
   $= 0.514$

   and the lever-arm parameter for the internal moment at the section

   $j = 1 - \dfrac{0.514}{3} = 0.829$

   It can be shown that at working stress condition

   $f_c = \dfrac{2M}{kjbd^2} = \dfrac{2 \times 1119.6 \times 10^6}{0.514 \times 0.829 \times 300 \times 630^2} = 44.1$ MPa (nominal value only)

   Therefore, the triangular stress block (as shown below) gives

   $F_c^* = f_c\, kdb/2 = 44.1 \times 0.514 \times 630 \times 300/2 = 2\,142\,069.3$ N $= 2142.1$ kN

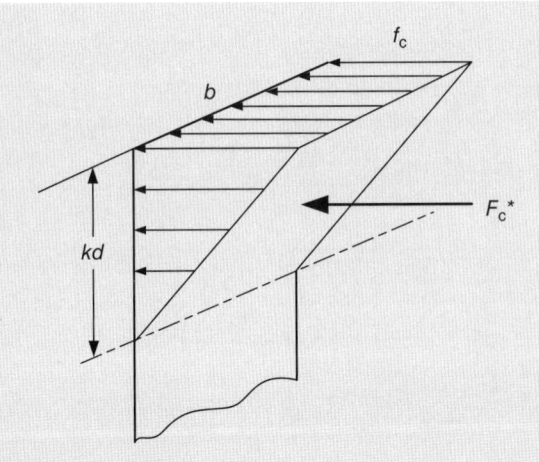

For $f'_c = 20$ MPa, $\alpha_2 = \gamma = 0.85$ and

Equation 3.6(1): $k_u = \dfrac{600 \times \dfrac{42}{630}}{600 - 500} = 0.4$

Equation 7.4(10): right-hand side $= (0.85 \times 20 \times 0.85 \times 0.4 \times 300 \times 630 + 3080 \times 500) \times 10^{-3}$
$$= 2632.42 > F_c^*,\text{ which is acceptable.}$$

From Figure 7.4(2):

$\bar{y} = 700 - 2 \times 42 = 616$ mm

$\bar{x} = 300 - 2 \times 42 = 216$ mm

Equation 7.4(6): $\Delta A_{st} = [0.5 \times 500 \times 0.2377 \times 2 (216 + 616) \times \cot^2 30°]/500$
$$= 593 \text{ mm}^2$$

And Equation 7.4(9): $\Delta A_{sc} = (593 \times 500 - 2\,142\,100)/500$, which is negative.

Hence, $\Delta A_{sc} = 0$.

5. Check adequacy of bending reinforcement.

So that AS 3600–2009, Clause 8.3.6 is complied with, the given cross-section less ΔA_{st} should be able to resist $M_A^* = 1119.6$ kNm.

$A_{st} = 6160 - 593 = 5567; p_t = 0.0295$

$A_{sc} = 3080; p_c = 0.0163$

$p_t - p_c = 0.0132$

Equation 3.6(3): $(p_t - p_c)_{\text{limit}} = \dfrac{0.85 \times 600 \times 0.85 \times 20 \times \dfrac{42}{630}}{(600 - 500) \times 500} = 0.01156 < p_t - p_c,$

and A_{sc} would yield at ultimate.

Equation 3.6(6): $a = \dfrac{(5567 - 3080) \times 500}{0.85 \times 20 \times 300} = 243.8$

Further, $k_u = \dfrac{a}{\gamma d} = \dfrac{243.8}{0.85 \times 630} = 0.46$ and $k_{uo} = \dfrac{0.46 \times 630}{658} = 0.44$

Thus, Equation 3.4(20)a with b: $\phi = 0.713$ and finally

Equation 3.6(8): $\phi M_{uo} = 0.713 \times \left[5567 \times 500 \times \left(630 - \dfrac{243.8}{2}\right) \right.$

$$\left. + 3080 \times 500 \times \left(\dfrac{243.8}{2} - 42\right) \right] \times 10^{-6}$$

$$= 1096.1 \text{ kNm}$$

That is, ϕM_{uo} = 1096.1 kNm ≈ M_A^* = 1119.6 kNm, which is acceptable.

Since the moment capacity of the section (see Figure 7.4(2)) less ΔA_{st} is very close to the design moment (M_A^*), the existing section does not need an increase in A_{st} to resist torsion.

6. Transverse reinforcement details and final design

    From (b) Step 3, above, we have A_{sw}/s_w = 0.2377 mm^2/mm. Use N10 closed ties and the required spacing s_w = 78/0.2377 = 328.1 mm.

    For the 10 m span, the number of N10 ties = 10 000/328.1 = 30.5, and the existing number of N10 ties as A_{sv} = 10 000/300 = 33.3.

    These make a total of 64 N10 ties @ s = 10 000/(64 − 1) = 158.7 mm. Say, use 65 N10 ties @ 155 mm, covering a total length of 64 × 155 = 9.92 m < 10 m, which is acceptable.

    Note that N10 ties @ 155 mm complies with the spacing requirements given in Equation 7.4(11).

7. The final design

    Details of the final design are as shown in Figure 7.4(3).

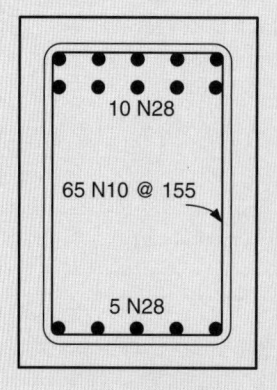

Figure 7.4(3) Final design details for the example design problem

7.5 SUMMARY

The origin and nature of torsion are discussed in detail together with the interaction of torsional moment and transverse shear force. Considering the relevant Standard recommendations, formulas are derived for the ultimate strength design of reinforced concrete beams for torsion. Applications of these design formulas are illustrated via fully worked examples.

7.6 PROBLEMS

1. Repeat the example in Section 7.4.4, but with R10 closed ties @ 100 mm.
2. In the example in Section 7.4.4, if the existing shear reinforcement comprises N12 stirrups @ 300 mm, what are the required A_{sw}/s_w and its distribution?
3. The T-beam detailed in Figure 7.6(1) is subjected to a pure torsion $T^* = 20$ kNm. Is torsional reinforcement required? If so, what is the maximum spacing s allowed for the N12 ties serving as torsional shear reinforcement? Assume $f'_c = 20$ MPa.

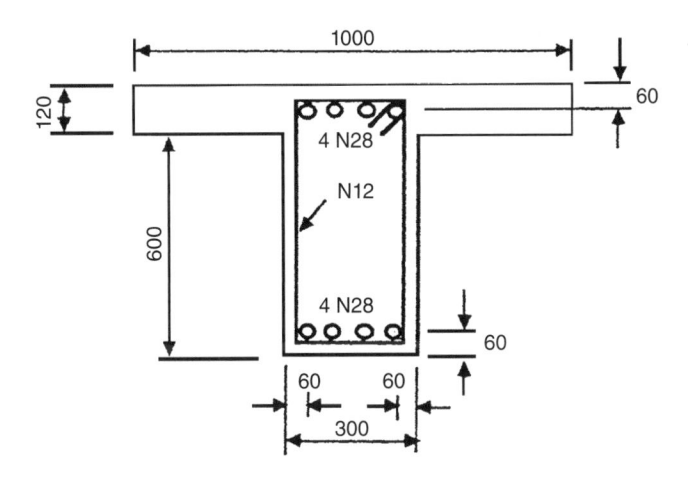

Figure 7.6(1) Details of the T-beam for Problem 3

Note: all dimensions are in mm.

4. If the beam in Problem 1 in Section 6.6 is subjected to a torsion $T^* = 5$ kNm in addition to the critical transverse shear force as indicated in Figure 6.6(1), check whether the use of R10 ties @ 125 mm is adequate.
5. If the beam in Figure 6.6(7) for Problem 9 in Section 6.6 is subjected to a torsion $T^* = 10$ kNm in addition to the specified transverse shear force, check whether the use of R12 ties @ 100 mm is adequate.

6. The reinforced concrete section detailed in Figure 7.6(2) is subjected to a torsion $T^* = 5$ kNm in addition to a shear $V^* = 150$ kN. Taking $f'_c = 32$ MPa, check whether R10 ties @ 125 mm are adequate for supporting this combined shear and torsion. Substantiate your answer with calculations.

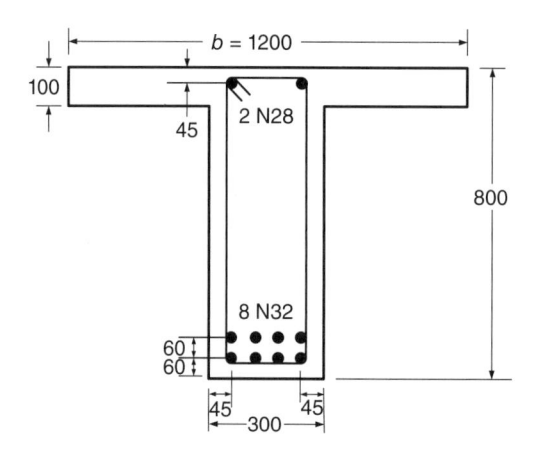

Figure 7.6(2) Details of the T-beam for Problem 6

Note: all dimensions are in mm.

7. The reinforced concrete T-section detailed in Figure 7.6(3) is subjected to a pure torsion $T^* = 60$ kNm. Taking $f'_c = 32$ MPa, answer the questions below following necessary calculations:
   (a) Is the concrete section reinforceable or does it need to be enlarged?
   (b) If so, what is the required torsional reinforcement?

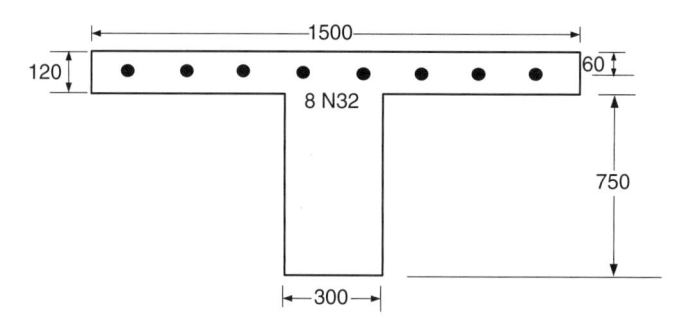

Figure 7.6(3) Details of the reinforced concrete T-section for Problem 7

Note: all dimensions are in mm.

8. For the three sections shown in Figure 7.6(4) and based on the AS 3600–2009 limitations, determine the maximum factored torque (T^*) in pure torsion that can be applied to each section without using torsional reinforcement.
   Assume $f'_c = 25$ MPa, $f_{sy} = 500$ MPa and R10 stirrups.

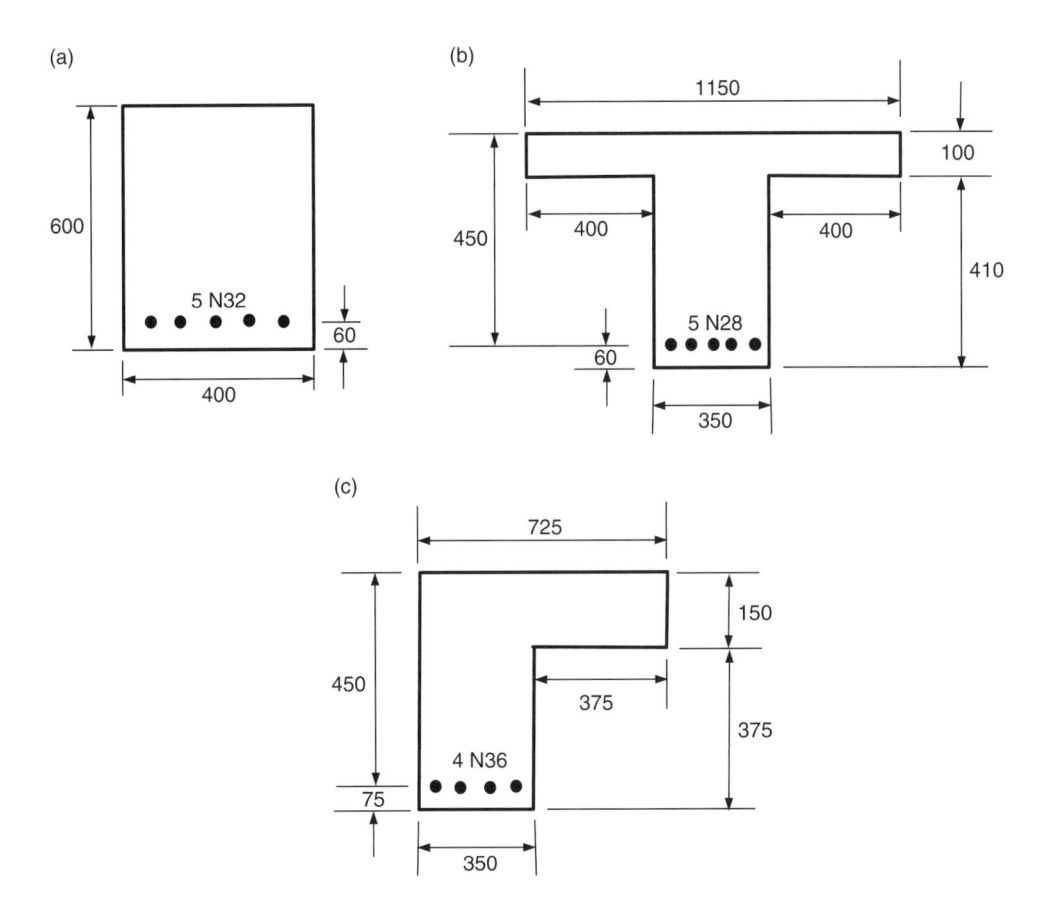

Figure 7.6(4) Details of the sections for Problem 8

Note: all dimensions are in mm.

170 PART 1 REINFORCED CONCRETE

9. What minimum total theoretical depth is needed for the beam shown in Figure 7.6(5) if no torsional reinforcement is to be used? The cross-section is not shown, but it is rectangular, with $b = 500$ mm, and the depth to be determined. The concentrated load is located at the end of the cantilever 200 mm to one side of the beam centre line and $f'_c = 25$ MPa.

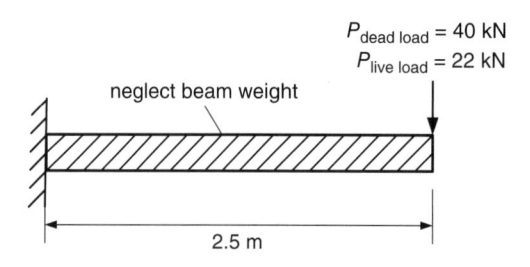

[Figure: cantilever beam, 2.5 m long, fixed at left end, with $P_{\text{dead load}} = 40$ kN and $P_{\text{live load}} = 22$ kN at the free end; neglect beam weight.]

Figure 7.6(5) Loading details of cantilever beam for Problem 9

10. What minimum total theoretical depth is needed for a rectangular beam subjected to a pure torsion $T^* = 15$ kNm, if no torsional reinforcement is to be provided? Take the width of the rectangular cross-section as $b = 500$ mm and $f'_c = 32$ MPa.

11. What minimum total theoretical depth is needed for the beam segment, AB, shown in Figure 7.6(6) if no torsional reinforcement is to be used? For the rectangular cross-section of the beam, $b = 450$ mm, and the depth is to be determined. The 25 kN is a concentrated live load and $f'_c = 40$ MPa. The beam weight may be neglected.

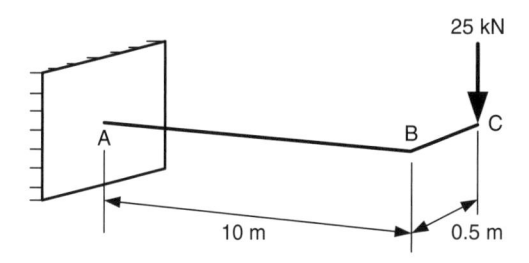

[Figure: beam fixed at A, horizontal segment AB of length 10 m, with a short perpendicular extension BC of 0.5 m, and a 25 kN downward load applied at C.]

Figure 7.6(6) Loading details of beam for Problem 11

BOND AND STRESS DEVELOPMENT

8.1 INTRODUCTION

8.1.1 General remarks

An assumption is made in the analysis and design of beams in bending that plane sections normal to the beam axis remain plane after bending (see Section 3.3.1). This enables the establishment of the compatibility condition, which is fundamental to the derivation of bending analysis and design equations. The compatibility condition can be maintained only if there is a complete bond between the concrete and the longitudinal steel bars.

Inadequate bond strength will lead to premature failure of the beam in bending or in shear; it will invalidate the analysis and design equations for reinforced concrete.

8.1.2 Anchorage bond and development length

The concept of anchorage bond can be readily explained using the free-body diagram of a beam segment illustrated in Figure 8.1(1).

If at section a-a (Figure 8.1(1)), the steel bar of an area A_b is to be able to develop the yield strength (f_{sy}), the resultant of the ultimate bond stress (u) around and along the embedded length must be at least equal to the tensile force (T). Otherwise, the bar would be 'pulled out' of the concrete, resulting in bond failure.

By considering the equilibrium of the horizontal forces, we have

$$\pi d_b l_b \, u = A_b f_{sy} \qquad \text{Equation 8.1(1)}$$

from which

$$l_b = \frac{A_b f_{sy}}{\pi d_b u} = \frac{d_b f_{sy}}{4u} \qquad \text{Equation 8.1(2)}$$

where l_b is the basic development length, d_b is the diameter of the bar and u is the apparent bond strength.

172 PART 1 REINFORCED CONCRETE

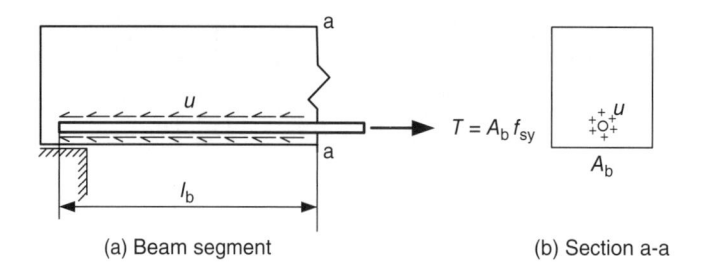

Figure 8.1(1) Free-body diagram showing bond stress in a beam section

8.1.3 Mechanism of bond resistance

Amongst other topics, the book by Park and Paulay (1975) contains a wealth of information from published literature on bond and the reader is referred to that all-encompassing book for an in-depth study. Only a brief discussion on the nature of bond is given here, so that the reader can better understand the design equations recommended in AS 3600–2009 (the Standard).

For plain reinforcing bars of Class N or L, bond resistance is provided by:
- cohesion or chemical bond between mortar and the bar surface
- the wedging action of dislodged sand particles at the interface between steel and concrete
- friction between steel and concrete.

Cohesion, wedging action and friction exist for plain bars as well as for deformed ones. They are represented collectively by v_a in Figure 8.1(2), which shows the mechanism of a deformed bar that provides bond. In addition, there is the bearing stress (f_b) at the 'compression' side of the ribs and the shear stress (v_c) in concrete at the level of the top of the ribs. Needless to say, bond strength is higher, or the development length is shorter for deformed bars than for plain bars of the same size.

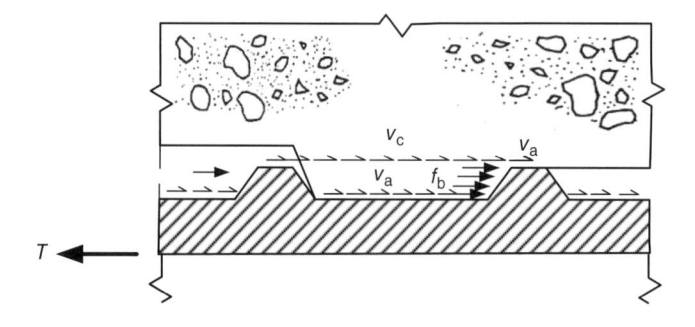

Figure 8.1(2) The mechanism of bond development in the case of a deformed bar

A hook (135° or 180° bend) or a cog (90° bend) is more effective than a straight bar in providing anchorage bond. As Figure 8.1(3) clearly illustrates, in addition to the bond stress, there are mechanical actions that help to enhance the bond resistance. The bearing stress (f_b) adds to the resistance against T, while the normal stress (f_n) also increases the friction between concrete and steel, which in turn will improve the bond resistance.

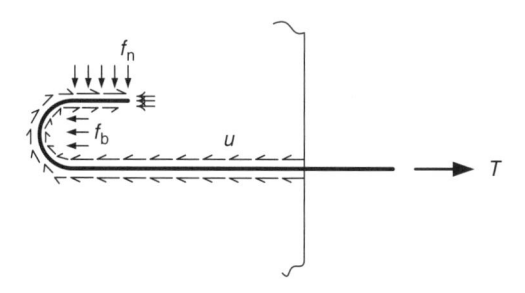

Figure 8.1(3) Enhancement of bond resistance for a hook

8.1.4 Effects of bar position

Figure 8.1(2) indicates that bond is provided by a combination of v_a, v_c and f_b, respectively, the bond between mortar and steel surface, the concrete shear and the bearing stress. It is obvious that the values of v_a, v_c and f_b are higher for concrete with a higher characteristic strength (f'_c). On average, f'_c is the strength representative of all concrete in a given beam. However, because of the phenomenon of 'weeping' in fresh concrete (the tendency of water to separate from the other constituent materials and rise to the top surface), some water is trapped under the horizontally placed steel bars. As a result, at that locality the water/cement ratio is higher than elsewhere. This means that the concrete strength (or f'_c) is lower there as well. This in turn leads to a lower bond strength at that locality. The strength reduction is particularly significant for bars near the top of a concrete beam, slab or wall. The design equations given in the next section reflect this observation.

8.2 DESIGN FORMULAS FOR STRESS DEVELOPMENT

This section discusses only deformed and plain bars in tension and compression, and as fitments (i.e. stirrups and ties). When necessary, the reader should refer to Clauses 13.1.4 and 13.1.7 of the Standard, respectively, for the stress development design of headed bars in tension and welded mesh.

8.2.1 Basic and refined development lengths for a bar in tension

Deformed bars

For deformed bars of all sizes, the basic design development length (according to Clause 13.1.2.2 of the Standard) is

$$L_{\text{sy.tb}} = \frac{0.5k_1k_3f_{\text{sy}}d_{\text{b}}}{k_2\sqrt{f'_c}} \geq 29k_1d_{\text{b}} \qquad \text{Equation 8.2(1)}$$

where $k_1 = 1.3$ for a horizontal bar with more than 300 mm of concrete cast below the bars, or
$k_1 = 1.0$ for all other bars
$k_2 = (132 - d_{\text{b}})/100$ in which d_{b} is in millimetres
$k_3 = 1.0 - 0.15(c_{\text{d}} - d_{\text{b}})/d_{\text{b}}$ but $0.7 \leq k_3 \leq 1.0$, in which the concrete cover (c_{d}) in millimetres is defined in Figure 8.2(1), which is a copy of Figure 13.1.2.3(a) of the Standard, and

$$f'_c \leq 65\text{Mpa}$$

The value of $L_{\text{sy.tb}}$ shall be:
- multiplied by 1.5 for epoxy-coated bars
- multiplied by 1.3 when lightweight concrete is used
- multiplied by 1.3 for all structural elements built with slip form.

Lower values of the design development length may be obtained from the more accurate formula given below (from Clause 13.1.2.3 of the Standard), that is

$$L_{\text{sy.t}} = k_4 k_5 L_{\text{sy.tb}} \qquad \text{Equation 8.2(2)}$$

where

$$k_4 = 1.0 - K\lambda \text{ but } 0.7 \leq k_4 \leq 1.0 \qquad \text{Equation 8.2(3)}$$

$$k_5 = 1.0 - 0.04\rho_{\text{p}} \text{ but } 0.7 \leq k_5 \leq 1.0 \qquad \text{Equation 8.2(4)}$$

In Equation 8.2(3), $K =$ a factor that accounts for the position of the bars being anchored with respect to the transverse reinforcement – see Figure 8.2(2), which is a reproduction of Figure 13.1.2.3(B) in the Standard, and

$$\lambda = \left(\sum A_{\text{tr}} - \sum A_{\text{tr.min}}\right)/A_s \qquad \text{Equation 8.2(5)}$$

where
- $\sum A_{\text{tr}} =$ sum of cross-sectional area of the transverse reinforcement along the development length ($L_{\text{sy.t}}$)
- $\sum A_{\text{tr.min}} =$ sum of cross-sectional area of the minimum transverse reinforcement, which may be taken as $0.25A_s$ for beams or columns and 0 (zero) for slabs or walls
- $A_s =$ cross-sectional area of a single bar of diameter d_{b} being anchored.

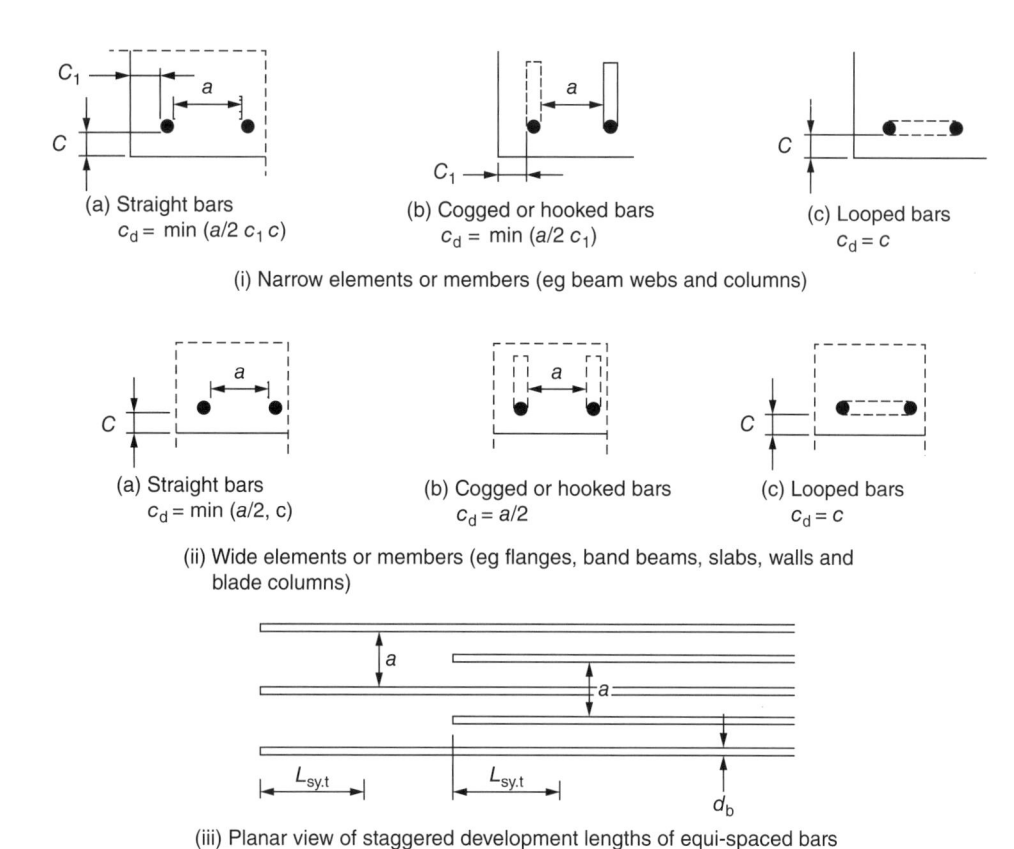

(a) Straight bars
$c_d = \min(a/2, c_1, c)$

(b) Cogged or hooked bars
$c_d = \min(a/2, c_1)$

(c) Looped bars
$c_d = c$

(i) Narrow elements or members (eg beam webs and columns)

(a) Straight bars
$c_d = \min(a/2, c)$

(b) Cogged or hooked bars
$c_d = a/2$

(c) Looped bars
$c_d = c$

(ii) Wide elements or members (eg flanges, band beams, slabs, walls and blade columns)

(iii) Planar view of staggered development lengths of equi-spaced bars

NOTE: For wide elements or members (such as band beams, slabs, walls and blade columns) edge cover (c_1) should be ignored

Figure 8.2(1) Values of c_d

Source: Standards Australia (2009). AS 3600–2009 *Concrete Structures*. Standards Australia, Sydney, NSW. © Standards Australia Limited. Copied by Cambridge University Press with the permission of Standards Australia and Standards New Zealand under Licence 1712-c038.

$K = 0.1$ $K = 0.05$ $K = 0$

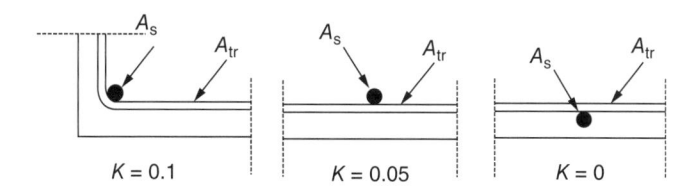

Figure 8.2(2) Values of K for different bar positions

Source: Standards Australia (2009). AS 3600–2009 *Concrete Structures*. Standards Australia, Sydney, NSW. © Standards Australia Limited. Copied by Cambridge University Press with the permission of Standards Australia and Standards New Zealand under Licence 1712-c038.

In Equation 8.2(4), ρ_p = transverse compressive pressure (in MPa), at the ultimate limit state along the development length and perpendicular to the plane of splitting.

Note that in all cases

$$k_3 k_4 k_5 \geq 0.7 \qquad \text{Equation 8.2(6)}$$

In a situation where only the development of $\sigma_{st} < f_{sy}$ is required, the design development length obtained from Equations 8.2(1) and 8.2(2) may be proportionally reduced.

Or

$$L_{st} = L_{sy.t} \sigma_{st} / f_{sy} \geq 12 d_b \qquad \text{Equation 8.2(7)}$$

Plain bars

For plain bars in tension

$$L_{sy.t} = 1.5 \text{ times the length obtained using Equation 8.2(1) but} \geq 450 \text{ mm} \qquad \text{Equation 8.2(8)}$$

8.2.2 Standard hooks and cog

As discussed in Section 8.1.3, a hook or a cog is more effective than a straight bar for bond resistance. Standard hooks and the cog as recommended in Clause 13.1.2.7 of the Standard are detailed in Figure 8.2(3).

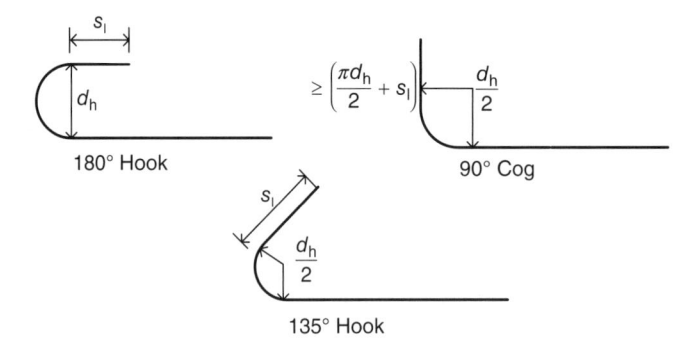

Figure 8.2(3) Standard hooks and cog

Note in Figure 8.2(3) that the hooks and the cog all have the same total length.

Further, as in Clause 17.2.3.2 of the Standard:

- $d_h \geq 5d_b$ for reinforcement
- $d_h \geq 3d_b$ for 500L fitments
- $d_h \geq 3d_b$ for R250N fitments
- $d_h \geq 4d_b$ for D500N fitments
- $d_h \geq 5d_b$ for epoxy-coated and galvanised bars with $d_b \leq 16$ mm and $d_h \geq 8d_b$ for such bars with $d_b \geq 20$ mm
  s_l has to be greater than the greater of $4d_b$ and 70 mm.
- But for cogs of all sizes $d_h \leq 8d_b$.

Equation 8.2(9)

When using the standard hook or cog, the development length of a bar is $L_{sy.t}/2$, or $L_{st}/2$ as the case may be, measured from the outer edge of the hook or cog. This is illustrated in Figure 8.2(4) using a 180° hook.

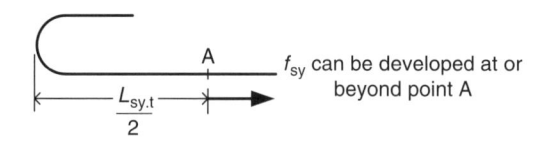

Figure 8.2(4) Development length of a bar with hook

8.2.3 Deformed and plain bars in compression
Deformed bars

For a bar in compression, the end bearing stress helps to enhance the bond strength. The 'basic' length for stress development up to f_{sy} is given in Clause 13.1.5.2 of the Standard as

$$L_{sy.cb} = 0.22 d_b f_{sy}/\sqrt{f'_c} \geq \text{the greater of } 0.0435 f_{sy} d_b \text{ and } 200 \text{ mm} \quad \text{Equation 8.2(10)}$$

A comparison with Equation 8.2(1) indicates, as it should be, that this development length is shorter than that for a bar of the same size in tension. For obvious reasons, hooks or cogs are not considered effective for bars in compression. The shorter or 'refined' development length is

$$L_{sy.c} = k_6 L_{sy.cb} \quad \text{Equation 8.2(11)}$$

where transverse reinforcement with at least three bars, transverse to and outside the bar being developed, is provided within $L_{sy.cb}$, and when $\Sigma A_{tr}/s \geq A_s/600$

$$k_6 = 0.75 \quad \text{Equation 8.2(12)}$$

and is equal to 1.0 for all other cases.

Note that $\sum A_{tr}$ and A_s are as defined for Equation 8.2(4).

The development length for a compressive stress (σ_{sc}), less than the yield stress (f_{sy}), may be calculated as

$$L_{sc} = L_{sy.c} \frac{\sigma_{sc}}{f_{sy}} \geq 200 \text{ mm} \qquad \text{Equation 8.2(13)}$$

Plain bars

The development length for plain bars in compression is twice the calculated value of $L_{sy.c}$ and $L_{sy.cb}$ for a deformed bar.

8.2.4 Bundled bars

Two types of bundled bars are specified in Clause 13.1.7 of the Standard:
1. three-bar bundle, for which

$$L_{sy.t} = 1.2 L_{sy.t} \text{ of the largest bar in the bundle} \qquad \text{Equation 8.2(14)}$$

2. four-bar bundle, for which

$$L_{sy.t} = 1.33 L_{sy.t} \text{ of the largest bar in the bundle} \qquad \text{Equation 8.2(15)}$$

8.3 SPLICING OF REINFORCEMENT

To comply with the Standard, lapped, welded or mechanical splices may be used in design. Only lapped splices for which the recommendations are based on stress development analysis are discussed here. For details of welded and mechanical splices, the reader is referred to Clause 13.2.6 of the Standard. The reader should also take note of the general splicing requirements as specified in Clause 13.2.1 of the Standard.

Four types of lapped splices are described in the following sections.

8.3.1 Bars in tension

For deformed bars in wide elements such as flanges, slabs, walls, and so forth, which may be in contact or otherwise, the lap length as in Clause 13.2.2 of the Standard is

$$L_{sy.t.lap} = k_7 L_{sy.t} \geq 29 k_1 d_b \qquad \text{Equation 8.3(1)}$$

where $L_{sy.t}$ is calculated using Equation 8.2(1) noting that the lower limit of $29 k_1 d_b$ does not apply, and k_7 shall be taken as 1.25, unless A_s provided is at least twice A_s required, and no more than half of the reinforcement at the section is spliced, in which case k_7 may be taken as 1.

In narrow elements, such as beam webs and columns, the tensile lap length is

$$L_{sy.t.lap} \geq \text{the larger of } 29k_1d_b, k_7L_{sy.t} \text{ and } (L_{sy.t} + 1.5s_b) \qquad \text{Equation 8.3(2)}$$

where s_b is the clear distance between bars of the lapped splice, noting that if s_b does not exceed $3d_b$, then s_b may be taken as zero.

8.3.2 Bars in compression

The minimum lap length for deformed bars in compression is taken to be the development length in compression ($L_{sy.c}$), which is given as appropriate in 1, 2 or 3 below:

1. The development length in compression is that given in Equation 8.2(11), but not less than $40d_b$.
2. In compression members with stirrups or ties where at least three sets of such fitments are present over the length of the lap and $A_{tr}/s \geq A_b/1000$, a lap length of 0.8 times the value given in 1.
3. In helically tied compression members, if at least three turns of the helix are present over the length of the lap and $A_{tr}/s \geq nA_b/6000$, a lap length of 0.8 times the value given in 1 is appropriate, where n denotes the number of bars uniformly spaced around the helix.

Note that in all cases

$$L_{sy.c} \geq 300 \text{ mm} \qquad \text{Equation 8.3(3)}$$

8.3.3 Bundled bars

For three-bar bundles

$$L_p \geq 1.2 \text{ times the lap splice length for single bars} \qquad \text{Equation 8.3(4)}$$

and for four-bar bundles

$$L_p \geq 1.33 \text{ times the lap splice length for single bars} \qquad \text{Equation 8.3(5)}$$

where in each case the lap splice length is for the largest bar in the bundle. Note that, for obvious reasons, individual bar splices within a bundle shall not overlap.

8.3.4 Mesh in tension

For welded mesh in tension, the two outermost transverse wires of one sheet of mesh overlap the two outermost transverse wires of the sheet being lapped. Figure 13.2.3 of the Standard shows the splicing details and it is reproduced in Figure 8.3(1).

180 PART 1 REINFORCED CONCRETE

(a) $s_1 = s_2$ (b) $s_1 < s_2$

Figure 8.3(1) Welded mesh in tension

Source: Standards Australia (2009). AS 3600–2009 *Concrete Structures*. Standards Australia, Sydney, NSW. © Standards Australia Limited. Copied by Cambridge University Press with the permission of Standards Australia and Standards New Zealand under Licence 1712-c038.

8.4 ILLUSTRATIVE EXAMPLES

8.4.1 Example 1

Problem

Given N28 bars in a simply supported beam as shown in Figure 8.4(1) and $f'_c = 25$ MPa, determine the position from the support beyond which the yield stress can be developed in the bars, if

(a) the bars are extended straight into the support
(b) standard 180° hooks are used.

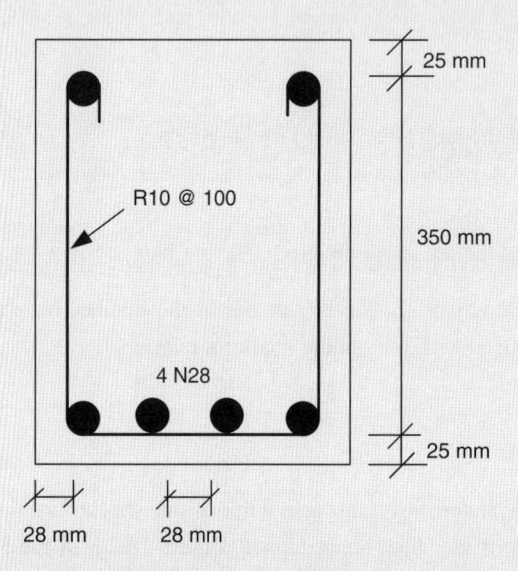

Figure 8.4(1) Details of the simply supported beam in Example 1

Solution

(a) Straight bars

For Equation 8.2(1)

$k_1 = 1.0$

$k_2 = (132 - 28)/100 = 1.04$

$k_3 = 1.0 - 0.15\,(28/2 - 28)/28 > 1.0$; use 1.0.

Thus Equation 8.2(1):

$$L_{sy.tb} = \frac{0.5 \times 1.0 \times 1.0 \times 500 \times 28}{1.04 \times \sqrt{25}} = 1346 \text{ mm} > 29 \times 1.0 \times 28 = 812 \text{ mm},$$

which is acceptable.

(b) Hooks

$L_{sy.t} = 1346/2 = 673$ mm

The points beyond which yield stress can be developed in the two cases are illustrated in Figure 8.4(2).

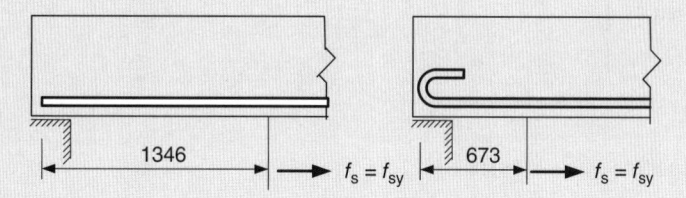

Figure 8.4(2) Development lengths for straight bars and hooks

8.4.2 Example 2

Problem

The stirrups specified for the beam in Figure 8.4(1) are made of R10 bars, each of which is required to develop f_{sy} at mid-depth of the section. Determine and sketch the dimensional details of the stirrup.

Solution

For Equation 8.2(1)

$k_1 = 1.0$

$k_2 = (132 - 10)/100 = 1.22$

$k_3 = 1.0 - 0.15(25 - 10)/10 = 0.78$

Thus, Equation 8.2(1): $L_{sy.tb} = 0.5 \times 1.0 \times 0.78 \times 250 \times 10/(1.22 \times \sqrt{25})$

$$= 160 < 29 \times 1.0 \times 10 = 290$$

Use 290 mm.

Equation 8.2(8): $L_{sy.t} = 1.5 \times 290 = 435$ mm

Since each end of the stirrup takes the form of a hook, we have $L_{sy.t}/2 = 435/2 = 218$ mm.

However, it can be seen in Figure 8.4(1) that at mid-depth the available development length is only 175 mm. It is therefore necessary to extend the hook by a minimum of $2 \times (218 - 175) = 86$ mm.

The details of the stirrup are given in Figure 8.4(3). In Equation 8.2(9), $d_h = 30$ mm and $s_l = 70$ mm.

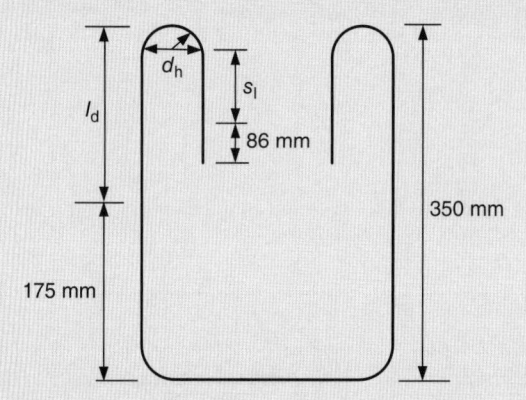

Figure 8.4(3) Dimensional details of the stirrup in Example 2

8.5 SUMMARY

The concepts of anchorage bond and development length as well as the mechanism of bond resistance are discussed in detail. Incorporating the relevant Standard recommendations, formulas are derived for the stress development design of deformed and plain bars in tension, in compression and as fitments. Applications of these design formulas are demonstrated using worked examples.

8.6 PROBLEMS

1. Figure 8.6(1) shows details of a cantilever beam. At the support section (a-a), all the bars are required to develop the yield stress, 500 MPa. Taking $f'_c = 25$ MPa, design and detail the support reinforcement if
   (a) straight bars are used
   (b) standard cogs are provided.
2. The cross-section of a beam subjected to combined bending, shear and torsion is detailed in Figure 7.4(2) in Section 7.4.4. The required torsional reinforcement is N12 ties at 150 mm. Assuming that it is necessary to develop f_{sy} at all sections of the tie, design the tie, with $f'_c = 20$ MPa.

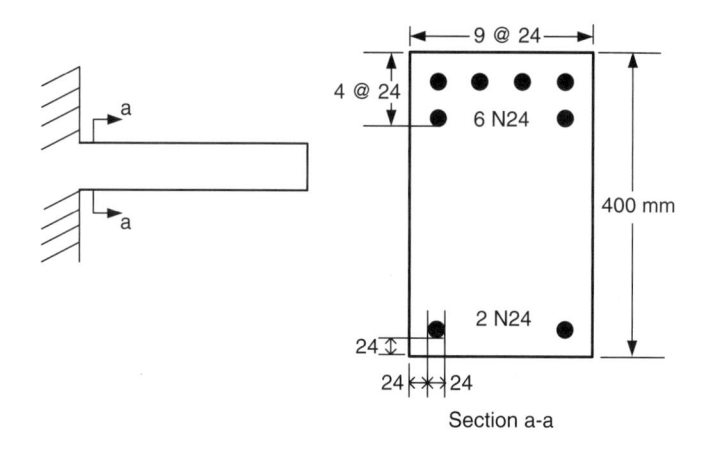

Figure 8.6(1) Details of a cantilever beam

Note: all dimensions are in mm.

3. For the cantilever beam shown in Figure 3.9(7)a for Problem 12 in Section 3.9, the maximum moment occurs at the support or root section. Assuming that every bar at the support section (see Figure 3.9(7)b) yields at the ultimate state, compute the minimum embedded lengths of the top and bottom bars if they are extended straight into the support. The concrete strength $f'_c = 32$ MPa.
4. Refer to the beam given in Problem 6 in Section 6.6 (see Figure 6.6(5)). If only the three bars in the bottom layer are extended 300 mm straight beyond the support face, is the design adequate for bond strength consideration?
5. The N20 bottom bars shown in Figure 8.6(2) are epoxy coated. Assuming normal weight concrete and $f'_c = 32$ MPa, determine the required development lengths as per AS 3600–2009 for the tension bars.

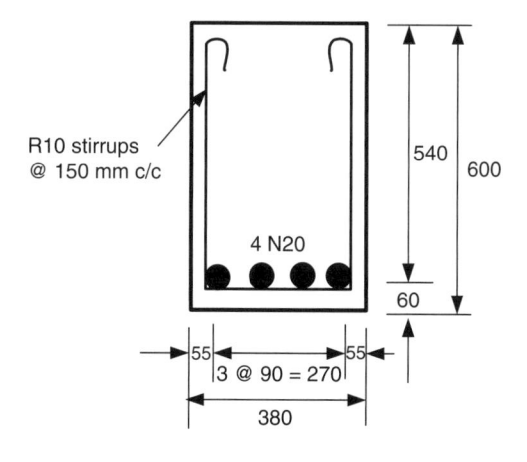

Figure 8.6(2) Details of the section for Problem 5

Note: all dimensions are in mm.

6. The cross-section of a reinforced lightweight concrete cantilever beam is shown in Figure 8.6(3). Compute development lengths for the N24 top bars, in accordance with AS 3600–2009, if $f'_c = 25$ MPa.

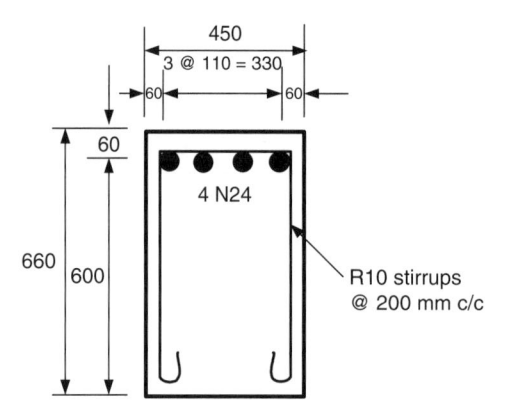

Figure 8.6(3) Cross-section of the cantilever beam for Problem 6

Note: all dimensions are in mm.

SLABS 9

9.1 INTRODUCTION

9.1.1 One-way slabs

Theoretically, all slabs are two-dimensional systems with bending in two orthogonal directions or two-way bending. In practice, however, some slabs are supported on parallel and continuous (line) supports in one direction only. Figure 9.1(1)a shows a typical slab of this type, in which the line supports take the form of a continuous wall or a girder, which is supported on discrete columns. Note that the coordinate system adopted here is different from that used in the earlier chapters. This is to conform to AS 3600–2009 (the Standard) convention.

If such a slab is subjected to a uniformly distributed load only, then bending mainly occurs in the direction perpendicular to the support – the x-direction in Figure 9.1(1)a. The bending in the y-direction may be ignored. Such a slab is referred to as a one-way slab.

Since bending occurs only in the x-direction, the entire slab structure can be seen as a very wide beam running continuously in the x-direction. For analysis and design, it is convenient to consider a strip of a unit width, for example, 1 metre. This is illustrated in Figure 9.1(1)b. The strip can be further assumed to be continuous over standard (beam) supports. However, such an assumption is valid only if the column-supported girders are rigid enough to prevent significant bending in the y-direction.

The 1-metre wide continuous strip can be treated as a beam of the same width. Consequently, the forces in this one-way strip include only the bending moment (M_x) – or simply M – and transverse shear (V) as depicted in Figure 9.1(1)c. A simplified method for computing these forces in a multispan strip is provided in Clause 6.10.2 of the Standard.

The use of a 1-metre wide strip is convenient for design purposes because the computed steel reinforcement area is given in mm^2/m. Then the required spacing of a given size bar can be obtained directly from Table 2.3(2) in Section 2.3.

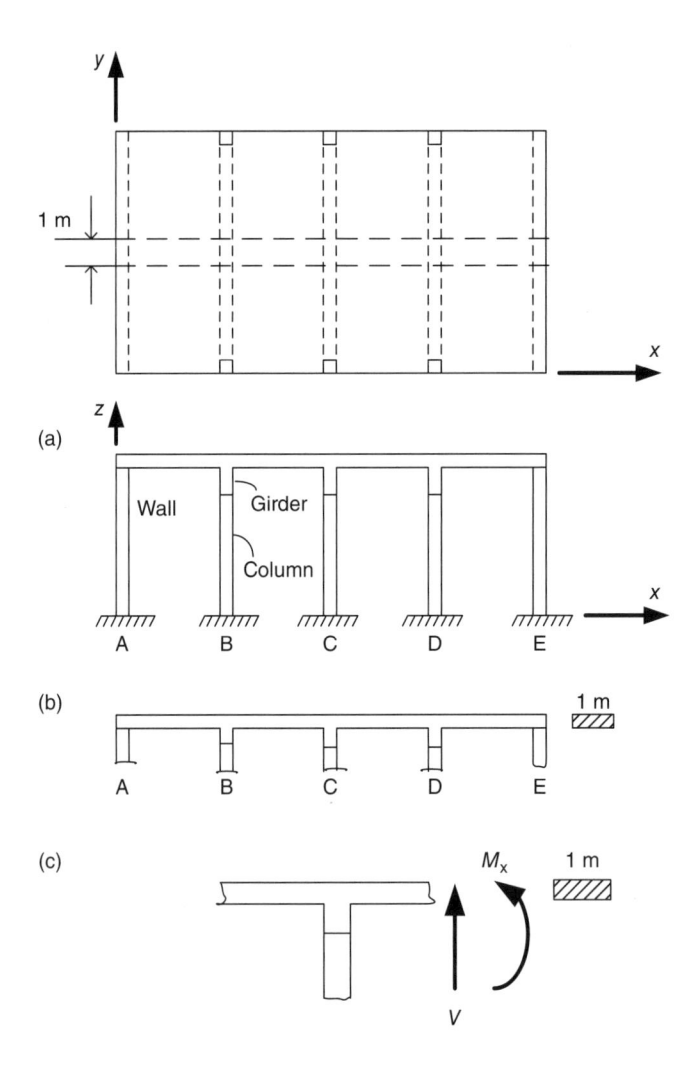

Figure 9.1(1) One-way slab and a typical 1-metre wide strip

9.1.2 Two-way slabs

Apart from those that can be treated as one-way slabs, all other slab structures are treated as two-way slabs. These include:
- slabs supported on four sides, as depicted in Figure 9.1(2)a
- beam-and-slab systems, Figure 9.1(2)b
- flat slabs, Figure 9.1(2)c
- flat plates, Figure 9.1(2)d.

Note that in the beam-and-slab system shown in Figure 9.1(2)b, girders are taken to be much heavier than beams. While the girders provide continuous support for the slab, the beams form part of the slab. In the classical theory of plates, such a beam-and-slab system is referred to as an orthotropic slab. In reinforced concrete design, a beam and the associated part of the slab (running in the same direction) together form a T-beam (see Section 4.2). Depending on their rigidity, the girders constitute a grillage system, as shown in Figure 7.1(1)b. For multistorey buildings, together with the columns they form a three-dimensional framework (see Section 10.1). The grillage and framework may be analysed and designed separately from the slab systems they are assumed to support.

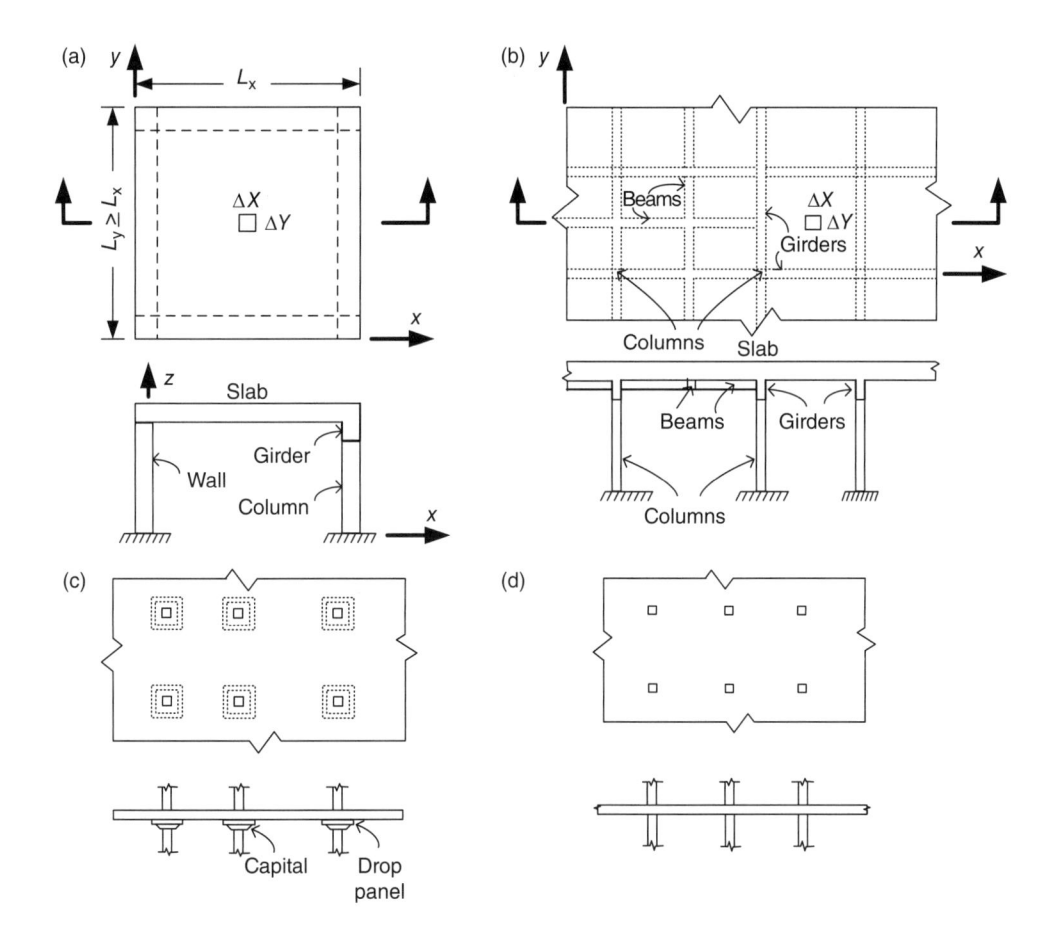

Figure 9.1(2) Examples of two-way slabs: **(a)** slab supported on four sides; **(b)** beam-and-slab system; **(c)** flat slab; and **(d)** flat plate

By definition, flat slabs and flat plates are slabs supported on columns only; there may be spandrel or edge beams constructed around the perimeter of the slab, but there are no interior beams or girders. The only difference between a flat slab and a flat plate is that there are drop panels or capitals (or both) in the former – but none in the latter – which help reduce the punching shear in the vicinity of the slab–column connection.

According to classical plate theory, in addition to the bending moments M_x and M_y in the x and y directions, respectively, there are the twisting moments (M_{xy} and M_{yx}) as well as the transverse plate shears (Q_x and Q_y). These are defined in Figure 9.1(3)a. The unit for plate moments is kNm/m or Nmm/mm; for plate shears, it is kN/m or N/mm. In general, for reinforced concrete floor slabs, it is required to design for M_x and M_y only. Special reinforcement detailing is sometimes necessary to cover twisting moments at exterior slab corners (see Section 9.3.3). The reinforcement for M_x and M_y is illustrated in Figure 9.1(3)b. Plate shears are generally insignificant except for areas in the vicinity of the column heads or under severe concentrated loads, where high punching shears exist.

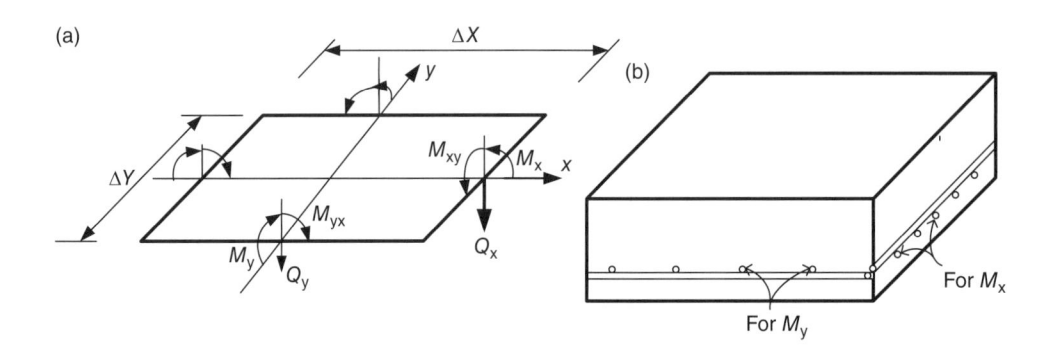

Figure 9.1(3) Definition of bending moments (M_x and M_y), twisting moments (M_{xy} and M_{yx}) and transverse plate shears (Q_x and Q_y) in classical plate theory

Clause 6.10.3 of the Standard gives bending moment coefficients for calculating M_x and M_y in a slab supported on four sides and reinforced by N and L bars, and mesh. Clause 6.10.4 presents a simplified method of analysis for both the plate bending moments and the transverse shears. For a multistorey building comprising two-way slabs and columns, an 'idealised frame method' may be used for computing the forces in both the structural components. This is given in Clause 6.9 of the Standard. Note that Clause 6.7 allows the use of plastic analysis for the strength design of beams, frames and slabs, and Clause 6.7.3.2 also stipulates the requirements in yield line analysis of slabs.

9.1.3 Effects of concentrated load

A concentrated load produces severe bending moments and transverse shears in slabs. The simplified methods recommended in Clauses 6.10.2–6.10.4 cover mainly the overall effects of uniformly distributed loads. Even the idealised frame approach (Clause 6.9) may not be used to analyse these localised forces. However, for one-way slabs, the allowable design strip widths to support a given concentrated load are specified in Clause 6.9.

Standard texts on the theory of plates like that by Timoshenko and Woinowsky-Kreiger (1959) carry in-depth discussions on this and related topics. Widely available computer software packages, on the other hand, are readily applicable for the detailed analysis of the deflection and the main forces in slabs under concentrated loads and other discrete loading patterns. The reader is also referred to the text by Park and Gamble (2000) on the effects of concentrated loads in yield line analysis. Note again that the provisions for the use of this plastic method of analysis are available in Clause 6.7.3.2.

9.1.4 Moment redistribution

For the convenience of the designer, the Standard recommends some simplified but proven procedures in determining bending moments in:
- continuous beams and one-way slabs (Clause 6.10.2)
- two-way slabs supported on four sides (Clause 6.10.3)
- two-way slabs having multiple spans (Clause 6.10.4).

The latter two procedures normally result in unbalanced negative moments over interior supports. That is, the moment at one side of the common support is different from that at the other side, as illustrated in Figure 9.1(4).

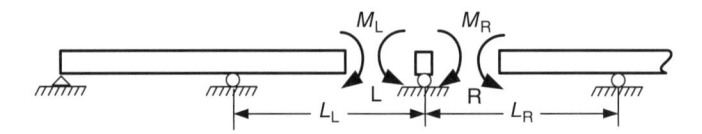

Figure 9.1(4) Example of an unbalanced moment at an interior support

For equilibrium

$$M_L = M_R$$

If such is not the case, the balance or $(M_R - M_L)$ has to be redistributed to sections L and R according to the bending stiffness of spans L_L and L_R. The moments (after such redistribution) on the left and right of the support are

$$M_{LB} = M_L + \frac{(M_R - M_L)S_L}{S_L + S_R} \qquad \text{Equation 9.1(1)}$$

and

$$M_{RB} = M_R + \frac{(M_R - M_L)S_R}{S_L + S_R} \qquad \text{Equation 9.1(2)}$$

respectively. The stiffnesses of the left and right spans are

$$S_L = \frac{(E_c I_{ef})_L}{L_L} \qquad \text{Equation 9.1(3)}$$

and

$$S_R = \frac{(E_c I_{ef})_R}{L_R}$$

Equation 9.1(4)

respectively. E_c is given in Equation 2.2(5) or 2.2(6), and I_{ef} may be estimated using Equation 5.3(11) for an endspan and Equation 5.3(12) for an interior span. Comparing Equations 9.1(1) and 9.1(2) shows that $M_{LB} = M_{RB}$.

For beams and slabs where class N bars or mesh are used as the main reinforcement, the design may be based on M_{LB} or M_{RB}, or the larger of M_L and M_R. If Class L reinforcement is used instead, the design has to be based on the larger of M_L and M_R. These requirements are specified in Clauses 6.10.3.2(b) and 6.10.4.3.

For redistribution of elastically determined moments at interior supports of statically indeterminate members or structures, the Standard stipulates additional ductility requirements. For details, the reader is referred to Clauses 6.2.7.1 and 6.2.7.2 of the Standard.

9.2 ONE-WAY SLABS

9.2.1 Simplified method of analysis

As discussed in Section 9.1.1, the behaviour of a one-way slab may be represented by a 1-metre wide strip running in the direction of bending. Figure 9.2(1) shows a typical one-way slab together with its representative strip.

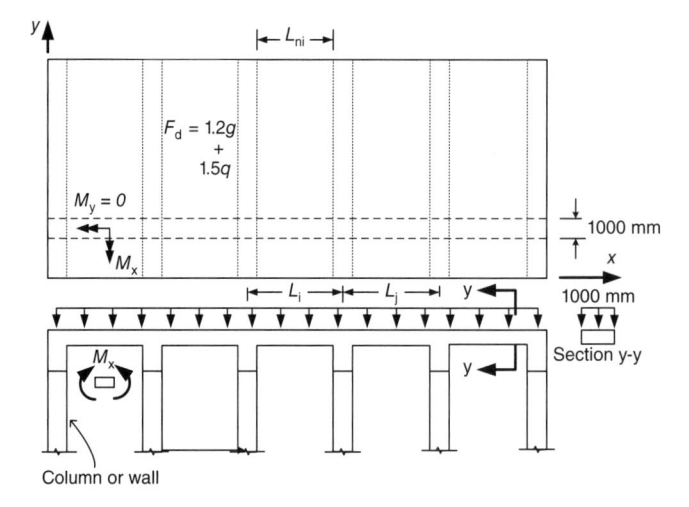

Figure 9.2(1) A typical one-way slab with its representative strip

Under a given loading condition, the strip or beam moments and shears can be determined using any structural analysis procedure. For simplicity, Clause 6.10.2 recommends an approximate method provided that the following conditions are satisfied:
- The ratio of any two adjacent spans

$$L_i/L_j \leq 1.2 \text{ or } \geq 1/1.2 \qquad \text{Equation 9.2(1)}$$

where L_i and L_j are the spans – see Figure 9.2(1).
- Loading is essentially uniformly distributed.
- Live load (q) is not greater than two times the dead load (g) or

$$q \leq 2g \qquad \text{Equation 9.2(2)}$$

- The slab cross-section is uniform (i.e. I is constant).
- If Class L bars or mesh are used as main reinforcement, the design strength of the slab shall be reduced by not less than 20%.

The design moment at the supports and at midspan can be computed as

$$M^* = \alpha F_d L_n^2 \qquad \text{Equation 9.2(3)}$$

where in general

$$F_d = 1.2g + 1.5q \qquad \text{Equation 9.2(4)}$$

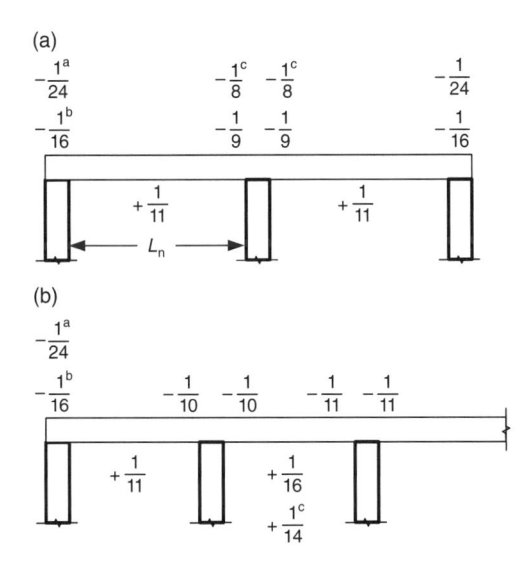

Figure 9.2(2) Values of moment coefficient (α) for **(a)** two-span continuous slabs and **(b)** slabs with more than two spans
a For slabs and beam on girder support
b For continuous beam support only
c Where Class L reinforcement is used

L_n is the clear span and the values of α are given in Figure 9.2(2). Note that the simplified method does not cover effects other than those due to g and q, which are gravity loads. In Figures 9.2(2)a and b, the slab is assumed to be built integrally with the exterior supports or some moment continuity is assumed.

The design shear at the supports and at midspan can be calculated as

$$V^* = \beta F_d L_n \qquad \text{Equation 9.2(5)}$$

where β is given in Figure 9.2(3). Clause 8.2.5(ii) states that no shear reinforcement is required if $V^* \leq \phi V_{uc}$.

Equations 9.2(3) and 9.2(5) are not valid for loading other than g and q.

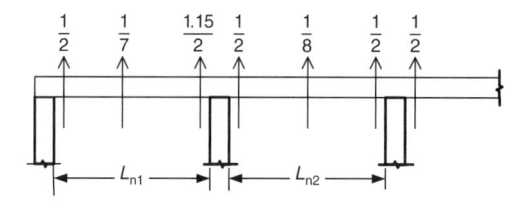

Figure 9.2(3) Values for shear coefficient, β

9.2.2 Reinforcement requirements

For the 1-metre wide strip, the design for moments is identical to that of a beam of the same width. Implicitly, the steel ratio (p_t) is also subject to the following limitation – see Equation 3.5(5):

$$p_t \geq 0.20(D/d)^2 f'_{ct.f}/f_{sy} \qquad \text{Equation 9.2(6)}$$

The simplified analysis is subject to one condition – the detailing of the positive and negative reinforcement follows that prescribed in Clause 9.1.3.1. This requirement is deemed to be complied with if the reinforcement arrangement is in agreement with that given in Figure 9.1.3.2 of the Standard. A version of this layout is presented in Figure 9.2(4). Further, for crack control, the bar spacing as per Clause 9.4.1(b) is

$$s \leq \text{the lesser of } 2.0D \text{ and } 300 \text{ mm} \qquad \text{Equation 9.2(7)}$$

As for beams, the clear minimum reinforcement spacing for slabs is specified qualitatively in Clause 9.1.5 simply as that which allows for the proper placing and compaction of concrete.

In addition to the bending steel, proper reinforcement has to be provided in the 'secondary' direction, which is 90° to the span. This is to control cracking due to shrinkage and temperature effects. For this and other crack control measures, the reader should consult Clause 9.4.3 of the Standard for details. Relevant extracts from this clause are presented in Section 9.4.7 for two-way slabs and are also applicable for one-way slabs.

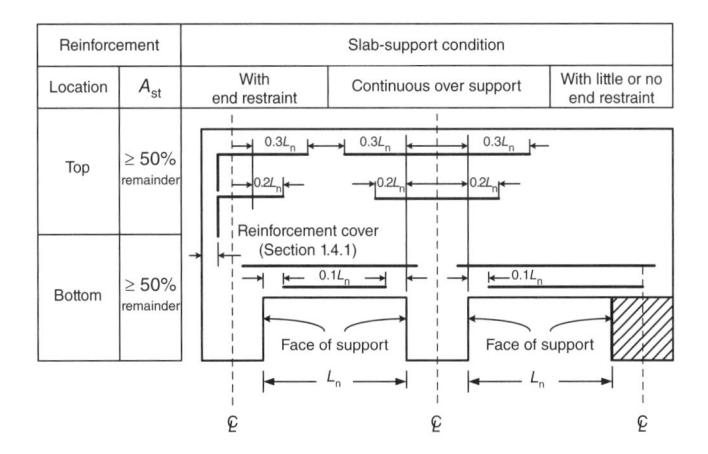

Figure 9.2(4) Reinforcement arrangement for one-way and two-way slabs

Source: Standards Australia (2009). AS 3600–2009 *Concrete Structures*. Standards Australia, Sydney, NSW. © Standards Australia Limited. Copied by Cambridge University Press with the permission of Standards Australia and Standards New Zealand under Licence 1712-c038.

9.2.3 Deflection check

The immediate deflection of an elastic slab is easily determined using commercially available computer packages for plate structures. However, for reinforced concrete slabs where the deflection is affected by cracking, and aggravated by creep and shrinkage, a reliable, rigorous analysis method is yet to be recognised.

Similar to beam design, the Standard recommends a simple deflection check for one-way slabs and two-way flat slabs of essentially uniform thickness and with live-load intensity (q) not greater than the dead load (g). For the design to satisfy the serviceability requirement, the effective slab thickness (d) has to be greater than an empirical limit or

$$d \geq \frac{L_{ef}}{k_3 k_4 \sqrt[3]{\frac{(\Delta/L_{ef})E_c}{F_{d.ef}}}} \quad \text{Equation 9.2(8)}$$

where
 L_{ef}, the effective span, is defined in Section 5.2.2
 E_c is in MPa
 Δ/L_{ef}, the deflection limitation, is given in Table 5.2(4)
 E_c and $F_{d.ef}$ are computed using Equations 2.2(5) or 2.2(6), and 5.5(7) or 5.5(8), respectively.

Further,

$$k_3 = 1.0 \qquad \text{Equation 9.2(9)}$$

for a one-way slab;

$$k_3 = 0.95 \qquad \text{Equation 9.2(10)}$$

for a two-way flat slab without drop panels;

$$k_3 = 1.05 \qquad \text{Equation 9.2(11)}$$

for a two-way flat slab with drop panels having the dimensions as shown in Figure 9.2(5);

$$k_4 = 1.4 \qquad \text{Equation 9.2(12)}$$

for simply supported slabs; and for continuous slabs with adjacent spans satisfying Equation 9.2(1) and where no endspan is longer than an interior span

$$k_4 = 1.75 \qquad \text{Equation 9.2(13)}$$

for an endspan; and

$$k_4 = 2.1 \qquad \text{Equation 9.2(14)}$$

for an interior span.

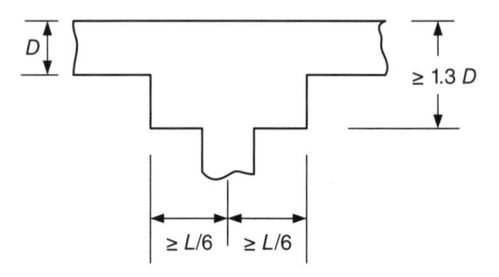

Figure 9.2(5) Details of a drop panel

9.2.4 Design example

Problem

A multistorey reinforced concrete and brick structure as part of an office building may be idealised as shown in Figure 9.2(6). Design the typical floor as a one-way slab. Take $f'_c = 25$ MPa, $D = 210$ mm, $q = 4$ kPa and the floor finish to be 0.5 kPa. Exposure classification A1 and fire resistance period of 60 minutes are assumed. The effects of wind may be ignored.

Figure 9.2(6) Details of a design: multistorey reinforced concrete and brick building

Solution

The design is to be carried out on a 1-metre wide strip.

First, the loading must be considered. Assuming the total steel reinforcement is 0.5% by volume in the slab, Equation 2.4(1) gives

$$\rho_w = 24 + 0.6 \times 0.5 = 24.3 \text{ kN/m}^3$$

Thus, the dead load

$$g = 1 \times 0.21 \times 24.3 + \text{floor finish} = 5.1 + 0.5 = 5.6 \text{ kN/m}$$

and the live load $q = 4$ kN/m

The design load according to Equation 9.2(4) is

$$F_d = 1.2 \times 5.6 + 1.5 \times 4 = 12.72 \text{ kN/m}$$

Following Figure 9.2(2), the 'beam' moment coefficients α for the various sections are shown in Figure 9.2(7) together with the M^* values.

Bending design

We may now proceed to design the main steel reinforcement using 500 N12 bars.

Table 1.4(2): cover = 20 mm for exposure classification A1
Table 5.4.2(A) of AS 3600–2009: cover $\geq$ 10 mm for fire resistance period of 60 minutes
Thus, as shown in Figure 9.2(8),

$$d = 210 - 20 - 6 = 184 \text{ mm}$$

196 PART 1 REINFORCED CONCRETE

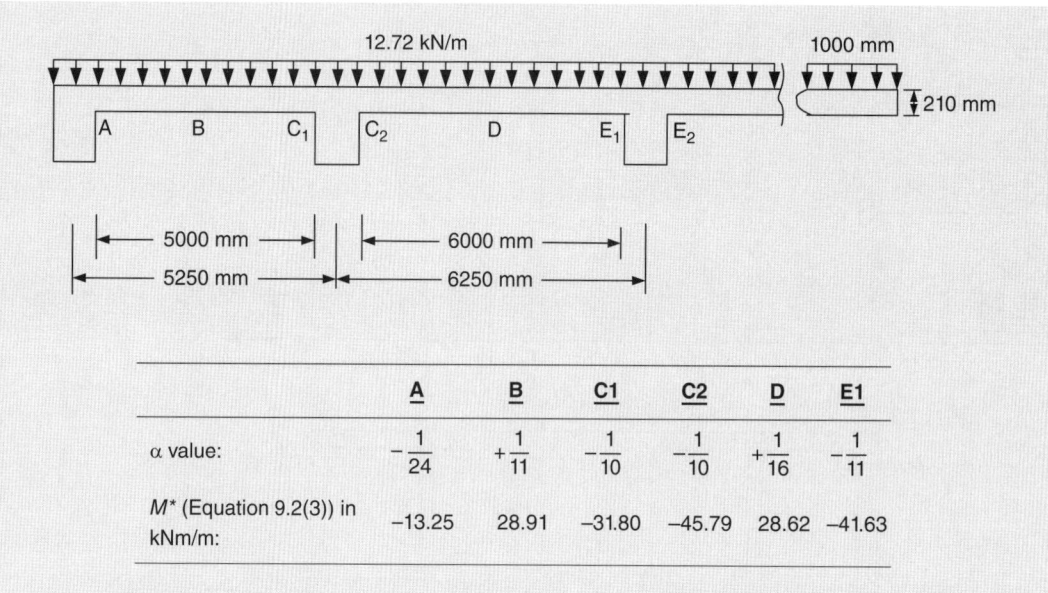

	A	**B**	**C1**	**C2**	**D**	**E1**
α value:	$-\dfrac{1}{24}$	$+\dfrac{1}{11}$	$-\dfrac{1}{10}$	$-\dfrac{1}{10}$	$+\dfrac{1}{16}$	$-\dfrac{1}{11}$
M^* (Equation 9.2(3)) in kNm/m:	−13.25	28.91	−31.80	−45.79	28.62	−41.63

Figure 9.2(7) Moment coefficients (α) and moments (M^*) at various sections of the slab

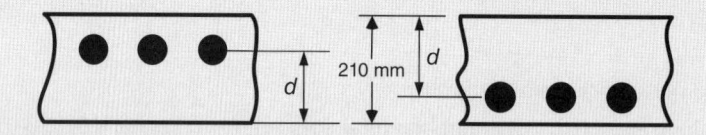

Figure 9.2(8) Definition of d for top and bottom bars in the slab

Beam section A

As shown in Figure 9.2(7),

$$M^* = -13.25 \text{ kNm/m}$$

Equation 3.3(2)a: $\alpha_2 = 1.0 - 0.003 \times 25 = 0.925$
but $0.67 \leq \alpha_2 \leq 0.85$; thus, $\alpha_2 = 0.85$

Equation 3.5(7): $\xi = \dfrac{0.85 \times 25}{500} = 0.0425$

Assuming $\phi = 0.8$,
   Equation 3.5(6):

$$p_t = 0.0425 - \sqrt{0.0425^2 - \dfrac{2 \times 0.0425 \times 13.25 \times 10^6}{0.8 \times 1000 \times 184^2 \times 500}} = 0.00099$$

Now check for ϕ.

Equation 3.3(2)b: $\gamma = 1.05 - 0.007 \times 25 = 0.875$; but $0.67 < \gamma < 0.85$
Thus,
$\gamma = 0.85$

Equation 3.4(8): $k_u = \dfrac{0.00099 \times 500}{0.85 \times 0.85 \times 25} = 0.027$

For a single layer of steel, we have $d = d_o$ and $k_{uo} = k_u = 0.027$. Then
Equations 3.4(20)a with b: $\phi = 0.8$, which is acceptable.
But

Equation 3.5(5): $p_{t.min} = 0.20 \left(\dfrac{D}{d}\right)^2 \dfrac{f'_{ct.f}}{f_{sy}} = 0.20 \times \left(\dfrac{210}{184}\right)^2 \times \dfrac{0.6 \times \sqrt{25}}{500}$
$= 0.001563$

Since $p_t < p_{t.min}$, use $p_{t.min}$ or

$A_{st} = p_{t.min}\, bd = 0.001563 \times 1000 \times 184 = 288\ \text{mm}^2/\text{m}$

From Table 2.3(2) we have for N12 bars at a spacing of $s = 300$ mm

$A_{st} = 377\ \text{mm}^2/\text{m}\ (> 288\ \text{mm}^2/\text{m})$

This is acceptable, as

Equation 9.2(7): $s \leq$ the lesser of $2.0D = 2 \times 210 = 420$ mm and 300 mm, and thereby $s = 300$ mm.

Thus, use N12 @ 300 mm ($A_{st} = 377\ \text{mm}^2/\text{m} > 288\ \text{mm}^2/\text{m}$) for the top bars.

Beam section B

As per Figure 9.2(7),

$M^* = 28.91$ kNm/m

Following the same process, we have

$p_t = 0.00219 > p_{t.min}$, which is acceptable

$A_{st} = 403\ \text{mm}^2/\text{m}$

for N12 at 275 mm, $A_{st} = 411\ \text{mm}^2/\text{m}$, which is acceptable.
Thus, use N12 @ 275 mm for the bottom bars.

Beam section C1
As shown in Figure 9.2(7),

$M^* = -31.80$ kNm/m

Following the same process, we have

$p_t = 0.00242 > p_{t.min}$, which is acceptable
$A_{st} = 446$ mm^2/m

for N12 bars at 250 mm, $A_{st} = 452$ mm^2/m, which is acceptable. Thus, use N12 @ 250 mm for the top bars.

Beam section C2
As shown in Figure 9.2(7),

$M^* = -45.79$ kNm/m

In view of the heavier moment, we may use N16 bars. This gives $d = 182$ mm. Similarly, we have

$p_t = 0.00361 > p_{t.min}$, which is acceptable
$A_{st} = 657$ mm^2/m

for N16 bars at 300 mm, $A_{st} = 670$ mm^2/m, which is acceptable. This also satisfies Equation 9.2(7).
Thus, use N16 @ 300 mm for the top bars.

Beam section D
As shown in Figure 9.2(7),
$M^* = 28.62$ kNm/m

Using, say, N12 bars, $d = 184$ mm. Then we have:
$p_t = 0.00217 > p_{t.min}$, which is acceptable
$A_{st} = 400$ mm^2/m
for N12 bars at 275 mm, $A_{st} = 411$ mm^2/m.
Thus, use N12 @ 275 mm for the bottom bars.

Beam section E1
As shown in Figure 9.2(7),

$M^* = -41.63$ kNm/m

In view of the heavier moment, we may use N16 bars. This gives $d = 182$ mm. Similarly, we have

$p_t = 0.00327 > p_{t.min}$, which is acceptable
$A_{st} = 596$ mm^2/m

for N16 bars at 300 mm, $A_{st} = 670$ mm^2/m, which is acceptable.

Also,

Equation 9.2(7): $s \leq$ the lesser of $2.0D = 2 \times 210 = 420$ mm and 300 mm, and thereby $s = 300$ mm

Thus, use N16 @ 300 mm ($A_{st} = 670$ mm²/m > 596 mm²/m) for the top bars.

Shear design

It is apparent in Figure 9.2(3) that the shear force is a maximum at section C1 of continuous strip if all spans are equal. For our case, $L_{n2} > L_{n1}$ and the maximum shear occur in section C2.

Equation 9.2(5) thus gives

$$V^* = 0.5 \times 12.72 \times 6 = 38.16 \text{ kN}$$

Equation 6.3(5)a: $\beta_1 = 1.56$

Equation 6.3(4): $V_{uc} = \left[1.56 \times 1 \times 1 \times 1000 \times 182 \times \sqrt[3]{25} \times \sqrt[3]{\dfrac{670}{1000 \times 182}}\right] \times 10^{-3}$

$$= 128.2 \text{ kN}$$

and

$$\phi V_{uc} = 0.7 \times 128.2 = 89.7 \text{ kN} > V^* = 38.16 \text{ kN}$$

This is acceptable. Note that if $\phi V_{uc} < V^*$, a thicker slab should be used.

Shrinkage and temperature steel (in y-direction)

For unrestrained edge conditions, Equation 9.4(14) gives:

$$A_{s.\,min} = 1.75 \times 1000 \times 210 \times 10^{-3} = 368 \text{ mm}^2/\text{m}$$

for N12 @ 300 mm, $A_{st} = 377$ mm²/m (> 368 mm²/m).

Thus, use N12 @ 300 mm for the bottom bars.

Serviceability check

Figure 9.2(7) shows that, for the exterior span, Equation 5.2(4) requires that

$$L_{ef} = \text{the lesser of } L \ (5250 \text{ mm}) \text{ and } L_n + D \ (5210 \text{ mm})$$

Thus,
$L_{ef} = 5210$ mm

For total deflection:
Table 5.2(4): $(\Delta/L_{ef}) = 1/250$

Equation 5.4(3): $k_{cs} = 2 - 1.2 \dfrac{A_{sc}}{A_{st}} = 2.0$

Table 1.3(1): for office building $\psi_s = 0.7$ and $\psi_l = 0.4$.

Then **Equation 5.5(8)** gives

$$F_{d.ef} = (1+2) \times 5.6 + (0.7 + 2 \times 0.4) \times 4 = 22.8 \text{ kN/m}^2 = 22.8 \times 10^{-3} \text{ MPa}$$

Equations 9.2(9) and **9.2(13)** give $k_3 = 1.0$ for one-way slab and $k_4 = 1.75$ for an endspan, respectively.

From Table 2.2(1), for $f_c' = 25$ MPa, $E_c = 26\,700$ MPa

Finally, **Equation 9.2(8)** gives

$$d_{min} = \frac{5210}{1 \times 1.75 \times \sqrt[3]{\frac{(1/250) \times 26700}{22.8 \times 10^{-3}}}} = 177.9 \text{ mm} < d = 184 \text{ mm}$$

This is acceptable.

For the interior span, $L_{ef} = 6210$ mm and $k_4 = 2.1$. These lead to

$$d_{min} = 176.7 \text{ mm} < d = 182 \text{ mm}$$

This is also acceptable.

Drawings

The design results are presented in Figure 9.2(9).

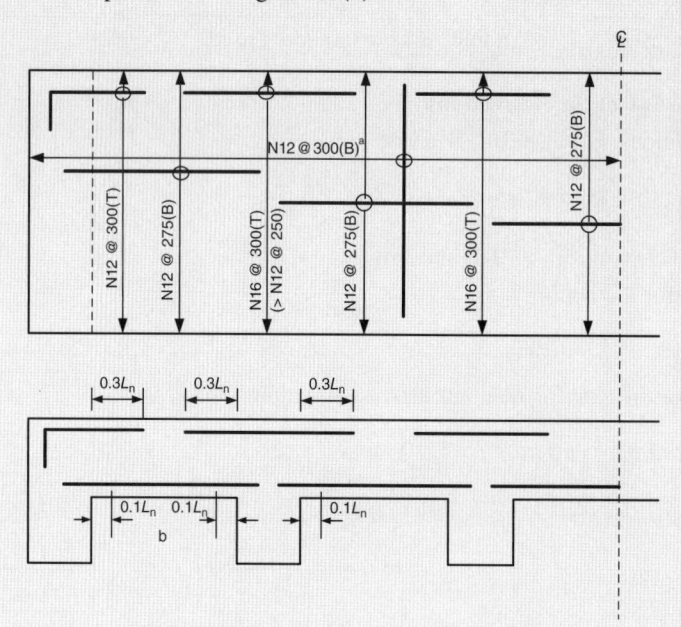

Figure 9.2(9) Reinforcement layout for the design slab
a Or half the amount provided near the top and bottom faces may also use steel fabrics
b $\leq$ 50% may be curtailed

9.3 TWO-WAY SLABS SUPPORTED ON FOUR SIDES

9.3.1 Simplified method of analysis

A typical slab supported on four sides is given in Figure 9.1(2)a. Since the beam-and-slab system as shown in Figure 9.1(2)b is supported on four sides by heavy girders (which in turn are supported by columns), each panel of the system may be designed individually.

A simplified method for analysing the bending moments (M_x and M_y) for slabs subjected to gravity loads only is recommended in Clause 6.10.3.2. Further, the loads (g and q) have to be essentially uniform for the simplified method to be valid. Clause 6.10.3.1(e) also stipulates that if Class L bars and mesh are used as main reinforcement, the slab must be continuously supported on walls. This is to account for the low ductility of the Class L reinforcement.

Positive moments

The slab to be analysed should be so oriented that

$$L_y > L_x$$

This is illustrated in Figure 9.3(1). For the purpose of analysing the positive moments, the slab is divided into two parts: the central region and the area surrounding the central region.

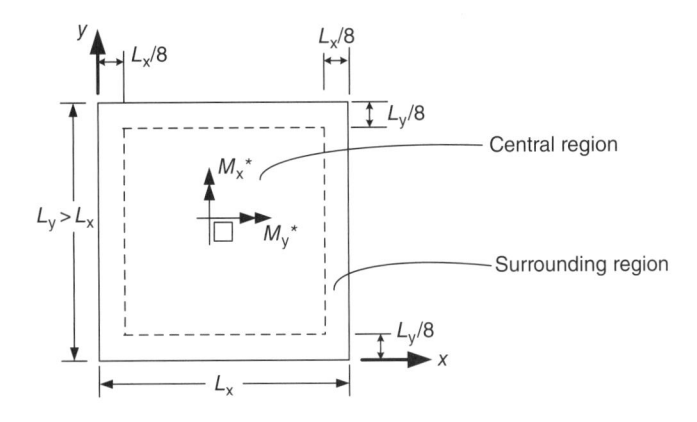

Figure 9.3(1) Orientation of a two-way slab for moment determination

The central region

For the central region, the moment

$$M_x^* = \beta_x F_d L_x^2 \quad \text{Equation 9.3(1)}$$

and

$$M_y^* = \beta_y F_d L_x^2 \qquad \text{Equation 9.3(2)}$$

where L_x is the shorter span, the design load (F_d) is given in Equation 9.2(4), and the moment coefficients (β_x and β_y) are listed in Tables 6.10.3.2(A) and (B) in the Standard. The tables apply to two different situations – Table 6.10.3.2(A) covers slabs with Class N main reinforcement, and Table 6.10.3.2(B) is for slabs with Class N bars, or Class L mesh as main reinforcement, where no moment redistribution is allowed at the serviceability or ultimate states. The two sets of coefficients are presented in a slightly different way in Tables 9.3(1) and 9.3(2), respectively. Note that the edge conditions referred to in these two tables are illustrated in Figure 9.3(2).

Table 9.3(1) Values of β_x and β_y for slabs supported on four sides[a]

Edge condition[b]	Short-span coefficients (β_x) Values of L_y/L_x								Long-span coefficients (β_y) for all values of L_y/L_x
	1.0	1.1	1.2	1.3	1.4	1.5	1.75	≥2.0	
1. Four edges continuous	0.024	0.028	0.032	0.035	0.037	0.040	0.044	0.048	0.024
2. One short edge discontinuous	0.028	0.032	0.036	0.038	0.041	0.043	0.047	0.050	0.028
3. One long edge discontinuous	0.028	0.035	0.041	0.046	0.050	0.054	0.061	0.066	0.028
4. Two short edges discontinuous	0.034	0.038	0.040	0.043	0.045	0.047	0.050	0.053	0.034
5. Two long edges discontinuous	0.034	0.046	0.056	0.065	0.072	0.078	0.091	0.100	0.034
6. Two adjacent edges discontinuous	0.035	0.041	0.046	0.051	0.055	0.058	0.065	0.070	0.035
7. Three edges discontinuous (one long edge continuous)	0.043	0.049	0.053	0.057	0.061	0.064	0.069	0.074	0.043
8. Three edges discontinuous (one short edge continuous)	0.043	0.054	0.064	0.072	0.078	0.084	0.096	0.105	0.043
9. Four edges discontinuous	0.056	0.066	0.074	0.081	0.087	0.093	0.103	0.111	0.056

a   Only for slabs with Class N bars as main reinforcement

b   See Figure 9.3(2).

Table 9.3(2) Values of β_x, β_y, α_x and α_y for slabs supported on four sides[a]

Edge condition[b]	Short-span coefficients (β_x and α_x) Values of L_y/L_x									Long-span coefficients (β_y and α_y) for all values of L_y/L_x
	1.0	1.1	1.2	1.3	1.4	1.5	1.75	2.0	> 2.0	
1. Four edges continuous	0.021 2.310	0.025 2.220	0.029 2.140	0.032 2.100	0.034 2.060	0.036 2.030	0.039 2.000	0.041 2.000	0.042 2.000	0.020 2.690
2. One short edge discontinuous	0.027 2.200	0.030 2.140	0.033 2.100	0.035 2.060	0.037 2.040	0.039 2.020	0.041 2.000	0.042 2.000	0.042 2.000	0.024 2.290
3. One long edge discontinuous	0.024 2.220	0.028 2.170	0.034 2.090	0.038 2.030	0.043 1.970	0.047 1.930	0.056 1.860	0.061 1.810	0.070 1.800	0.028 2.460
4. Two short edges discontinuous	0.032 2.090	0.035 2.050	0.037 2.030	0.038 2.010	0.039 2.000	0.040 2.000	0.042 2.000	0.042 2.000	0.042 2.000	0.024 —
5. Two long edges discontinuous	0.024 —	0.028 —	0.035 —	0.042 —	0.049 —	0.056 —	0.071 —	0.085 —	0.125 —	0.039 2.310
6. Two adjacent edges discontinuous	0.031 2.130	0.036 2.070	0.041 2.010	0.046 1.960	0.050 1.920	0.053 1.890	0.060 1.830	0.064 1.800	0.070 1.800	0.034 2.130
7. Three edges discontinuous (one long edge continuous)	0.039 2.040	0.044 1.970	0.048 1.930	0.052 1.890	0.055 1.860	0.058 1.840	0.063 1.800	0.066 1.800	0.070 1.800	0.035 —
8. Three edges discontinuous (one short edge continuous)	0.033 —	0.039 —	0.047 —	0.054 —	0.061 —	0.067 —	0.082 —	0.093 —	0.125 —	0.046 2.120
9. Four edges discontinuous	0.044 —	0.052 —	0.059 —	0.066 —	0.073 —	0.079 —	0.091 —	0.100 —	0.125 —	0.049 —

a Only for slabs with Class N bars, and Class L mesh as main reinforcement

b See Figure 9.3(2).

— = not applicable

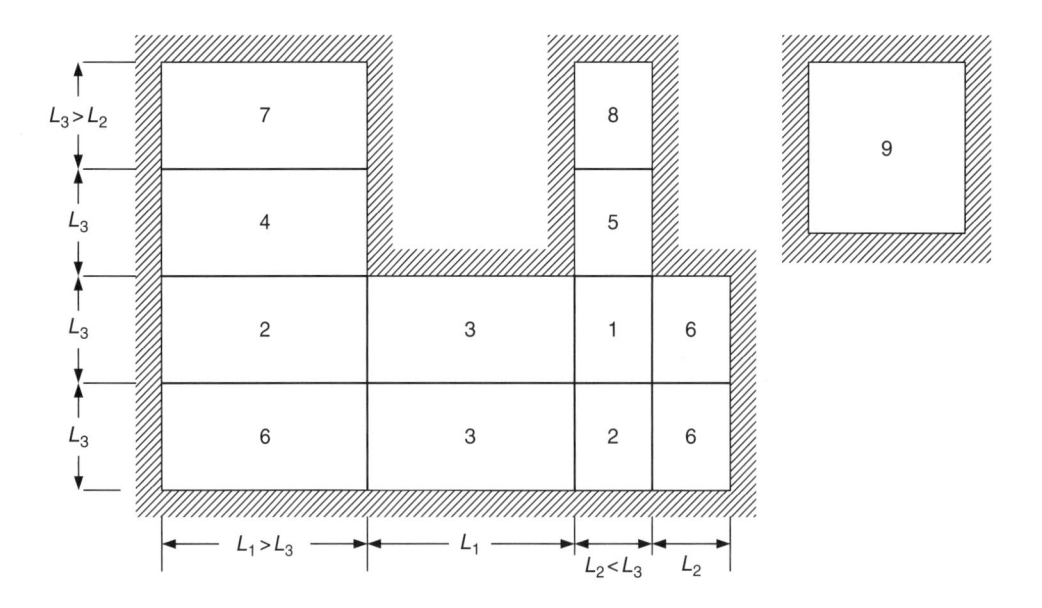

Figure 9.3(2) Definition of edge conditions for two-way slabs

The surrounding region

For the surrounding region, it is only necessary to provide the minimum steel in both the x and y directions. Or, according to Clause 9.1.1 (b),

$$p_{t.\min} = 0.19(D/d)^2 f'_{ct.f}/f_{sy} \qquad \text{Equation 9.3(3)}$$

It may be noted in Equations 9.3(1) and 9.3(2) that the positive bending moments are given in terms of the shorter span (L_x). This is because the bending rigidity of the shorter span is higher, thereby carrying a larger portion of the loading than the longer span (L_y). This phenomenon is reflected in Table 9.3(1) in which $\beta_y < \beta_x$ for all cases in which $L_y/L_x > 1$.

Negative moments

After determining the positive M_x^* and M_y^*, we can proceed to compute the negative bending moments at the exterior and interior supports. The negative moment at the exterior support (or the discontinuous edge) of span 1 in Figure 9.3(3) is given as

$$\overline{M}_{x1}^* = -\lambda_e M_{x1}^* \qquad \text{Equation 9.3(4)}$$

for the x-direction and

$$\overline{M}_{y1}^* = -\lambda_e M_{y1}^* \qquad \text{Equation 9.3(5)}$$

CHAPTER 9 SLABS 205

for the y-direction, where M_{x1}^* and M_{y1}^* are given in Equations 9.3(1) and 9.3(2), respectively. Also, $\lambda_e = 0.5$ if the midspan moments M_{x1}^* and M_{y1}^* are calculated based on Table 9.3(1), and $\lambda_e = 0.8$ if they are based on Table 9.3(2). The positions and directions of the support moments are illustrated in Figure 9.3(3).

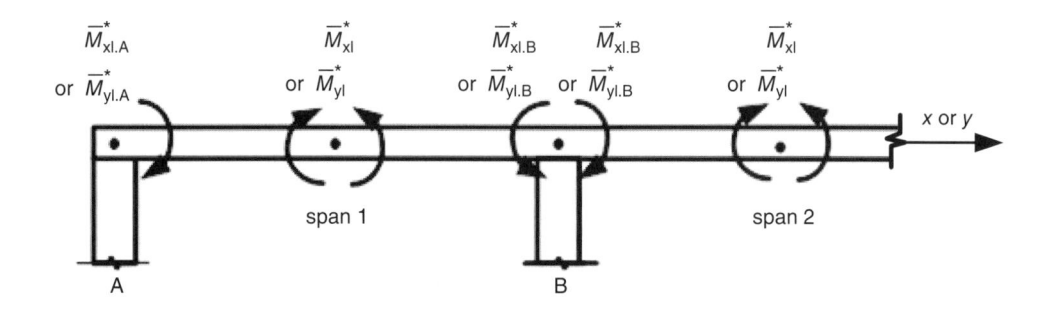

Figure 9.3(3) Position and direction of the support moments

For the interior support B of span 1 in Figure 9.3(3)

$$\overline{M}_{x1.B}^* = -\lambda_{ix} M_{x1}^* \qquad \text{Equation 9.3(6)}$$

and

$$\overline{M}_{y1.B}^* = -\lambda_{iy} M_{y1}^* \qquad \text{Equation 9.3(7)}$$

Similarly, for the interior span 2

$$\overline{M}_{x2.B}^* = -\lambda_{ix} M_{x2}^* \qquad \text{Equation 9.3(8)}$$

and

$$\overline{M}_{y2.B}^* = -\lambda_{iy} M_{y2}^* \qquad \text{Equation 9.3(9)}$$

In Equations 9.3(6) to (9), $\lambda_{ix} = \lambda_{iy} = 1.33$ if the midspan positive moments are calculated based on Table 9.3(1), and $\lambda_{ix} = \alpha_x$, $\lambda_{iy} = \alpha_y$ if they are based on Table 9.3(2), in which α_x and α_y are also given in Table 9.3(2).

When considering slabs with Class N main reinforcement, any unbalanced moment at an interior support (e.g. if there was an unbalanced moment at support B in Figure 9.3(3)) may be redistributed. Alternatively, the slab would have to be reinforced on both sides of the support for the larger of the two support moments.

It should be noted that Equations 9.3(1) to 9.3(9) are valid for analysing the gravity loads (g and q) only. Also, at exterior corners, the slab shall be reinforced in accordance with Clause 9.1.3.3 to prevent uplifts due to the twisting moments (see Section 9.3.3).

Slab shear to be carried by supporting beams or walls

The uniformly distributed load on the slab is carried to the four supporting beams or walls by the bending action of the slab. For simplicity, Clause 6.10.3.4 states that the distribution of the load to the four supporting beams or walls, or the shear forces they carry, may be determined according to the process illustrated in Figure 9.3(4).

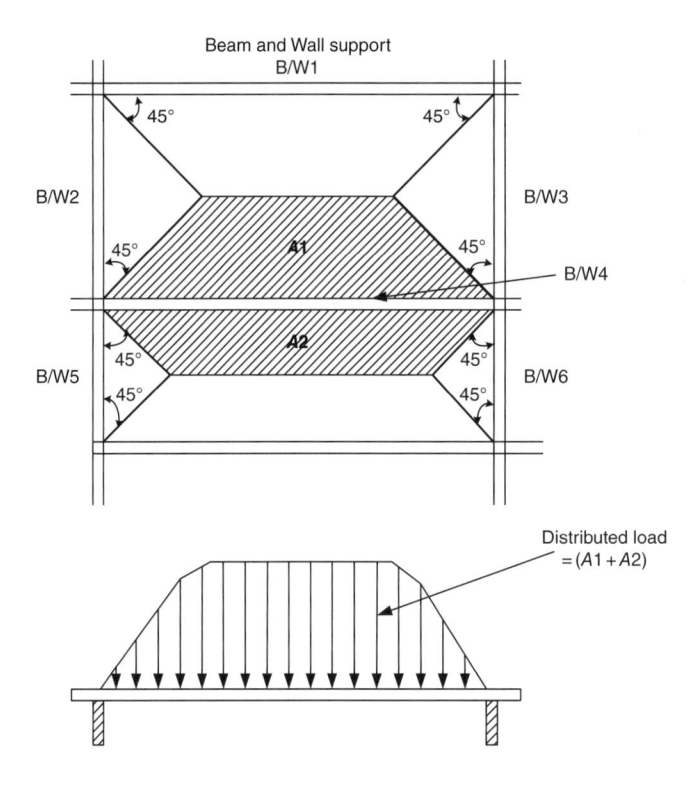

Figure 9.3(4) Load distribution to the beams or walls supporting the slabs
B/W = beam or wall

Note, however, that when one of the four edges is discontinuous, the loads distributed to the three continuous edges shall be increased by 10% and that for the discontinuous edge shall be decreased by 20%. For cases where two or three adjacent edges are discontinuous, it is advisable to conduct a rigorous elastic analysis to determine the respective shear forces.

Note also that as per Clause 9.7, the longitudinal shear check procedure, as discussed in Section 6.4, is also applicable for slab designs.

9.3.2 Reinforcement requirements for bending

For Equations 9.3(1) to 9.3(9) to be valid, the positive and negative bending reinforcements have to be arranged in accordance with the details specified in Clause 9.1.3.1. Further, to comply with this clause, the arrangement in each of the two directions shall follow that illustrated in Figure 9.2(4) (for one-way slab) subject to the following three modifications.

Positive moments
For a typical panel where $L_y > L_x$

$$L_{ny} = L_{nx} \quad \text{Equation 9.3(10)}$$

in which L_n is the clear span of the slab as depicted in Figure 9.2(4). Equation 9.3(10) indicates that the shorter span should be adopted for either of the two spans as the basis for measuring the extents or curtailment points for the bending reinforcement.

Negative moments over continuous supports
A typical layout of a multispan two-way slab is shown in Figure 9.3(5).

The clear span to be used in conjunction with Figure 9.2(4) in the x and y directions can be determined as described below.

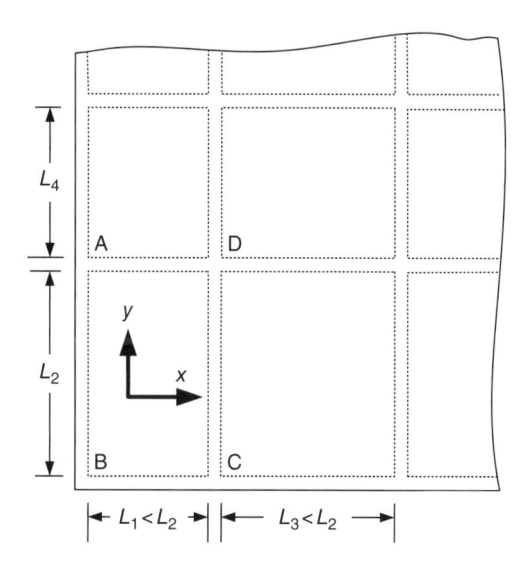

Figure 9.3(5) Typical layout of a multispan two-way slab

Reinforcement in the x-direction
For reinforcement in the x-direction

$$L_{nx} = L_1 \text{ if } L_1 > L_3 \quad \text{Equation 9.3(11)}$$

Otherwise,

$$L_{nx} = L_3 \qquad \text{Equation 9.3(12)}$$

Reinforcement in the y-direction

For reinforcement in the y-direction

$$L_{ny} = L_3 \text{ if } L_2 > L_4 \qquad \text{Equation 9.3(13)}$$

Otherwise,

$$L_{ny} = L_4 \qquad \text{Equation 9.3(14)}$$

In the above equations, L_1 and L_3 are the adjacent clear spans in the x-direction; L_2 and L_4 are in the y-direction.

Negative moments over discontinuous supports

For the typical panel ABCD shown in Figure 9.3(5), the extents of the negative reinforcement over the discontinuous supports AB and BC are governed by the clear span L_n as shown in Figure 9.2(4) where

$$L_n = L_1 \text{ if } L_1 < L_2 \qquad \text{Equation 9.3(15)}$$

Otherwise,

$$L_n = L_2 \qquad \text{Equation 9.3(16)}$$

The minimum extent of the negative steel in either direction is shown in Figure 9.3(6).

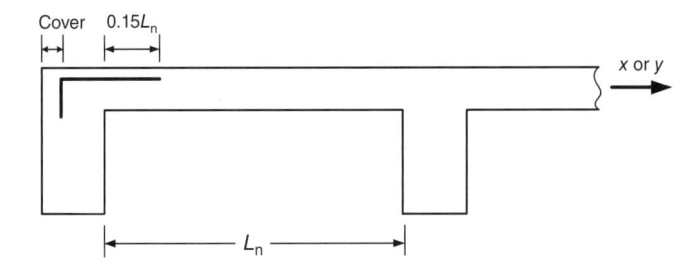

Figure 9.3(6) Minimum extent of negative steel over discontinuous support

9.3.3 Corner reinforcement

The twisting moments (M_{xy} and M_{yx}) shown in Figure 9.1(3)a have the effect of producing uplifts of the slab at the corners of the floor. To be valid, the uplifts must be prevented for the moment analysis in Section 9.3.1. If uplifting is prevented that way, cracking could occur in the vicinity of the slab corners. To reinforce against such cracks, corner reinforcements are required at all exterior corners as shown in Figure 9.3(7). Further explanations are given for the next two types of corner reinforcement.

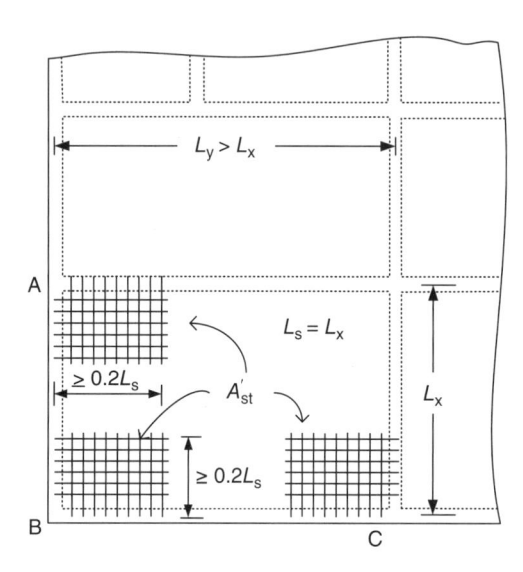

Figure 9.3(7) Corner reinforcement to prevent corner cracking

Exterior corner B

For the exterior corner B

$$A'_{st} = 0.75 A_{st} \qquad \text{Equation 9.3(17)}$$

where A'_{st} is the total area for each of the four layers of reinforcement – the top and bottom layers in the x and y directions, and A_{st} is the maximum positive steel for the midspan.

Exterior corners A and C

For the exterior corners A and C

$$A'_{st} = 0.5 A_{st} \qquad \text{Equation 9.3(18)}$$

Note that for all exterior corners, the existing reinforcement for bending and shrinkage and temperature effects may be considered as part of A'_{st}.

9.3.4 Deflection check

A slab subjected to uniform load must have an effective thickness greater than d_{min} – as given in Equation 9.2(8) – to satisfy the deflection serviceability design requirement. For slabs supported on four sides, the k_4 values are recommended in Table 9.3.4.2 of the Standard for different L_y/L_x ratios. The values are reproduced in Table 9.3(3). For all cases $k_3 = 1$.

Table 9.3(3) Values of k_4 for slabs supported on four sides

Edge condition type[a]	k_4 Values of L_y/L_x			
	1.00	1.25	1.50	2.00
1	3.60	3.10	2.80	2.50
2	3.40	2.90	2.70	2.40
3	3.40	2.65	2.40	2.10
4	3.20	2.80	2.60	2.40
5	3.20	2.50	2.00	1.60
6	2.95	2.50	2.25	2.00
7	2.70	2.30	2.20	1.95
8	2.70	2.10	1.90	1.60
9	2.25	1.90	1.70	1.50

a   See Figure 9.3(2) for definition.

Source: Standards Australia (2009). AS 3600–2009 *Concrete Structures*. Standards Australia, Sydney, NSW. © Standards Australia Limited. Copied by Cambridge University Press with the permission of Standards Australia and Standards New Zealand under Licence 1712-c038.

9.3.5 Crack control

The Standard in Clause 9.4 provides stringent requirements for crack control in two-way slabs. Clause 9.4.1 specifies, for slabs in bending, the minimum A_{st} and the maximum bar spacing in the two orthogonal directions. It also limits the tensile reinforcement stresses (σ_{scr} and $\sigma_{scr.1}$) at a cracked section under various service load combinations to some given maximum values. Note that σ_{scr} and $\sigma_{scr.1}$ may be calculated using the transformed section approach described in Appendix A.

To control cracking due to shrinkage and temperature effects, specifications in terms of minimum A_{st} are detailed in Clause 9.4.3. Also, the reader should be aware of the qualitative recommendations in Clauses 9.4.4 and 9.4.5 for crack control in the vicinity of restraints, and at slab openings and discontinuities, respectively.

9.3.6 Design example

Problem

The corner slab shown in Figure 9.3(8) is part of a hotel complex. Sides AB and BE are supported on brick walls, and sides AC and CE on reinforced concrete spandrel beams. The beams are supported at C by a column cast monolithically with the slab.

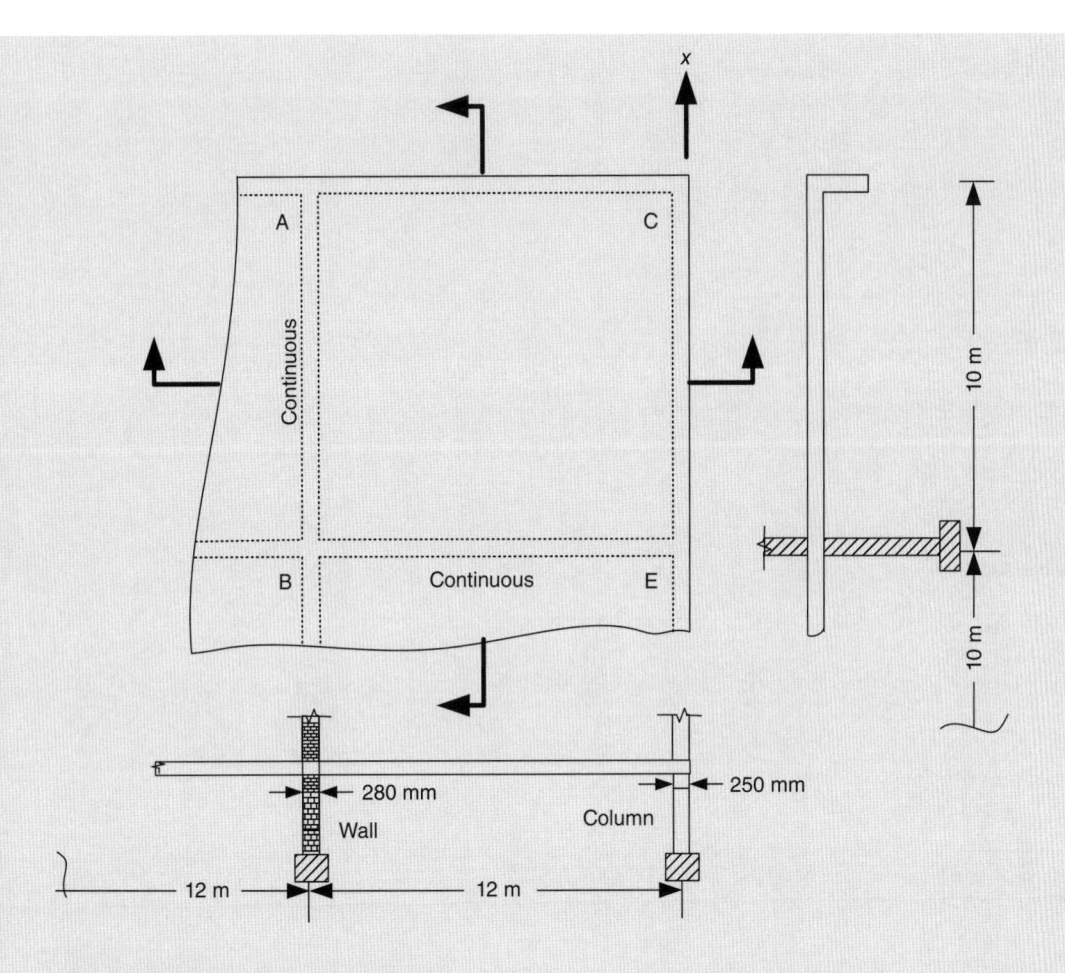

Figure 9.3(8) Corner slab in a hotel complex

Design the slab in accordance with the Standard, given $f'_c = 25$ MPa, exposure classification A1 and fire-resistance period = 90 minutes. Use N bars only.

Solution

Loading
Table 3.1 of AS/NZS 1170.1–2002 gives $q = 4$ kPa. Assuming $D = 300$ mm and the total amount of steel in the slab is 0.5% by volume, Equation 2.4(1) gives

$$\rho_w = (24 + 0.6 \times 0.5) = 24.3 \text{ kN/m}^3$$

Then
$$g = 0.3 \times 24.3 = 7.29 \text{ kPa}$$

Also, the design load is

$$F_d = 1.2g + 1.5q = 14.75 \text{ kPa}$$

Design moments
Now, $L_x = 10$ m and $L_y = 12$ m yield $L_y/L_x = 1.2$. With the two adjacent edges being continuous, Table 9.3(1) gives

$$\beta_x = 0.046 \text{ and } \beta_y = 0.035$$

Positive moments

Equation 9.3(1): $M_x^* = 0.046 \times 14.75 \times 10^2 = 67.85$ kNm/m

Equation 9.3(2): $M_y^* = 0.035 \times 14.75 \times 10^2 = 51.63$ kNm/m

Negative moments
About edge AB

Equation 9.3(7): $M_y^* = -1.33 \times 51.63 = -68.67$ kNm/m

About edge CE

Equation 9.3(5): $M_y^* = -0.5 \times 51.63 = -25.82$ kNm/m

About edge BE

Equation 9.3(6): $M_x^* = -1.33 \times 67.85 = -90.24$ kNm/m

About edge AC

Equation 9.3(4): $M_x^* = -0.5 \times 67.85 = -33.93$ kNm/m

Minimum effective depth
The shorter effective span is $L_{ef} = 10000$ mm. For total deflection,

Table 5.2(4): $(\Delta/L_{ef}) = 1/250$

Equation 5.4(3): $k_{cs} = 2 - 1.2\dfrac{A_{sc}}{A_{st}} = 2.0$

Table 1.3(1): $\psi_s = 0.7$ and $\psi_l = 0.4$

Then,

Equation 5.5(8): $F_{def} = (1.0 + 2) \times 7.29 + (0.7 + 2 \times 0.4) \times 4$
$$= 27.87 \text{ kN/m}^2 = 27.87 \times 10^{-3} \text{ MPa}$$

From Section 9.3.4, we have $k_3 = 1.0$ and by interpolation from Table 9.3(3)

$k_4 = 2.50 + [(0.05/0.25) \times (2.95 - 2.5)] = 2.59$

Equation 2.2(5): $E_c = 25\,280$ MPa

Finally, Equation 9.2(8) gives

$$d_{min} = \dfrac{10\,000}{1 \times 2.59 \times \sqrt[3]{\dfrac{(1/250) \times 25\,280}{27.87 \times 10^{-3}}}} = 251.3 \text{ mm}$$

Since D = 300 mm, use N16 bars. From Table 1.4(2), the concrete cover is 20 mm for exposure classification A1 and from Table 5.5.3(A) of the Standard, only 15 mm is needed for fire resistance. Thus

$d_x = 300 - 20 - 8 = 272$ mm $> d_{min} = 251.3$ mm

This is acceptable. Now

$d_y = 272 - 16 = 256$ mm $> d_{min} = 251.3$ mm

This is also acceptable. Finally, the average

$d_{average} = 272 - 8 = 264$ mm $> d_{min} = 251.3$ mm

This, too, is acceptable.

Bending design

Equation 3.3(2)a: $\alpha_2 = 1.0 - 0.003 \times 25 = 0.925$

but $0.67 \leq \alpha_2 \leq 0.85$ thus; $\alpha_2 = 0.85$ and

Equation 3.5(7): $\xi = \dfrac{0.85 \times 25}{500} = 0.0425$

Assuming $\phi = 0.8$, Equation 3.5(6) yields the p_t values for all the design sections. The results are summarised in Table 9.3(4).

Table 9.3(4) Results for all of the design sections

Location	M_x^*	M_y^*	p_t	A_{st}	N16 @ s	A_{st} (actual)	Top or bottom	x or y direction
Central region	67.85	–	0.00236	641.9	300	670	B	x
Central region	–	51.63	0.002	512	300	670	B	y
Outside the central region			0.002	528 ($d_{average}$)	300	670	B	x and y
About AB	–	–68.67	0.0027	691.2	275	731	T	y
About CE	–	–25.82	0.00156[a]	399.4	300	670	T	y
About BE	–90.24	–	0.0032	870.4	225	893	T	x
About AC	–33.93	–	0.00139[a]	378.1	300	670	T	x

a $p_{t.min}$ values, since the minimum reinforcement requirements govern in these cases. See Equation 9.3(3).

Note: check for ϕ. Take the largest p_t value of 0.0032, Equation 3.4(8) gives $k_u = 0.0886$ and for $d = d_o$, $k_{uo} = k_u = 0.0886$ and Equations 3.4(20) a with b give $\phi = 0.8$. Hence, for all cases, the assumption of $\phi = 0.8$ is acceptable.

Corner reinforcement

Corner C
At corner C, the uplift is prevented. Thus
Equation 9.3(17): $A'_{st} = 0.75 \times 670 = 502.5$ mm²/m

The existing top and bottom reinforcement in the x and y directions exceed 502.5 mm²/m. Therefore, no additional reinforcement is required over corner C.

Corners A and E
Equation 9.3(18): $A'_{st} = 0.5 \times 670 = 335$ mm²/m

All existing top and bottom reinforcement in the x and y directions exceed 335 mm²/m. Therefore, no additional reinforcement is required over corners A and E.

Reinforcement layout
A sketch of the reinforcement arrangement is given in Figure 9.3(9).

Main steel	Corner steel
(1) N16@300 mm (B)	No additional reinforcement required over corners A, C and E.
(2) N16@300 mm (T)	
(3) N16@225 mm (T)	
(4) N16@275 mm (T)	
(5) N16@300 mm (B)	
(6) N16@300 mm (B)	

B = bottom; T = top

Figure 9.3(9) Reinforcement layout for the design slab

Note: all dimensions are in mm.

9.4 MULTISPAN TWO-WAY SLABS

9.4.1 General remarks

A building floor slab, as shown in Figure 9.1(2)b, is supported on four sides by a system of walls or heavy girders on columns. Thus, it may be designed following the procedure described in Section 9.3 even if it is continuous over several spans in the x or y direction. This is because the wall or girder supports are rigid enough to isolate the slab and render it into a system of single panels.

Multispan slabs that cannot be 'isolated' into single panels have to be analysed and designed in a different way. Such slabs may be solid with or without drop panels; they may incorporate ribs in two directions (which are generally referred to as waffle slabs) or they may be beam-and-slab systems without heavy girders. See Figure 9.l(2)b for definitions of these terms.

There are two methods recommended in the Standard for the analysis of the multispan slab moments:

1. the simplified method, in which each floor of the building is analysed separately (Clause 6.10.4)
2. the 'idealised frame' method, which is applied to situations where the effects of columns or the behaviour of the multistorey structure is taken into consideration (Clause 6.9).

As the floor slabs are basically supported on columns, the punching shear would be high. Special design considerations are necessary.

The simplified method of analysis is discussed in this section and the idealised frame approach is presented in Section 9.5. Section 9.6 details the checking and design procedure for punching shear, and illustrative examples are given where necessary. The inadequate performance of the Standard's punching shear design procedure is discussed in Section 9.6.11. Finally, a practical design example for a flat plate building is included in Section 9.7.

It should be noted that a multistorey building may be analysed using a computer-based method such as the finite element procedure. The use of such rigorous analyses is mentioned in Clause 6.4 of the Standard.

Also, the crack control requirements summarised in Section 9.3.5 must be met in the design of multispan slabs.

9.4.2 Design strips

Figure 9.4(1) illustrates the plan view of a column-and-slab floor system (i.e. flat plate or flat slab), which can be considered as a multispan two-way slab, spanning both the x and y directions. Note that at the near edge it is unsupported (or a free edge); at the far edge it is supported on a wall.

For analysis and design purposes using either the simplified or idealised frame method the slab may be divided into two different types of design strips:

1. edge design strip (either with a free or wall-supported edge)
2. interior design strip.

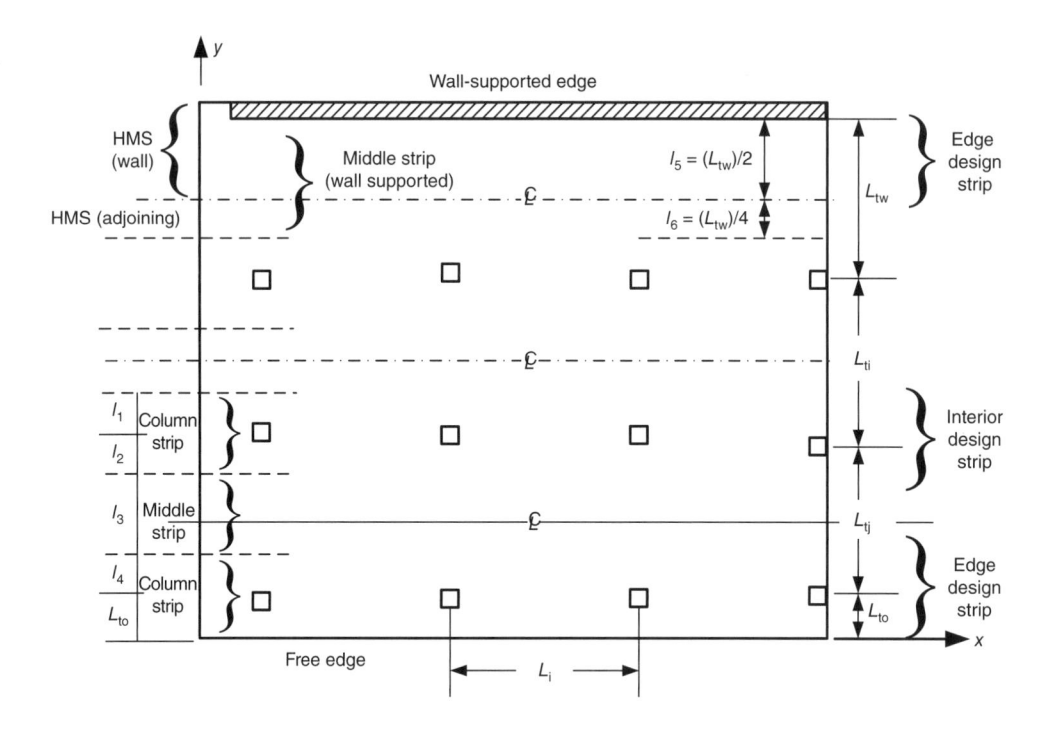

Figure 9.4(1) Plan view of a column-and-slab floor system identifying the design strips
HMS = half middle strip

Each of these design strips can be regarded as an independent unit for the analysis of bending moments. For reinforcement design, it may be treated as a wide beam spanning multiple (column) supports.

Each of the free-edged and interior design strips consists of one column strip and the half middle strip(s); the design strip with a wall-supported edge stands on its own. For a given design strip under gravity loads, the total moment at a given section may be calculated using the simplified formulas recommended in the Standard (see Section 9.4.4 of this book). This total moment can be distributed according to some formulas to the column strip and the two half middle strips. The distribution of moment is to approximate the true behaviour of the slab.

All the strips are defined geometrically in Figure 9.4(1). Note that for a typical span (L_j) of a design strip

$$l_1 = L_{ti}/4 \qquad \text{Equation 9.4(1)}$$

$$l_2 = L_{tj}/4 \qquad \text{Equation 9.4(2)}$$

noting that $l_1 + l_2 < L_j/2$; otherwise, $l_1 + l_2 = L_j/2$ and

$$l_4 = L_{tj}/4 \qquad \text{Equation 9.4(3)}$$

noting that $l_4 + L_{to} < L_i/2$; otherwise $l_4 + L_{to} < L_i/2$ and

$$l_3 = L_{tj} - l_2 - l_4 \qquad \text{Equation 9.4(4)}$$

and

$$l_5 = L_{tw}/2 \qquad \text{Equation 9.4(5)}$$

Finally

$$l_6 = L_{tw}/4 \qquad \text{Equation 9.4(6)}$$

The above design strips run in the x-direction, the design of which produces the reinforcement in the x-direction. The design of the other series of strips in the y-direction leads to the reinforcement in that direction.

9.4.3 Limitations of the simplified method of analysis

The total static moment of a given span of a given design strip may be computed using a simple formula. However, the following structural and geometrical conditions must prevail:
- There is a minimum of two continuous spans in each of the x and y directions.
- The column grid should be rectangular but may be with a maximum of 10% 'offset'.
- For any panel

$$0.5 \leq L_i/L_{ti} \leq 2 \qquad \text{Equation 9.4(7)}$$

- If $L_i < L_j$

$$(L_j - L_i)/L_j \leq 0.3333 \qquad \text{Equation 9.4(8)}$$

- All horizontal forces are resisted by shear walls or braced frames.
- Vertical loads are essentially uniformly distributed.
- The live load (q) is not greater than twice the dead load (g), or

$$q \leq 2g \ (\text{in kN/m}^2) \qquad \text{Equation 9.4(9)}$$

- Class L bars or mesh are not used as the main bending reinforcement.

9.4.4 Total moment and its distribution

For a given span of a design strip, the total static moment (M_o) as per Clause 6.10.4.2 is given as

$$M_o = F_d L_t L_o^2 / 8 \qquad \text{Equation 9.4(10)}$$

where the design load (F_d) in kN/m² is obtained using Equation 9.2(4), L_t is the width of the design strip – see Figure 9.4(1) – and L_o is the span which may be determined as

$$L_o = L - 0.7(a_{sl} + a_{sr})$$ **Equation 9.4(11)**

in which L, a_{sl} and a_{sr} are defined in Figure 6.1.2(B) of the Standard, reproduced in Figure 9.4(2).

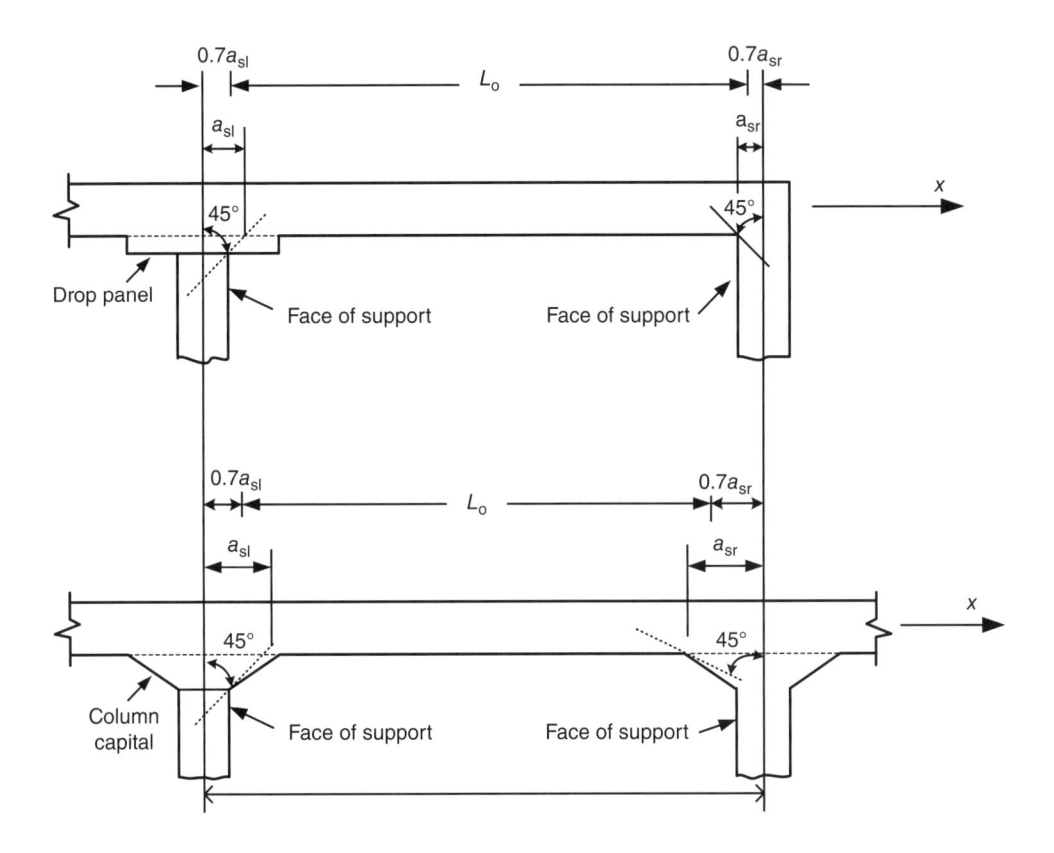

Figure 9.4(2) Definitions of L_o
Source: Standards Australia (2009). AS 3600–2009 *Concrete Structures*. Standards Australia, Sydney, NSW.
© Standards Australia Limited. Copied by Cambridge University Press with the permission of Standards Australia and Standards New Zealand under Licence 1712-c038.

For an interior span, the total moment (M_o) is to be distributed to the two supports and midspan of the design strip as illustrated in Figure 9.4(3). For an exterior span, the distribution factors are as shown in Figure 9.4(4).

Each of the bending moments shown in Figure 9.4(4) is the total value to be taken by a section of the design strip in question. The value (in kNm) is to be distributed to the column strip and the two half middle strips according to a simple procedure described in Clause 6.9.5.3 of the Standard. The procedure is identical for a multispan two-way slab and for a slab

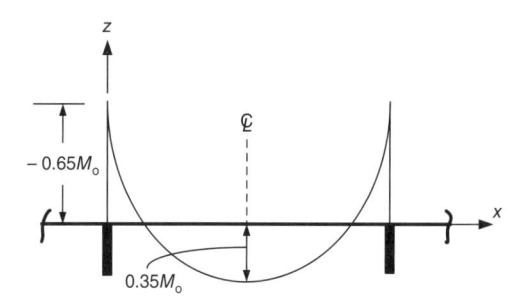

Figure 9.4(3) Distribution of total moment (M_o) for an interior span

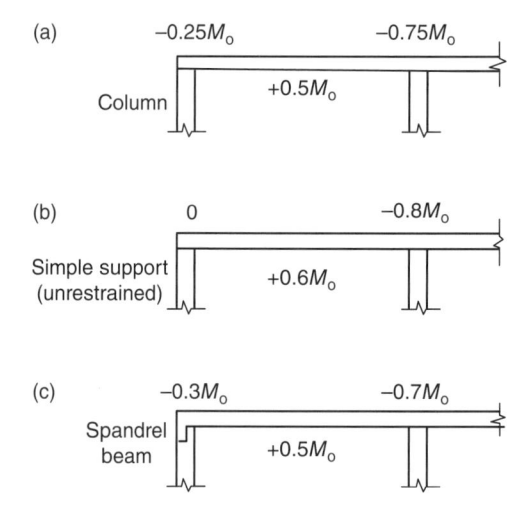

Figure 9.4(4) Distribution of total moment (M_o) for an exterior span

in the idealised frame; details are given in Section 9.5.3 in this book. Further, and as specified in Clause 6.10.4.3, the distribution factors for cases with the exterior edge fully restrained are –0.65, +0.35 and –0.65 for the exterior, midspan and interior support moments, respectively. The corresponding factors for beam-and-slab construction (see Figure 9.1(2)b) are –0.15, +0.55 and –0.75.

9.4.5 Punching shear

In punching shear design, the moment to be transferred from the slab to the column support is referred to as M_v^*. For the multispan two-way slab, M_v^* for an interior support should be taken as the unbalanced moment at that support (i.e. $\sum M$ at the joint). However,

$$M_v^* \geq 0.06\left[(1.2g + 0.75q)\, L_t L_o^2 - 1.2gL_t L_o'^2\right] \qquad \text{Equation 9.4(12)}$$

where L_o' is the shorter L_o for the adjoining spans.

For an exterior support, M_v^* is the actual moment.

The design procedure for punching shear recommended in Clause 9.2 may be applicable to slabs as part of a multispan two-way slab or of an idealised frame. The procedure is presented in Section 9.6.

9.4.6 Reinforcement requirements

The arrangement of reinforcement for continuous slabs supported on walls or girders is described in Section 9.3.2.

For flat plates where the slabs are supported on columns, the extent and curtailment of the reinforcement in the x and y directions are depicted in Figure 9.1.3.4 of the Standard. The arrangement is also illustrated in Figure 9.4(5) with the same elaborations. Note that, for adjacent spans that are unequal, L_n for the top bars in Figure 9.4(5) is taken to be the longer of the two spans.

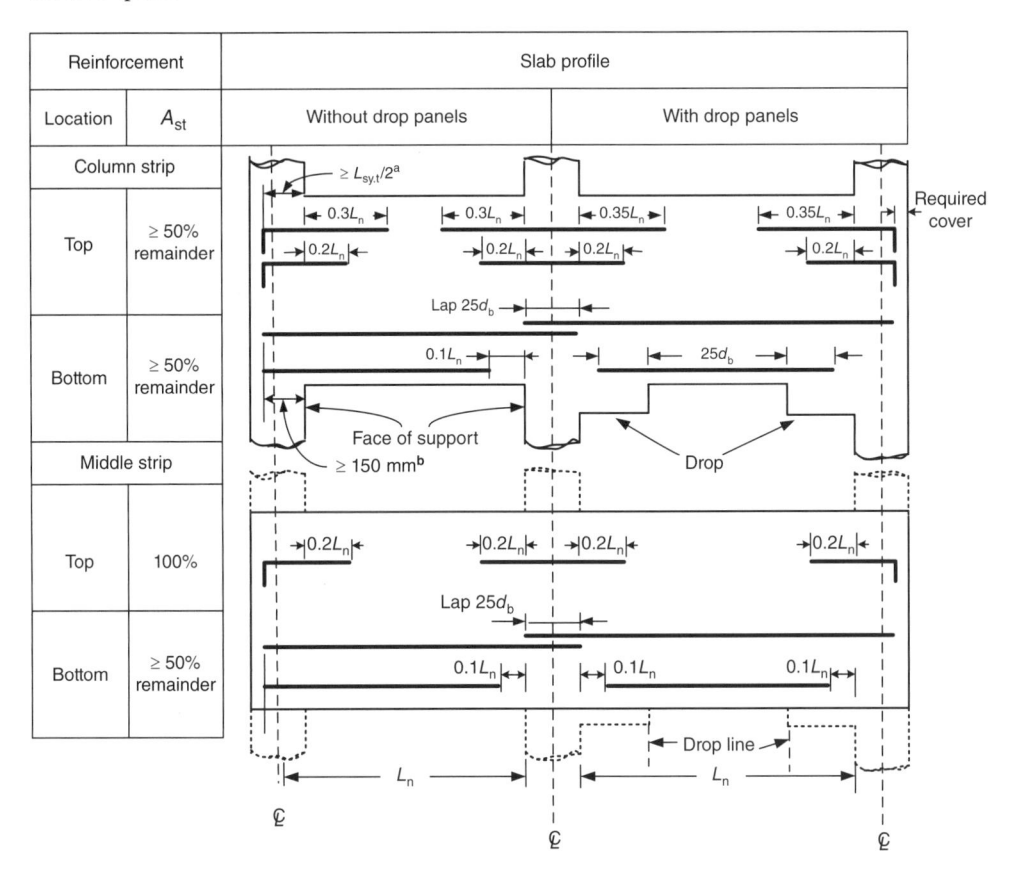

Figure 9.4(5) Reinforcement arrangement for column-and-slab structures

a $L_{sy.t}$ for straight bars (see Section 8.2.1)
b Applies to all bottom bars perpendicular to the discontinuous edge. For slabs without spandrel beams, extend all the way to the edge of the slab (less cover)

As per Clause 9.1.1(a), the minimum steel ratio is

$$p_{t.\min} = 0.24(D/d)^2 f'_{ct.f}/f_{sy} \qquad \text{Equation 9.4(13)}$$

9.4.7 Shrinkage and temperature steel

For reinforcement in the secondary direction or direction parallel to the major bending axis, Clause 9.4.3.3 specifies that the shrinkage and temperature steel area is

$$A_{s.\min} = 1.75bD \times 10^{-3} \qquad \text{Equation 9.4(14)}$$

where the slab is free to expand or contract in the secondary direction in which b and D are given in mm.

On the other hand, Clause 9.4.3.4 stipulates that for a slab that is restrained, or not free to expand or contract in the secondary direction

$$\text{(a) } A_{s.\min} = 6.0\,bD \times 10^{-3} \qquad \text{Equation 9.4(15)}$$

where a strong degree of crack control is required,

$$\text{(b) } A_{s.\min} = 3.5bD \times 10^{-3} \qquad \text{Equation 9.4(16)}$$

where a moderate degree of crack control is required and

$$\text{(c) } A_{s.\min} = 1.75\,bD \times 10^{-3} \qquad \text{Equation 9.4(17)}$$

where a minor degree of crack control is required.

Clause 9.4.3.2 further states that, in the primary or major bending direction, no additional reinforcement is required if

- the existing reinforcement is greater than as specified in Equation 9.3(3) or 9.4(13), as appropriate
- 75% of the steel area as obtained in Equation 9.4(15), (16) or (17) (whichever is required) is provided in the said direction.

9.5 THE IDEALISED FRAME APPROACH

9.5.1 The idealised frame

A multistorey three-dimensional column-and-slab building structure – as shown in Figure 9.1(2)c and d – may be divided into a series of 'idealised frames' in the x-direction and another series in the y-direction (though this is subject to some limitations). These 'plane' frames basically consist of columns and slabs. The latter can be treated as wide beams in a structural analysis. These two series of frames may be analysed and designed individually, and then recombined to form the three-dimensional structure to be constructed. Note that Class L main reinforcement is not allowed in a design based on the idealised frame approach (Clause 6.9.5.1).

Figure 9.5(1) shows a typical column-and-slab building system and the idealised frames in the x–z and y–z planes. A comparison with Figure 9.4(1) indicates that the 'thickness' of an idealised frame is identical to the width of a design strip. Further, each frame basically consists of the column strip and two half middle strips. All the strip dimensions can also be calculated using Equations 9.4(1) to 9.4(6).

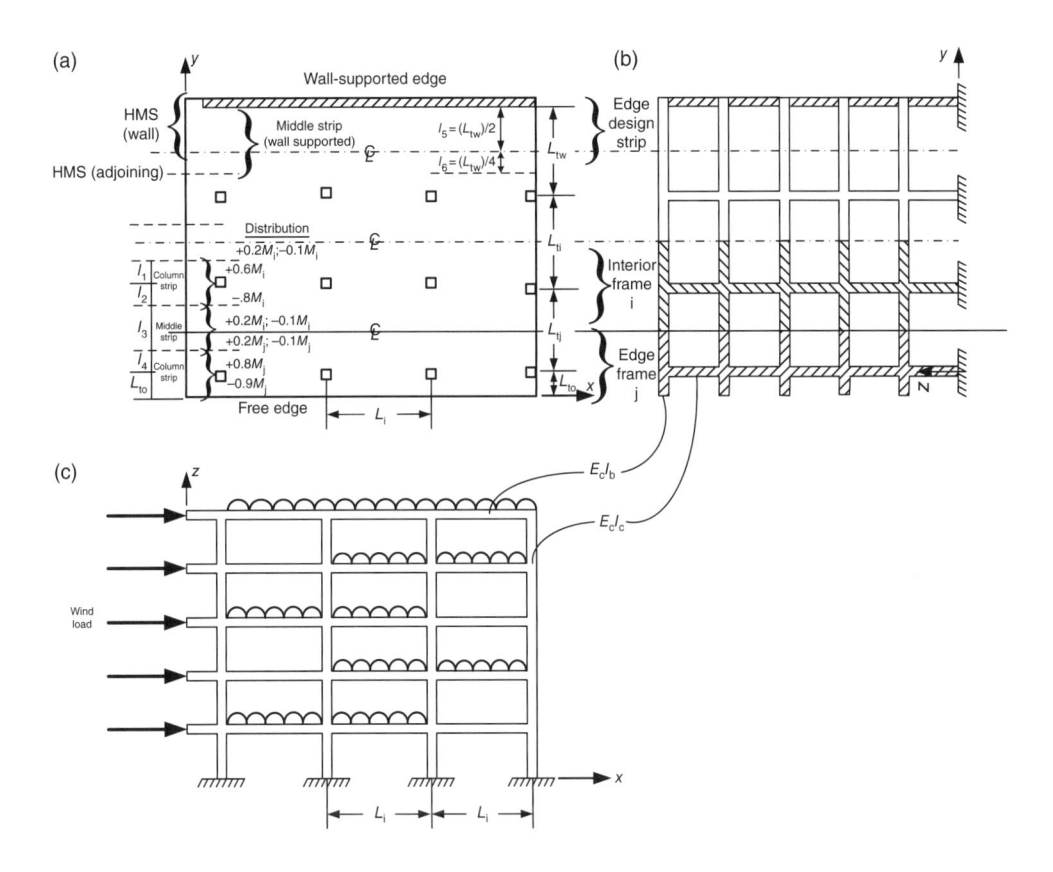

Figure 9.5(1) Typical idealised frames: **(a)** plan; **(b)** elevation, *yz*; and **(c)** elevation, *xz*
HMS = half middle strip

It should be remembered that the slabs in the column-and-slab system may be solid or they may be of the beam-and-slab type, including systems with thickened slab bands or band-beam slabs. They may also incorporate ribs in the two orthogonal directions such as a waffle slab. Other permissible slab systems are specified in Clause 6.9.5.1.

9.5.2 Structural analysis

The bending moments, shear forces and axial loads in each of the idealised frames can be analysed using the traditional linear structural analysis method. In view of the high degree of indeterminacy, a computer-based procedure will be necessary. Because of the widespread availability of structural analysis packages on desktop computers or notebooks, analysis of the entire frame is preferable to treating one floor at a time as permitted in Clause 6.9.3.

The analysis of the frame in its entirety should include all possible loading combinations as applicable (see Section 1.3). For vertical loads, the following load combinations are considered the minimum:

- If the live-load pattern is fixed, use the factored live load – Equation 1.3(2).
- If the live load is variable but is not greater than three-quarters of the dead load, use the factored live load on all spans.
- If the live load is variable and exceeds three-quarters of the dead load, use the following for the floor under consideration:
  – three-quarters of the factored live load on two adjacent spans
  – three-quarters of the factored live load on adjacent spans
  – factored live load on all spans.

As input to a linear structural analysis, the moment of inertia of each frame member has to be determined. For columns, it is straightforward:

$$I_c = \frac{b_c D_c^3}{12}$$ Equation 9.5(1)

where b_c and D_c are the width and the overall depth of the column section. If the horizontal member is a solid slab, b_b is the width of the slab or the 'thickness' of the idealised frame and D_b is the overall slab thickness. That is,

$$I_b = \frac{b_b D_b^3}{12}$$ Equation 9.5(2)

If the horizontal member includes n parallel flanged beams, then

$$I_b = \sum_{i=1}^{n} I_{bi}$$ Equation 9.5(3)

where I_{bi} is the moment of inertia of flanged beam i. Note that when calculating I_{bi}, the effective flange width rules apply, as discussed in Section 4.2.2.

To estimate the overall deflected shapes, the full bending stiffness may be assumed for each of the idealised frame members. For individual beams and slabs, Clause 6.2.5 stipulates that the deflection calculation shall take into account the effects of cracking, tension stiffening, shrinkage and creep. For detailed discussions, the reader is referred to Chapter 5 for beam deflections, and Sections 9.2.3 and 9.3.4 for one-way and two-way slabs, respectively. To analyse the frame for the bending moments, and the shear and axial forces to be used for the strength design, Clause 6.3.2(b) recommends some reductions. Alternatively,

$$EI = 0.8 E_c I_c$$ Equation 9.5(4)

for columns and

$$EI = 0.4\,E_c I_b \qquad \text{Equation 9.5(5)}$$

for the horizontal or 'beam' members. In both equations, E_c may be computed using Equation 2.2(5) or 2.2(6), whichever is appropriate for the case.

9.5.3 Distribution of moments

After analysing the idealised frame for the moments in the horizontal members (which have been treated as beams in the structural analysis), these moments have to be distributed appropriately to the column and half middle strips. This is necessary because, in reality, the bending moment is not uniform over the width of the slab. Instead, the closer the strip is to the column, the higher the moment. This is, to some extent, similar to the stress distribution diagram shown in Figure 4.2(1)b. For obvious reasons, no distribution of axial forces is necessary for the vertical members or columns.

The distribution of bending moments to the column strip is in accordance with Clause 6.9.5.3 of the Standard. The distribution factors for the column strips and the two half middle strips within the 'thickness' of the frame or the width of the design strip are given in Table 9.5(1).

The table shows that the Standard recommends a range of values for each of the distribution factors. Note that for each design strip, the distribution factors sum to unity. Consequently, after distributing the moment to the column strip the remainder is equally shared by the two half middle strips. Figure 9.5(1) shows examples of the application of these distribution factors.

Table 9.5(1) Distribution factors

Bending moment location	Column strip	Each half middle strip
Negative moment at an interior support	0.60–1.00	0.00–0.20
Negative moment at an exterior support with or without spandrel beam	0.75–1.00	0.00–0.125
Positive moment at all spans	0.50–0.70	0.15–0.25

For the design strip with a wall-supported edge shown in Figure 9.5(1), the bending moment it carries is equal to twice the value assigned to the adjoining half middle strip that forms part of the adjacent first interior frame.

Figure 9.5(1)a shows two examples of the moment distribution procedure (where the average values of the distribution factors are used): one for the interior frame i and another for the exterior frame j. Note that the M_i and M_j are the total bending moments resulting from the structural analysis of the idealised frame as a whole.

The application of the idealised frame approach is demonstrated in the design example given in Section 9.7. For the design of torsional moments and to account for openings in the slabs, the reader is referred to Clauses 6.9.5.4 and 6.9.5.5, respectively.

9.6 PUNCHING SHEAR DESIGN

For a column-and-slab structure, high shear force exists in the vicinity of the column heads. The so-called punching shear phenomenon is complicated by the existence of the bending moments transferred from the slab to the column. This is especially true at the perimeter of the floor where there are edge beams or spandrels. Considerable research work has been done with the aim of developing an accurate prediction procedure for the punching shear strength of flat plate structures (e.g. Falamaki and Loo 1992; Loo and Falamaki 1992).

This section presents the design equations for punching shear in detail (from the Standard) followed by an illustrative example. The accuracy of the Standard procedure is discussed in Section 9.6.11.

9.6.1 Geometry and definitions

Figure 9.6(1) illustrates a portion of a column-and-slab system including two slab–column connections. The forces at each of the connections include the axial forces (N^*) for the top and bottom columns, and the moments (M^*) from the connecting slab or slabs as the case may be. After a structural analysis of the idealised frame, we obtain the values of N and M for the individual load cases. By applying the load combination equations given in Section 1.3, the design values (N^*) and (M^*) can be computed.

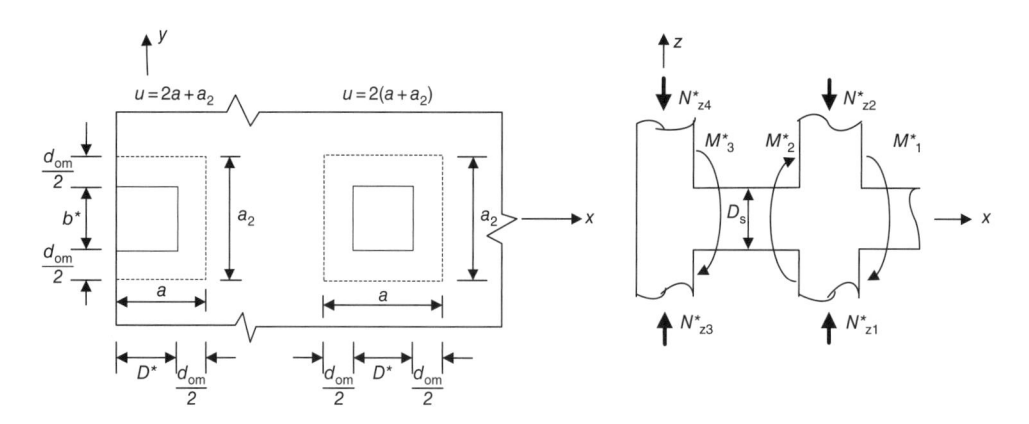

Figure 9.6(1) Examples of slab–column connections

Note that in Figure 9.6(1), $d_{om} = (d_{ox} + d_{oy})/2$, where d_{ox} and d_{oy} are the distances between the extreme compressive fibre of the concrete to the centre of gravity of the outer layer of tensile reinforcement running in the x and y directions, respectively. See Figure 6.3(1) for an example of this. Also, D^* and b^* are the overall depth and width, respectively, of the column head (or drop panel if it exists) and a and a_2 are the dimensions of the critical shear perimeter. The design forces

$$N_z^* = N_{z\text{ bottom}}^* - N_{z\text{ top}}^* \qquad \text{Equation 9.6(1)}$$

$$M_v^* = M_{\text{right}}^* + M_{\text{left}}^* \qquad \text{Equation 9.6(2)}$$

where the moments are positive if they are of the same direction. Otherwise, the larger moment is given the positive sign and M_v^* is the difference of the two moments.

9.6.2 Drop panel and shear head

The use of drop panels (and capitals) at column heads has the indirect effect of increasing the shear strength of the slab. For aesthetic or economic reasons, a shear head consisting of steel sections may be embedded in the slab on top of the column. It may be used instead of, or in conjunction with, the drop panel. Practical design examples of shear heads may be found in some existing textbooks (e.g. Darwin, Dolan and Nilson 2016). The existence of a shear head increases the shear strength of the slab in the area where it is located. This increase is reflected in the basic shear strength calculations discussed in the next section.

9.6.3 The basic strength

The ultimate shear strength of a slab that is not transferring bending moment to the column is referred to as the basic punching shear strength or V_{uo}. It is unlikely that such a pure punching shear condition exists in reality at the ultimate state. But the value of V_{uo} is an important parameter for computing the ultimate shear strength of the slab.

For a slab without a shear head

$$V_{uo} = u d_{om} f_{cv} \qquad \text{Equation 9.6(3)}$$

where d_{om} is the mean value of the shear effective depth (d_o) for the slab – see Figure 6.3(1) – around the critical shear perimeter (u), in which

$$u = 2a + a_2 \qquad \text{Equation 9.6(4)}$$

for an edge column position and

$$u = 2(a + a_2) \qquad \text{Equation 9.6(5)}$$

for an interior column location. Note that, for other column locations, the formulas for u are given in Appendix B. In Equation 9.6(3)

$$f_{cv} = 0.17(1 + 2/\beta_h)\sqrt{f_c'} \leq 0.34\sqrt{f_c'} \qquad \text{Equation 9.6(6)}$$

in which

$$\beta_h = D^*/b^*$$ Equation 9.6(7)

noting that $D^* \geq b^*$ – see Figure 9.6(1). Otherwise interchange the symbols. For a slab with a shear head

$$V_{uo} = 0.5ud_{om}\sqrt{f'_c} \leq 0.2ud_{om}f'_c$$ Equation 9.6(8)

9.6.4 The ultimate strength

The ultimate punching shear strength (V_u) of a slab is given as

$$V_u = \frac{V_{uo}}{1 + \frac{uM_v^*}{8N_z^* ad_{om}}}$$ Equation 9.6(9)

It is apparent in Equation 9.6(9) that in addition to the basic shear strength and the geometry of the slab–column connection, the slab shear strength is influenced by the net column axial force and the unbalanced slab moment.

For a given design, if

$$\phi V_u \geq N_z^*$$ Equation 9.6(10)

where $\phi = 0.7$, then the punching shear strength of the slab is adequate. Otherwise, torsion strip or spandrel beam has to be provided. Alternatively, the slab thickness has to be increased. The following section shows how to determine the new d_{om} for the slab. Sections 9.6.6 and 9.6.7 discuss the design of torsion strips and spandrel beams, respectively.

9.6.5 Minimum effective slab thickness

If a given slab is found to be inadequate to resist the punching shear at the column-slab connection, the value of d_{om} may be increased such that

$$\phi V_u = N_z^*$$ Equation 9.6(11)

Substituting Equation 9.6(9) into the above equation gives

$$\frac{N_z^*}{\phi} = \frac{V_{uo}}{1 + \frac{uM_v^*}{8N_z^* a d_{om}}}$$ Equation 9.6(12)

Equations 9.6(3) and 9.6(8) show that V_{uo} is a function of d_{om} and can be given in the following general form

$$V_{uo} = \zeta d_{om}$$ Equation 9.6(13)

where for slab without a shear head

$$\zeta = uf_{cv}$$ Equation 9.6(14)

in which f_{cv} is given in Equation 9.6(6). For slabs with shear heads use

$$\zeta = 0.5u\sqrt{f'_c}$$ Equation 9.6(15)

Substituting Equation 9.6(13) into Equation 9.6(12) leads to

$$\frac{N_z^*}{\phi} = \frac{\zeta d_{om}}{1 + \dfrac{u M_v^*}{8N_z^* a\, d_{om}}}$$ Equation 9.6(16)

Rearranging the terms in the equation yields

$$\left(8\phi N_z^* a\zeta\right) d_{om}^2 - \left(8N_z^{*2} a\right) d_{om} - u N_z^* M_v^* = 0$$

which may be rewritten as

$$\alpha' d_{om}^2 - \beta' d_{om} - \gamma' = 0$$ Equation 9.6(17)

where

$$\alpha' = 8\phi N_z^* a\zeta$$ Equation 9.6(18)

$$\beta' = 8N_z^{*2} a$$ Equation 9.6(19)

and

$$\gamma' = u N_z^* M_v^*$$ Equation 9.6(20)

in which a is a length of the critical shear perimeter as shown in Figure 9.6(1). Solving Equation 9.6(17) yields the minimum d_{om} required to resist the punching shear or

$$d_{om.min} = \frac{\beta' + \sqrt{\beta'^2 + 4\alpha'\gamma'}}{2\alpha'}$$ Equation 9.6(21)

9.6.6 Design of torsion strips

A torsion strip is a width of the slab running in the direction of the axis of bending for M_v^*. For the case in Figure 9.6(1), the torsion strip runs in the y-direction. The width of the strip should be as wide as possible, but not greater than a, which is the length of the critical shear perimeter in the x-direction (i.e. the slab bending direction).

For a torsion strip to be effective, Clause 9.2.5 states that the minimum torsional shear reinforcement shall be

$$\left(\frac{A_{sw}}{s}\right)_{min} = \frac{0.2 y_1}{f_{sy.f}}$$

Equation 9.6(22)

where A_{sw} is the area of the reinforcing bar making up the closed tie, s is the spacing of the ties, $f_{sy.f}$ is the yield strength of the bar and

$$y_1 = a - \text{minimum concrete cover} - \frac{d_{bar}}{2}$$

Equation 9.6(23)

where a is defined in Figure 9.6(1).

Figure 9.6(2) shows that z_1 is equal to the slab thickness, less the top and bottom covers to the bar centre. For values of clear cover see Section 1.4.1. Further, for Equation 9.6(22) to be valid, $y_1 > z_1$, which is always the case for torsion strips because a is always greater than the slab thickness (D_s) as evident in Figure 9.6(1). However, for spandrel beams y_1 can be smaller than z_1, depending on the width and depth ratios of the beam (see Section 9.6.7).

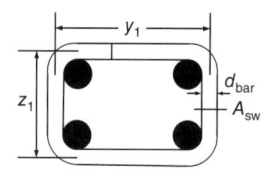

Figure 9.6(2) Suggested dimensions of a closed tie

For a torsion strip reinforced with the minimum torsional reinforcement as given in Equation 9.6(22), the corresponding ultimate punching shear strength as per Clause 9.2.4(b) is taken as

$$V_{u.min} = \frac{1.2 V_{uo}}{1 + \frac{u M_v^*}{2 N_z^* a^2}}$$

Equation 9.6(24)

If for a given column-and-slab connection, $\phi V_{u.min} \geq N_z^*$, then the use of $(A_{sw}/s)_{min}$ is adequate. On the other hand, if $\phi V_{u.min} < N_z^*$, the required torsional reinforcement according to Clause 9.2.4(d) is computed as

$$\left(\frac{A_{sw}}{s}\right) = \frac{0.2 y_1}{f_{sy.f}} \left(\frac{N_z^*}{\phi V_{u.min}}\right)^2$$

Equation 9.6(25)

Note that, for all cases, Clause 9.2.4(d) also stipulates that

$$N_z^* \leq 3\phi V_{u.min} \sqrt{D_s/a}$$

Equation 9.6(26)

If this equation cannot be satisfied, increase the slab thickness accordingly.

Following the design process for shear and torsional reinforcement given in Chapters 6 and 7, we can decide on which bar size to use (i.e. A_{sw}), and then compute the spacing (s), or vice versa. The detailing of reinforcement is discussed in Section 9.6.8.

9.6.7 Design of spandrel beams

For edge columns where the punching shear is heavy enough to require torsional reinforcement, it is more effective to provide spandrel beams along the edge of the slab. The use of spandrel beams also increases the rigidity of the multistorey system in resisting horizontal forces. Normally, the dimensions of a spandrel beam are first chosen by the designer. For a section with width b_w and depth D_b, where $D_b > b_w$ as illustrated in Figure 9.6(3), Clause 9.2.5 specifies that the minimum effective torsional reinforcement is

$$\left(\frac{A_{sw}}{s}\right)_{min} = \frac{0.2 y_1}{f_{sy.f}} \qquad \text{Equation 9.6(27)}$$

where y_1 is the larger of the two centre-to-centre dimensions of the closed tie. As shown in Figure 9.6(3), y_1 is D_b, less the top and bottom covers to the centres of the bar. In case of $b_w > D_b$, then y_1 is b_w less the covers, or z_1.

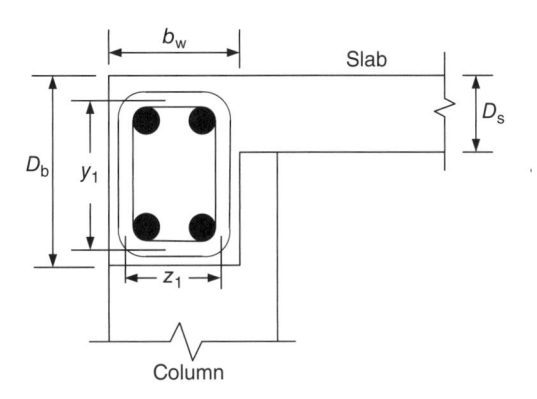

Figure 9.6(3) A spandrel beam section with width (b_w) and depth (D_b), where $D_b > b_w$

With the reinforcement prescribed in Equation 9.6(27), the ultimate punching shear strength ($V_{u.min}$) as per Clause 9.2.4(c) is given as

$$V_{u.min} = \frac{1.2 V_{uo}(D_b/D_s)}{1 + \dfrac{u M_v^*}{2 N_z^* a\, b_w}} \qquad \text{Equation 9.6(28)}$$

where M_v^* and N_z^* are the design moment and column axial force, respectively (see Section 9.6.1), and V_{uo} and u are as defined in Section 9.6.3.

For a given slab–column connection, if

$$\phi V_{u.min} > N_z^* \qquad \text{Equation 9.6(29)}$$

where $\phi = 0.7$, then the design is adequate against punching shear failure. The beam requires only the minimum amount of steel as specified in Equation 9.6(27). If Equation 9.6(29) cannot be satisfied, the required amount of torsional steel is

$$\left(\frac{A_{sw}}{s}\right)_{min} = \frac{0.2y_1}{f_{sy.f}} \left(\frac{N_z^*}{\phi V_{u.min}}\right)^2 \qquad \text{Equation 9.6(30)}$$

where y_1 is defined in Figure 9.6(3) and $V_{u.min}$ is given in Equation 9.6(28). Note that in all cases

$$N_z^* \leq 3\phi V_{u.min} \sqrt{X/Y} \qquad \text{Equation 9.6(31)}$$

where $X = b_w$, and $Y = D_b$, if $b_w < D_b$, – otherwise interchange the values. Revise the spandrel beam dimensions if Equation 9.6(31) cannot be satisfied.

With Equation 9.6(30) in hand, we can either specify A_{sw} and compute s, or vice versa subject to the requirements specified in Section 9.6.8.

9.6.8 Detailing of reinforcement

The dimensions of a closed tie are given in Figures 9.6(2) and 9.6(3) for a torsion strip and a spandrel beam, respectively. Clause 9.2.6 specifies the following requirements:
- The ties have to be provided along the torsion strip or spandrel beam – the y-direction in Figure 9.6(1) – for a distance of not less than $L_t/4$ from the face of the column on one or both sides of the column, as applicable. For an example, see Figure 9.5(1). The first tie is to be located at $0.5s$ or less from the face of the column.
- For torsion strips,

$$s \leq \text{the greater of 300 mm and } D_s \qquad \text{Equation 9.6(32)}$$

and for spandrel beams,

$$s \leq \text{the greater of 300 mm and } D_b \qquad \text{Equation 9.6(33)}$$

- At least one longitudinal bar has to be provided at each corner of the tie.

9.6.9 Summary

The design for punching shear may be summarised in the following steps:
1. Compute V_u using Equation 9.6(9).
2. If $\phi V_u \geq N_z^*$, the slab–column connection is adequate against punching shear. Otherwise,
   - increase the value of d_{om} (Section 9.6.5),
   - provide a torsion strip (Section 9.6.6) or
   - provide a spandrel beam (Section 9.6.7).

3. For cases requiring a torsion strip, ensure that Equation 9.6(26) prevails. If not, increase the slab thickness. For cases with a spandrel, Equation 9.6(31) must be satisfied. If it is not, then enlarge the beam dimensions.

Also note that the designer should be aware of the shortcomings of this Standard design procedure, as discussed in Section 9.6.11.

9.6.10 Illustrative example

Problem

An interior idealised frame is illustrated in Figure 9.6(4)a with the details of connection A shown in Figure 9.6(4)b. Given $M_v^* = 120$ kNm, $N_z^* = 1500$ kN, $D_s = 350$ mm and $d = 300$ mm. Is the connection adequate in resisting the punching shear? Design a torsion strip as necessary. Assume $f'_c = 25$ MPa and that a shear head is provided. Use N12 bars for A_{sw}.

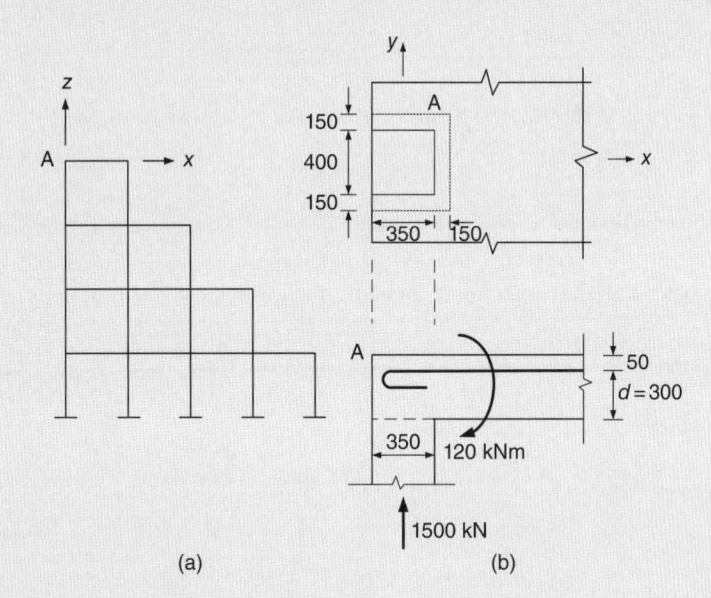

Figure 9.6(4) An interior idealised frame with the details of its connection at A

Note: all dimensions are in mm.

Solution

Punching shear strengths

Equation 9.6(4): $u = 2 \times (350 + 150) + (400 + 150 + 150) = 1700$ mm

Equation 9.6(8): $V_{uo} = 0.5 \times 1700 \times 300 \times \sqrt{25} \times 10^{-3} = 1275$ kN

Equation 9.6(9): $V_u = \dfrac{1275}{1 + \dfrac{1700 \times 120 \times 10^6}{8 \times 1500 \times 10^3 \times 500 \times 300}} = 1145$ kN

But

$\phi V_u = 0.7 \times 1145 = 802$ kN $< N_z^* = 1500$ kN

Thus, the slab–column connection has insufficient punching shear strength.

Torsion strip

Equation 9.6(23): $y_1 = 500 - 25 - 6 = 469$ mm

Equation 9.6(22): $\left(\dfrac{A_{sw}}{s}\right)_{min} = \dfrac{0.2 \times 469}{500} = 0.1876$ mm^2/mm

Equation 9.6(24): $V_{u.min} = \dfrac{1.2 \times 1275}{1 + \dfrac{1700 \times 120 \times 10^6}{2 \times 1500 \times 10^3 \times 500^2}} = 1202.83$ kN

But

$\phi V_{u.min} = 0.7 \times 1202.83 = 841.98$ kN $< N_z^* = 1500$ kN

Thus, $(A_{sw}/s)_{min}$ is inadequate and the required amount of torsional steel, according to Equation 9.6(25), is

$\left(\dfrac{A_{sw}}{s}\right) = \dfrac{0.2 \times 469}{500}\left(\dfrac{1500}{841.98}\right)^2 = 0.5954$

Equation 9.6(26): the right-hand side $= 3 \times 841.98 \times \sqrt{350/500} = 2113.35$ kN
$> N_z^* = 1500$ kN

This is acceptable.

Detailing

Using N12 closed ties, Table 2.3(1) gives $A_{sw} = 113$ mm^2. Thus, the spacing is

$s = \dfrac{113}{0.5954} = 189.8$ mm

We can now take $s = 180$ mm, and from Equation 9.6(32) we see that s is less than the greater of 300 mm and $D_s = 350$ mm. This is acceptable. The layout of the torsion strip is given in Figure 9.6(5).

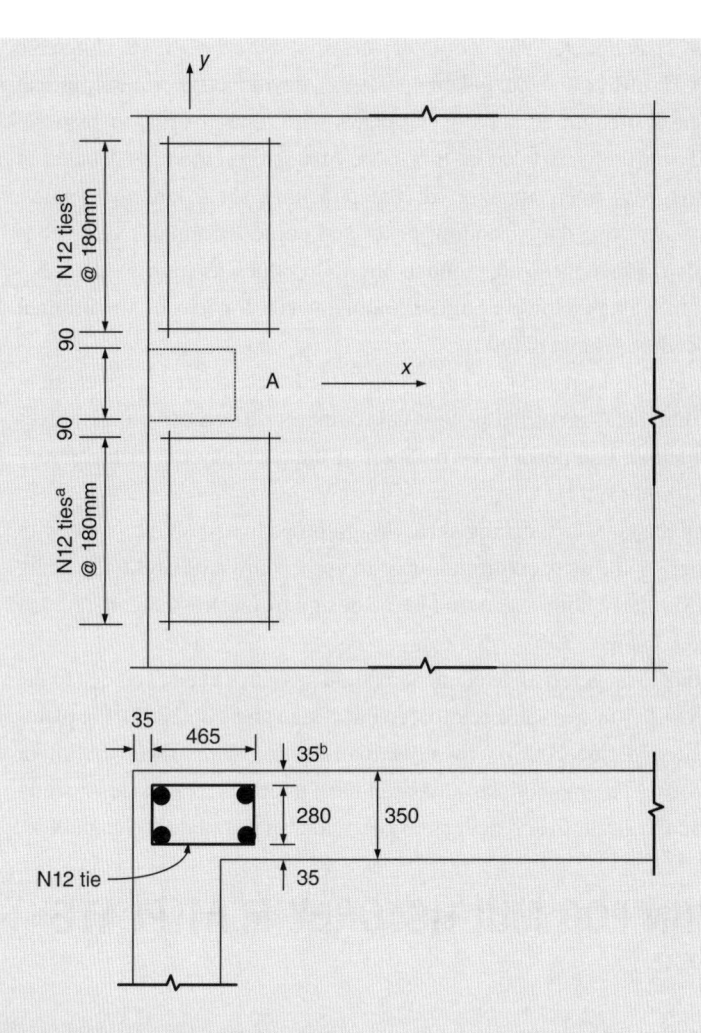

Figure 9.6(5) Reinforcement details of the torsion strip
a Over a distance not less than a quarter of the respective centre to centre span of the next column
b The cover of 35 mm is inadequate for B1 exposure classification.
Note: all dimensions are in mm.

9.6.11 Semi-empirical approach and layered finite element method

Punching shear failure of flat plates at edge and column locations is a complex problem, particularly where the slab edges are strengthened by torsion strips or spandrel beams. The empirical formulas recommended in the Standard have rendered this complex problem amenable to analysis and design by manual calculations. This is a significant achievement, although the shortcomings of this simplified approach ought not to be overlooked.

Falamaki and Loo (1992) conducted the laboratory tests of a series of 18 half-scale models. Based on the test results, Loo and Falamaki (1992) developed a semi-empirical method for the punching shear strength analysis of flat plates with torsion strips or spandrel beams. A comparative study has indicated that, at times, the Standard method (as detailed in Section 9.6) could lead to gross overestimation of the punching shear strength values, especially for connections where the slab, spandrel, beam and corner columns meet. It has been shown that the Standard procedure can produce strength values as high as 2.75–3.5 times that of the experimental values. However, these findings were doubted by the original author of the Standard procedure (Rangan 1993).

This doubt was not lifted until Guan and Loo (1997) published their layered finite element method of cracking and failure analysis of flat plates. This computer-based, nonlinear, three-dimensional analysis provided an independent verification of the accuracy of the laboratory test results of Falamaki and Loo (1992).

In view of the above, the Standard's punching shear design procedure should be used with care. If in doubt, the designer should seek confirmation of the procedure's reliability by using some alternative methods. Similarly, this is also true in the design of prestressed concrete flat plates as found by Loo and Chiang (1996).

It may be worth noting that the layered finite element method has since been extended to the failure analysis of flat plates with studded shear reinforcement (Guan and Loo 2003), flat plates with openings (Garda, Guan and Loo 2006), as well as load-bearing walls with or without openings (Guan and Doh 2007). The method has also been used to conduct an in-depth study of the effects of openings on the strength and deflection behaviour of flat plates (Guan 2009).

9.7 SLAB DESIGN FOR MULTISTOREY FLAT PLATE STRUCTURES

The main purpose of this section is to illustrate the complete design process for a slab panel in a multistorey flat plate system, including serviceability considerations. However, it excludes punching shear checks, for which the reader should refer to Section 9.6.

9.7.1 Details and idealisation of a three-storey building

The centre-line dimensions of a three-storey car park building of flat plate construction are given in Figure 9.7(1). The three-dimensional structure may be divided into two series of idealised frames, one in the x–z plane and another in the y–z plane. Figures 9.7(1)b and c show the first interior idealised frames in these two vertical planes, respectively.

As an illustrative example, the complete design of slab I of the three-bay frame in the x–z plane as shown in Figure 9.7(1)c is considered here. The dimensions of the member cross-sections are given below:

- column a–a: 350 mm × 350 mm
- column b–b: 400 mm × 400 mm

- column c–c: 450 mm × 450 mm
- slab d–d: 400 mm thick (all sections).

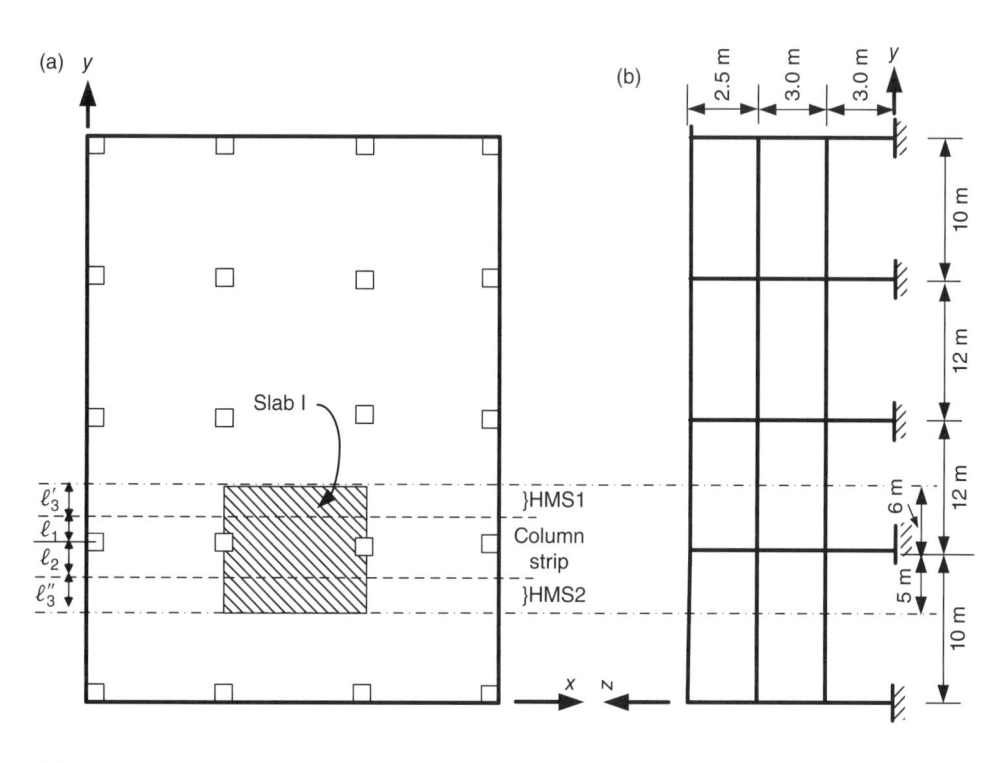

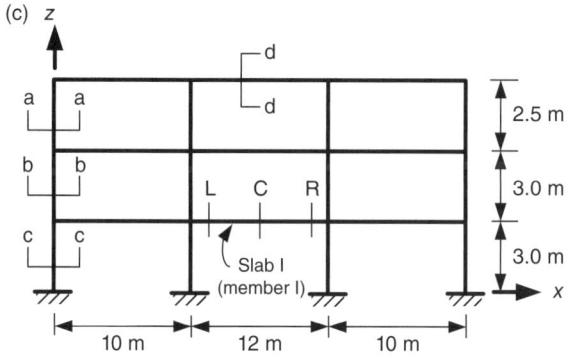

Figure 9.7(1) Centre-line dimensions of a three-storey car park building of flat plate construction: **(a)** plan; **(b)** side elevation (y–z plane); and **(c)** elevation (x–z plane)

HMS = half middle strip

9.7.2 Loading details

In addition to dead load, the design is required to carry the live load (for the car park) and wind load. Details are as follows.

For dead load (*g*):
- self-weight
- superimposed dead load of 3 kPa.

For live load (for car park as per AS/NZS 1170.1–2002 *Structural Design Actions Part 1: Permanent, Imposed and Other Actions*):

q_1: all panels, 5 kPa

q_2: alternate panels, 5 kPa

q_3: adjacent panels, 5 kPa.

And for ultimate wind load (assumed):

w_{u1} = left wind, 5 kPa

w_{u2} = right wind, 5 kPa

w_{u3} = ballooning effect, 5 kPa.

For self-weight, assume a total of 1.25% steel by volume in the slab. Then according to Equation 2.4(1), $\rho_w = 24 + 0.6 \times 1.25 = 24.75$ kN/m³ and the dead load on the slab is

$$g_s = 24.75 \times 0.4 = 9.90 \text{ kPa}$$

For the columns, assuming 1% by volume of steel content, the self-weight is

$$\text{Equation 2.4(1): } g_c = 24 + 0.6 \times 1 = 24.6 \text{ kN/m}^3$$

The directions and other details of the live loads and wind loads are illustrated in Figures 9.7(2) and 9.7(3), respectively. Note that the thickness of the idealised frame or the width of the design strip is 5 + 6 = 11 m.

9.7.3 Load combinations

In addition to dead load (*g*), three types of live load and three types of wind load are given. For a complete loading consideration, the application of Equations 1.3(2), (4) and (5) leads to 15 load combinations. These are detailed in Table 9.7(1).

9.7.4 Material and other specifications

For this example, the following specifications are assumed:
- $f'_c = 25$ MPa
- $E_c = 26\,700$ MPa, from Table 2.2(1) for $f'_c = 25$ MPa
- Exposure classification B1 applies.
- Fire-resistance period = 90 minutes
- Use N bars only.
- For each span, design the midspan section as well as the left and right support sections.

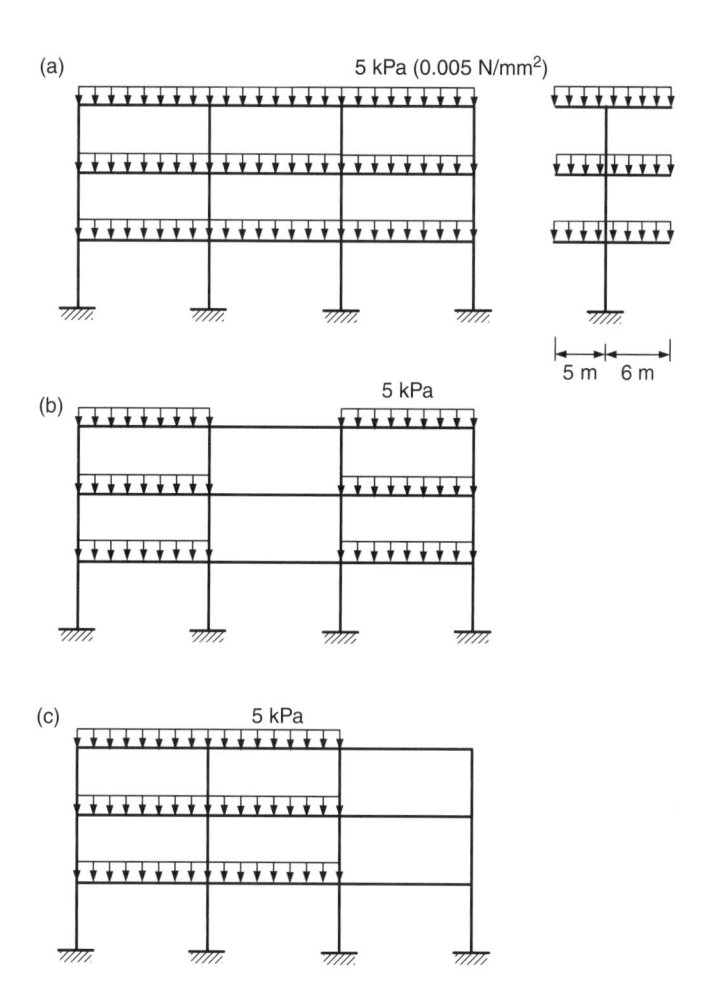

Figure 9.7(2) Live-load cases: **(a)** all panels loaded; **(b)** alternate panels loaded; and **(c)** adjacent panels loaded

9.7.5 Structural analysis and moment envelopes

The three-storey, three-bay idealised frame (Figure 9.7(2)) is a highly statically indeterminate structure. This, together with the multiple load cases to be analysed, makes it virtually impossible to carry out the structural analysis manually. Instead, the structural analysis module (ANALY2) of the personal computer flat plate design package (Loo 1988) was used to analyse the idealised frame under the 15 loading cases as specified in Table 9.7(1).

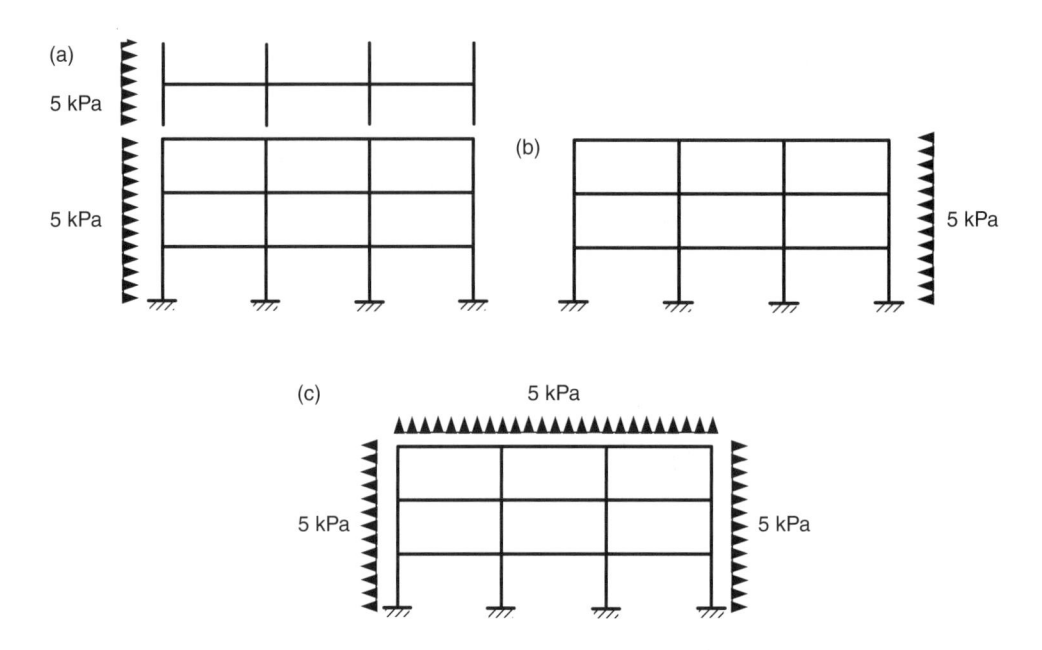

Figure 9.7(3) Wind load cases: **(a)** left wind; **(b)** right wind; and **(c)** ballooning

Table 9.7(1) Load combinations for dead load, live load and wind load

Load combination	Combination[a]	Reference
1	$1.2g + 1.5q_1$	
2	$1.2g + 1.5q_2$	Equation 1.3(2)
3	$1.2g + 1.5q_3$	
4	$0.9g + w_{u1}$	
5	$0.9g + w_{u2}$	Equation 1.3(5)
6	$0.9g + w_{u3}$	
7	$1.2g + w_{u1} + 0.4q_1$	
8	$1.2g + w_{u2} + 0.4q_1$	Equation 1.3(4)
9	$1.2g + w_{u3} + 0.4q_1$	
10	$1.2g + w_{u1} + 0.4q_2$	
11	$1.2g + w_{u2} + 0.4q_2$	Equation 1.3(4)
12	$1.2g + w_{u3} + 0.4q_2$	
13	$1.2g + w_{u1} + 0.4q_3$	
14	$1.2g + w_{u2} + 0.4q_3$	Equation 1.3(4)
15	$1.2g + w_{u3} + 0.4q_3$	

a   As per Table 1.3(1), $\psi_c = 0.4$ for parking area

For the bending design of the slabs, it is necessary to obtain the moment envelope for each of the spans. Table 9.7(2) tabulates the total moments at sections L, C and R – see Figure 9.7(1)c – of span I under the 15 different load combinations as specified in Table 9.7(1). A search through the three columns of moment values leads to the conclusion that the design (total) moments for the three sections are:

- $M^* = -3053$ kNm for section L
- $M^* = 1749$ kNm for section C
- $M^* = -2973$ kNm for section R.

Graphically, they are presented in Figure 9.7(4).

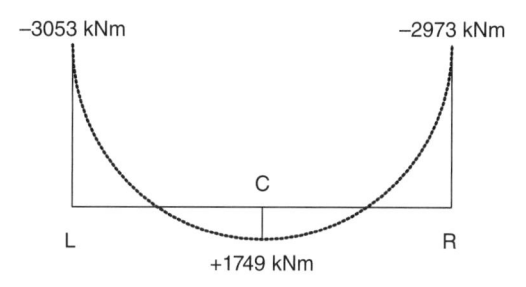

Figure 9.7(4) Design moments for sections L, C and R of span I

Table 9.7(2) Moment envelope – span I

Load combination	M^* (kNm) at section L	M^* (kNm) at section C	M^* (kNm) at section R
1	−2973	1628	**−2973**
2	−2247	855	−2247
3	**−3053**	**1749**	−2651
4	−2496	1451	−2806
5	−2269	1210	−2580
6	−2521	1489	−2706
7	−1150	702	−1415
8	−2395	1238	−2130
9	−2202	1032	−1937
10	−2417	1270	−2045
11	−1415	702	−1150
12	−1288	697	−1288
13	−2268	1233	−2268
14	−2074	1027	−2074
15	−2289	1265	−2182

9.7.6 Design strips and design moments

The total moment for each section obtained from the structural analysis of the idealised frame is over the thickness of the frame or the entire width of the design strip. As shown in Figure 9.7(1) the width is 11 m. The moment given in Figure 9.7(4) for each of the sections L, C and R is to be distributed to the column strip and the two half middle strips according to the distribution factors given in Table 9.5(1).

The various strip widths are computed as described as follows.

Design strips
Column strip

Equation 9.4(1): $l_1 = L_{ti}/4 = 12/4 = 3$ m

Equation 9.4(2): $l_2 = L_{tj}/4 = 10/4 = 2.5$ m

Thus, the column strip width

$$l_1 + l_2 = 5.5 \text{ m} < L_i/2 = 12/2 = 6 \text{ m}$$

Use $l_1 + l_2 = 5.5$ m

Half middle strip 1 (HMS1) – north of column

$$l'_3 = L_{ti}/2 - l_1 = 12/2 - 3 = 3 \text{ m}$$

Half middle strip 2 (HMS2) – south of column

$$l''_3 = L_{tj}/2 - l_2 = 10/2 - 2.5 = 2.5 \text{ m}$$

The sum of all the strip widths is

$$l_1 + l_2 + l'_3 + l''_3 = 3 + 2.5 + 3 + 2.5 = 11 \text{ m}$$

which is the total width of the slab.

The total moments to be distributed at sections L, C and R are illustrated in Figure 9.7(4). Following Table 9.5(1) for distribution factors and taking the midrange values we have:
- for the column strip 0.8 and 0.6 for the negative and positive moments, respectively
- for the half middle strips, the respective factors are 0.1 and 0.2.

We can now proceed to compute the design moments for each of the strips.

Design moments
Column strip
For the column strip (with $b = 5500$ mm, $D = 400$ mm):

> Section L: $M^* = -3053 \times 0.8 = -2442.4$ kNm
> C: $M^* = 1749 \times 0.6 = 1049.4$ kNm
> R: $M^* = -2973 \times 0.8 = -2378.4$ kNm $= -2442.4$ kNm (for symmetry)

Half middle strip 1 (HMS1) – north of column
For the half middle strip 1 (with $b = 3000$ mm, $D = 400$ mm):

> Section L: $M^* = -3053 \times 0.1 = -305.3$ kNm
> C: $M^* = 1749 \times 0.2 = 349.8$ kNm
> R: $M^* = -2973 \times 0.1 = -297.3$ kNm $= -305.3$ kNm (for symmetry)

Half middle strip 2 (HMS2) – south of column
For the half middle strip 2 (with $b = 2500$ mm, $D = 400$ mm):

> Section L: $M^* = -305.3$ kNm
> C: $M^* = 349.8$ kNm
> R: $M^* = -297.3$ kNm $= -305.3$ kNm (for symmetry)

9.7.7 Design of column and middle strips

The overall thickness of the slab has been given as 400 mm. Each of the column and middle strips can be designed as a wide beam using the restricted design process discussed in Section 3.5.2. Two equations need to be used repeatedly. They are:

$$\xi = \frac{\alpha_2 f'_c}{f_{sy}} \qquad \text{Equation 3.5(7)}$$

and

$$p_t = \xi - \sqrt{\xi^2 - \frac{2\xi M^*}{\phi b d^2 f_{sy}}} \qquad \text{Equation 3.5(6)}$$

As per Equation 9.4(13)

$$p_{t.min} \geq \frac{0.24(D/d)^2 f'_{ct.f}}{f_{sy}} \qquad \text{Equation 9.7(1)}$$

For exposure classification B1 with $f'_c = 25$ MPa, Table 1.4(2) indicates that a clear cover of 60 mm is required. This is greater than the 25 mm cover required for the 90-minute fire-resistance period as given in Table 5.4.2(B) of the Standard. Thus, using N28 bars for both the positive and negative steel, or the bottom and top reinforcement,

$$d = 400 - 60 - 28/2 = 326 \text{ mm}$$

Note that with $d = 326$ mm for bending in the x–z plane, the corresponding d for the y–z plane has to be $d = 326 - 28 = 298$ mm. These are illustrated in Figure 9.7(5).

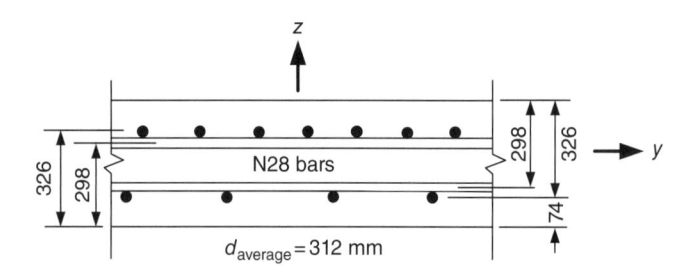

Figure 9.7(5) Definitions of d for the top and bottom reinforcement layers in the slab

Note: all dimensions are in mm.

In practice, it is economical to adopt the larger d for the plane having a larger moment in a majority of the sections. Alternatively, it is acceptable to use the average d value (312 mm in this case). In the present example, let us assume that the bending in the x–z plane is more severe. Thus, we adopt

$$d = 326 \text{ mm}$$

Note that this value is acceptable for fire-resistance considerations since Table 5.4.1 of the Standard indicates that, for a period of 90 minutes, the minimum effective or actual thickness (D) is 100 mm for insulation. For structural adequacy on the other hand, the minimum D required is 200 mm as per Table 5.4.2(B) of the Standard. Both values are smaller than the available $D = 400$ mm.

With d now calculated and using Equations 3.5(6) and 3.5(7), the bending reinforcement can readily be computed.

The column strip

$$(b = 5500 \text{ mm}, \ d = 326 \text{ mm})$$

Section L

$M^* = 2442.4$ kNm. Consider a 1-metre width of the column strip.

$$M^* = 2442.4/5.5 = 444.1 \text{ kNm/m}$$

For $f'_c = 25$ MPa, $\alpha_2 = 0.85$ and $\xi = \dfrac{0.85 \times 25}{500} = 0.0425$.

Assuming $\phi = 0.8$:

$$p_t = 0.0425 - \sqrt{0.0425^2 - \dfrac{2 \times 0.0425 \times 444.1 \times 10^6}{0.8 \times 1000 \times 326^2 \times 500}} = 0.012197$$

Equation 9.7(1): $p_{t.min} = 0.0022$
Equation 3.4(6): $p_{all} = 0.01445$
As $p_{t.min} < p_t$ and $p_t < p_{all}$, which is acceptable. Therefore,

$$p_t = 0.012197$$

and

$$A_{st} = p_t bd = 0.012197 \times 1000 \times 326 = 3976.22 \text{ mm}^2/\text{m}$$

Provide N28 @ 150 mm top bars ($A_{st} = 4107$ mm²/m).

Section C

$$M^* = 1049.4 \text{ kNm}$$

Consider a 1-metre width of the column strip.

$$M^* = 1049.4/5.5 = 190.8 \text{ kNm/m}$$
$$p_t = 0.00475 > p_{t.min} = 0.0022$$
$$p_t = 0.00475 < p_{all} = 0.01445$$

This is acceptable. Therefore,

$$p_t = 0.00475$$

and

$$A_{st} = 0.00475 \times 1000 \times 326 = 1548.5 \text{ mm}^2/\text{m}$$

Provide N28 @ 300 mm bottom bars ($A_{st} = 2053$ mm²/m). Also,
Equation 9.2(7): $s =$ the lesser of $2.0D$ and 300 mm $= 300$ mm
Thus, provide N28 @ 300 mm bottom bars ($A_{st} = 2053$ mm²/m).

Section R

The reinforcement is the same as Section L for reason of symmetry.

Therefore, provide N28 @ 150 mm top bars (A_{st} = 4107 mm²/m).>

Half middle strip 1 (HMS1)

$$(b = 3000 \text{ mm}, d = 326 \text{ mm})$$

Sections L and R

$$M^* = 305.3 \text{ kNm}$$

Consider a 1-metre width.

$$M^* = 305.3/3 = 101.77 \text{ kNm/m}$$
$$p_t = 0.00247 > p_{t.min} = 0.0022$$
$$p_t = 0.00247 < p_{all} = 0.01445$$

This is acceptable. Therefore,

$$p_t = 0.00247$$

and

$$A_{st} = 0.00247 \times 1000 \times 326 = 805.22 \text{ mm}^2/\text{m}$$

Provide N28 @ 300 mm top bars (A_{st} = 2053 mm²/m).

Section C

$$M^* = 349.8 \text{ kNm}$$

Consider a 1-metre width.

$$M^* = 349.8/3 = 116.6 \text{ kNm/m}$$
$$p_t = 0.00284 > p_{t.min} = 0.0022$$
$$p_t = 0.00284 < p_{all} = 0.01445$$

This is acceptable. Therefore,

$$p_t = 0.00284$$

and

$$A_{st} = 0.00284 \times 1000 \times 326 = 925.84 \text{ mm}^2/\text{m}$$

Provide N28 @ 300 mm bottom bars (A_{st} = 2053 mm²/m).

Half middle strip 2 (HMS2)

$$(b = 2500 \text{ mm}, d = 326 \text{ mm})$$

Sections L and R

$$M^* = 305.3 \text{ kNm}$$

Consider a 1-metre width.

$$M^* = 305.3/2.5 = 122.12 \text{ kNm/m}$$
$$p_t = 0.00298 > p_{t.min} = 0.0022$$
$$p_t = 0.00298 < p_{all} = 0.01445$$

This is acceptable. Therefore,

$$p_t = 0.00298$$

and

$$A_{st} = 0.00298 \times 1000 \times 326 = 971.48 \text{ mm}^2/\text{m}$$

Provide N28 @ 300 mm top bars (A_{st} = 2053 mm²/m).

Section C

$$M^* = 349.8 \text{ kNm}$$

Consider a 1-metre width.

$$M^* = 349.8/2.5 = 139.92 \text{ kNm/m}$$
$$p_t = 0.00343 > p_{t.min} = 0.0022$$
$$p_t = 0.00343 < p_{all} = 0.01445$$

This is acceptable. Therefore,

$$p_t = 0.00343$$

and

$$A_{st} = 0.00343 \times 1000 \times 326 = 1118.18 \text{ mm}^2/\text{m}$$

Provide N28 @ 300 mm bottom bars (A_{st} = 2053 mm²/m).

The design of the bending reinforcement is now completed. However, note that, in practice, it is acceptable to combine the total moments of the two adjacent half middle strips from the two adjoining idealised frames. Based on this combined total moment, the A_{st} for the middle strip can be computed accordingly.

It should also be pointed out that in all the bending design calculations, $\phi = 0.8$ is assumed. This, as in Equation 3.4(20)a, requires $k_{uo} \leq 0.36$. It may be shown that this criterion is satisfied throughout.

9.7.8 Serviceability check – total deflection

According to Equation 9.2(8), the minimum d, which is deemed to satisfy the serviceability requirement is given as:

$$d_{min} = \frac{L_{ef}}{k_3 k_4 \sqrt[3]{\frac{(\Delta/L_{ef})E_c}{F_{def}}}} \qquad \text{Equation 9.7(2)}$$

In our case:

L_{ef} = the lesser of L = 12 000 mm and $L_n + D$ = 12 000 – 450 + 400 = 11 950 mm.
Therefore L_{ef} = 11 950 mm
Δ/L_{ef} = 1/250 (for total deflection as in Table 5.2(4))
k_3 = 0.95 (for a two-way flat slab without drop panels – Equation 9.2(10))
k_4 = 2.1 (for interior span – Equation 9.2(14))
ψ_s = 0.7 and ψ_l = 0.4 (for parking area – Table 1.3(1))
$k_{cs} = 2 - 1.2 \dfrac{A_{sc}}{A_{st}} = 2$ (for $A_{sc} = 0$ – Equation 5.4(3))

$$F_{def} = (1.0 + k_{cs})g \text{ (including self-weight)} + (\psi_s + k_{cs}\psi_1)q$$
$$= (1.0 + 2) \times 12.90 + (0.7 + 2 \times 0.4) \times 5 = 46.2 \text{ kN/m}^2$$
$$= 46.2 \times 10^{-3} \text{ MPa (as per Equation 5.5(8))}$$

Thus, Equation 9.7(2) gives

$$d_{min} = 453.0 \text{ mm} > 326 \text{ mm, which is definitely not acceptable!}$$

Revision of the design is necessary and there are four immediate options:
1. Increase d (not advisable in this case since D = 400 mm is already heavy).
2. Use f'_c > 25 MPa (not effective as E_c increases proportionally to $\sqrt{f'_c}$).
3. Use drop panels in accordance with Figure 9.2(5); then k_3 = 1.05 and d_{min} = 409.9 > 326 mm. This is unacceptable for bending in the x–z plane and still inadequate in the y–z plane, where d = 298 mm. It is also an expensive option.
4. Use A_{sc}.

Use, say, $A_{sc} = A_{st}$; this means providing an equal number of top bars at all the centre sections (sections C) in the column and half middle strips. Thus,

$$k_{cs} = 2 - 1.2 = 0.8$$

Then F_{def} = 28.32 × 10^{-3} MPa, and the required effective depth according to Equation 9.7(2) is

$$d_{min} = 384.8 \text{ mm} > 326 \text{ mm}$$

Even after providing the maximum amount of permissible A_{sc}, the design does not meet the deflection requirement. Therefore, a detailed plate analysis should be conducted using a numerical approach, such as a finite element method. Following this, the design is to be revised as necessary.

9.7.9 Reinforcement detailing and layout

Since the column strip attracts a higher bending moment, Clause 9.1.2 of the Standard stipulates that at least 25% of the negative (top) reinforcement must be distributed over a distance $\overline{D}$ where

$$\overline{D} = b_{\text{column}} + 2D_s \qquad \text{Equation 9.7(3)}$$

In the present case, the total number of negative N28 bars is (37 + 10 + 9) = 56, 25% of which is 14 bars. $\overline{D}$ = 450 + 2 × 400 = 1250 mm. The required spacing s = 89.3 mm. Use, say 90 mm, therefore:

$$14 \text{ N28 @ 90 mm over } \overline{D} \text{ (top)}$$

The remaining width of the column strip is (5500 − 1250) = 4250 mm, which is required to accommodate (37 − 14) = 23 bars. That is, s = 4250/23 = 184.8 mm. For the portion north of the column, the number of bars is approximately

$$n = \left(3000 - \frac{1250}{2}\right)/184.8 = 12.85, \text{ say 13 bars, therefore 13 N28 @ 180 mm (top).}$$

For the portions south of the column, n = 23 − 13 = 10 bars and the spacing

$$s = \left(2500 - \frac{1250}{2}\right)/10 = 187.5 \text{ mm, say 185 mm, therefore 10 N28 @ 185 mm (top).}$$

A sketch of the reinforcement layout incorporating all the design results is presented in Figure 9.7(6). At the border between this panel and the adjoining ones, the final reinforcement arrangements also depend on the outcome of the design for those adjoining panels.

9.7.10 Comments

This section presents the bending design process for the slab panel represented by member I of the three-storey flat plate structure (Figure 9.7(1)). This procedure will be similar for all the other panels in the idealised frame, and for other idealised frames parallel to the x–z plane and parallel to the y–z plane.

The design for punching shear is not included in the present example. Details of the checking and design process may be found in Section 9.6.

It should by now be obvious that considerable manual effort is required to design just one slab panel out of a total of nine, as illustrated in Figure 9.7(1). This does not even include the punching shear and column design. To do all this work manually is quite uninviting. In an attempt to automate the entire design process, a personal computer-based integrated design package called UNIDES-FP was developed (see Loo 1988).

It should be emphasised that this example is for illustrative purposes. The 10-metre endspans and 12-metre interior spans are considered to be large for reinforced concrete flat

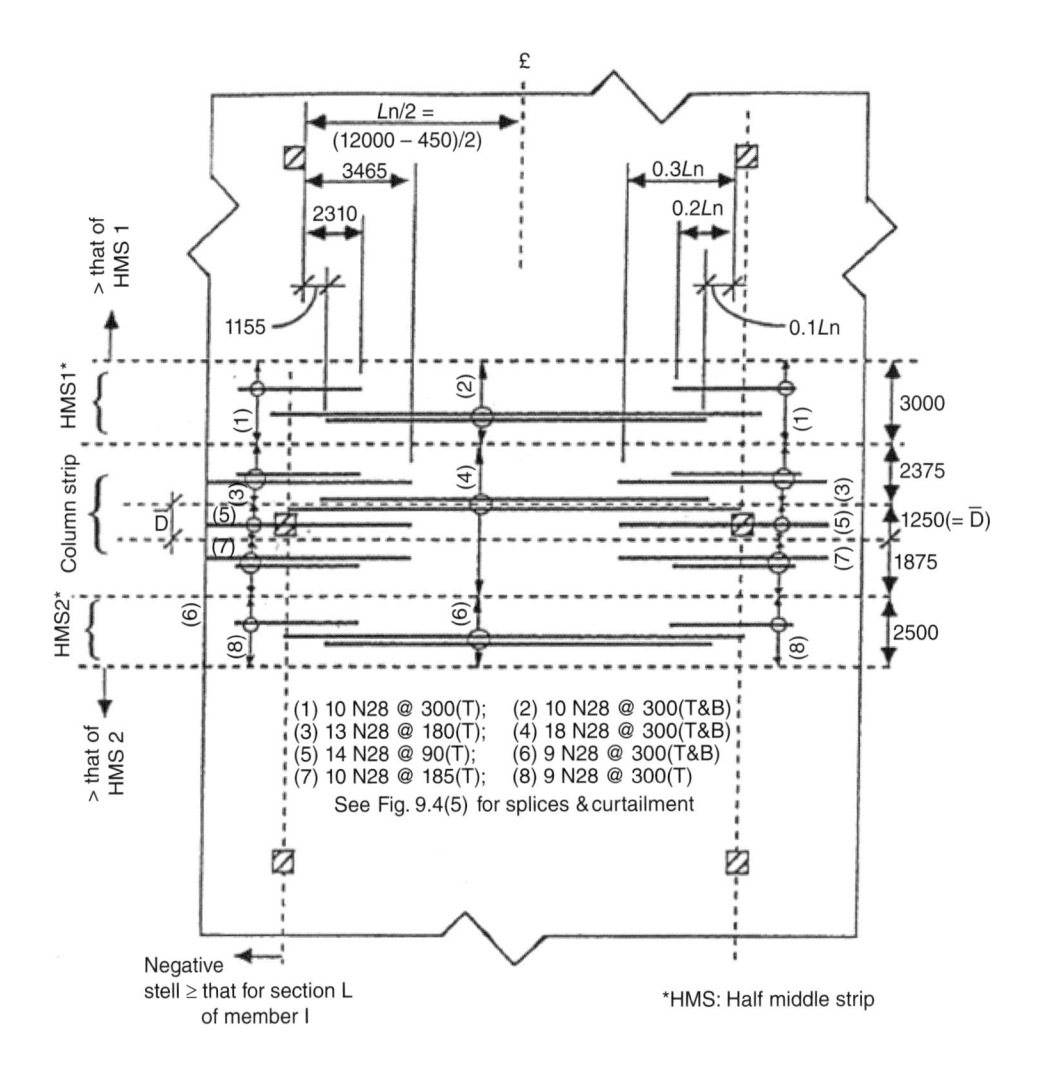

Figure 9.7(6) Reinforcement layout for the design slab

Note: in practice, separate drawings should be provided for top and bottom reinforcement, each in x and y directions. All dimensions are in mm.

plate construction. The 400 mm thick slab is no doubt heavy and thus uneconomical. In practice, it is better to use more spans that are shorter or adopt a different structural system.

9.8 SUMMARY

Single and multispan concrete building slabs can be categorised as a one-way or two-way system. In accordance with relevant Standard recommendations, the analysis and design procedures are developed for these systems covering various support conditions. Both

CHAPTER 9 SLABS 251

strength and serviceability are considered in these procedures. Their applications are demonstrated using worked examples.

The punching shear of column-and-slab structures is discussed including the design of torsion strips and spandrel beams. For the structural analysis of multistorey multibay buildings, the 'idealised frame' approach as provided by the Standard is explained. As a capstone, the complete design of a typical slab panel in a three-storey flat plate system is presented.

9.9 PROBLEMS

1. Slab A in Figure 9.9(1) is 150 mm thick and is to carry a live load of 4 kPa. Compute the central region positive moments M_x^* and M_y^* as well as the negative moments, $\overline{M}_{x1}^*$ and $\overline{M}_{y1}^*$, at both the interior and exterior supports for the slab, namely those at AB, BC, CD and DA. Assume $L_1 = 4.2$ m, $L_3 = 3.0$ m and $\rho_w = 24$ kN/m³.

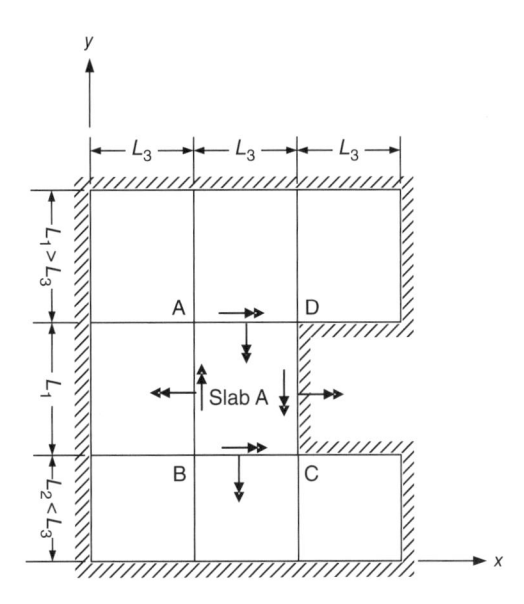

Figure 9.9(1) Support conditions and orientation of Slab A

2. The slab in a flat plate structure discussed in Section 9.6.10 has been found to be inadequate in punching shear.
   (a) What is the required minimum effective depth of the slab?
   (b) Design the torsional reinforcement for a spandrel beam with dimensions $b = 300$ mm and $D_b = 500$ mm.

3. Figure 9.9(2) shows the dimensional details of a column-and-slab connection of a flat plate structure. The structure is to be designed for 15 different ultimate loading conditions. The structural analysis results for the bending moment at Section L (M_L^*) and the top and bottom column forces (N_{zT}^* and N_{zB}^*, respectively) are presented in Table 9.9(1).

Check the adequacy of the design for punching shear and design, as necessary, the required A_{sw}/s for a torsion strip. You may provide a shear head only if it is necessary. Assume $f_c' = 25$ MPa and use N12 bars for the ties.

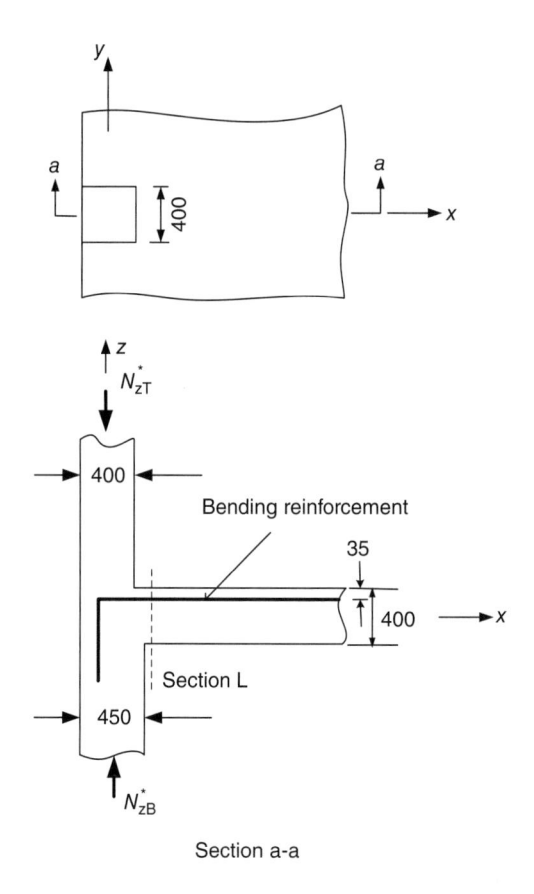

Figure 9.9(2) Dimensional details of a column-and-slab connection of a flat plate structure

Note: all dimensions are in mm.

Table 9.9(1) Bending moments at section L and the top and bottom column forces for various loading conditions

Loading condition	M^*_L (kNm)	N^*_{zT} (kN)	N^*_{zB} (kN)
1	−626	−2012	−3043
2	−685	−2116	−3198
3	−599	−1996	−3017
4	−201	−1672	−2489
5	−219	−1703	−2535
6	−193	−1668	−2482
7	28	−927	−1362
8	−845	−1720	−2642
9	−863	−1751	−2688
10	−839	−1715	−2634
11	−620	−990	−1539
12	−518	−1330	−2195
13	−535	−1360	−2240
14	−510	−1325	−2187
15	−291	−600	−1090

4. Repeat the example in Section 9.7 with the following new specifications:
   - f'_c = 32 MPa
   - the endspans in the x and y directions = 8 m (instead of 10 m)
   - the interior spans = 10 m (instead of 12 m).
5. A reinforced concrete slab for an assembly area with movable seats, spans between 200 mm-wide beams at 4.5-metre centres over five bays. The discontinuous ends of the slab are unrestrained. The superimposed dead load due to finishes and services is 2 kPa. Take f'_c = 20 MPa, exposure classification as A1 and use N12 bars for reinforcement.
   Design the slab as a one-way slab for the endspan and at the penultimate support. Take D = 200 mm and the effects of wind may be ignored.
6. A reinforced concrete flat plate floor has 450 mm square columns at 6-metre centres in both directions and supports a factored load of 12.5 kPa. For an interior bay, compute the positive and negative moments of the column strip and of the middle strip for an endspan, interior span and interior support.

254   PART 1   REINFORCED CONCRETE

7. The typical slab–column connection of a reinforced concrete flat plate floor is shown in Figure 9.9(3). The floor has 500 mm diameter columns at 6-metre centres in both directions and supports a factored load of 12.5 kPa. For an interior column, compute the minimum required mean value of shear effective depth for the slab (d_{om}) to resist punching shear if $f_c' = 25$ MPa.

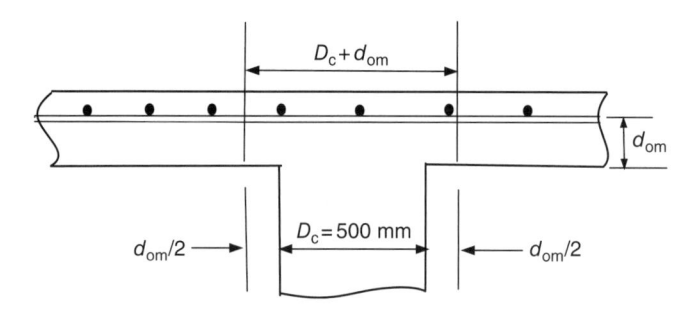

Figure 9.9(3) Details of a typical slab–column connection of the flat plate structure for Problem 7

8. The 450 mm square interior column of a flat plate floor supports a factored shear force $N_z^* = 440$ kN and an unbalanced moment $M_v^* = 50$ kNm. The mean value of shear effective depth for the slab (d_{om}) is 180 mm and $f_c' = 25$ MPa. Determine if the punching shear strength of the slab is adequate.

9. Design a 3.5-metre simply supported slab as part of a residential building to carry a uniformly distributed dead load (excluding self-weight) of 1.5 kPa and a uniformly distributed live load of 3 kPa. Take $f_c' = 20$ MPa, $D = 180$ mm and use N12 bars for the reinforcement. You may assume exposure classification A1 and ignore the effects of wind.

10. The cross-section of a continuous one-way solid slab in an office building is shown in Figure 9.9(4). The slabs are supported by beams that span 3.3 m between simple supports. Design the continuous slab and provide detailed drawings for the reinforcement layout. Take $f_c' = 25$ MPa, $D = 150$ mm, $q = 4$ kPa and the floor finish to be 0.5 kPa. Exposure classification A1 and fire resistance period of 60 minutes are assumed. The effects of wind may be ignored.

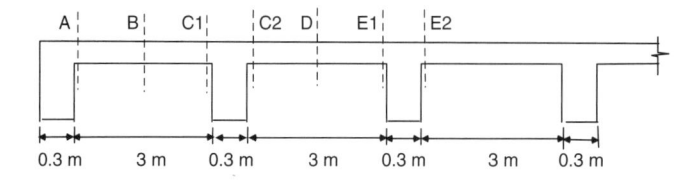

Figure 9.9(4) Cross-section of a continuous one-way solid slab for Problem 10

COLUMNS 10

10.1 INTRODUCTION

Columns exist in all conventional building structures. Whereas beams, slabs or even trusses may be used to span the floors, columns carry loads vertically, floor by floor, down to the foundations. Even in specialised systems such as shear-wall, shear-core and framed-tube structures, columns are used to support parts of the floor areas.

Figure 10.1(1)a shows a portion of a three-dimensional building frame. For the purposes of discussion on the role of columns, the frame may be taken as representative of other popular building systems, such as multistorey flat slabs, as well as beam/slab and column structures. At each level, the floor spans in both the x and z directions. As a result, bending occurs in both the x–y and y–z planes. Thus, for a typical column, AB, the forces acting at the top end, or joint A, include:

- N, the axial force equal to the portion of the vertical load (from the floor immediately above) to be carried by column AB plus the axial load transmitted by the column above (i.e. column CA)
- M_x, the bending moment about the z-axis
- M_z, the bending moment about the x-axis.

These are illustrated in Figure 10.1(1)b. Note that a similar set of end forces also exists at the bottom end (joint B).

These three-dimensional forces are statically indeterminate, and computer-based structural analysis procedures are normally relied on to determine their values. Because of the interaction between axial force and bending moments, the analysis and design of reinforced concrete columns, subjected to either uniaxial or biaxial bending, are considerably more complicated than the treatments for beams. In addition, the stability of slender columns and problems associated with column side-sway need to be considered. Consequently, analysis and design formulas for columns are nonlinear and the process is cumbersome. It is impossible to carry out a direct rigorous design of a reinforced concrete column. In practice, with the use of column interaction diagrams, a trial-and-error approach based on a preliminary column section can be used. A computer may also be programmed to do this rather tedious work.

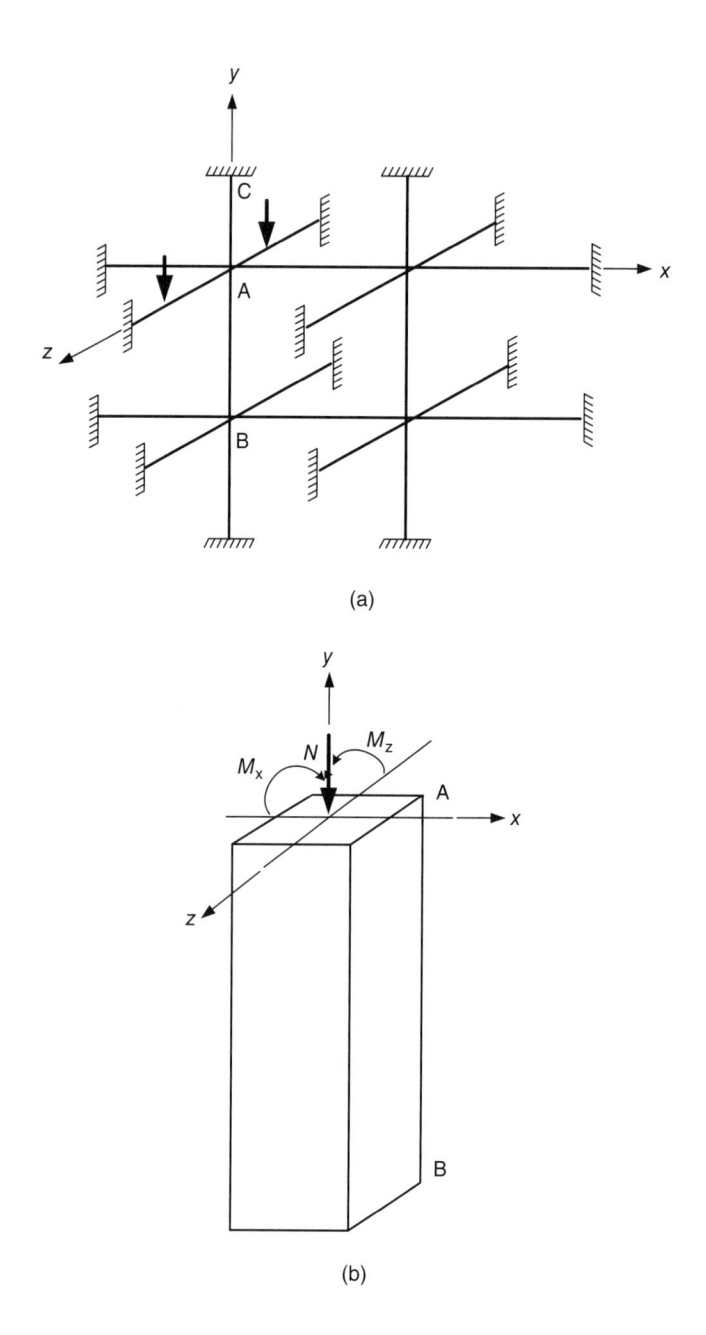

Figure 10.1(1) Forces acting on a column

In this chapter, the basic strength equations for centrally loaded columns and columns subjected to uniaxial bending are given in detail, together with the procedure for constructing the column interaction diagram. As advanced topics, two methods of analysis for arbitrary sections are included: the iterative approach and the semi-graphical procedure.

A preliminary design procedure is presented, along with the AS 3600–2009 recommendations on:
- the capacity reduction factor (ϕ)
- the moment magnifiers for slender columns, with or without side-sways
- the interaction of axial load and bending moments (M_x and M_z)
- reinforcement requirements.

Finally, an indirect but rigorous design process is summarised.

10.2 CENTRALLY LOADED COLUMNS

Figure 10.2(1) shows a centrally loaded column or one that is loaded at the 'plastic centre'. Note that the plastic centre coincides with the geometric centre only for symmetrical sections that are symmetrically reinforced with respect to the x and z axes.

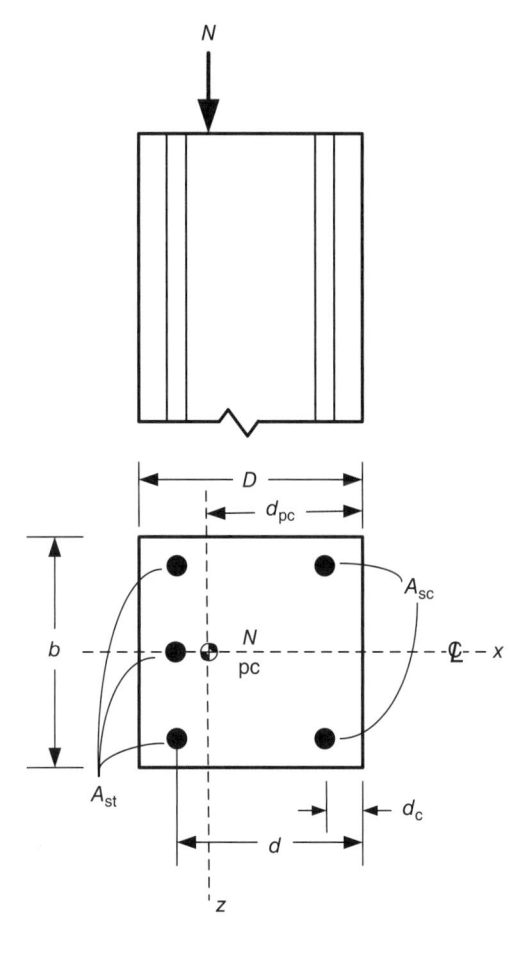

Figure 10.2(1) A centrally loaded column (loaded at the plastic centre [pc])

For the section illustrated, the ultimate strength or 'squash' load is

$$N_{uo} = \alpha_1 f'_c (A_g - A_s) + A_s f_{sy}$$ Equation 10.2(1)

where, as in Clause 10.6.2.2 of AS 3600–2009 (the Standard),

$$\alpha_1 = 1.0 - 0.003 f'_c \quad \text{but} \quad 0.72 \leq \alpha_1 \leq 0.85$$ Equation 10.2(1)a

A_g is the gross column section, and

$$A_s = A_{st} + A_{sc}$$ Equation 10.2(2)

noting that $\varepsilon_{cu} = 0.0025$ in this case.

The reliable load is

$$N' = \phi N_{uo}$$ Equation 10.2(3)

in which ϕ is the capacity reduction factor. For columns, the value of ϕ is between 0.6 and 0.8, depending upon the failure conditions. This is discussed in detail in Section 10.5.

In Figure 10.2(1), the position of the plastic centre can be obtained by considering the equilibrium of moments with respect to (say) the right edge of the section, or

$$d_{pc} = \frac{\alpha_1 f'_c (A_g - A_s) D/2 + f_{sy}(A_{sc} d_c + A_{st} d)}{N_{uo}}$$ Equation 10.2(4)

in which N_{uo} is given in Equation 10.2(1).

Although a centrally loaded column hardly exists in practice, its ultimate strength in compression is an important parameter in column analysis and design calculations.

Note also that for the squash load condition illustrated in Figure 10.2(1), the ultimate strains for steel and concrete are set as 0.0025.

10.3 COLUMNS IN UNIAXIAL BENDING

10.3.1 Strength formulas

Figure 10.3(1)a shows a column section symmetrically reinforced and symmetrically loaded with respect to the *x*-axis. The moment (M_x) in Figure 10.3(1) has been transformed to be the product of the axial load (N_u) and the eccentricity e' with respect to the plastic centre. That is,

$$M_{ux} = N_u e'$$ Equation 10.3(1)

Under the eccentric load, the strain and stress diagrams are detailed in Figure 10.3(1), parts b and c, respectively. At the ultimate state, the compressive forces in concrete and steel, respectively, are

$$C_1 = \alpha_2 f'_c \gamma k_u db \qquad \text{Equation 10.3(2)}$$

and

$$C_2 = A_{sc}(f_{sy} - \alpha_2 f'_c) \qquad \text{Equation 10.3(3)}$$

where all the symbols used here carry the same significance as those adopted for beams in bending. Note that in Equation 10.3(3) the negative $\alpha_2 f'_c$ is to compensate for the concrete area taken by the compression steel (A_{sc}). The equation is valid only if the compression steel yields.

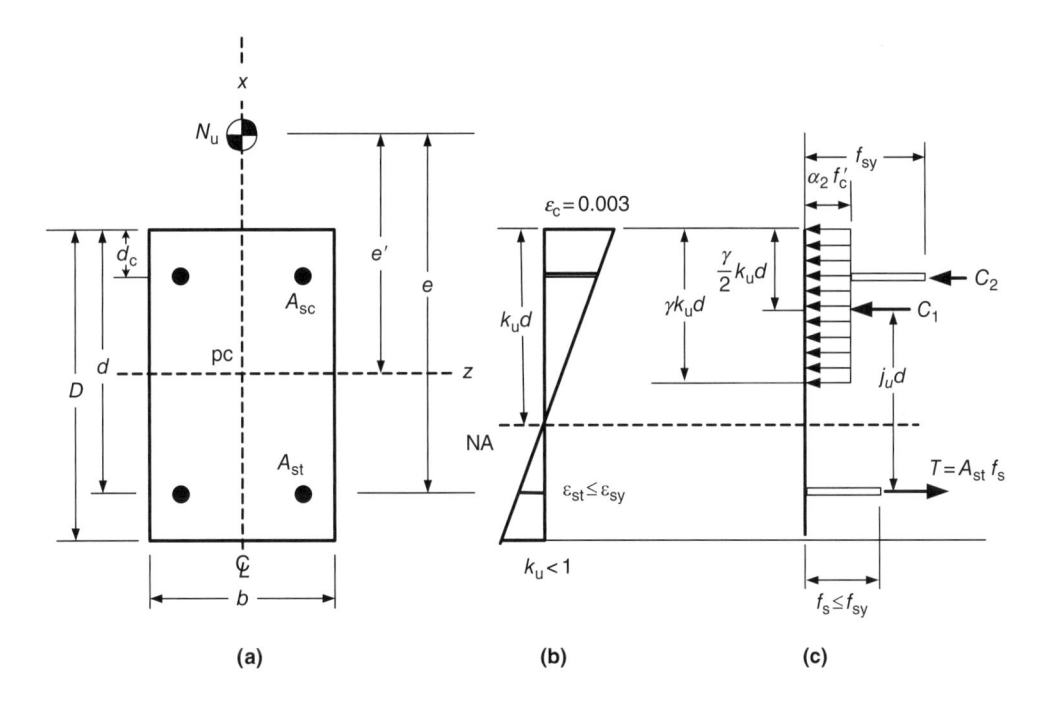

Figure 10.3(1) Stress and strain diagrams for a symmetrically reinforced and symmetrically loaded column section

As the tension steel may not yield at ultimate, the force

$$T = A_{st} f_s \qquad \text{Equation 10.3(4)}$$

where $f_s \leq f_{sy}$ is the steel stress.

Considering the equilibrium of axial forces leads to

$$N_u = C_1 + C_2 - T \qquad \text{Equation 10.3(5)}$$

Substituting Equations 10.3(2), 10.3(3) and 10.3(4) into 10.3(5) gives

$$N_u = \alpha_2 \gamma k_u f'_c bd + A_{sc}(f_{sy} - \alpha_2 f'_c) - A_{st} f_s \qquad \text{Equation 10.3(6)}$$

By taking moment about the level of A_{st}, we obtain

$$eN_u = C_1 j_u d + C_2(d - d_c) \qquad \text{Equation 10.3(7)}$$

where

$$j_u d = (1 - \gamma k_u/2)d \qquad \text{Equation 10.3(8)}$$

Substituting Equations 10.3(2), 10.3(3) and 10.3(8) into Equation 10.3(7) leads to

$$eN_u = \alpha_2 f'_c \gamma k_u b d^2 (1 - \gamma k_u/2) + A_{sc}(f_{sy} - \alpha_2 f'_c)(d - d_c) \qquad \text{Equation 10.3(9)}$$

Equations 10.3(6) and 10.3(9) are the two basic formulas for the strength analysis of a column under the combined effect of axial load (N) and moment about the z-axis (M_x). The equations are of course equally applicable to uniaxial bending about the x-axis (M_z). At ultimate, the compression steel (A_{sc}) is assumed to be yielding under the combined axial and bending compressive stress. Similar to beams in bending, however, the tension steel may or may not yield at the ultimate state. This gives rise to the various failure conditions as discussed in the next section.

10.3.2 Tension, compression, decompression and balanced failure

Like beams in bending, all columns fail as a result of the concrete strain in the compression zone attaining the ultimate value, that is,

$$\varepsilon_c = \varepsilon_{cu} = 0.003 \qquad \text{Equation 10.3(10)}$$

Again, at the ultimate state, the tension steel strain (ε_{st}) may be greater or less than the yield strain (ε_{sy}). If $\varepsilon_{st} > \varepsilon_{sy}$ (see Figure 10.3(2)), the failure is referred to as a tension failure. In this case, Equation 10.3(6) becomes

$$N_u = \alpha_2 \gamma k_u f'_c b d + A_{sc}(f_{sy} - \alpha_2 f'_c) - A_{st} f_{sy} \qquad \text{Equation 10.3(11)}$$

If, on the other hand, $\varepsilon_{st} < \varepsilon_{sy}$ at the ultimate state, it is referred to as compression failure. This condition is depicted in Figure 10.3(3).

If at the ultimate state $\varepsilon_{st} = \varepsilon_{sy}$, it is referred to as balanced failure. The position of the neutral axis is governed by $k_{uB} d$, where according to Equation 3.4(4)

$$k_{uB} = \frac{600}{600 + f_{sy}} \qquad \text{Equation 10.3(12)}$$

which is identical to the equation derived in Chapter 3 for bending analysis. Even if the column is purposely designed, the likelihood of a balanced failure eventuating in practice is

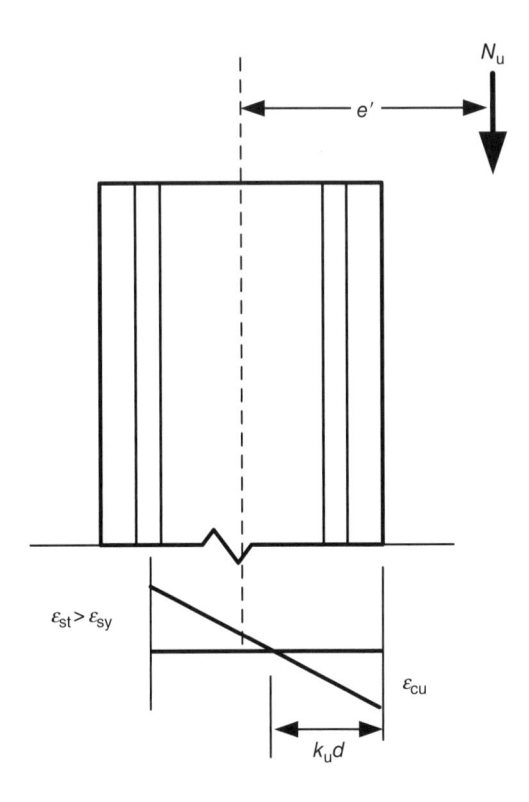

Figure 10.3(2) Tension failure ($\varepsilon_{st} > \varepsilon_{sy}$)

minute. However, the 'balanced' ultimate axial strength (N_{uB}) and the 'balanced' ultimate moment

$$M_{uB} = N_{uB} e'_B \qquad \text{Equation 10.3(13)}$$

are important quantities in the analysis and design of columns.

In Equation 10.3(13), e'_B is the eccentricity of N_{uB} with respect to the plastic centre. By definition, for a compression failure,

$$e' < e'_B \qquad \text{Equation 10.3(14)a}$$

and for a tension failure,

$$e' > e'_B \qquad \text{Equation 10.3(14)b}$$

Within compression failure mode, the Standard, for the first time, also defines the so-called 'decompression' failure point. This mode prevails when, in Figure 10.3(1)b, the concrete strain in the extreme tension fibres is equal to zero, creating a threshold beyond which tensile stress does not exist in the section, or

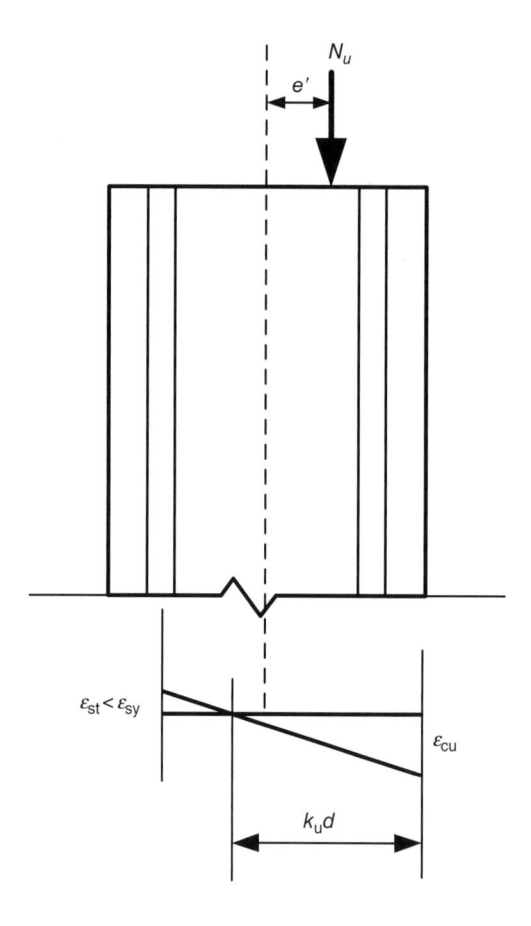

Figure 10.3(3) Compression failure ($\varepsilon_{st} < \varepsilon_{sy}$)

$$N_{u.dc} = \alpha_2 \gamma f'_c bD + A_{sc}(f_{sy} - \alpha_2 f'_c) + A_{st} f_s \qquad \text{Equation 10.3(15)}$$

where

$$f_s = \varepsilon_s E_s = 600\left(\frac{D}{d} - 1\right) \qquad \text{Equation 10.3(16)}$$

Clause 10.6.2.4 of the Standard states that the column strengths between the squash load condition (or N_{uo}) and the decompression mode are for cases where the neutral axis lies outside the section. Such can be determined by a linear interpolation process which is discussed in Section 10.3.5.

10.3.3 Interaction diagram

For a given column section subjected to an axial load (N_u) at an eccentricity (e') giving an ultimate moment of $M_u = e'N_u$, the failure mode and strength depend upon the combined effect of N_u and M_u. The interaction diagram of a column prescribes all the combinations of N_u and M_u that can cause failure to the column. The procedure in detail for constructing an interaction diagram is illustrated using the following example.

Illustrative example

Problem

A column section subjected to bending in the x–y plane is detailed in Figure 10.3(4). Construct the interaction diagram.

Take $f'_c = 25$ MPa, $f_{sy} = 500$ MPa, $\alpha_2 = \gamma = 0.85$ and $A_{sc} = A_{st} = 2$ N32 bars @ 804 mm² = 1608 mm².

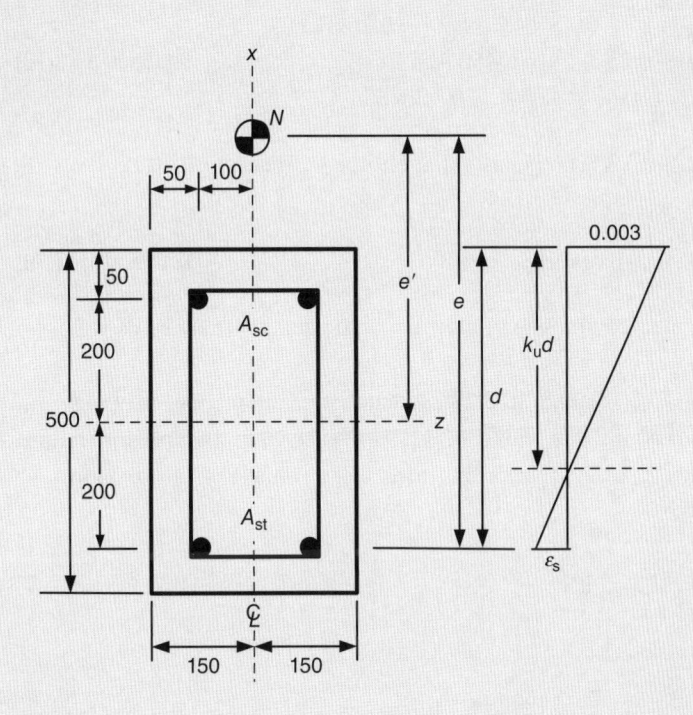

Figure 10.3(4) Details of the example column section subjected to bending in the x–y plane

Note: all dimensions are in mm.

Solution

The following are given: $b = 300$ mm; $d = 450$ mm; $d_c = 50$ mm.

Since $A_{sc} = A_{st} = 1608$ mm², we have $p_t = p_c = 0.01191$.
For $f'_c = 25$ MPa, $\alpha_2 = 0.85$ as per Equation 3.3(2)a.

Under different combinations of N_u and M_u, the column could fail in one of three modes. For compression failure, Equation 10.3(6) for the squash load capacity yields:

$$N_u = 0.85 \times 0.85 k_u \times 25 \times 300 \times 450 + 1608 \times (500 - 0.85 \times 25) - 1608 f_s$$

where f_s is given in Equation 3.4(13), or

$$f_s = 600(1 - k_u)/k_u$$

We then have

$$N_u = [2.438 k_u + 0.770 - 0.965(1 - k_u)/k_u] \times 10^3 \,(\text{kN}) \qquad \text{Equation 10.3(17)}$$

The moment equation (Equation 10.3(9)) becomes

$$\begin{aligned} eN_u &= 0.85 \times 0.85 k_u \times 25 \times 300 \times 450^2 \times (1 - 0.85 k_u/2) \\ &\quad + 1608 \times (500 - 0.85 \times 25) \times (450 - 50) \\ &= [10.97 k_u (1 - 0.425 k_u) + 3.079] \times 10^2 \,(\text{kNm}) \end{aligned} \qquad \text{Equation 10.3(18)}$$

For tension failure, Equation 10.3(18) remains valid, but for N_u Equation 10.3(17) gives

$$N_u = [2.438 k_u - 0.034] \times 10^3 \,(\text{kN}) \qquad \text{Equation 10.3(19)}$$

Note that either the tension or compression failure strength equations may be used for the balanced failure analysis.

With Equations 10.3(17), 10.3(18) and 10.3(19) in hand, the interaction curve can be obtained by appropriately varying the value of k_u. However, for compression failure $k_u > k_{uB}$; for tension failure $k_u < k_{uB}$; according to Equation 10.3(12) $k_{uB} = 0.5454$.

- N_{uo} (i.e. $e' = 0$)

    Equation 10.2(1): $N_{uo} = 4727.2$ kN

- M_{uo} (i.e. $N_u = 0$)

    Equation 3.6(15)a or b: $M_{uo} = 328.5$ kNm

- Balanced failure (i.e. $k_{uB} = 0.5454$).

    Equation 10.3(17): $N_{uB} = 1295.3$ kN

    Equation 10.3(18): $eN_{uB} = 767.5$ kNm

But from Figure 10.3(4),

$$\begin{aligned} M_{uB} &= e'_B N_{uB} = N_{uB}(e - 0.2) = eN_{uB} - 0.2 N_{uB} = 767.5 - 0.2 \times 1295.3 \\ &= 508.4 \text{ kNm} \end{aligned}$$

- Compression failure (i.e. $k_u > k_{uB}$).

    The variable k_u may be given some appropriate values and N_u and M_u can be computed using Equations 10.3(17) and 10.3(18), as for the balanced failure case given above.

    For $k_u = 1$

    > Equation 10.3(17): $N_u = 3208.0$ kN
    > Equation 10.3(18): $eN_u = 938.7$ kNm
    > $M_u = eN_u - 0.2N_u = 297.1$ kNm

    For $k_u = 0.9$

    > Equation 10.3(17): $N_u = 2857.0$ kN
    > Equation 10.3(18): $eN_u = 917.6$ kNm
    > $M_u = 346.2$ kNm

    For $k_u = 0.8$

    > Equation 10.3(17): $N_u = 2479.2$ kN
    > Equation 10.3(18): $eN_u = 887.1$ kNm
    > $M_u = 391.3$ kNm

    For $k_u = 0.7$

    > Equation 10.3(17): $N_u = 2063.0$ kN
    > Equation 10.3(18): $eN_u = 847.3$ kNm
    > And $M_u = 434.7$ kNm

- Tension failure (i.e. $k_u < k_{uB}$ and $f_s = f_{sy}$)

    For tension failure, Equations 10.3(19) and 10.3(18), respectively, should be used for computing N_u and eN_u. Note that in Equations 10.3(6) and 10.3(9), yielding of A_{sc} is assumed. For this to be valid, k_u must be greater than a certain lower limit.

    Considering $k_u = 0.4$ as the limiting value,

    > Equation 10.3(19): $N_u = 941.2$ kN
    > Equation 10.3(18): $eN_u = 672.1$ kNm
    > $M_u = 483.9$ kNm

    For $k_u = 0.5$ ($< k_{uB} = 0.5454$),

    > Equation 10.3(19): $N_u = 1185.0$ kN
    > Equation 10.3(18): $eN_u = 739.8$ kNm
    > $M_u = 502.8$ kNm

- Decompression mode

    For the threshold beyond which no tensile stress exists in the section, Equation 10.3(16) gives

    $$f_s = 600\left(\frac{500}{450} - 1\right) = 66.67 \text{ MPa}.$$

    Then from Equation 10.3(15),

    $$N_{u.dc} = [0.85 \times 0.85 \times 25 \times 300 \times 500 + 1608(500 - 0.85 \times 25) + 1608 \times 66.67] \times 10^{-3}$$
    $$= 3586.4 \text{ kN}$$

    Further,

    Equation 10.3(8): $j_u d = (450 - 0.85 \times 500/2) = 237.5$ mm

    Equation 10.3(9): $eN_{u.dc} = [0.85 \times 0.85 \times 25 \times 300 \times 500 \times 237.5$
    $$+ 1608(500 - 0.85 \times 25) \times (450 - 50)] \times 10^{-6} = 951.4 \text{ kNm}$$

    and

    $$M_{u.dc} = eN_{u.dc} - 0.2 N_{u.dc} = 951.4 - 0.2 \times 3586.4 = 234.1 \text{ kNm}$$

Finally, with all the above coordinates of N_u and M_u, the interaction diagram can be drawn. This is shown in Figure 10.3(5).

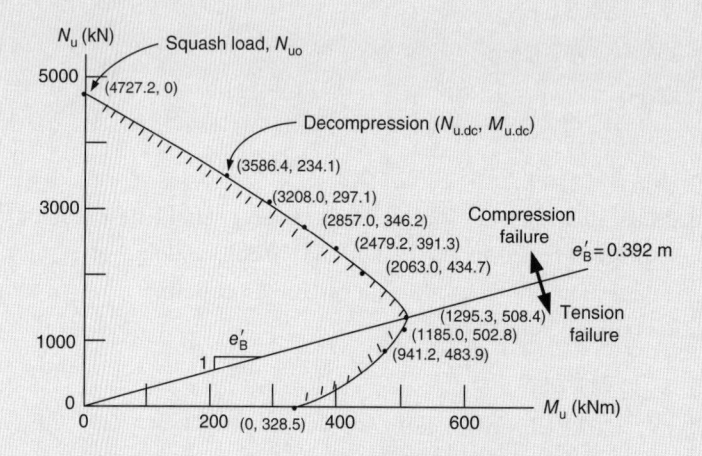

Figure 10.3(5) Interaction diagram for the example column section

Since the interaction curve describes all the combinations of N_u and M_u that can cause failure to the column, the zone enclosed by the curve and the vertical and horizontal axes is referred to as the safe zone. In practice, the safe zone is smaller as a result of the capacity reduction factor (ϕ) being applied. This is discussed in Section 10.5.

With a given combination of N^* and M^* outside the safe zone, the column could fail in compression if $e' < e'_B$ or in tension if $e' > e'_B$.

It may be obvious in Figure 10.3(5) that, for compression failure, $N_u > N_{uB}$ and $M_u < M_{uB}$; and for tension failure, $N_u < N_{uB}$, but depending on the values of e' and p_t, M_u can be greater or less than M_{uB}.

For bending in the y–z plane, another interaction diagram can be constructed in a similar manner.

10.3.4 Approximate analysis of columns failing in compression

It is observed in Figure 10.3(5) that the top portion of the interaction curve is close to a straight line and that, unlike columns failing in tension, the shape or nature of the curve is unchanged. Thus, for compression failure, approximate equations can be derived by replacing the top part of the interaction curve by a straight line connecting N_{uo} with the point (N_{uB}, M_{uB}) as shown in Figure 10.3(6). The procedure given below is recognised to have been first presented by Lorell (1968).

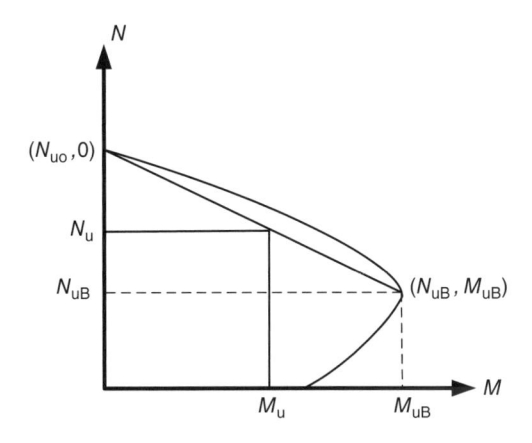

Figure 10.3(6) Approximation of interaction diagram for columns failing in compression

By considering the two similar triangles, we obtain

$$\frac{N_u - N_{uB}}{N_{uo} - N_{uB}} = \frac{M_{uB} - M_u}{M_{uB}} = 1 - \frac{M_u}{M_{uB}} \qquad \text{Equation 10.3(20)}$$

from which

$$M_u = \frac{N_{uo} - N_u}{N_{uo} - N_{uB}} M_{uB} \qquad \text{Equation 10.3(21)}$$

But $M_u = e' N_u$ and $M_{uB} = e'_B N_{uB}$. Substituting these equations into Equation 10.3(21) and rearranging terms gives

$$N_u = \frac{N_{uo}}{1 + \left(\dfrac{N_{uo}}{N_{uB}} - 1\right)\dfrac{e'}{e'_B}} \qquad \text{Equation 10.3(22)}$$

Equations 10.3(21) and 10.3(22) may also be found in the book by Warner, Rangan and Hall (1989). Note too that the two equations always give conservative results, which is obvious in Figure 10.3(6).

Illustrative example

Problem

For the column shown in Figure 10.3(4), calculate the approximate values of N_u and M_u for the case with $e' = 0.0925$ m (i.e. for $k_u = 1$).

Solution

From the example in Section 10.3.3, we have

$N_{uB} = 1295.3$ kN

$M_{uB} = 508.4$ kNm

$e'_B = 0.392$ m,

and

$N_{uo} = 4727.2$ kN. Thus

Equation 10.3(22): $N_u = \dfrac{4727.2}{1 + \left(\dfrac{4727.2}{1295.3} - 1\right)\dfrac{0.0925}{0.392}} = 2906.8$ kN

Equation 10.3(21): $M_u = \dfrac{4727.2 - 2906.8}{4727.2 - 1295.3} \times 508.4 = 269.7$ kNm

Note that the approximate formulas have underestimated the values of N_u and M_u by 9.4% and 9.2%, respectively.

10.3.5 Strengths between decompression and squash points

As indicated in Section 10.3.2, for cases where the neutral axis lies outside the section, the column strength or the N_u and M_u combination may be obtained by linear interpolation between the decompression and squash points (as identified in Figure 10.3.(5) for a given column section).

Substituting $N_{u.dc}$ and e'_{dc} for N_{uB} and e'_B, respectively, in Equation 10.3(22) gives

$$N_u = \frac{N_{uo}}{1 + \left(\dfrac{N_{uo}}{N_{u.dc}} - 1\right) \dfrac{e'}{e'_{dc}}}$$

Equation 10.3(23)

where e' is the given eccentricity of N_u (see Figure 10.3(1)), and

$$e'_{dc} = M_{u.dc}/N_{u.dc}$$

Equation 10.3(24)

Similarly, replacing N_{uB} and M_{uB}, respectively, in Equation 10.3(21) with $N_{u.dc}$ and $M_{u.dc}$ leads to

$$M_u = \frac{N_{uo} - N_u}{N_{uo} - N_{u.dc}} \cdot M_{u.dc}$$

Equation 10.3(25)

10.4 ANALYSIS OF COLUMNS WITH AN ARBITRARY CROSS-SECTION

Analysis equations for circular columns are available in Appendix A of the earlier Australian Standard, AS 1480–1982 *SAA Concrete Structures Code*. The topic has also been discussed by Warner, Rangan and Hall (1989). For details, the reader should consult these publications.

The analysis of a more general case of columns with an arbitrary cross-section is discussed here. Two methods are included: the iterative approach and the semi-graphical procedure. Numerical examples are given to elaborate on the details.

10.4.1 Iterative approach

An arbitrarily shaped column section is shown in Figure 10.4(1)a. The section is reinforced by m layers of steel, designated as $A_{s1}, A_{s2}, \ldots, A_{si}$ and A_{sm} as shown in Figure 10.4(1). Because of symmetry, the eccentric load produces bending in the x–y plane only. The strain distribution, concrete stress block and steel forces are illustrated in Figures 10.4(1)b, c and d.

For the column, the ultimate load (N_u) for a specified eccentricity along the x-axis (or e_{given}) can be obtained following the steps described below.

Step 1 – Assume a value of d_{NA} with $\varepsilon = \varepsilon_{cu} = 0.003$.
Step 2 – Determine ε_{si} for the steel layer i by proportion (see Figure 10.4(1)b).
  Then compute for $i = 1$ to m.

$$T_i = E_s \varepsilon_{si} A_{si} \leq f_{sy} A_{si}$$

Equation 10.4(1)

where T_i is positive if it is in compression.
Step 3 – Compute the concrete stress resultant.

$$C_c = \alpha_2 f'_c A'$$

Equation 10.4(2)

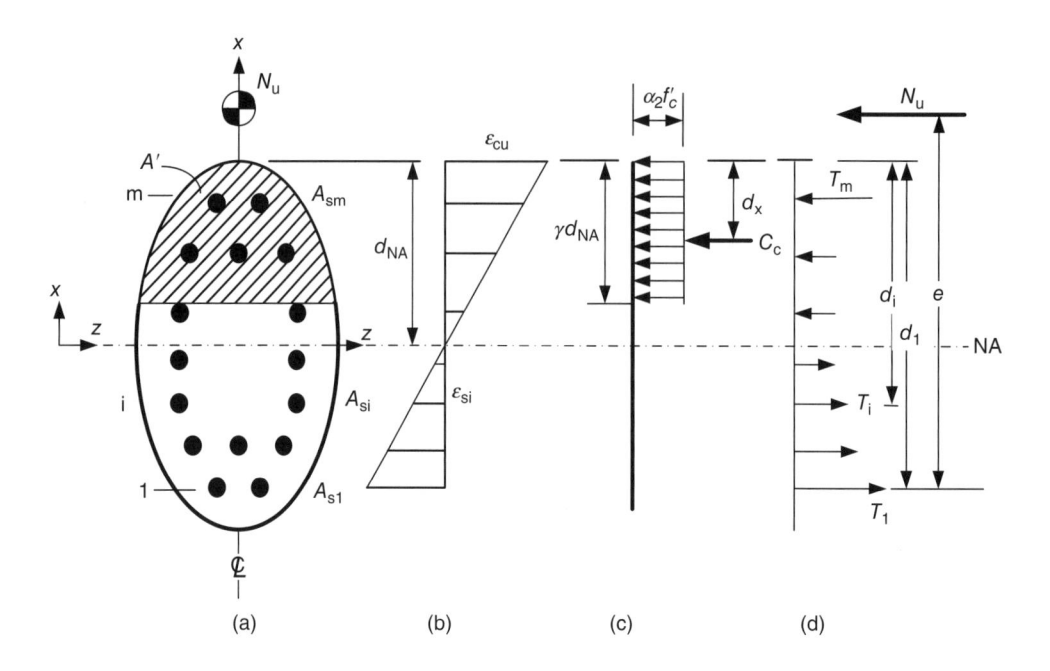

Figure 10.4(1) An arbitrarily shaped column section: **(a)** the section; **(b)** strains; **(c)** concrete stress; and **(d)** steel forces

where

$$A' = \int dA'$$

Equation 10.4(3)

can be determined numerically or by integration. The position of C_c from the top or the extreme compression fibres (see Figure 10.4(1)c) is given as

$$d_x = \frac{\int x dA'}{\int dA'}$$

Equation 10.4(4)

Step 4 – Compute the ultimate load.

$$N_u = C_c + \sum_{i=1}^{m} T_i$$

Equation 10.4(5)

By taking moment about the level of A_{s1} and rearranging terms, we obtain

$$e_{computed} = \frac{1}{N_u}\left\{C_c(d_1 - d_x) + \sum_{i=1}^{m}[T_i(d_1 - d_i)]\right\}$$

Equation 10.4(6)

in which the symbols are defined in Figures 10.4(1)c and d.

Step 5 – Carrying out the iterative process as summarised in Figure 10.4(2) will eventually lead to the correct value of N_u.

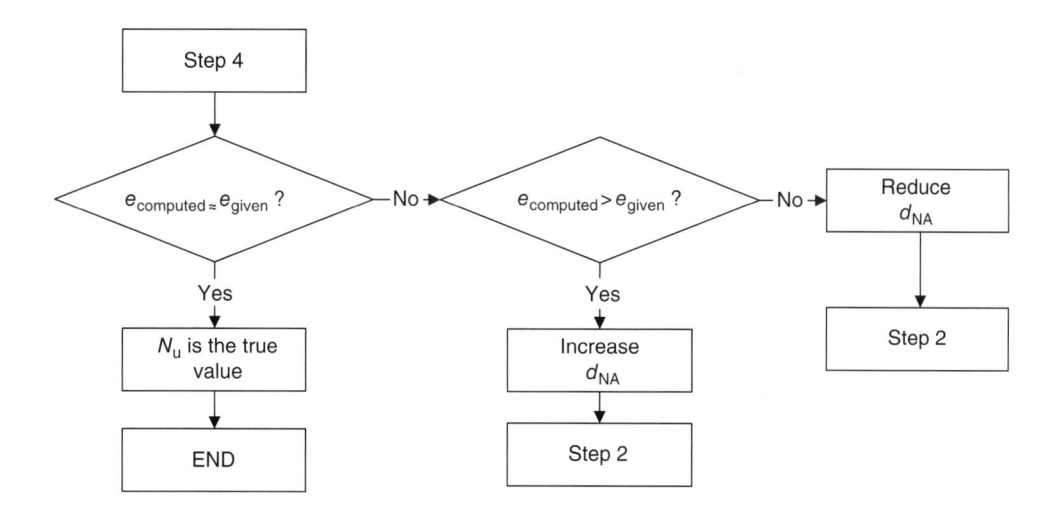

Figure 10.4(2) Iterative process for calculating N_u

10.4.2 Illustrative example of iterative approach

Problem

An irregular-shaped column section is shown in Figure 10.4(3). Compute N_u using the iterative procedure. Take $f'_c = 32$ MPa and $f_{sy} = 500$ MPa.

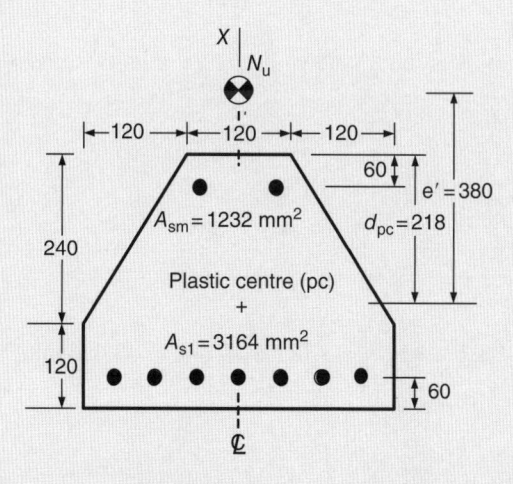

Figure 10.4(3) An irregular-shaped column section

Note: all dimensions are in mm.

Solution

Equation 3.3(2)a: $\alpha_2 = 0.85$

Equation 3.3(2)b: $\gamma = 1.05 - 0.007 \times 32 = 0.826$

And there are two layers of steel or $m = 2$.

Trial 1

Step 1

Assume $d_{NA} = 240$ mm.

Step 2

For the top steel layer 2 (or A_{sm} with $m = 2$ in this case)

$$\varepsilon_{s2} = 0.003 \times (240 - 60)/240 = 0.00225 < \varepsilon_{sy} = 0.0025$$

that is, $f_{s2} = 200\,000\varepsilon_{s2} = 450$ MPa

$$T_2 = 450 \times 1232 \times 10^{-3} = 554.4 \text{ kN}$$

For the bottom steel layer 1 (or A_{s1}),

$$\varepsilon_{s1} = 0.003 \times (240 - 300)/240 = -0.00075 \text{ (tension)}$$

and

$$f_{s1} = 200\,000\varepsilon_{s1} = -150 \text{ MPa}$$
$$T_1 = -150 \times 3164 \times 10^{-3} = -474.6 \text{ kN (tension)}$$

Step 3

The height of the concrete area in compression as shown in Figure 10.4(4) is given as

$$\gamma d_{NA} = 0.826 \times 240 = 198.2 \text{ mm}$$

and

$$c = (198.2/240) \times 120 = 99.1 \text{ mm}$$

Thus, the concrete area

$$A' = 120 \times 198.2 + 2 \times 0.5 \times 198.2 \times 99.1 = 23\,784 + 19\,642 = 43\,426 \text{ mm}^2$$

Using Equation 10.4(4),

$$d_x = (23\,784 \times 198.2/2 + 19\,642 \times 198.2 \times 2/3)/43\,426 = 114 \text{ mm}$$

Equation 10.4(2): $C_c = 0.85 \times 32 \times 43\,426 \times 10^{-3} = 1181$ kN

Figure 10.4(4) Determination of neutral axis (NA) for the irregular-shaped column section

Note: all dimensions are in mm.

Step 4

Equation 10.4(5): $N_u = 1181 + 554.4 - 474.6 = 1260.8$ kN

Equation 10.4(6): $e_{computed} = [1181 \times (300 - 114) + 554.4 \times (300 - 60)]/1260.8$
$= 279.8$ mm

Thus, the eccentricity

$e'_{computed} = 279.8 - (300 - 218) = 197.8$ mm

Step 5

$e'_{computed} < e'_{given} (= 380$ mm$)$

Thus, reduce d_{NA} and repeat the process.

Trial 2

Step 1

Assume $d_{NA} = 210$ mm.

Step 2

$\varepsilon_{s2} = 0.00214 < \varepsilon_{sy}$, therefore

$T_2 = 527.3$ kN

$\varepsilon_{s1} = -0.001286$, therefore

$T_1 = -813.8$ kN

Step 3

$\gamma d_{NA} = 173.5$ mm

Accordingly,

$$A' = 35\,871 \text{ mm}^2$$
and $d_x = 98.9$ mm
Thus, $C_c = 975.7$ kN

Step 4

$N_u = 975.7 + 527.3 - 813.8 = 689.2$ kN, and

$e_{\text{computed}} = 468.3$ mm, therefore

$e'_{\text{computed}} = 386.3$ mm

Step 5

$e'_{\text{computed}} > e'_{\text{given}}$

but the value is quite close. Thus, increase d_{NA} to 210.5 mm and try again.

Trial 3

Step 1

$$d_{\text{NA}} = 210.5 \text{ mm}$$

Step 2

$T_2 = 528.5$ kN

$T_1 = -807.2$ kN

Step 3

$C_c = 978.7$ kN

Step 4

$N_u = 700.0$ kN, and

$e_{\text{computed}} = 462.1$ mm

Or, $e'_{\text{computed}} = 380.1$ mm

Step 5

$e'_{\text{computed}} \cong e'_{\text{given}}$

Therefore, we take $N_u = 700.0$ kN.

10.4.3 Semi-graphical method

For column sections having shapes not easily handled numerically, some of the steps in the iterative approach may be executed graphically on graph paper.

With the cross-sectional details of the column illustrated in Figure 10.4(1)a and the trial strain (stress) diagram in Figure 10.4(1)b drawn to scale, Steps 2, 3 and 4 in the iterative procedure may be modified as follows.

Step 2

The strains (or stresses) for the m layers of steel bars (i.e. ε_{si} for $i = 1$ to m) may be measured directly from a scaled drawing.

Step 3

Equations 10.4(3) and 10.4(4), respectively, may be rewritten as

$$A' = \sum_{j=1}^{n} \Delta A_j = \sum_{j=1}^{n} b_j t_j \qquad \text{Equation 10.4(7)}$$

and

$$d_x = \frac{\sum_{j=1}^{n} x_j \Delta A_j}{\sum_{j=1}^{n} \Delta A_j}$$

Or

$$d_x = \frac{\sum_{j=1}^{n} b_j t_j x_j}{\sum_{j=1}^{n} b_j t_j} \qquad \text{Equation 10.4(8)}$$

where the various symbols are defined in Figure 10.4(5), and b_j, t_j as well as x_j can be measured directly from the scaled drawing.

Step 4

Equation 10.4(5) may be rewritten as

$$N_u = \sum_{j=1}^{n} \alpha_2 f'_c b_j t_j + \sum_{i=1}^{m} A_{si} f_{si} \qquad \text{Equation 10.4(9)}$$

where f_{si} is as measured from the steel strain (stress) diagram.

Taking moment about the level of the plastic centre (pc) gives

276 PART 1 REINFORCED CONCRETE

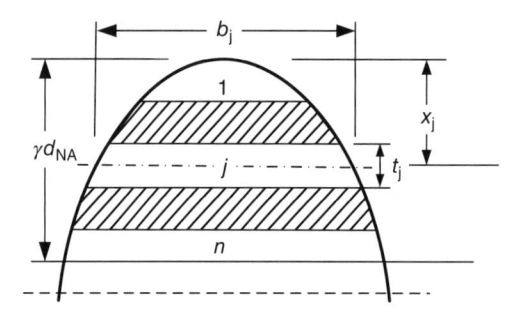

Figure 10.4(5) Scaled drawing for the iteration in graphical method

$$e'N_u = \sum_{j=1}^{n} \alpha_2 f'_c b_j t_j (d_{pc} - x_j) + \sum_{i=1}^{m} A_{si} f_{si} (d_{pc} - d_i) \qquad \text{Equation 10.4(10)}$$

where d_{pc} is the distance between the extreme compression fibres and the plastic centre.

And, with Equations 10.4(9) and 10.4(10),

$$e'_{conputed} = e'N_u/N_u \qquad \text{Equation 10.4(11)}$$

Similarly, Step 5 in Section 10.4.1 will eventually lead to the correct N_u.

10.4.4 Illustrative example of semi-graphical method

Problem

A rectangular column section, reinforced with four layers of steel (a total of twelve 20-mm diameter bars) is subjected to uniaxial bending in the x–y plane. Details of the section are shown in Figure 10.4(6). Take $f'_c = 32$ MPa, $f_{sy} = 500$ MPa, $\alpha_2 = 0.85$ and $\gamma = 0.826$. Determine N_u using the semi-graphical method.

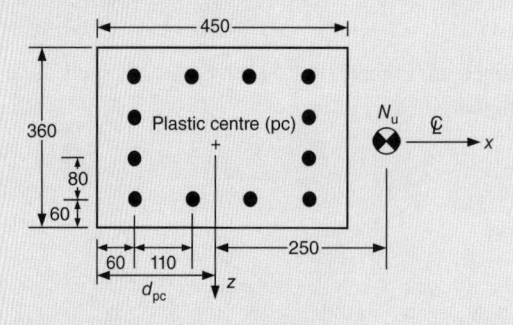

Figure 10.4(6) Details of the example rectangular column section

Note: all dimensions are in mm.

Solution

There are four layers of steel (or $m = 4$) and for the symmetrical section, the position of the plastic centre $d_{pc} = 225$ mm.

The scaled drawing of the cross-section including the reinforcement positions is given in Figure 10.4(7), together with the steel stress diagram. Note that for the rectangular section $b_j = 360$ mm, which is constant, there is no need to draw the entire cross-section. The drawing of the full section is mandatory for an irregular section.

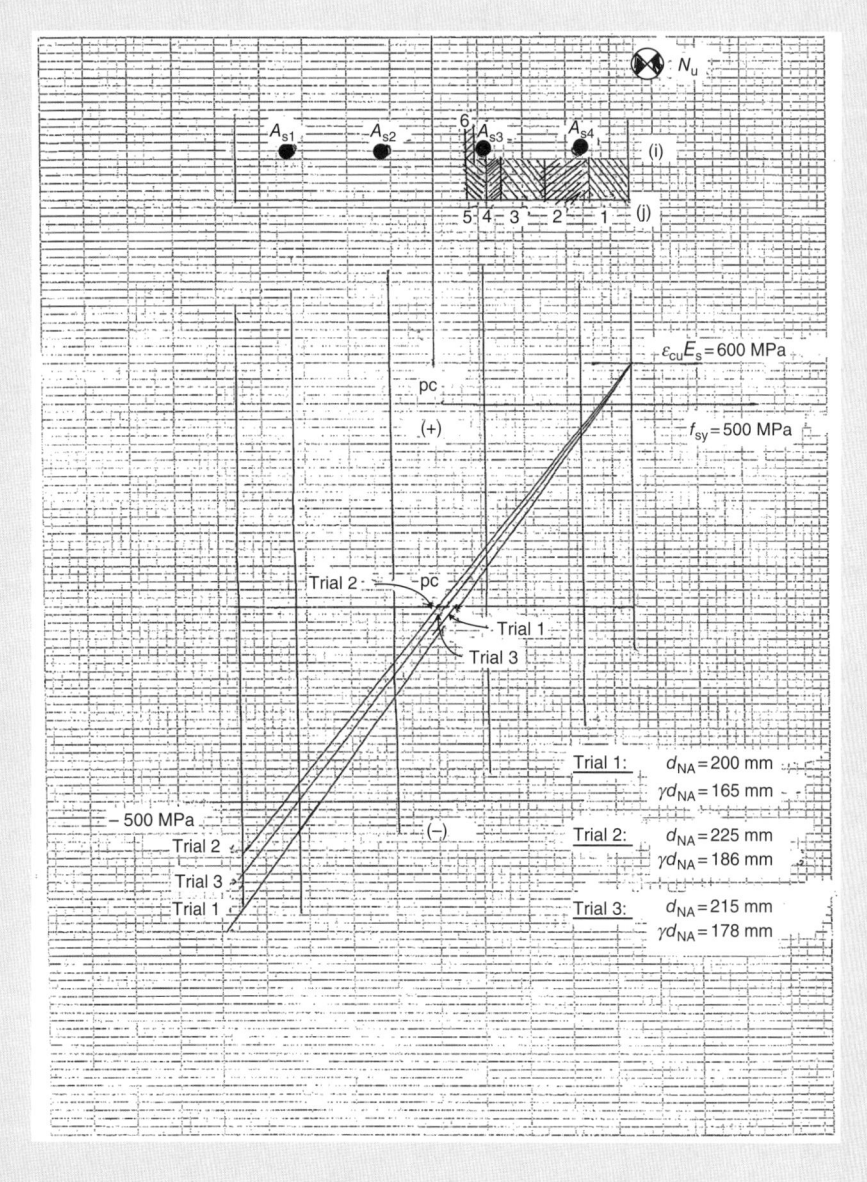

Figure 10.4(7) Scaled drawing on graph paper for the semi-graphical method example

Table 10.4(1) Calculations for semi-graphical method example

1	2	3	4	5	6	7	8	9	10
Trial	Element	A_{si}	f_{si}	$(d_{pc} - d_i)$	$A_{si}f_{si}$	ΔA_{cj}	$(d_{pc} - x_j)$	$\alpha_2 f'_c \Delta A_{cj}$	$\Delta(e'N_u)$
1	$i = 1$	1256	−500	225 − 390 = −165	−628 kN				103.6 kNm
	$i = 2$	628	−240	225 − 280 = −55	−150.7				8.3
	$i = 3$	628	+85	+55	+53.4				2.9
	$i = 4$	1256	+410	+165	+515.0				85.0
	$j = 1$					$50 \times 360 = 18000$	$225 − 25 = 200$	489.6 kN	97.9
	$j = 2$					$18000 − 1240 = 16760$	$225 − 75 = 150$	455.9	68.4
	$j = 3$					18000	$225 − 125 = 100$	489.6	49.0
	$j = 4$					$15 \times 360 = 5400$	$225 − 158 = 67$	146.9	9.8
	Σ				**−210.3**			**1582.0**	**424.9**
2	$i = 1$	1256	+60	−165	75.4				−12.4
	$i = 2$	628	+90	−55	56.5				−3.1
	$i = 3$	628	+60	+55	37.7				2.1
	$i = 4$	1256	+30	+165	37.7				6.2
	$j = 5$					$21 \times 360 − 620 = 6940$	$225 − 186 + 11 = 50$	188.8	9.4
	Σ				**−3.0**			**1770.8**	**427.1**
3	$i = 1$	1256	−40	−165	−50.2				8.3
	$i = 2$	628	−30	−55	−18.8				1.0
	$i = 3$	628	−20	+55	−12.6				−0.7
	$i = 4$	1256	−15	+165	−18.8				−3.1
	$j = 6$					$−8 \times 360 = −2880$	$225 − 182 = 43$	−78.3	−3.4
	Σ				**−103.4**			**1692.5**	**429.2**

The working is given in detail in Table 10.4(1). For Trial 1, assume $d_{NA} = 200$ mm. Then the two sums in Equation 10.4(9) are given, respectively, in columns 9 and 6 of Table 10.4(1). We have $N_u = 1582.0 - 210.3 = 1371.7$ kN.

Equation 10.4(10) is represented by column 10 of Table 10.4(1) and the sum $e'N_u = 424.9$ kNm.

Thus

$e'_{computed} = 424.9/1371.7 = 0.31$ m, which is greater than e'_{given} (= 0.25 m).
For Trial 2, $d_{NA} = 225$ mm. This gives $N_u = 1770.8 - 3.0 = 1767.8$ kN, and
$e'_{computed} = 427.1/1767.8 = 0.24$ m, which is less than e'_{given}.
For Trial 3, $d_{NA} = 215$ mm. This gives $N_u = 1692.5 - 103.4 = 1589.1$ kN, and
$e'_{computed} = 429.2/1589.1 = 0.27$ m, which is also greater than $e'_{given} = 0.25$ m.
Thus, take N_u = greater than 1589.1 kN, say 1600 kN.

10.5 CAPACITY REDUCTION FACTOR

The column analysis methods discussed in the preceding sections produce the ultimate strength, N_u and M_u, in one way or another. In accordance with Equation 1.1(1),

$$\phi N_u \geq N^*$$ Equation 10.5(1)

and

$$\phi M_u = \phi e' N_u \geq e' N^*$$ Equation 10.5(2)

where ϕ is the capacity reduction factor.

The value of ϕ for column strength depends on the failure conditions. This is summarised in Figure 10.5(1).

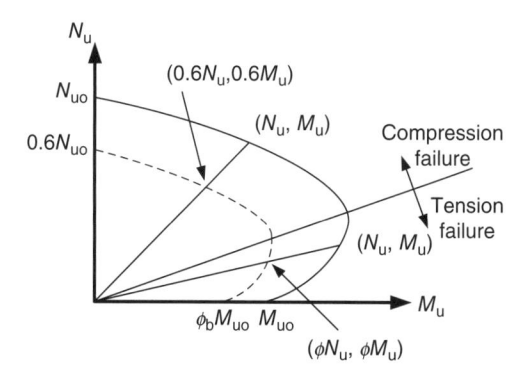

Figure 10.5(1) Failure conditions and their effects on the values of ϕ

For compression failure or $N_u > N_{uB}$

$$\phi = 0.6 \qquad \text{Equation 10.5(3)}$$

For tension failure or $N_u < N_{uB}$

$$\phi = 0.6 + (\phi_b - 0.6)\left(1 - \frac{N_u}{N_{uB}}\right) \qquad \text{Equation 10.5(4)}$$

where ϕ_b is for bending as specified in Equations 3.4(20)a, b and c.

10.6 PRELIMINARY DESIGN PROCEDURE

An approximate method of proportioning column sections has been presented by Warner, Rangan and Hall (1989). It is described here with minor modifications.

10.6.1 Design steps

For given values of N^* and M^*, the following steps may be used to obtain the column size and the reinforcement:

1. Ignore M^* and assume $\phi = 0.6$. Equation 10.2(1) yields

$$\overline{b}\overline{D} = A_g \approx \frac{N^*}{0.6\left[\alpha_2 f'_c + f_{sy}(p_t + p_c)\right]} \qquad \text{Equation 10.6(1)}$$

2. Increase $\overline{b}$ and $\overline{D}$ each by 20%, that is

$$b = 1.2\overline{b} \quad \text{and} \quad D = 1.2\overline{D} \qquad \text{Equation 10.6(2)}$$

3. Ignore N^* and compute the dimensions.

$$\overline{b}\,\overline{d}^2 = \frac{M^*}{0.8(0.9 f_{sy} p_t)} \qquad \text{Equation 10.6(3)}$$

4. Reduce $\overline{b}$ and $\overline{d}$ each by 10%, or

$$b = 0.9\overline{b} \quad \text{and} \quad d = 0.9\overline{d} \qquad \text{Equation 10.6(4)}$$

5. Choose the larger of the section from Equations 10.6(2) and 10.6(4).
6. Adopt the values for p_t and p_c, then $A_{st} = p_t bd$ and $A_{sc} = p_c bd$.

Note that the Standard, in general, specifies a minimum $p = p_t + p_c = 1\%$ and a maximum $p = 4\%$ (see Section 10.10).

To ensure the adequacy of the design, check the capacity of the preliminary section using Equations 10.3(17) and 10.3(18) or, more accurately, Equations 10.3(6) and 10.3(9), applying ϕ appropriately. A practical design procedure is discussed in Section 10.11.

10.6.2 Illustrative example

Problem

Given $M_g = 34$ kNm, $M_q = 38.3$ kNm; $N_g = 340$ kN, $N_q = 220$ kN where subscripts g and q, respectively indicate dead- and live-load effects. Assume $f'_c = 32$ MPa and $f_{sy} = 500$ MPa, and proportion a symmetrically reinforced square section.

Solution

By invoking Equation 1.3(2), we have

$$M^* = 1.2 \times 34 + 1.5 \times 38.3 = 98.25 \text{ kNm}$$

and

$$N^* = 1.2 \times 340 + 1.5 \times 220 = 738 \text{ kN}$$

1. Assuming $p_t = p_c = 0.008$, Equation 10.6(1) gives

    $$\overline{b}\,\overline{D} = (738 \times 10^3)/[0.6 \times (0.85 \times 32 + 500 \times 0.016)] = 34\,943.2 \text{ mm}^2$$

    from which $\overline{b} = \overline{D} = 187$ mm.

2. Use $b = D = 225$ mm, say.
3. Equation 10.6(3) leads to

    $$\overline{b}\,\overline{d}^2 = (98.25 \times 10^6)/[0.8 \times (0.9 \times 500 \times 0.008)] = 34.1 \times 10^6 \text{ mm}^3$$

    Take $\overline{b} = 1.1\,\overline{d}$ and we have $\overline{d} = 315$. This means

    $$b = D = 0.9 \times 1.1 \times 315 = 312 \text{ mm} \cong 320 \text{ mm, say.}$$

4. Thus, the details of the preliminary column section are

    $b = D = 320$ mm and
    $A_{st} = A_{sc} = 0.008 \times 0.9 \times 320^2 = 738$ mm^2

    or use 4 N24 bars, one each at the four corners, giving a total of 1808 mm^2, which is greater than $2 \times 738 = 1476$ mm^2.

10.7 SHORT COLUMN REQUIREMENTS

A column shall be deemed to be short if it satisfies one of the following two requirements:
1. For a braced column

$$L_e/r \leq 25, \text{ or } \leq \alpha_c (38 - f'_c/15)(1 + M_1^*/M_2^*), \text{ whichever is the greater where}$$

$$\alpha_c = \sqrt{2.25 - 2.5N^*/0.6N_{uo}} \quad \text{for} \quad N^*/0.6N_{uo} \geq 0.15, \text{ or}$$

$$\alpha_c = \sqrt{1/(3.5H^*/0.6N_{uo})} \quad \text{for} \quad N^*/0.6N_{uo} < 0.15$$

2. For an unbraced column $L_e/r \leq 22$.

Note that in 1 and 2 above, the radius of gyration $r = D/3$ or $b/3$, as the case may be for a rectangular section, and for a circular column with diameter D, $r = D/4$.

M_1^*/M_2^* is the ratio of the smaller to the larger of the design bending moments at the ends of the column (see Figure 10.7(1)). The ratio is taken to be negative when the column is bent in single curvature and positive when the column is bent in double curvature. When the absolute value of M_2^* is less than or equal to $0.05DN^*$, the ratio shall be taken as -1.0.

L_e is the effective length which may be taken as:
- L_u for a braced column restrained by a flat slab floor
- $0.9L_u$ for a braced column restrained by a beam

in which the unsupported column length (L_u) is defined in Figure 10.7(1). For a more rigorous analysis of L_e, the reader should refer to Clause 10.5.3 of the Standard.

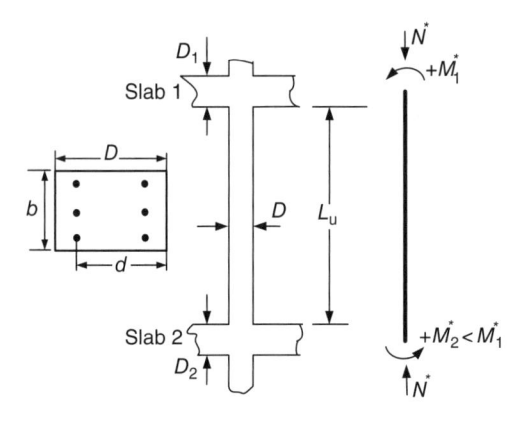

Figure 10.7(1) Definitions of variables for a short column

10.8 MOMENT MAGNIFIERS FOR SLENDER COLUMNS

For slender columns, the interaction between axial force and lateral deflection leads to an increase in the bending moments. Depending on the length of the column, these secondary effects can be significant. Clause 10.4 of the Standard covers both braced and unbraced slender columns.

10.8.1 Braced columns

A braced column exists in a rectangular frame where side-sway is prevented by bracing of one form or another. Thus, lateral deflection for a column not subjected to lateral load is due to the rotation of the top and bottom joints only. For such a case, the moment magnifier is taken to be

$$\delta_b = \frac{k_m}{(1 - N^*/N_c)} \geq 1.0 \qquad \text{Equation 10.8(1)}$$

where

$$k_m = 0.6 - 0.4\frac{M^*_2}{M^*_1} \qquad \text{Equation 10.8(2)}$$

but

$$0.4 \leq k_m \leq 1.0 \qquad \text{Equation 10.8(3)}$$

and

$$N_c = \left(\frac{\pi^2}{L_e^2}\right)[182d_o(\phi M_{uB})/(1 + \beta_d)] \qquad \text{Equation 10.8(4)}$$

in which $\phi = 0.6$, M_{uB} is obtained using Equation 3.6(15)a or 3.6(15)b with $k_u = k_{uB} = 0.545$, and $\beta_d = g/(g + q)$ is the ratio of the dead and live loads to be carried by the column, and $\beta_d = 0$ when $L_e/r \leq 40$ and $N^* \leq M^*/2D$. As in Clause 10.5.2 of the Standard, the radius of gyration, $r = D/3$, or $b/3$ as the case may be for a rectangular section, and for a circular column with diameter D, $r = D/4$. Other variables in these equations are defined in Figure 10.7(1). Note that M^*_2 is positive if it has the same direction as M^*_1.

In Equation 10.8(4) the effective column length

$$L_e = kL_u \qquad \text{Equation 10.8(5)}$$

where L_u is the unsupported length (see Figure 10.7(1)), and the values of the effective length factor (k) for various common braced and unbraced columns are given in Figure 10.5.3(A) of the Standard, which is reproduced as Figure 10.8(1). For columns with non-rigid end restraints, such as those forming part of a framed structure, the reader is referred to Clause 10.5.4 of the Standard for the process in determining k. The design moment for the column shown in Figure 10.7(1) is therefore

$$M^* = \delta_b M^*_1 \qquad \text{Equation 10.8(6)}$$

where M^*_1 is the larger of the two end moments. However, if the sections at different heights of the column are to be designed separately, the design moment at every section should be increased accordingly. This is done by applying Equation 10.8(6) to all the design moments for the column under consideration.

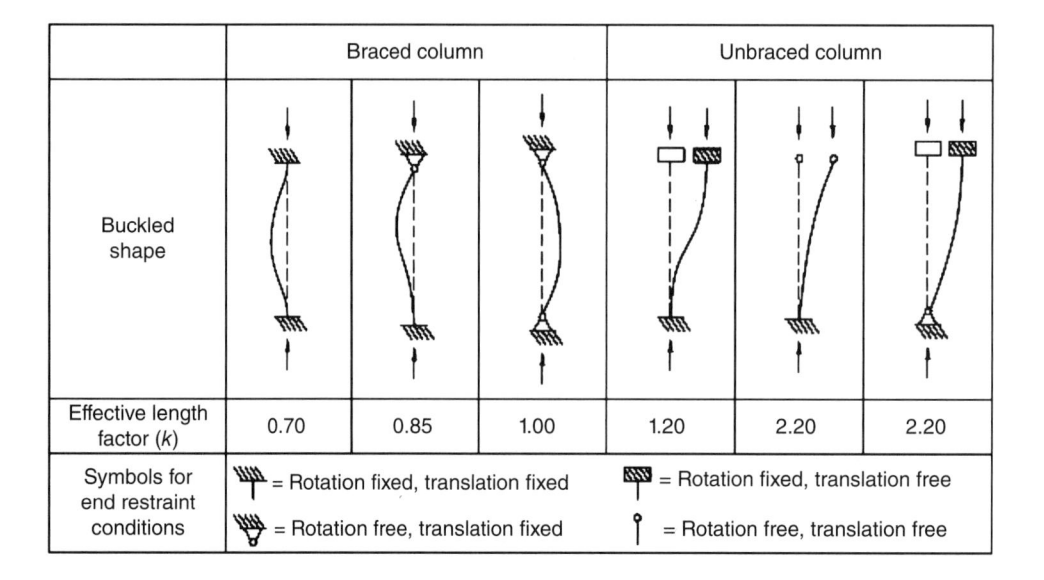

	Braced column			Unbraced column		
Buckled shape						
Effective length factor (k)	0.70	0.85	1.00	1.20	2.20	2.20
Symbols for end restraint conditions	= Rotation fixed, translation fixed = Rotation free, translation fixed			= Rotation fixed, translation free = Rotation free, translation free		

Figure 10.8(1) Effective length factors for columns with simple end restraints

Source: Standards Australia (2009). AS 3600–2009 *Concrete Structures.* Standards Australia, Sydney, NSW. © Standards Australia Limited. Copied by Cambridge University Press with the permission of Standards Australia and Standards New Zealand under Licence 1712-c038.

10.8.2 Unbraced columns

If a storey of a building frame to be designed is unbraced, then in addition to joint rotations, side-sways will occur. For each column in the group of m columns in an unbraced frame, the Standard recommends that the (swayed) magnifier (δ_s) is given as

$$\delta_s = \frac{1}{\left[1 - \sum_{i=1}^{m} N_i^* / \sum_{i=1}^{m} N_{ci}\right]} \geq \delta_b \qquad \text{Equation 10.8(7)}$$

where N_i^* is the design axial load for column i and N_c for column i can be determined using Equation 10.8(4).

In design, δ_s is to be compared with δ_b, computed using Equation 10.8(1) for a given column. The larger value should be adopted as the magnifier to be used in Equation 10.8(6).

As an alternative to Equation 10.8(7), the Standard recommends a more rigorous approach that also involves the computation of the elastic buckling load of the entire frame. Comparatively, this is a much more tedious method and the reader is referred to Clause 10.4.3(b) of the Standard for details.

10.9 BIAXIAL BENDING EFFECTS

Figure 10.9(1) shows a column subjected to axial compression and biaxial bending. The design values N^*, M_x^* and M_z^* are normally obtained by applying the relevant load combination equations described in Section 1.3.1 to the structural analysis results for N, M_x and M_z. If the column is part of a three-dimensional structure, it is unlikely that either M_x^* or M_z^* will be zero. In the event that such is indeed the case, Clause 10.1.2 of the Standard stipulates the minimum design values for short columns to be

$$M_x^* = N^*(0.05D) \qquad \text{Equation 10.9(1)}$$

and

$$M_z^* = N^*(0.05b) \qquad \text{Equation 10.9(2)}$$

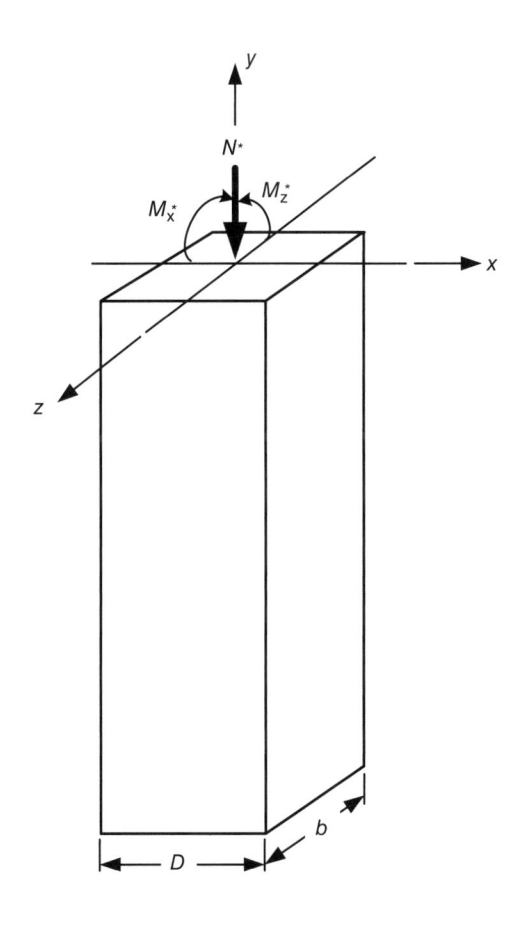

Figure 10.9(1) A column subjected to axial compression and biaxial bending

For a column subjected to any given values of N^*, M_x^* and M_z^*, the adequacy of the design must be checked using the interaction formula

$$\left(\frac{M_x^*}{\phi M_{ux}}\right)^{\alpha_n} + \left(\frac{M_z^*}{\phi M_{uz}}\right)^{\alpha_n} \leq 1.0 \qquad \text{Equation 10.9(3)}$$

where M_x^* and M_z^*, respectively, are about the z and x axes.

$$\alpha_n = 0.7 + 1.7\frac{N^*}{0.6N_{uo}}, \quad \text{but} \quad 1 \leq \alpha_n \leq 2 \qquad \text{Equation 10.9(4)}$$

ϕM_{ux} and ϕM_{uz} can be obtained from the interaction diagrams for uniaxial bending about the principal axes as shown in Figure 10.9(2), and ϕN_{uo} in either of the interaction diagrams is $0.6N_{uo}$.

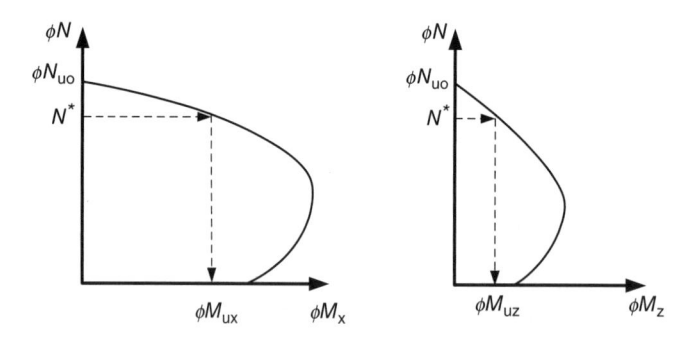

Figure 10.9(2) Interaction diagrams for uniaxial bending about the principal axes

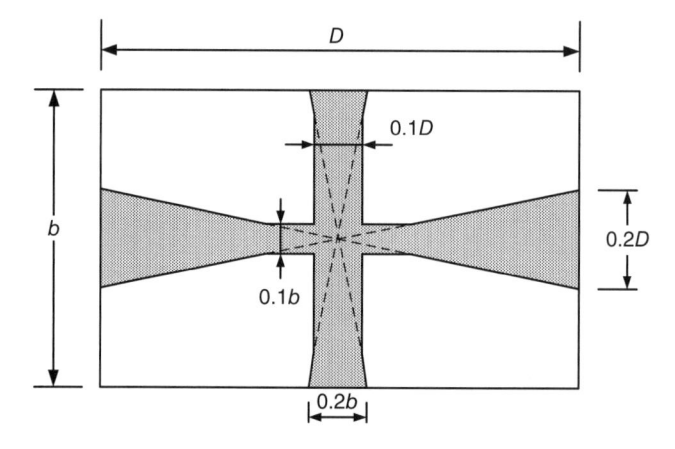

Figure 10.9(3) Limitation for the line of action of the resultant axial force in a rectangular column

For slender columns, the values of M_x^* and M_z^* in Equation 10.9(3) should be suitably magnified using the magnifiers described in Section 10.8.

Further and as in Clause 10.6.3 of the Standard, for a rectangular column where $D/b < 3.0$, which is subjected simultaneously to N^*, M_x^* and M_y^*, the section may be designed for N^* with each M_x^* and M_z^* considered separately, provided that the line of action of the resultant force falls within the shaded area of the section shown in Figure 10.9(3).

10.10 REINFORCEMENT REQUIREMENTS

10.10.1 Limitations and bundled bars

In general, the total cross-sectional area of the longitudinal reinforcement (A_s) as in Clause 10.7.1 should be such that

$$0.01A_g \leq A_s \leq 0.04A_g \qquad \text{Equation 10.10(1)}$$

where A_g is the gross area of the column section.

Bundled bars may be used to avoid congestions, but there should not be more than four bars in a bundle. Each bundle should be tied together in contact.

10.10.2 Lateral restraint and core confinement

To prevent buckling of the longitudinal reinforcement, lateral restraints must be provided using circular and noncircular ties, or helices.

Columns where $f'_c \leq 50$ MPa

Clause 10.7.3.1(a) of the Standard stipulates that each bundle of bars, each corner bar and all bars spaced at centres of more than 150 mm need to be restrained at the bend of the ties or by a helix. For spacing at 150 mm or less, at least every alternate bar should likewise be restrained. Figure 10.10(1) shows some typical examples. For more details, the reader should consult Clause 10.7 of the Standard. The bar size of the ties or helices must be not less than:

- 6 mm for up to N20 single bars
- 10 mm for N24 to N36 single bars
- 12 mm for bundled bars.

No design is necessary for s, the spacing of ties or pitch of a helix (see Figure 10.10(1)) as long as it does not exceed the smaller of:

- D_c and $15d_b$, for single bars, or
- $0.5D_c$ and $7.5d_b$, for bundled bars

where d_b, is the smallest bar diameter in the column, and D_c is the smaller column dimension if rectangular or the column diameter if circular.

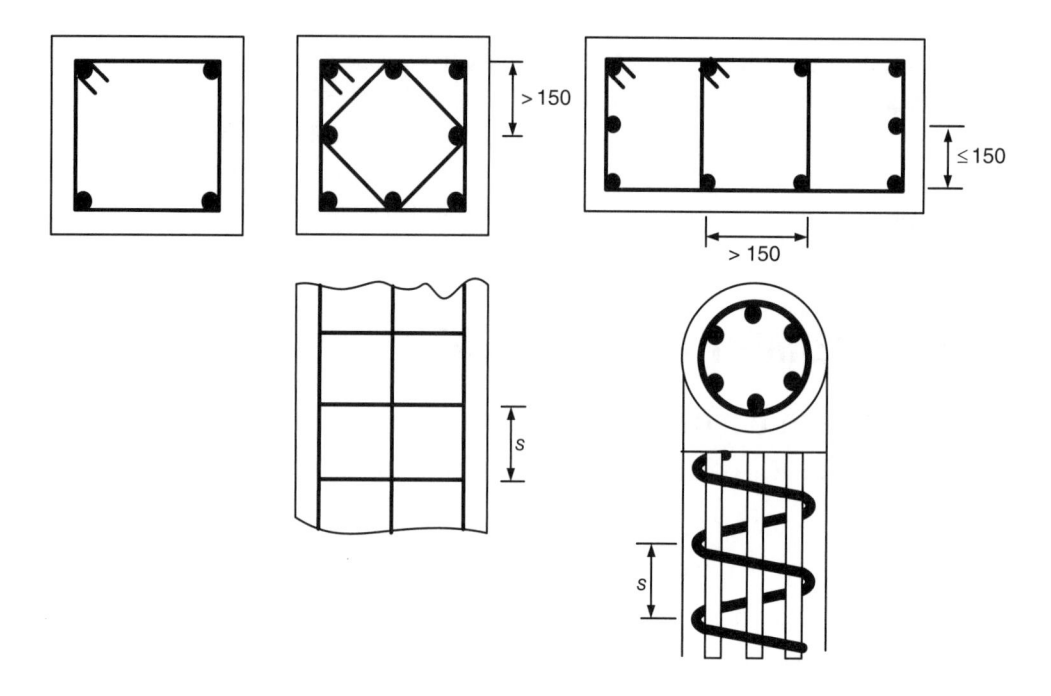

Figure 10.10(1) Examples of ties and helices for restraining longitudinal bars in columns
Note: all dimensions are in mm.

Columns where $f'_c > 50$ MPa

For high-strength concrete columns, Clause 10.7.3.1(b) of the Standard specifies an elaborate process for concrete confinement to which the reader should refer for details.

10.10.3 Recommendations

A very brief discussion has been given here on column reinforcement. Detailed practical guidance on all aspects of design requirements and detailing of both longitudinal and lateral reinforcements can be found in the Standard. The reader is advised to consult Clause 10.7 of the Standard while carrying out practical design work.

Note that the transmission of column axial force through floor systems also requires consideration in practice. Quantitative design provisions are given in Clause 10.8 of the Standard.

10.11 COMMENTS

The reader should by now appreciate that all the equations given in this chapter, except the approximate expressions discussed in Section 10.6, are executable only if the sectional details of the column are given. However, the objective of design is to determine these details. This underlies the fact that reinforced concrete column design is an indirect process: that is, assume a section then check all the equations to ensure that the section is adequate and at the same time not over-sized. If not, assume another section and try again until the section is adequate. No doubt such a trial and error approach is tedious and highly undesirable, even with the aid of a computer. For practical projects of any size, the following steps are recommended:

1. Conduct structural analyses under the load combinations specified in Section 1.3.1. These give the values of N^*, M_x^* and M_z^* for each column.
2. Based on these design axial forces and bending moments, estimate a range of column sizes which form a 'pool' from which a suitable section is later selected for each set of N^*, M_x^* and M_z^*. The preliminary design procedure described in Section 10.6 may be helpful to produce this pool.
3. For each section in the pool, construct the interaction diagrams for uniaxial bending about the x and z axes. Follow the procedure given in Section 10.3.3. Modify the interaction diagrams using the appropriate capacity reduction factors (see Section 10.5).
4. The design or selection process can now begin by systematically going through the following checking process for each of the columns or for each set of N^*, M_x^* and M_z^*:
   - Select the smallest section in the pool.
   - Check uniaxial bending capacity using the interaction diagrams obtained in (3) for the section.
   - Compute the moment magnifiers δ_b and δ_s as described in Section 10.7; magnify the design moments M_x^* and M_z^* accordingly.
   - With N^* and the magnified M_x^* and M_z^*, check Equation 10.9(3) for adequacy of the section against the combined effects of compression and biaxial bending.
   - If the section passes all these checks, then it is satisfactory; proceed to the next column and repeat the selection process and if any of the above checks is negative, go back to the pool and select the next larger section and try again until all checks are satisfactory.

The accuracy of this indirect design process depends on the number of sections available in the pool; the more sections in the pool, the better the design outcome, and the resulting section is not only safe, but will not be unduly over-designed. This indirect process is tedious, but fortunately the pool of sections and the corresponding interaction diagrams may be built up gradually by an individual designer or in a design office – they may be used repeatedly for column design in future projects.

This indirect procedure has been adopted in an integrated computerised design package for multistorey flat plate structures. Its operational details may be found in the user manual for the package (Loo 1988).

It should also be noted that the Cement and Concrete Association of Australia (CCAA) in conjunction with Standards Australia (SA) has in the past published a compilation of design charts (HB71–2002). An updated design manual, HB71–2011, which complies with the Standard, has since been published.

10.12 SUMMARY

The basic strength equations for centrally loaded columns and columns subjected to uniaxial bending are developed, complemented by the procedure for constructing the column interaction diagram. As advanced topics, an iterative approach and a semi-graphical procedure for the analysis of arbitrary-shaped column sections are also discussed. In addition, a preliminary design procedure is presented, together with the Standard specifications for the capacity reduction factor (ϕ), the moment magnifiers for slender columns, biaxial bending, and reinforcement requirements. Worked examples are given to illustrate the important aspects of column analysis and design. As a supplement, a rigorous column design process is expounded.

10.13 PROBLEMS

1. The symmetrically reinforced triangular column illustrated in Figure 10.13(1) has a 100 mm × 100 mm square void at the centre of the section. Compute the squash load N_{uo} for the section. The reinforcement comprises six N20 bars and f'_c = 80 MPa.

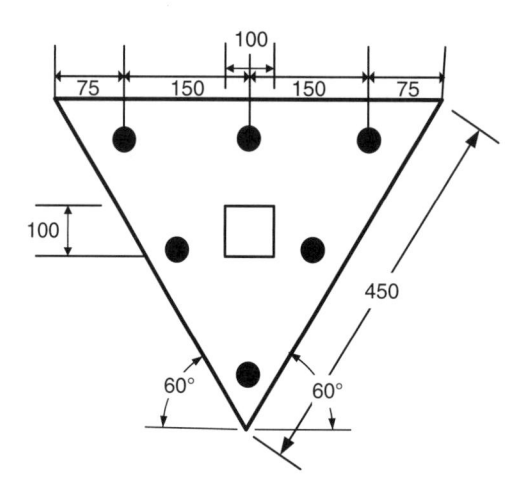

Figure 10.13(1) Details of symmetrically reinforced triangular column

Note: all dimensions are in mm.

2. The 450 mm diameter circular column detailed in Figure 10.13(2) has a 75 mm diameter void in the middle of the section. Compute the squash load N_{uo} for the section. The reinforcement comprises eight N20 bars and f'_c = 80 MPa.

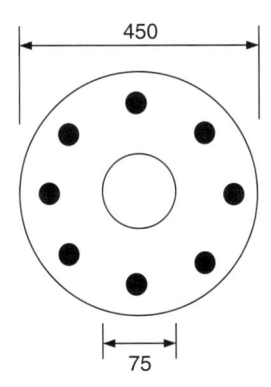

Figure 10.13(2) Details of circular column

Note: all dimensions are in mm.

3. For the column section detailed in Figure 10.3(4), construct the interaction diagram for bending about the x-axis. On the same graph, superimpose the strength-reduced interaction diagram.
4. Details of an asymmetrically reinforced rectangular column section are given in Figure 10.13(3). The column is subjected to uniaxial bending about the x-axis. For the given eccentricity, compute ϕN_u.

    Note that A_{st} = 4 N24 and A_{sc} = 3 N24. Take f'_c = 32 MPa. You may use the approximate approach.

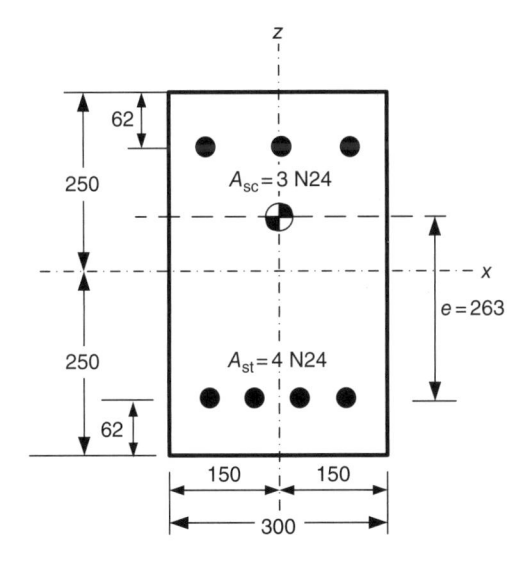

Figure 10.13(3) Details of asymmetrically reinforced rectangular column section

Note: all dimensions are in mm.

5. Figure 10.13(4) shows a triangular column section subjected to uniaxial bending with the main reinforcement consisting of six N20 bars.
   (a) Compute N_u using the semi-graphical method.
   (b) Check your result by the iterative approach.

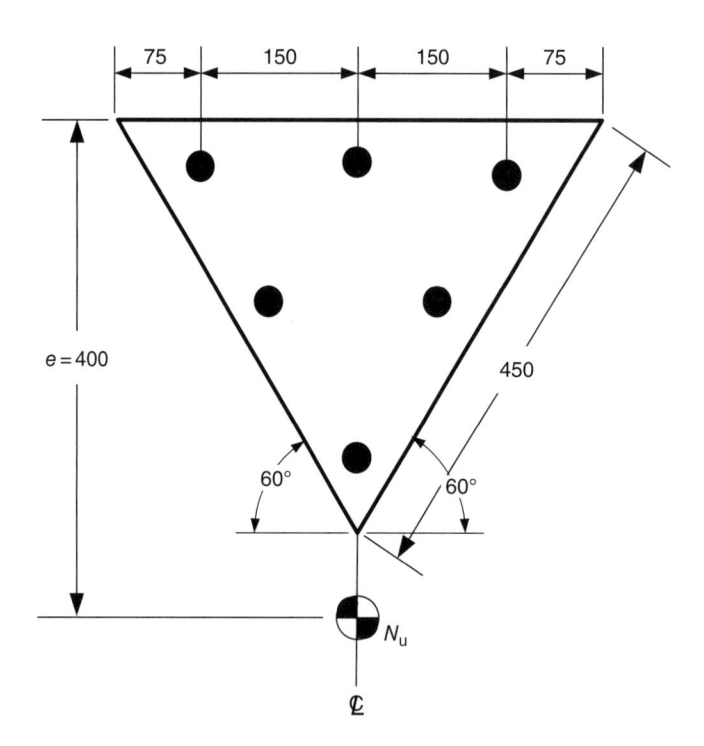

Figure 10.13(4) Details of triangular column section ($f'_c = 32$ MPa)

Note: all dimensions are in mm unless otherwise specified.

6. The elliptical section of a column detailed in Figure 10.13(5) is governed by the equation

$$\left(\frac{x}{300}\right)^2 + \left(\frac{y}{150}\right)^2 = 1$$

For the specified eccentricity, obtain the ultimate load (N_u) using the semi-graphical procedure. Take $f'_c = 50$ MPa and a main reinforcement consisting of eight N28 bars.

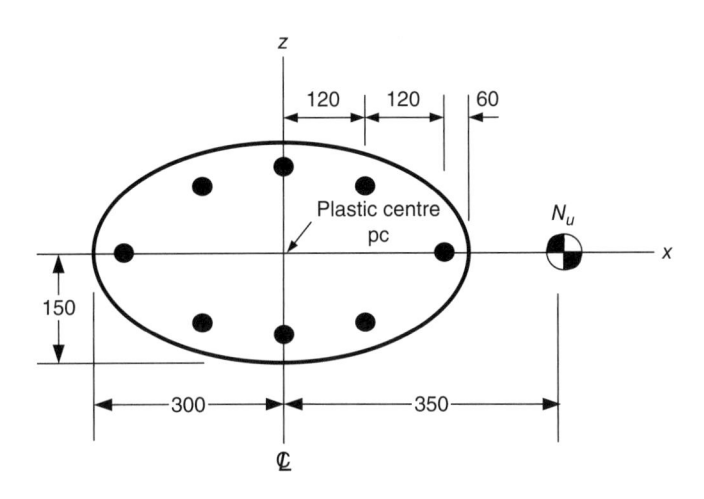

Figure 10.13(5) Details of elliptical column section

Note: all dimensions are in mm.

7. You are given the following service load data for a column subjected to uniaxial bending:
    - dead load, $g = 380$ kN
    - dead-load moment, $M_g = 38$ kNm
    - live load, $q = 250$ kN
    - live-load moment, $M_q = 44$ kNm.

    Assuming $f'_c = 65$ MPa and using N bars only, proportion a (preliminary) square section that is to be symmetrically reinforced with $p_t + p_c = 1.5\%$.

8. Figure 10.13(6) shows the cross-section and loading details of a square column reinforced with four N32 bars. The raw data for plotting the interaction diagrams for bending about the x and z axes are shown in Table 10.13(1). Assume $f'_c = 25$ MPa and that the column is short (stubby). Can the column withstand the biaxial loading? Support your answer with all necessary computations.

294 PART 1 REINFORCED CONCRETE

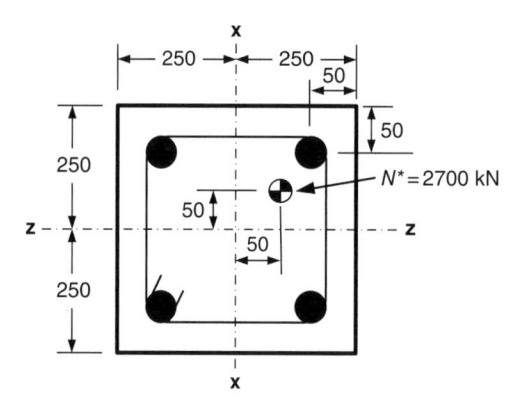

Figure 10.13(6) Details of square column

Note: all dimensions are in mm.

Table 10.13(1) Data for plotting interaction diagrams for bending about the x and z axes for Problem 8

	N_u (kN)	M_{ux} and M_{uz} (kNm)
Concentric load	6524.5	0.0
$k_u = 1$	4670.0	360.0
$k_u = 0.7$	3039.4	533.9
$k_u = k_{uB} = 0.545$	2404.4	579.0
$k_u = 0.5$	1998.0	562.9
$k_u = 0.333$	1319.3	501.3
Pure bending	0.0	263.7

9. Design a square tied column to support an axial dead load, $g = 550$ kN and an axial live load, $q = 800$ kN. Initially assume that 2% longitudinal steel is desired and take $f'_c = 32$ MPa. Use N bars for reinforcement.
10. Design a round spiral column to support an axial dead load, $g = 1050$ kN and an axial live load, $q = 1325$ kN. Initially assume that approximately 2% longitudinal steel is desired and take $f'_c = 32$ MPa. Use N bars for reinforcement.
11. A short column section is detailed in Figure 10.13(7). Considering uniaxial bending only:
    (a) show that the column would sustain a compression failure (i.e. $e' < e'_B$)
    (b) compute the ultimate load N_u.
    Note that $A_{st} = A_{sc} = 4$ N28. Take $f'_c = 32$ MPa. You may use the approximate approach.

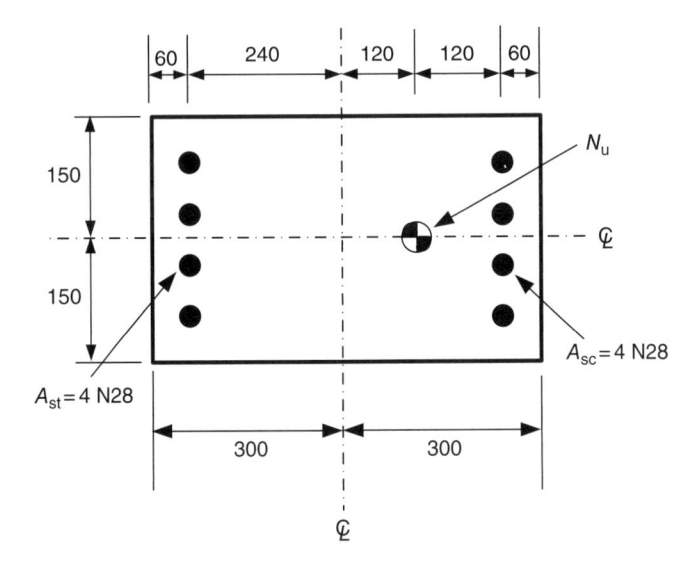

Figure 10.13(7) Details of a short column section

Note: all dimensions are in mm.

12. For the asymmetrically reinforced rectangular column section shown in Figure 10.13(8), assuming $f'_c = 32$ MPa:
    (a) compute the value of the ultimate section capacity N_{uo}
    (b) find the balanced failure values for axial load (N_{uB}) and bending moment (M_{uB}) for bending about the z-axis.

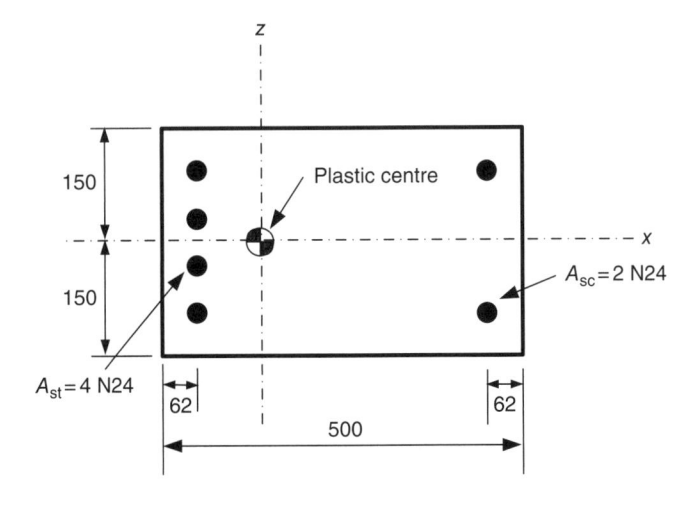

Figure 10.13(8) Details of column section for Problem 12

Note: all dimensions are in mm.

11 WALLS

11.1 INTRODUCTION

In the past, when design provisions were lacking, wall panels were usually taken as nonload-bearing. Due to the increased acceptance of precast techniques in building construction and a trend towards reinforced concrete core structures, walls have now become popular as load-bearing elements.

A wall is a vertical planar continuum with a thickness much smaller than its height or length. If a wall is short, with a length of the same order as the thickness, it can be treated as a column. In fact, AS 3600–2009 (the Standard) defines a wall as an element wider than three times its thickness. Otherwise, the element is considered to be a column.

Walls are normally supported at the bottom end by a floor system and at the top end by a roof structure or another floor. Or a wall may be freestanding. Depending on the chosen structural system, walls may be supported on either or both sides by interconnecting walls or other structural elements. Consequently, a wall may act like a column, a beam cantilevered at one end or a slab standing vertically. Figure 11.1(1)a depicts a wall under vertical in-plane and lateral bending loads; Figure 11.1(1)b shows a wall that is under combined in-plane vertical (axial) and horizontal (shear) forces. At times a wall may be subjected to simultaneous axial, bending and shear forces.

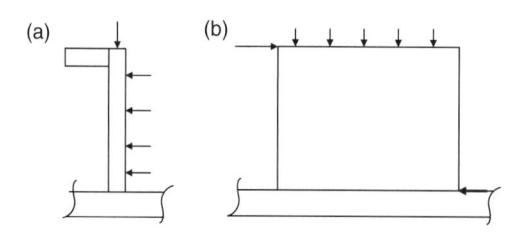

Figure 11.1(1) Wall panels under in-plane and lateral loads: **(a)** axial (in-plane) and lateral forces and **(b)** axial and in-plane shear forces

Walls in tilt-up construction and low-rise structures are relatively easy to analyse, based on forces acting statically on the element. In multistorey buildings, shear walls are present in elevator shafts, central cores or stairwells; they may also exist as load-bearing planar continuums. In addition to resisting horizontal forces from wind and earthquake loading, shear walls also carry in-plane vertical forces due to dead and live loading. With the axial, bending and shear forces in a wall computed using an appropriate structural analysis procedure, the provisions in Section 11 of the Standard may be used to design a wall for strength and serviceability.

The Standard's strength design provisions for walls are highlighted in this chapter. Also summarised are relevant recommendations given in ACI 318–2014. Doh and Fragomeni (2004) have proposed a strength design formula covering walls in both one-way and two-way actions. This new formula is included in the discussion here and illustrative examples are provided to demonstrate the application of the three design procedures.

11.2 STANDARD PROVISIONS

The major design recommendations given in the Standard may be summarised as follows:
- As in Clause 11.1 in the Standard, walls that are unbraced, or braced and subjected to simultaneous in-plane and lateral or bending loads must be designed as slabs (Chapter 9) or columns (Chapter 10) as appropriate.
- If designed as a column, vertical reinforcement must be provided in both faces of a wall. Lateral restraint requirements for vertical reinforcement may be relaxed if the conditions specified in Clause 11.7.4 of the Standard are met.
- For a wall to be designed as a slab, the stress at the mid-height section due to the design action effect caused by the combined axial and in-plane bending loads must be less than the lesser of $0.03\,f'_c$ and 2 MPa. Further, if designed as a slab, the ratio of effective height (H_{we}) (see Section 11.3.1) to thickness must not exceed 50, or 40 where fire resistance is also a design consideration (see Clause 5.7.3 of the Standard).
- Clause 11.2 of the Standard specifies that braced walls subjected to vertical in-plane or axial loads only may be designed using the simplified method given in Clause 11.5 (which is discussed in Section 11.3.1). If subjected to horizontal in-plane shear forces, braced walls must be designed in accordance with the procedure given in Clause 11.6 (see Section 11.4).
- As stipulated in Clause 11.3 of the Standard, a wall is considered braced if it forms part of a structure that does not rely on out-of-plane strength and stiffness of the wall and its connection to the rest of the said structure is capable of transmitting:
  – any calculated load effects
  – 2.5% of the total vertical load that the wall is designed to carry at the lateral support level, but not less than 2 kN per metre length of the wall.

11.3 WALLS UNDER VERTICAL LOADING ONLY

The Standard provides two methods for the axial strength design of concrete walls. Clause 11.5 specifies a simplified design formula that can be used when certain loading and bracing restrictions are met. Otherwise, the Standard allows any wall to be designed as a column as discussed in Chapter 10.

The Standard presents and recognises wall panels in two-way action that are supported laterally on three and four sides. Figures 11.3(l)a, b and c, respectively, depict the anticipated failure modes of wall panels in one-way action, and in two-way action with either three or four sides supported.

(a) One-way action   (b) Two-way action with three sides supported laterally   (c) Two-way action with all four sides supported laterally

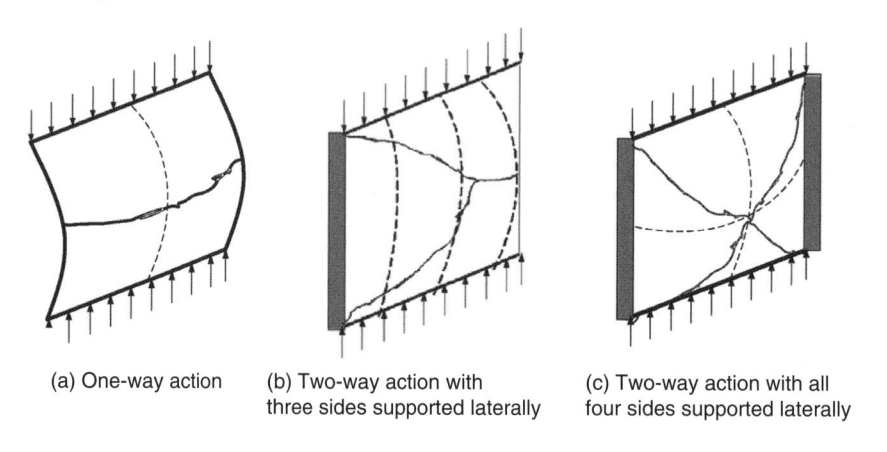

Figure 11.3(1) Behaviour of vertically loaded wall panels

11.3.1 Simplified method

Clause 11.5 of the Standard describes a simplified procedure in which the design axial strength per unit length of a braced wall in compression is given as

$$\phi N_u = \phi(t_w - 1.2e - 2e_a)0.6f'_c \qquad \text{Equation 11.3(1)}$$

where
   $\phi = 0.6$ is the capacity reduction factor
   e = eccentricity of the load applied to the top of the wall and measured at right angles to the plane of the wall, which must be determined as follows:
1. Due to the load applied by the floor (or roof) supported by the top of the wall, being designed must be taken as:
   - for a discontinuous floor, one-third of the depth of the bearing area measured from the span face of the wall
   - for a cast in situ concrete floor continuous over the wall, zero.
2. Due to the aggregated load from all floors above the floor at the top of the wall being designed may be taken to be zero.

Note that the resultant eccentricity of the total load from 1 and 2 must be calculated, but taken as not less than $0.05t_w$.

An additional eccentricity due to the deformation of the wall is

$$e_a = H_{we}^2/2500t_w \qquad \text{Equation 11.3(2)}$$

where
- t_w = the thickness of the wall
- H_{we} = the effective wall height (as in Clause 11.4 of the Standard) is taken to be kH_w, in which k is determined for various support conditions as follows:

1. For one-way action with top and bottom floors providing the lateral support at both ends:
   - $k = 0.75$ where restraint against rotation is provided at both ends
   - $k = 1.0$ where no restraint against rotation is provided at one or both ends.
2. For two-way action with three sides supported laterally by floors and intersecting walls

$$k = \frac{1}{1 + \left(\frac{H_w}{3L_1}\right)^2} \qquad \text{Equation 11.3(3)}$$

but not less than 0.3 or greater than what is obtained in condition 1.

3. For two-way action with four sides supported laterally by floors and intersecting walls

$$k = \frac{1}{1 + \left(\frac{H_w}{L_1}\right)^2} \quad \text{where } H_w \leq L_1 \qquad \text{Equation 11.3(4)}$$

$$k = \frac{L_1}{2H_w} \quad \text{where } H_w > L_1 \qquad \text{Equation 11.3(5)}$$

in which H_w is the unsupported height of the wall and L_1, the horizontal length of the wall between the centres of lateral restraints. Note that Clause 11.4(c) of the Standard also covers walls with openings and the reader should refer to this as necessary.

Clause 11.5.1 of the Standard limits the application of Equation 11.3(1) to walls with a slenderness ratio $H_{we}/t_w \leq 30$. A minimum eccentricity of $0.05t_w$ must be designed for. The required minimum wall reinforcement ratios are $p_w = 0.0015$ in the vertical direction and $p_w = 0.0025$ in the horizontal direction. In computing p_w, the overall thickness t_w is used.

11.3.2 American Concrete Institute code provision

From the American Concrete Institute, ACI 318–2014 also recommends a simplified procedure for the design of walls under vertical loading, which incorporates the following empirical formula:

$$\phi N_u = \phi 0.55 f'_c A_g \left[1 - (kH/32t_w)^2\right] \qquad \text{Equation 11.3(6)}$$

where

$A_g = L \times t_w$, is the gross area of the wall panel section

H = the unsupported height of the wall

$k = 0.8$ and 1.0 for walls restrained and unrestrained against rotation, respectively, and 2.0 for walls not braced laterally

$\phi = 0.7$ for compression members.

The equation applies when either H/t_w or L/t_w is less than or equal to 25. The minimum allowable thickness is 100 mm. The resultant load must be in the 'middle third' of the overall thickness of the wall. This allows for a maximum eccentricity of $t_w/6$.

At first glance, Equations 11.3(6) and the simplified formula (or Equation 11.3(1)) appear to be different. An appropriate rearrangement of Equation 11.3(1), that is, the replacement of e by $t_w/6$ and the multiplication of L for length, reveals that the two equations are actually very similar. In general, Equation 11.3(6) gives a slightly higher capacity load (see Section 11.6.1). It is also not applicable to cases with $e > t_w/6$.

11.3.3 New design formula

Doh and Fragomeni (2004) have conducted research on normal and high-strength reinforced concrete walls with various support conditions. Wall panels with f'_c up to 80 MPa were tested to failure. Their extensive laboratory and analytical study has led to a new design formula, which is capable of analysing walls in both one-way and two-way actions. It also allows for higher slenderness ratios than those specified in the Standard and in ACI 318–2014. The formula takes a form similar to Equation 11.3(1), that is,

$$\phi N_u = \phi 2.0 (f'_c)^{0.7} (t_w - 1.2e - 2e_a) \qquad \text{Equation 11.3(7)}$$

where ϕ, t_w, e and e_a are as defined in Section 11.3.1. However, in Equation 11.3(2), for e_a, that is

$$e_a = H_{we}^2 / 2500 t_w \qquad \text{Equation 11.3(8)}$$

the effective wall height

$$H_{we} = \beta H \qquad \text{Equation 11.3(9)}$$

where for walls simply supported at top and bottom or walls in one-way action only

$$\beta = 1 \text{ if } H/t_w < 27 \qquad \text{Equation 11.3(10)}$$

$$\beta = \frac{18}{\left(\frac{H}{t_w}\right)^{0.88}} \text{ if } H/t_w \geq 27 \qquad \text{Equation 11.3(11)}$$

and for walls restrained laterally at all four sides or walls in two-way action

$$\beta = \alpha \frac{1}{1 + \left(\frac{H}{L}\right)^2} \quad \text{if } H \leq L \qquad \text{Equation 11.3(12)}$$

$$\beta = \alpha \frac{L}{2H} \quad \text{if } H > L \qquad \text{Equation 11.3(13)}$$

In Equations 11.3(12) and 11.3(13), the eccentricity parameter

$$\alpha = \frac{1}{1 - \frac{e}{t_w}}, \quad \text{if } H/t_w < 27 \qquad \text{Equation 11.3(14)}$$

and

$$\alpha = \frac{1}{1 - \frac{e}{t_w}} \cdot \frac{18}{\left(\frac{H}{t_w}\right)^{0.88}} \quad \text{if } H/t_w \geq 27 \qquad \text{Equation 11.3(15)}$$

Equation 11.3(7) in conjunction with Equations 11.3(8) to 11.3(15) has been shown to be superior to Equations 11.3(1) and 11.3(6). It is applicable to walls with slenderness ratios greater than 30, which is the conservative limit set for Equation 11.3(1), and it accounts for the effect of eccentricity up to $t_w/6$.

11.3.4 Alternative column design method

If the Standard stipulations on bracing, loading or slenderness cannot be complied with, a wall must be designed as a column (refer to Chapter 10). As a column, the amount of A_{sc} must not be less than $0.01A_g$. However, for a column with a larger area than is required for strength consideration, a reduced amount of A_{sc} may be used if $A_{sc} f_y > 0.15N^*$, but $A_{sc} \leq 0.04A_g$ (see Clause 10.7.1(a) of the Standard). Experience has indicated that the column design process normally leads to a much higher reinforcement content than is due to the simplified method for walls.

When treating a wall as a column, the upper limit of the slenderness ratio is $H_{we}/r = 120$, which is equal to $H_{we}/t_w = 36$ (assuming $r = 0.3t_w$). This limit is higher than the limit specified for the simplified method described in Section 11.3.1. As detailed in Chapter 10, the column design method is based on the moment magnifier principle, which has been recommended in the ACI 318 code for many years. The method is considered reliable, but it leads to a heavier section in terms of reinforcement.

11.4 WALLS SUBJECTED TO IN-PLANE HORIZONTAL FORCES

11.4.1 General requirements

If in-plane horizontal forces act in conjunction with axial forces, and compression exists over the entire section, the wall may be designed conservatively for horizontal shear only.

If tension exists in part of the section, the wall must be designed for in-plane bending as described in Chapter 10 and for horizontal shear in accordance with Section 11.4.2, or for combined in-plane bending and shear according to relevant clauses related to the design of non-flexural members in Section 12 of the Standard.

11.4.2 Design strength in shear

The Standard stipulates that, for a reinforced concrete wall, the critical section for maximum shear must be taken at a distance of $0.5L_w$ or $0.5H_w$ from the base, whichever is less, where L_w and H_w are the overall length and height of the wall, respectively.

The design strength in shear as in Clause 11.6 of the Standard is given as

$$\phi V_u = \phi(V_{uc} + V_{us}) < \phi V_{u.max} \qquad \text{Equation 11.4(1)}$$

where $\phi = 0.7$

$$V_{u.max} = 0.2 f'_c (0.8 L_w t_w) \qquad \text{Equation 11.4(2)}$$

and the ultimate strength, excluding shear reinforcement, is

$$V_{uc} = \left(0.66\sqrt{f'_c} - 0.21\frac{H_w}{L_w}\sqrt{f'_c}\right) 0.8 L_w t_w \text{ if } H_w/L_w \leq 1 \qquad \text{Equation 11.4(3)}$$

or

$$V_{uc} = \left[0.05\sqrt{f'_c} + \frac{0.1\sqrt{f'_c}}{\left(\dfrac{H_w}{L_w} - 1\right)}\right] 0.8 L_w t_w \text{ if } H_w/L_w > 1 \qquad \text{Equation 11.4(4)}$$

In either case, the ultimate shear strength of a wall without shear reinforcement is

$$V_{uc} \geq 0.17\sqrt{f'_c}(0.8 L_w t_w) \qquad \text{Equation 11.4(5)}$$

Further, in Equation 11.4(1) the contribution to shear strength of the shear reinforcement is

$$V_{us} = p_w f_{sy}(0.8 L_w t_w) \qquad \text{Equation 11.4(6)}$$

where p_w is the lesser of the vertical and horizontal reinforcement ratios of the wall (for $H_w/L_w \leq 1$) or the horizontal reinforcement ratio per vertical metre (for $H_w/L_w > 1$). In computing p_w, the overall wall thickness or overall depth is used, which is different from beams and slabs where the effective depth is specified.

11.4.3 American Concrete Institute recommendations

Equations 11.4(3) and 11.4(4) ignore the often significant effects of axial compression on the shear strength of a wall (Warner, Rangan, Hall and Faulkes 1998). ACI 318–2014, on the other hand, takes such effects into account by redefining $V_{u.max}$ and V_{uc} in Equation 11.4(1) as follows:

$$V_{u.max} = 0.83\sqrt{f'_c}(0.8L_w t_w) \quad \text{Equation 11.4(7)}$$

and

$$V_{uc} = 0.275\sqrt{f'_c}(0.8L_w t_w) + 0.2N^* \quad \text{Equation 11.4(8)}$$

or

$$V_{uc} = \left[0.05\sqrt{f'_c} + \frac{L_w\left(0.1\sqrt{f'_c} + 0.2\dfrac{N^*}{L_w t_w}\right)}{\left(\dfrac{M^*}{V^*} - \dfrac{L_w}{2}\right)}\right] 0.8L_w t_w \quad \text{Equation 11.4(9)}$$

whichever is less, noting that in either case, Equation 11.4(5) remains valid.

In Equations 11.4(8) and 11.4(9), the axial force on the wall (N^*) is taken as positive for compression and negative for tension. If $[(M^*/V^*) - (L_w/2)]$ is negative, Equation 11.4(9) is not applicable.

In addition, the shear strength due to the shear reinforcement in Equation 11.4(1) takes the form

$$V_{us} = A_{wh} f_{sy}(0.8L_w)/s \quad \text{Equation 11.4(10)}$$

where A_{wh} is the total horizontal reinforcement area within a given vertical distance (s).

Note that the strut-and-tie modelling approach (Chapter 13) may also be used to determine the shear strength of a wall subjected to in-plane vertical and horizontal forces.

11.5 REINFORCEMENT REQUIREMENTS

Clause 11.7.1 of the Standard stipulates that walls designed using the simplified method are required to have a minimum reinforcement ratio of:
- $p_w = 0.0015$ vertically
- $p_w = 0.0025$ horizontally.

Where there is no restraint against horizontal shrinkage or thermal movements, the horizontal reinforcement ratio can be reduced to zero for walls less than 2.5 m in overall length in the horizontal direction, or to 0.0015 for walls longer than this.

In ACI 318–2014, the same rules apply except that the vertical and horizontal p_w can be reduced to 0.0012 and 0.0020, respectively, if bars of less than 16 mm diameter or mesh are used.

As mentioned previously, walls designed as columns must comply with the requirement that the vertical reinforcement ratio is between 0.01 and 0.04.

For crack control of walls that are restrained from expanding or contracting horizontally due to shrinkage or temperature effects, Clause 11.7.2 of the Standard specifies that, for exposure classifications A1 and A2, the horizontal reinforcement ratio (p_w) shall not be less than:

- 0.0025 for minor control
- 0.0035 for moderate control
- 0.0060 for strong control.

For the more severe exposure classifications B1, B2, C1 and C2, the ratio is strictly 0.006.

As in Clause 11.7.3 of the Standard, the minimum clear distance between parallel bars, ducts and tendons must be sufficient to ensure that the concrete can be placed properly, but not less than $3d_b$ where d_b is the bar diameter. The maximum centre-to-centre spacing of parallel bars must be $2.5t_w$ or 350 mm, whichever is less. For walls thicker than 200 mm, vertical and horizontal reinforcement is to be provided in two layers, one near each face of the wall.

For walls designed as columns, lateral restraint of the vertical reinforcement may also be required. For details, the reader is referred to Clause 11.7.4 of the Standard.

11.6 ILLUSTRATIVE EXAMPLES

11.6.1 Example 1 – load-bearing wall

Problem

A rectangular wall is 300 mm thick with an unsupported height of 6 m and a length of 5 m. The vertical axial loading from the floor beams above acts with a 10 mm eccentricity. Assuming that the wall is restrained against rotation at the top and bottom ends by the floors, and that $f'_c = 32$ MPa and $f_{sy} = 500$ MPa, compute the design axial strength values for the following side-restraint conditions:

(a) without side restraints using the Standard procedure
(b) without side restraints using the ACI 318–2014 formula
(c) with restraints at both sides of the wall.

Solution

(a) Without side restraint – Standard procedure

Equation 11.3(1) is applicable in this case, for which $H_{we} = kH_w = 0.75 \times 6 = 4.5$ m, which is less than the wall length of 5 m.

Hence, adopt $H_{we} = 4.5$ m.

Since the eccentricity $e = 10$ mm $< 0.05 t_w = 0.05 \times 300 = 15$ mm, use $e = 15$ mm.

The slenderness ratio $H_{we}/t_w = 4500/300 = 15 < 30$, which is acceptable.

Equation 11.3(2): $e_a = (H_{we})^2/2500 t_w = (4500)^2/(2500 \times 300) = 27$ mm

With these values:

Equation 11.3(1): $\phi N_u = \phi(t_w - 1.2e - 2e_a) 0.6 f'_c$
$$= 0.6\,[300 - 1.2(15) - 2(27)] \times 0.6 \times 32 \times 5000 \times 10^{-3}$$
$$= 13\,133 \text{ kN}$$

(b) Without side restraint – ACI 318–2014 procedure

Apply Equation 11.3(6) for which

$L/t_w = 5000/300 = 16.7 < 25$, which is acceptable.

$t_w > 100$, which is acceptable.

$k = 0.8$ as the top and bottom ends are restrained against rotation.

Thus,

Equation 11.3(6): $\phi N_u = \phi 0.55 f'_c A_g [1 - (kH/32 t_w)^2]$
$$= 0.7 \times 0.55 \times 32 \times (300 \times 5000) \times \left[1 - \{(0.8 \times 6000)/(32 \times 300)\}^2\right] \times 10^{-3}$$
$$= 13\,860 \text{ kN}$$

Note that Equation 11.3(6) gives a 5.5% higher capacity than Equation 11.3(1). This is because the ACI 318–2014 equation does not consider the effect of eccentricity, or the additional eccentricity due to the secondary $P - \Delta$ effects.

(c) With both sides restrained laterally

The Standard's simplified method in the form of Equation 11.3(1) is also applicable when both sides are restrained laterally. The solution is obtained by using the appropriate input parameters for the wall in two-way action.

Since $H_w > L_1$,

Equation 11.3(5): $k = \dfrac{L_1}{2H_w} = \dfrac{5000}{2 \times 6000} = 0.4167$

That is, $H_{we} = kH_w = 0.4167 \times 6 = 2.5$ m

Similarly, the eccentricity $e = 10$ mm $< 0.05 t_w = 0.05 \times 300 = 15$ mm

Therefore, adopt $e = 15$ mm.

Equation 11.3(2): $e_a = (H_{we})^2/2500 t_w = (2500)^2/(2500 \times 300) = 8.3333$ mm

Then

Equation 11.3(1): $\phi N_u = \phi(t_w - 1.2e - 2e_a) 0.6 f'_c$
$$= 0.6\,[300 - 1.2(15) - 2(8.3333)]\,0.6 \times 32 \times 5000 \times 10^{-3}$$
$$= 15\,283 \text{ kN}$$

Comparing the results in cases (a) and (c) indicates that, with both sides of the wall restrained laterally, the design axial strength increases by $(15\,283 - 13\,133)/13\,133 \times 100\%$ = 16.4%.

11.6.2 Example 2 – tilt-up panel

Problem

A two-storey office building is to be constructed as an assembly of tilt-up panels. The (continuous) external panels are each 7.5 m high by 5 m long. They are designed to support the loading from the first floor as well as the roof. The first floor is 3.6 m above the base slab; the roof is a further 3.6 m above the first floor. The dead and live loads from the roof are 45 kN/m and 40.5 kN/m, respectively, along the wall. The dead and live loads from the first floor are 76.5 kN/m and 66 kN/m, respectively. Assume $f'_c = 32$ MPa and that the floor slab bearing into the panels is 25 mm.

Is a 150-mm thick concrete wall with minimum vertical and horizontal reinforcement adequate for the design?

Solution

(a) Design at 1.8 m (mid-height of wall between base and first floor)
   Assuming $\rho_w = 24$ kN/m³, wall dead load = $24 \times (7.5 - 1.8) \times 0.15 = 20.52$ kN/m
   Total dead load g = self-weight + roof dead load + floor dead load
   $\qquad = 20.52 + 45 + 76.5 = 142.02$ kN/m
   Total live load q = roof live load + floor live load = $40.5 + 66 = 106.5$ kN/m
   Design load = $(1.2 \times g) + (1.5 \times q) = 330.17$ kN/m

(b) Eccentricity of load at 1.8 m height
   Vertical load of first floor = $(1.2 \times 76.5) + (1.5 \times 66) = 190.8$ kN/m
   As in Clause 11.5.2 of the Standard, vertical load is assumed to act at one-third the depth of the bearing area from the span face of the wall. Thus, the eccentricity of the load above the first floor
   $e_f = (150/2) - (25/3) = 66.667$ mm $> e_{min} = 0.05t = 7.5$ mm, which is acceptable.
   At 1.8 m, the eccentricity of load, $e = 66.667 \times 190.8/330.17 = 38.525$ mm.
   Also,

   Equation 11.3(2): $e_a = (0.75 \times 3600)^2/(2500 \times 150) = 19.44$ mm

   Equation 11.3(1): $N_u = 0.6[150 - 1.2(38.525) - 2(19.44)] \times 0.6 \times 32$
   $\qquad = 747.53$ kN/m > 330.17 kN/m, which is satisfactory.

11.6.3 Example 3 – the new strength formula

Problem

Repeat the example in Section 11.6.1 using the new design formula for a wall:
(a) without side restraint
(b) with all four sides restrained laterally.

Solution

(a) Design strength without side restraint.
   Since $H/t_w < 27$

   Equation 11.3(10): $\beta = 1$. Hence,
   Equation 11.3(9): $H_{we} = \beta H = 6$ m and
   Equation 11.3(8): $e_a = (6000)^2/(2500 \times 300) = 48$ mm.

   Thus

   Equation 11.3(7): $\phi N_u = 0.6 \times 2 \times 32^{0.7}[300 - 1.2(10) - 2(48)] \, 5000 \times 10^{-3}$
   $= 13\,033$ kN

(b) Design strength with all sides restrained.
   For $H/t_w < 27$

   Equation 11.3(14): $\alpha = \dfrac{1}{1 - \dfrac{e}{t_w}} = \dfrac{1}{1 - \dfrac{10}{300}} = 1.0345$

   Hence, for $H = 6$ m $> L = 5$ m

   Equation 11.3(13): $\beta = \alpha \dfrac{L}{2H} = 1.0345 \dfrac{5000}{2 \times 6000} = 0.4310$

   Thus $H_{we} = 0.4310 \times 6000 = 2586$ mm
   and

   Equation 11.3(2): $e_a = (2586)^2/(2500 \times 300) = 8.917$ mm

   Finally,

   Equation 11.3(7): $\phi N_u = 0.6 \times 2 \times 32^{0.7}[300 - 1.2(10) - 2(8.917)] \, 5000 \times 10^{-3}$
   $= 18\,340$ kN

It should be noted that the new formula yields higher load-carrying capacities. In particular, when side restraints are considered, the capacity is about 40% greater than that allowed for a one-way-action wall. This shows that Equation 11.3(1) is unduly conservative when applied to walls with lateral restraints at all the four sides.

11.6.4 Example 4 – design shear strength

Problem

A reinforced concrete wall is required to resist vertical and horizontal in-plane forces in a multistorey building. The overall first-floor height and length of the wall are both 4.5 m. The wall is 250 mm thick with two 500 mm × 500 mm boundary columns. A structural analysis of the building gives $N^* = 6925$ kN, $V^* = 3750$ kN and the in-plane moment $M^* = 19\,300$ kNm.

Compute the shear strength of the wall and provide suitable reinforcement. Assume $f'_c = 50$ MPa and $f_{sy} = 500$ MPa. Compare the strength result with the ACI 318–2014 provisions.

Solution

The boundary columns serve to resist the in-plane bending moment in the wall, as well as the axial compression. The shear force due to horizontal load on the other hand is carried by the wall.

(a) Check the maximum shear capacity.

Equation 11.4(2): $V_{u.max} = 0.2 \times 50(0.8 \times 4500 \times 250) \times 10^{-3} = 9000$ kN

And $\phi V_{u.max} = 0.7 \times 9000 = 6300$ kN $> V^*$, which is acceptable.

(b) Calculate the shear strength of the wall.

Trial 1
Use N16 bars @ 200 mm in each face of the wall in both the vertical and horizontal directions.

Equation 11.4(3): $V_{uc} = \left[0.66\sqrt{50} - 0.21(4.5/4.5)\sqrt{50}\right] 0.8 \times 4500 \times 250 \times 10^{-3}$

$= 2863.8$ kN $> 0.17\sqrt{f'_c}(0.8 L_w t_w)$, which is acceptable.

The steel ratio, $p_w = 201/(200 \times 250) = 0.004$

Equation 11.4(6): $V_{us} = 0.004 \times 500 \times (0.8 \times 4500 \times 250) \times 10^{-3} = 1800$ kN

Then

Equation 11.4(1): $\phi V_u = 0.7(2863.8 + 1800) = 3264.7$ kN < 3750 kN, which is not acceptable.

Trial 2
Use N20 bars @ 220 mm.

The steel ratio, $p_w = 314/(220 \times 250) = 0.00571$

Equation 11.4(6): $V_{us} = 0.00571 \times 500 \times 0.8 \times 4500 \times 250 \times 10^{-3} = 2569.5$ kN

Equation 11.4(1): $\phi V_u = 0.7(2863.8 + 2569.5) = 3803.3$ kN > 3750 kN, which is acceptable.

(c) Calculate the shear strength using the ACI 318–2014 recommendations.

Equation 11.4(8): $V_{uc} = 0.275\sqrt{50}(0.8 \times 4500 \times 250) \times 10^{-3} + 0.2 \times 6925$
$= 3135.1$ kN

Equation 11.4(9): $V_{uc} = 3028$ kN < 3135.1 kN. Thus, adopt this value.

Then

Equation 11.4(10): $V_{us} = 314 \times 500 \times (0.8 \times 4500) \times 10^{-3}/220$
$= 2569.1$ kN

Finally,

Equation 11.4(1): $\phi V_u = 0.7(3028 + 2569.1) = 3918.0$ kN > 3750 kN, which is acceptable. Note that the ACI method gives a higher capacity load. This is to be expected as it also accounts for the beneficial effects of axial loading.

(d) Check related design requirements.

The wall panel should be checked for adequate crack control in both vertical and horizontal directions, each having a reinforcement ratio of 0.00571. As quoted in Section 11.5, this ratio is more than adequate for a moderate degree of control, which requires a minimum $p_w > 0.0035$ in exposure classification A1 or A2. It is just short of the required 0.006 for a strong degree of crack control.

Finally, the flexural strength of the wall must also be checked. This is done by adopting the rectangular stress block for the compression zone and applying the general column analysis given in Chapter 10. The moment capacity of the section ϕM_u must be greater than or equal to M^*.

11.7 SUMMARY

Strength design provisions for walls in the Standard and relevant recommendations of ACI 318–2014 are summarised. In addition, a sophisticated strength design formula is also discussed covering walls in both one-way and two-way actions. Illustrative examples are provided to demonstrate the application of these three design procedures.

11.8 PROBLEMS

1. Repeat Example 3 in Section 11.6.3 using the Standard approach (Equation 11.3(1)). Design the necessary reinforcement. Compare the strength results with those given in Section 11.6.3 and comment as appropriate.
2. A braced wall as part of a building structure is 8 m high and 10 m long and may be assumed to be restrained at all four sides. The action effect due to axial loads $N^* = 11050$ kN uniformly distributed over the top end with an eccentricity $e = 28$ mm, and that due to horizontal in-plane loading $V^* = 6780$ kN.

(a) Design the wall based on appropriate Standard provisions, including the reinforcement details.
(b) Check the design using relevant ACI 318–2014 formulas.

Assume f'_c = 25 MPa with A1 exposure classification and a moderate degree of crack control.

3. Design a concrete bearing wall using the ACI 318–2014 formula (Equation 11.3(6)) to support precast concrete roof beams spaced at 2.1 m intervals, as shown in Figure 11.8(1). The bearing width of each beam is 250 mm. The wall is considered to be laterally restrained top and bottom and is further assumed to be restrained against rotation at the footing; thus k = 0.8. Neglect wall weight. Take f'_c = 20 MPa, f_{sy} = 500 MPa, beam dead-load reaction g = 135 kN and beam live-load reaction q = 80 kN.

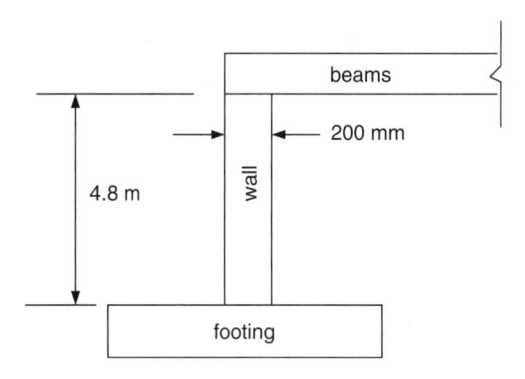

Figure 11.8(1) Details of concrete bearing wall

4. Repeat Problem 3 as above using the Standard approach (Equation 11.3(1)). Design the necessary reinforcement. Compare the strength results with those of Problem 3 and comment as appropriate.

5. A reinforced concrete wall is subjected to a total factored load N^* = 5000 kN uniformly distributed over its 5 m length with an eccentricity e = 15 mm. The wall is part of a 4 m high braced structure and may be taken as being restrained along all four sides. Assuming f'_c = 25 MPa, A1 exposure classification and a moderate degree of crack control, design the wall based on relevant Standard provisions. The solution must include the reinforcement details.

6. A 200 mm thick, braced shear wall as part of a multistorey in-situ reinforced concrete building is 5 m long and 3.2 m high. The wall is subjected to a total factored load N^* = 6600 kN uniformly distributed over its length. Of this load, 600 kN is applied from the floor immediately above at an eccentricity of 40 mm. Shear force acting on the wall, V^* = 1800 kN and the wall may be taken to be restrained along all four sides. Assume f'_c = 32 MPa with A2 exposure classification and a moderate degree of crack control. Using the Standard approach, check the adequacy of the wall assuming that it is doubly reinforced with N12@300 mm centres in both directions.

FOOTINGS, PILE CAPS AND RETAINING WALLS

12.1 INTRODUCTION

For a traditional building or bridge structure, the vertical forces in the walls, columns and piers are carried to the subsoil through a foundation system. The most common system consists of footings. In soft soil, because the bearing capacity is low, piles are needed to transfer the forces from the superstructure to deeper grounds where stiffer clay, sand layers or bed rock exist. The wall or column forces are each distributed to the piles or group of piles through a footing-like cap – a pile cap.

In civil and structural engineering, slopes often need be to cut to provide level grounds for construction. To ensure stability at and around the cuts or to meet similar requirements, the use of retaining walls for the disturbed soil and backfill is sometimes necessary.

The design of reinforced concrete footings and pile caps is generally governed by shear or transverse shear for wall footings, and transverse or punching shear for column footings and pile caps. Retaining walls behave like a cantilever system, resisting the horizontal pressures exerted by the disturbed soil or backfill (or both) by bending action.

The analysis of the forces acting above and below typical wall footings and their design are presented in Section 12.2. The treatments for footings supporting single and multiple columns are given in Section 12.3 whereas Section 12.4 deals with pile caps. Illustrative and design examples are given to highlight the application of the analysis and design procedures.

The horizontal soil pressure analysis and the design of retaining walls are discussed in Section 12.5, which also includes illustrative and design examples.

12.2 WALL FOOTINGS

12.2.1 General remarks

There are two main types of wall loading to be carried by the footing – the concentric loading as illustrated in Figure 12.2(1) and the eccentric loading as shown in Figure 12.2(2).

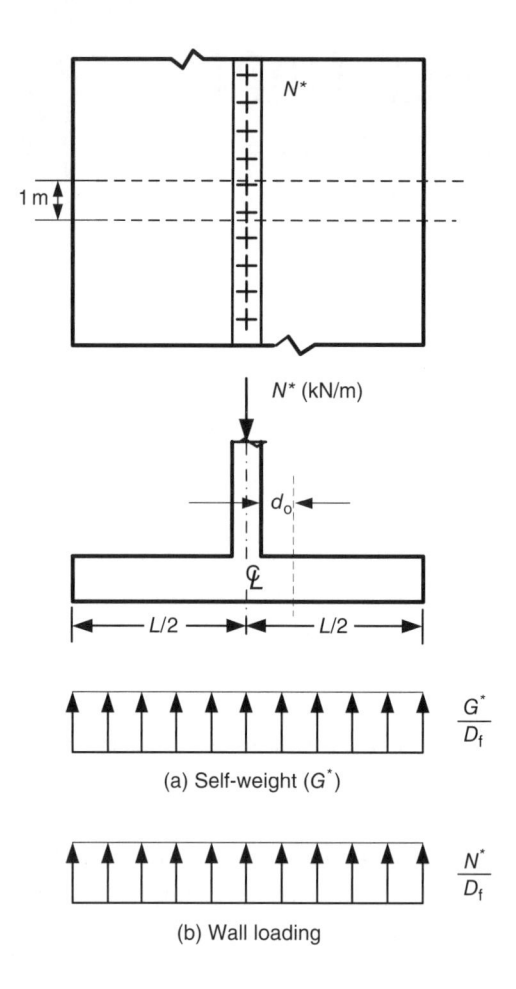

Figure 12.2(1) Concentric loading on wall and soil pressure profiles

Concentric loading leads to uniform pressure on the subsoil, which normally requires a symmetrical footing – see Figure 12.2(1). On the other hand, eccentric loading, depending on the ratio of M^* and N^*, produces two possible pressure distributions:
- the entire footing sustains pressure from the subsoil – see Figures 12.2(2)b(i) and b(ii)
- only part of the footing sustains the subsoil pressure – see Figure 12.2(2)b(iii).

In the above discussion, the self-weight of the footing is included in the loading.

For footings with prismatic cross-section, the subsoil pressure due to the self-weight is uniformly distributed.

Also, for an eccentric loading, an asymmetrical footing may be proportioned to produce a uniformly distributed subsoil pressure. This is illustrated in Figure 12.2(3).

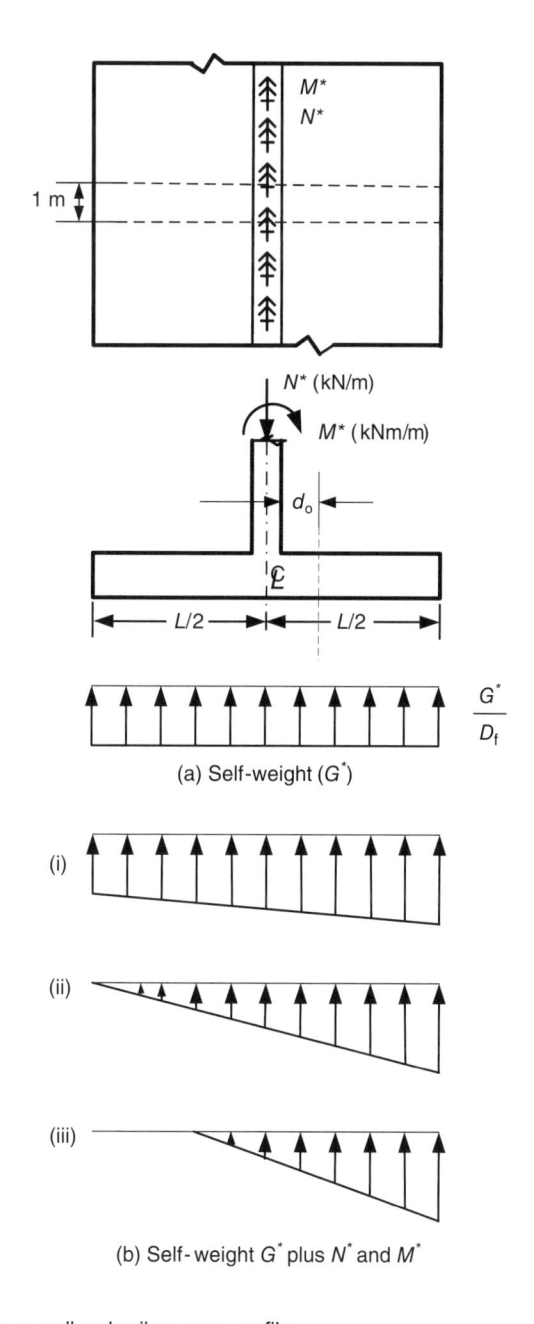

Figure 12.2(2) Eccentric loading on wall and soil pressure profiles

Note: the double-headed arrows indicate M^*, which follow the 'right-hand fingers rule'

The design of the more common eccentrically loaded wall footings is discussed in Section 12.2.2, followed by the concentrically loaded ones in Section 12.2.3. Section 12.2.4 presents the design of asymmetrical footings under eccentric loading. Numerical examples are given in Section 12.2.5.

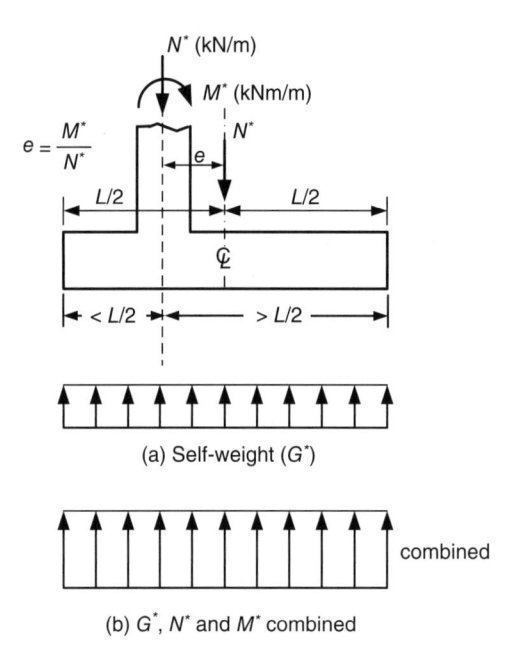

Figure 12.2(3) Asymmetrical wall footing

12.2.2 Eccentric loading

For the wall footing illustrated in Figures 12.2(4)a and b, the axial load and moment per unit length of the wall are N^*(kN/m) and M^*(kNm/m), respectively. The breadth of the footing (L) may be determined for a unit length (i.e. $b_w = 1$ m) of the wall by imposing two conditions.

Condition 1

The first condition occurs when, under the self-weight of the footing, the design force N^* (which includes the self-weight of the wall) and the superimposed loads plus the design moment M^* minus the maximum pressure on the subsoil at point B is

$$f_B = q_f \qquad \text{Equation 12.2(1)}$$

in which q_f is the allowable soil bearing capacity.

Condition 2

The second condition is when there is no uplift at A or

$$f_A = 0 \qquad \text{Equation 12.2(2)}$$

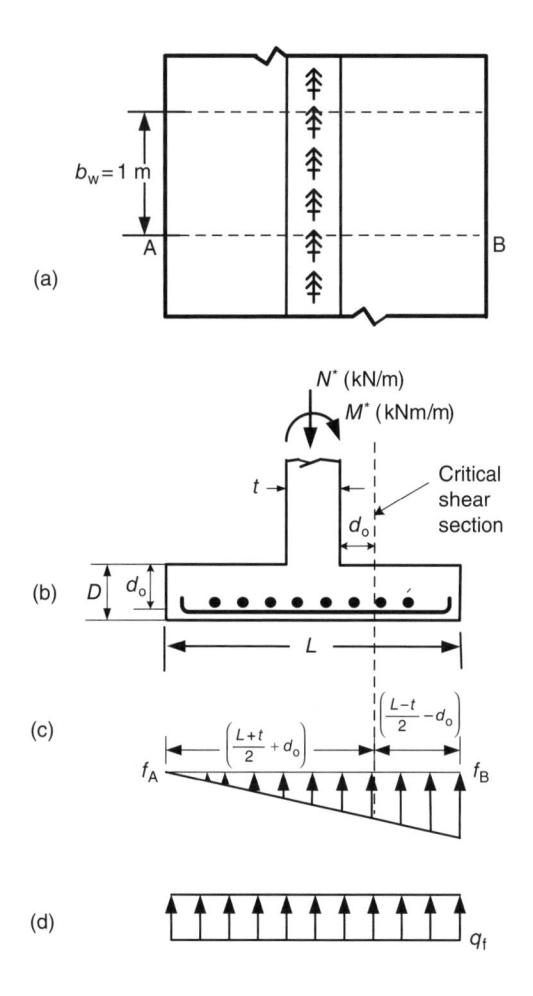

CHAPTER 12 FOOTINGS, PILE CAPS AND RETAINING WALLS 315

Figure 12.2(4) Wall footing under eccentric loading

Note that for condition 1, we can write

$$1.2\rho_w D + \frac{N^*}{L} + \frac{6M^*}{L^2} - q_f = 0 \quad \text{Equation 12.2(3)}$$

from which

$$L = \frac{0.5}{(6\rho_w D - 5q_f)}\left(2.24\sqrt{(5N^{*2} - 144\rho_w M^* D + 120M^* q_f)} - 5N^*\right) \quad \text{Equation 12.2(4)}$$

where 1.2 is the load factor for self-weight of the footing, ρ_w is the unit weight of the footing and D is the overall depth of the footing.

For condition 2, we have

$$1.2\rho_w D + \frac{N^*}{L} - \frac{6M^*}{L^2} = 0 \quad \text{Equation 12.2(5)}$$

from which

$$L = \frac{8.3 \times 10^{-3}}{\rho_w D}\left(2.24\sqrt{(5N^{*2} + 144\rho_w M^* D)} - 5N^*\right) \qquad \text{Equation 12.2(6)}$$

The design L is the larger of the two values from Equations 12.2(4) and 12.2(6). Note that D is unknown at this stage of the design process, but it can be computed by considering the shear strength of the concrete footing. The effective depth for shear consideration (d_o) may be estimated using Equation 6.3(4), which is reproduced below:

$$V_{uc} = \beta_1\beta_2\beta_3 b_w d_o f_{cv}\sqrt[3]{\frac{A_{st}}{b_w d_o}} \qquad \text{Equation 12.2(7)}$$

Similar to one-way slabs, wall footings may be designed as beams of a unit width. Considering the limit set in Equation 3.5(5) we may take

$$\frac{A_{st}}{b_w d_o} = 0.20\left(\frac{D}{d}\right)^2 \frac{f'_{ct.f}}{f_{sy}} \qquad \text{Equation 12.2(8)}$$

in view of the fact that footing design is governed by shear and that the required bending steel is usually less than the minimum value specified in AS 3600–2009 (the Standard).

Substituting Equation 12.2(8) into Equation 12.2(7) leads to

$$V_{uc} = \beta_1\beta_2\beta_3 b_w d_o f_{cv}\sqrt[3]{\frac{0.20 D^2 f'_{ct.f}}{d^2 f_{sy}}} \qquad \text{Equation 12.2(9)}$$

It may be immediately obvious that, for shear consideration, the empirical formulas recommended in the Standard are suitable only for beams or footings with a known or given depth. Therefore, to determine d_o, a trial and error approach is necessary. This may be carried out following these steps:

1. Assume D and determine L via Equations 12.2(4) and 12.2(6). Adopt the larger of the two values.
2. Select bars of a suitable size and determine d_o taking into consideration the exposure classification and other relevant factors. Then compute V^* at a section d_o from the appropriate face of the wall – see Figures 12.2(1)a and b.
3. Compute $V_{u.max}$ using Equation 6.3(1), and check adequacy of d_o in terms of the reinforceability of the section (noting that this criterion is seldom violated). Revise by increasing D (hence d_o) as necessary.
4. Compute V_{uc} using Equation 12.2(9) and $V_{u.min}$ using Equation 6.3(15).
5. Perform the four shear capacity checks (Cases A, B, C and D) using Equations 6.3(9), 6.3(10), 6.3(11) and 6.3(13), respectively, as detailed in Section 6.3.5. As necessary, compute $A_{sv.min}$ using Equation 6.3(12) and reinforce the footing accordingly. If the use

of shear reinforcement is to be avoided, d_o (hence D) should be increased such that Equation 6.3(9) or Equation 6.3(10) prevails. Then proceed to step 9. Otherwise go to step 6.

6. If $V^* > \phi V_{u.min}$, revise by assuming a larger d_o and repeat steps 1 to 4, or proceed to step 7.
7. Compute V_{us} with Equation 6.3(17) and choose A_{sv} with reference to the multiple arrangements of stirrups or ties. Then compute the spacing (s) using Equation 6.3(21).
8. Detail the (multiple) shear reinforcement for the footing.
9. With the adopted d_o determine d (note that $d = d_o$ if A_{st} is provided in a single layer). Then compute p_t using Equation 3.5(6) ensuring that Equation 3.5(8) prevails. Note that, in locating the critical bending section, the footing should be treated as a slab supported on a wall, referring to Figure 9.4(2) in computing the span.
10. Compute $A_{st} = p_t b_f d$ noting that if A_{st} is larger than $A_{st.min}$ as specified in Equation 12.2(8), the current V_{uc} is lower than the true value. Revise as necessary to obtain a more economical section.
11. Detail A_{st} ensuring that bond is adequate. Use hooks or cogs as necessary.
12. Provide shrinkage and temperature steel in the direction orthogonal to A_{st} or parallel to the length of the wall. The procedure described in Section 9.4.7 should be followed.

An application of this trial and error approach may be found in Section 12.2.5.

12.2.3 Concentric loading

Comparing Figures 12.2(1) and (2), it is easy to conclude that the concentric loading case is a special eccentric loading case with zero moment. If no moment is transmitted from the wall to the footing (i.e. the case depicted in Figure 12.2(2) with $M^* = 0$) then the pressure on the subsoil is uniformly distributed as illustrated in Figure 12.2(1). In such a case, Equation 12.2(3) is reduced to

$$1.2\rho_w D + \frac{N^*}{L} = q_f$$

or

$$L = \frac{N^*}{q_f - 1.2\rho_w D} \qquad \text{Equation 12.2(10)}$$

Note that Equation 12.2(5) does not exist under concentric loading.

Since the design of a concentrically loaded footing (i.e. with $M^* = 0$) is a special case of footings under eccentric loads, it follows similar steps to those enumerated in Section 12.2.2 and demonstrated numerically in Section 12.2.5.

12.2.4 Asymmetrical footings

The approach to designing an asymmetrical footing to produce uniformly distributed subsoil pressure is briefly discussed in Section 12.2.1 and illustrated in Figure 12.2(3). The method of proportioning the plan dimensions of such a footing is given below.

For the eccentrically loaded footing depicted in Figure 12.2(3), the concentric load per unit length of the wall is N^* and the corresponding moment is M^*. The combination of N^* and M^* may be replaced by N^* acting with an eccentricity (e) from the centre plane of the wall, where

$$e = \frac{M^*}{N^*}$$

Equation 12.2(11)

Thus, the centre line of the footing is located at a distance e to the centre plane of the wall. With e in hand, the breadth L and depth D of the footing can be determined following the procedure developed for concentrically loaded footings (Section 12.2.3). Note that, for both the transverse shear and bending design, the critical sections are located on the toe side of the footing (i.e. the right side of the symmetrical footing), shown in Figure 12.2(3). For a footing of a 1-metre run, the breadth (L) can be computed using Equation 12.2(10). A numerical example is given in Section 12.2.5 to illustrate the design process for such asymmetrical footings.

12.2.5 Design example

The example given in this section is for an asymmetrical footing under eccentric loading. It demonstrates the use of the equations and process developed in Section 12.2.2 for symmetrical wall footings under eccentric loading. However, as discussed in Sections 12.2.3 and 12.2.4, concentrically loaded footings and asymmetrical footings under eccentric loading may be treated as special cases of a symmetrical footing under eccentric loading. Thus, the example below also illustrates the design of the other two types of wall footings.

Numerical example

Problem

Given a reinforced concrete wall, 300 mm thick and subjected to a dead load $DL = 200$ kN/m and a live load $LL = 150$ kN/m, each having the same eccentricity of $e = 100$ mm from the centre plane of the wall. Take $f'_c = 25$ MPa, the soil bearing capacity $q_f = 250$ kPa; A2 exposure classification applies. Design an asymmetrical (strip) footing in such a way that the subsoil pressure is uniformly distributed. Use N16 or N20 bars in one layer and use no shear reinforcement.

Solution

Details of the footing of a 1-metre run are depicted in Figure 12.2(5). Note that the clear cover of 30 mm is in accordance with the recommendation given in Table 1.4(2). This assumes that the base of the footing is well prepared and compacted following excavation and before casting the concrete. If this cannot be assured, a larger cover ought to be used.

The design action per metre run (i.e. $b = 1000$ mm) is

$$N^* = 1.2 \times 200 + 1.5 \times 150 = 465 \text{ kN}$$

$M^* = 0$ at the centre line of the asymmetrical footing; see Figure 12.2(5).

The breadth of the footing as per Equation 12.2(10) and with an assumed $\rho_w = 24$ kN/m^3

$$L = \frac{465}{250 - 1.2 \times 24D} = \frac{465}{250 - 28.8D} \qquad \text{Equation (a)}$$

With Equation (a) in hand, the footing design may follow the trial and error process detailed in Section 12.2.2.

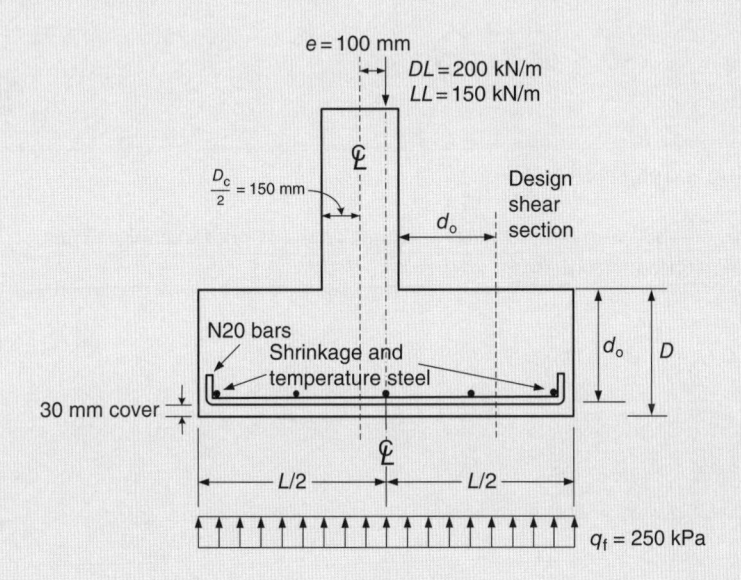

Figure 12.2(5) Details of the design asymmetrical wall footing

Trial 1

Assume $D = 600$ mm, and Equation (a) gives

$$L = \frac{465}{250 - 28.8 \times 0.6} = 1.998 \text{ m}$$

Use $L = 2.0$ m.

Step 1

For N20 bars with exposure classification A2,

$$d_o = 600 - 30 - 20/2 = 560 \text{ mm}$$

Step 2

The design shear force per metre run is

$$V^* = \left(\frac{L}{2} + 0.1 - 0.15 - d_o\right) \times (q_f - 0.9 \times 24D)$$

$$= (1.0 + 0.1 - 0.15 - 0.56) \times (250 - 0.9 \times 24 \times 0.6) = 92.45 \text{ kN}$$

Note that a load factor of 0.9 is applied to the self-weight of the footing to produce a more critical V^*.

Step 3

Equation 6.3(1) gives

$$V_{u.max} = 0.2 \times 25 \times 1000 \times 560 \times 10^{-3} = 2800 \text{ kN}$$

and

$$\phi V_{u.max} = 0.7 \times 2800 = 1960 \text{ kN} > V^*$$

The assumed section with $D = 600$ mm is more than adequate (i.e. it is reinforceable). Thus, this step will be omitted in subsequent trials.

Step 4

Equation 2.2(2): $f'_{ct.f} = 0.6\sqrt{f'_c} = 3.0$ MPa

Equation 6.3(5)b: $\beta_1 = 1.1\left(1.6 - \dfrac{560}{1000}\right) = 1.144 > 0.8$ — this is acceptable

Equations 6.3(6) and 6.3(8): $\beta_2 = \beta_3 = 1.0$

For N20 bars in one layer, $d = d_o = 560$ mm, and

Equation 12.2(9) gives

$$V_{uc} = 1.144 \times 1 \times 1 \times 1000 \times 560 \times \sqrt[3]{25} \times \sqrt[3]{0.20 \times (600/560)^2 \times (3/500)} \times 10^{-3}$$
$$= 208.45 \text{ kN}$$

Step 5

The specification was that shear reinforcement is not to be used. Thus, check Cases A and B (see Section 6.3.5) only by treating the footing as a beam (rather than slab or plate) in view of the very large D/L ratio. Since Case A is for $D \leq 250$ mm (or $b/2 = 500$ mm) only, this check may be ignored for this particular example where $D = 600$ mm.

For Case B, Equation 6.3(10) gives

$$0.5 \times 0.7 \times 208.45 = 72.96 \text{ kN} < V^* = 92.45 \text{ kN}$$

Thus, $D = 600$ mm is inadequate.

Trial 2

Assume $D = 650$ mm, and Equation (a) gives

$$L = \frac{465}{250 - 28.8 \times 0.65} = 2.011 \text{ m}$$

Use $L = 2.0$ m.

Step 1

$$d_o = 650 - 30 - 20/2 = 610 \text{ mm}$$

Step 2

$$V^* = (1.0 + 0.1 - 0.15 - 0.61) \times (250 - 0.9 \times 24 \times 0.65) = 80.23 \text{ kN}$$

Step 3

As explained in step 3 in trial 1, this step is skipped here.

Step 4

Equation 6.3(5)b: $\beta_1 = 1.1\left(1.6 - \dfrac{610}{1000}\right) = 1.089$

Equation 12.2(9):
$$V_{uc} = 1.089 \times 1 \times 1 \times 1000 \times 610 \times \sqrt[3]{25} \times \sqrt[3]{0.20 \times (650/610)^2 \times (3/500)} \times 10^{-3}$$
$$= 215.34 \text{ kN}$$

Step 5

Check Case B.

Equation 6.3(10): $0.5 \times 0.7 \times 215.34 = 75.37 \text{ kN} \approx V^* = 80.23 \text{ kN}$

In view of the tolerable difference (i.e. the section being under-designed by only 6.06%), $D = 650$ mm is acceptable. Proceed to the bending steel design in step 9, skipping steps 6–8 which are no longer required for this example.

Step 9

Design for A_{st}.

For a single layer of N20 bars, $d = d_o = 610$ mm and

Equation 3.3(2)a: $\alpha_2 = 0.85$

Equation 3.3(2)b: $\gamma = 0.85$

According to Figure 9.4(2) and for the wall-footing layout as shown in Figure 12.2(6), the location of the critical bending section is at L_b from the right edge where

$$L_b = \frac{L}{2} + e - 0.7a_s$$

where e is the eccentricity of the loading and a_s is half the wall thickness.
Thus,

$$L_b = 1 + 0.1 - 0.7 \times 0.15 = 0.995$$

With $\rho_w = 24$ kN/m^3 and

$$M^* = (250 - 24 \times 0.65 \times 1) \times 0.995^2/2 = 116.03 \text{ kNm}$$

Equation 3.5(7): $\xi = \dfrac{0.85 \times 25}{500} = 0.0425$

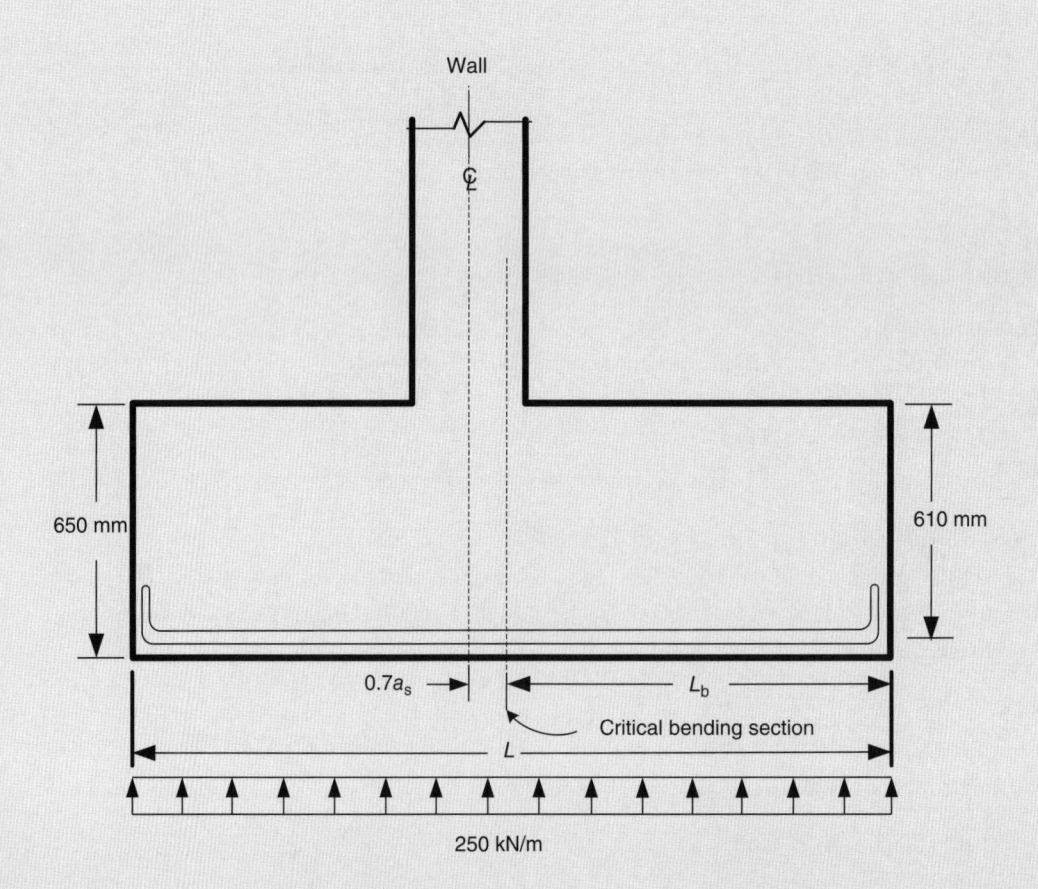

Figure 12.2(6) Defining critical bending section for the trial footing

Assuming $\phi = 0.8$, Equation 3.5(6) gives

$$p_t = 0.0425 - \sqrt{0.0425^2 - \frac{2 \times 0.0425 \times 116.03 \times 10^6}{0.8 \times 1000 \times 610^2 \times 500}} = 0.0007868$$

Check that the assumed $\phi = 0.8$ is correct.

$$k_u = k_{uo} = \frac{0.0007868 \times 500}{0.85 \times 0.85 \times 25} = 0.0218 \text{ from Equation 3.4(8)}$$

and then Equations 3.4(20)a with b: $\phi = 0.8$. Therefore, the assumption is correct.
Now, Equation 12.2(8):

$$p_{t.\min} = 0.20 \times (650/610)^2 \times \left(0.6 \times \sqrt{25}\right)/500 = 0.001363 > p_t$$

Therefore, $p_t = 0.001363$.

Step 10
For a 1-metre run of the footing

$$A_{st} = 0.001363 \times 1000 \times 610 = 831.3 \text{ mm}^2/\text{m}$$

Using Table 2.3(2), N20 bars @ 300 mm yield A_{st} = 1047 mm^2/m > 831.3 mm^2/m.
This is acceptable.

Also, for crack control purposes in this case, the maximum spacing is s = 300 mm as shown in Table 1.4(5). Therefore, use N16 bars and from Table 2.3(2), N16 @ 225 mm yields A_{st} = 893 mm^2/m > 831.3 mm^2/m. This is acceptable.

With the reduced bar size, d is increased and the stress development length is shortened accordingly. In consequence, no extra check is necessary for this bar size variation.
Since $A_{st} = A_{st.\min}$, V_{uc} = 215.34 kN is the true value as seen in step 4.

Step 11
Check the bond.
For Equation 8.2(1)

$$k_1 = 1.3; k_2 = (132 - 16)/100 = 1.16; k_3 = 1 - 0.15(32 - 16)/16 = 0.85$$

Equation 8.2(1) thus gives

$$L_{sy.tb} = (0.5 \times 1.3 \times 0.85 \times 500 \times 16)/\left(1.16 \times \sqrt{25}\right) = 762 \text{ mm}$$

Since the available $L_{sy.t} < (L_b - \text{cover}) = 0.995 - 0.03 = 0.965$ m on the right of the critical bending section, and $L_{sy.t} < (L - L_b - \text{cover}) = 0.975$ m on the left of the critical bending section, straight bars (without hooks or cogs) of length (2000 − 2 × 30) = 1940 mm @ 225 mm are adequate for bond.

In view of Equation 8.2(9), or the fact that the actual tensile stress in A_{st} or $\sigma_{st} < f_{sy}$, the bond strength is more than adequate, thereby rendering step 11 superfluous.

Step 12
Provide shrinkage and temperature steel.
Equation 9.4(14) gives

$$A_{s.min} = 1.75 \times 1000 \times 650 = 10^{-3} = 1137.5 \text{ mm}^2$$

Use six N16, $A_s = 1206 \text{ mm}^2$. This is acceptable. Note that $s \leq 300$ mm does not apply to shrinkage and temperature steel.

The final design is detailed in Figure 12.2(7).

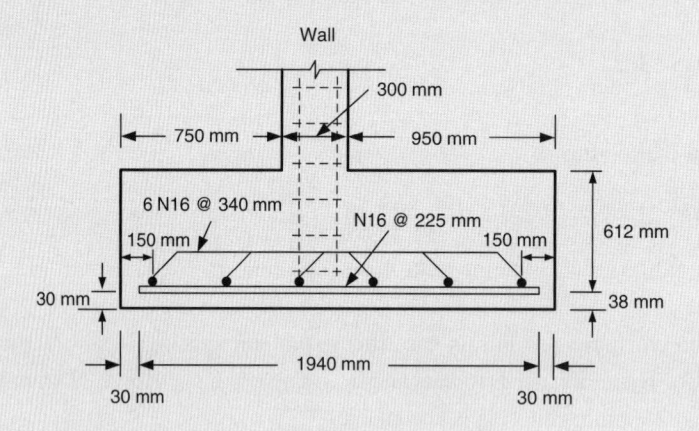

Figure 12.2(7) Details of the final design of the asymmetrical wall footing

Comments
In step 5 it was explained that, for $D = 650$ mm, the footing is under-designed in shear by 6.06%. However, for $p_t = p_{t.min} = 0.001363$ in step 9 and $p_{s.min} = 0.00175$ in step 12, the steel content of the footing is 0.311%. Then, from Equation 2.4(1), we have

$$\rho_w = 24 + 0.6 \times 0.311 = 24.187 \text{ kN/m}^3$$

As shown in Figure 12.2(7) the final $d = d_o = 612$ mm.
In step 2

$$V^* = (1.0 + 0.1 - 0.15 - 0.612) \times (250 - 0.9 \times 24.187 \times 0.65) = 79.72 \text{ kN}$$

Also, in step 4 with $d = d_o = 612$ mm, $V_{uc} = 215.57$ kN. Together, these calculations reduce the under-design to 5.66%. This review also indicates that for the (lightly reinforced) footings, Equation 2.4(1) may be ignored for simplification and ρ_w taken as 24.0 kN/m^3.

12.3 COLUMN FOOTINGS

12.3.1 General remarks

Similar to wall footings, column or pad footings may be subjected to concentric or eccentric loading. This class of footings may support more than one column and may also be subjected to biaxial bending actions.

The analysis and design of centrally and eccentrically loaded column footings are discussed in Sections 12.3.2 and 12.3.3, respectively. Section 12.3.4 covers footings supporting multiple columns, whereas Section 12.3.5 deals with biaxial actions. The reinforcement requirements as given in the Standard are highlighted in Section 12.3.6, which is followed by the design of an eccentrically loaded asymmetrical footing that achieves uniform subsoil pressure.

12.3.2 Centrally loaded square footings

For a centrally loaded square column ($b_c \times D_c$ where $b_c = D_c$) the footing is normally also square (i.e. $b_f = D_f$) as shown in Figure 12.3(1), in which case

$$b_f D_f = \frac{N^*}{q_f}$$

Equation 12.3(1)

where q_f is the allowable soil bearing capacity.

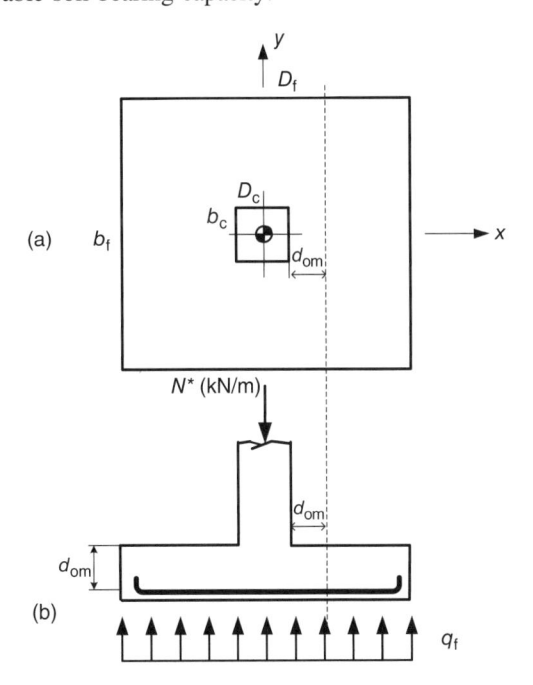

Figure 12.3(1) A centrally loaded square column footing

To determine d_o, we need to consider transverse and punching shear. For transverse shear, the process described in Section 12.2.3 should be followed. For punching shear consideration for an equivalent interior column, the procedure described in Section 9.6.3 may be used. Substituting Equation 9.6(5) to Equation 9.6(3) gives

$$V_{uo} = 2(D_c + b_c + 2d_{om})d_{om}f_{cv} \qquad \text{Equation 12.3(2)}$$

where f_{cv} is defined in Equation 9.6(6), or in our case

$$f_{cv} = 0.17\left(1 + \frac{2b_c}{D_c}\right)\sqrt{f'_c} \le 0.34\sqrt{f'_c} \qquad \text{Equation 12.3(3)}$$

and d_{om} is the mean value of the shear effective depth (d_o) around the critical shear perimeter. Thus, for a square column where $D_c = b_c$ we use

$$f_{cv} = 0.34\sqrt{f'_c} \qquad \text{Equation 12.3(4)}$$

Invoking Equation 9.6(10) we have

$$\phi V_{uo} = N^* \qquad \text{Equation 12.3(5)}$$

Substituting Equations 12.3(1) and 12.3(2) into Equation 12.3(5) with reference to Equation 12.3(4), we have

$$0.952 d_{om}(D_c + d_{om})\sqrt{f'_c} = b_f D_f q_f \qquad \text{Equation 12.3(6)}$$

which leads to

$$d_{om} = \frac{-b_F + \sqrt{b_F^2 + 4a_F c_F}}{2a_F} \qquad \text{Equation 12.3(7)}$$

where

$$a_F = 0.952\sqrt{f'_c}$$
$$b_F = 0.952 D_c \sqrt{f'_c}$$
$$c_F = b_f q_f D_f$$

The design shear effective depth of the footing is the larger of the values from Equation 12.3(7) (i.e. d_{om}) and the value (d_o) derived for cases where transverse shear is considered. Using the larger value, proceed to design and detail the bending reinforcement (A_{st}) following the procedure described in steps 9, 10 and 11 of Section 12.2.2. This A_{st} runs in the x-direction – see Figure 12.3(1). Repeat the bending design for A_{st} running in the y-direction.

12.3.3 Eccentric loading

For an eccentrically loaded column footing, the design may follow the approach described in Section 12.2.4 in which the footing dimensions (b_f and D_f) can be proportioned in such a way to produce a uniform pressure in the subsoil.

Illustrative example

Problem

Figure 12.3(2) details a rectangular column ($b_c \times D_c$) subjected to a combination of N^* and M^*. Proportion a suitable rectangular footing ($b_f \times D_f$) such that the uniform subsoil pressure is equal to the allowable soil bearing capacity (q_f). Discuss the procedure for the complete design.

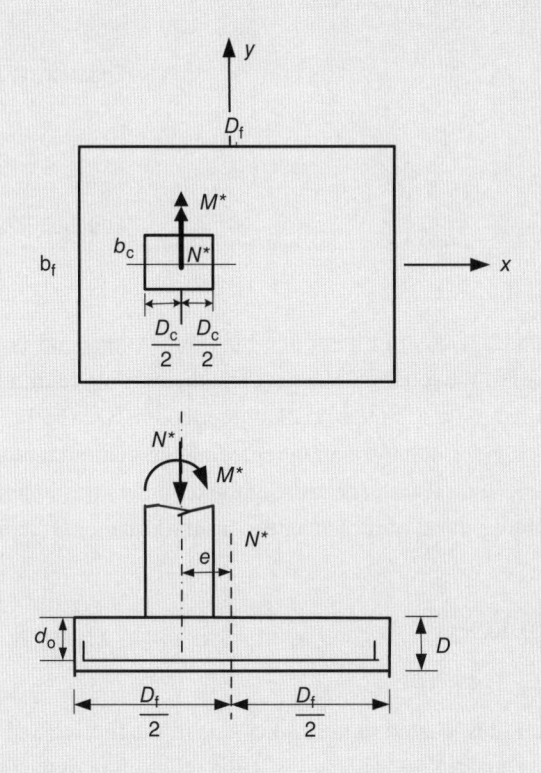

Figure 12.3(2) Footing for eccentric column loading

Solution

For the given N^* and M^*, the eccentricity of the equivalent N^* is

$$e = \frac{M^*}{N^*}$$

Equation 12.3(8)

The bearing area of the footing is

$$b_f D_f = \frac{N^*}{q_f - 1.2\rho_w D}$$
Equation 12.3(9)

where ρ_w is the unit weight of the footing.
But

$$\frac{b_f}{D_f} = \frac{b_c}{D_c}$$

or

$$b_f = \frac{b_c}{D_c} D_f$$
Equation 12.3(10)

Substituting Equation 12.3(10) into Equation 12.3(9) leads to

$$\frac{b_c}{D_c} D_f^2 = \frac{N^*}{q_f - 1.2\rho_w D}$$
Equation 12.3(11)

from which

$$D_f = \sqrt{\frac{D_c N^*}{b_c(q_f - 1.2\rho_w D)}}$$
Equation 12.3(12)

Equation 12.3(10) then yields b_f.

With Equations 12.3(8), 12.3(10) and 12.3(12) in hand, the plan dimensions and layout of the footing, the shear effective depth of the footing d_o or d_{om}, and the bending reinforcement can be obtained following an iterative process similar to that described in Section 12.2.5 for wall footings. Note that the Standard gives no information regarding the minimum reinforcement $p_{t.min}$ for pad footings. However, in view of shear being the dominant action and most column footing dimensions resembling those of a rectangular beam, Equation 12.2(9) may also apply

$$V_{uc} = \beta_1\beta_2\beta_3 b_w d_o f_{cv} \sqrt[3]{\frac{0.20 D^2 f'_{ct.f}}{d^2 f_{sy}}}$$
Equation 12.3(13)

The critical transverse shear section is located in the longer side of the footing and in this case, the right side. When deriving Equations 12.3(12) and (13), the self-weight of the footing, or D in Figure 12.3(2), is unknown. Therefore, an iterative process is needed to ensure that q_f is not exceeded. This may be done by ignoring the self-weight in computing b_f and D_f, but suitably increasing their values before proceeding to obtain d_o or d_{om}, and then checking Equation 12.3(9). Revise as necessary until either q_f is not exceeded or $b_f D_f$ is adequate. Also, a final check for punching shear strength is required by treating N^* as the design load from an edge column. The process given in Sections 9.6.3–9.6.5 may be used.

It should be recognised that, on certain construction sites, the length D_f may be restricted (e.g. where the footings are close to the boundary lines of the property). In such cases, footings with a stepped or trapezoidal plan shape may be mandatory. Such irregular plan shapes, as shown in Figure 12.3(3), can readily be obtained by ensuring that the geometric centre of the plan (cg) has the same eccentricity e from the column axis 0, where $e = M^*/N^*$.

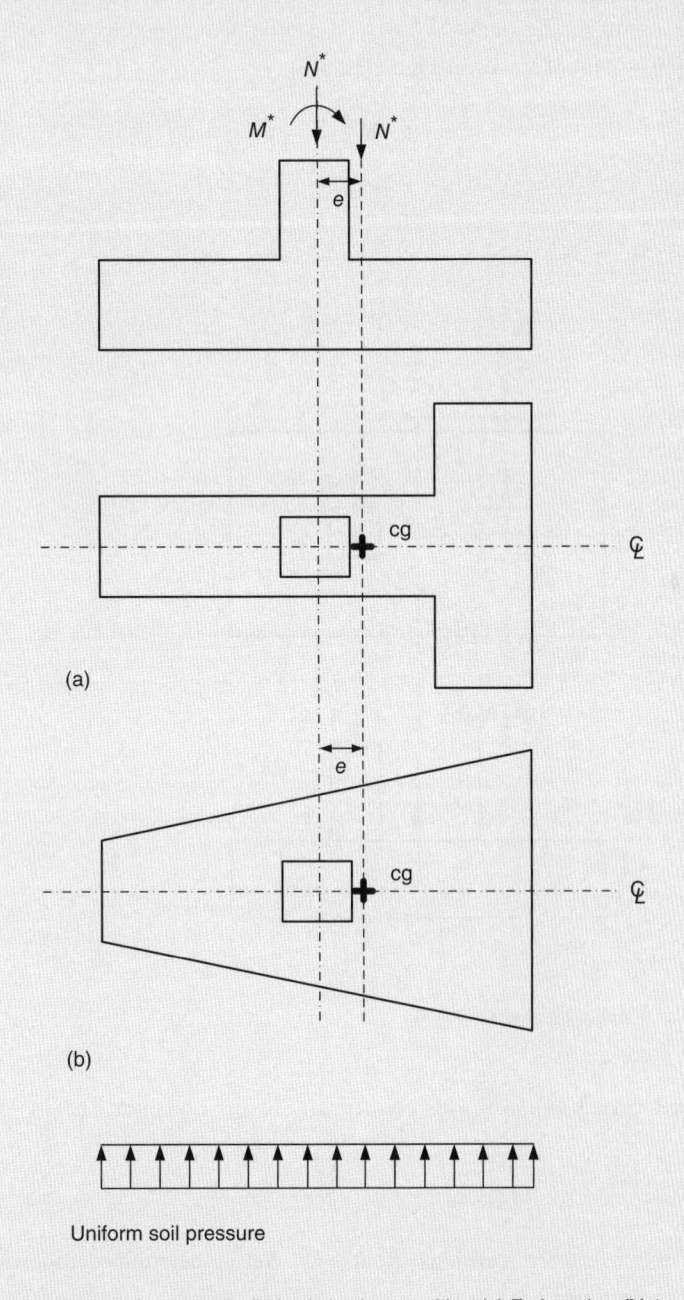

Figure 12.3(3) Asymmetrical footings for eccentrically loaded columns, either **(a)** T-shaped or **(b)** trapezoidal

12.3.4 Multiple columns

For single footings supporting more than one column, it is again desirable to proportion and lay out the footing area in such a way that the subsoil pressure is uniformly distributed.

Figure 12.3(4) shows a footing under two columns, each subjected to an eccentric loading, or a combination of N^* and M^*.

The position (i.e. cg) of the resultant $(N_1^* + N_2^*)$ can be determined by a simple static analysis. For the given system of forces in Figure 12.3(4)

$$\sum M_{\text{about y}} = 0$$

yields

$$M_1^* + M_2^* + N_2^*(e_1 + e_2) = (N_1^* + N_2^*)e_1 \qquad \text{Equation 12.3(14)}$$

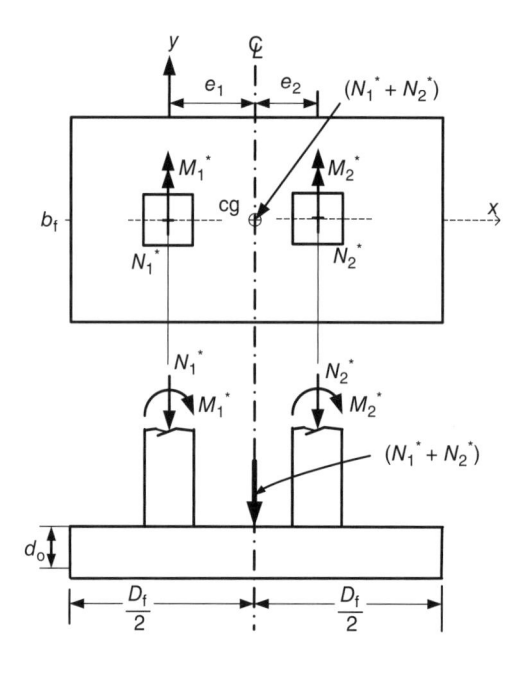

Figure 12.3(4) Footing for eccentrically loaded columns

Note that in Equation 12.3(14), $e_1 + e_2$ is prescribed. Solving Equation 12.3(14) gives

$$e_1 = \frac{N_2^* e_2}{N_1^*} \qquad \text{Equation 12.3(15)}$$

With e_1 known, the footing area dimensions (b_f and D_f) can be determined following the procedure discussed in Section 12.3.3.

The value of d_o can be obtained by considering the transverse and punching shear. Depending on the intensities of the column loads and the footing geometry, the most critical transverse shear section can be on either side of one or the other column. Similarly, the critical punching shear perimeter can be around either of the two columns.

12.3.5 Biaxial bending

For a typical column footing subjected to biaxial bending as shown in Figure 12.3(5) (where the longer side is designated D_f or $D_f > b_f$), the design may be carried out as follows:

1. Compute e_x and e_y so that the subsoil pressure is uniformly distributed where

$$e_x = M_x^*/N^*$$ Equation 12.3(16)

$$e_y = M_y^*/N^*$$ Equation 12.3(17)

2. Determine b_f and D_f following the illustrative example given in Section 12.3.3 where d_o is also determined. Since $D_f > b_f$, the d_o obtained is the governing value.
3. Calculate the bending reinforcement (A_{st}) along the x and y axes, using the usual process for restricted design (see Section 3.5.2).
4. Check to ensure that the bars have adequate lengths for stress development.

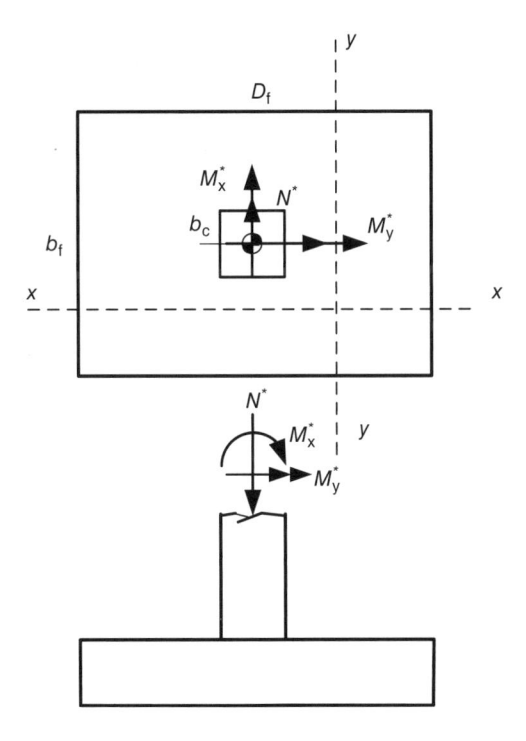

Figure 12.3(5) Footing for column under biaxial bending

In addition, the punching shear may also need to be considered if the critical shear perimeter is shorter than b_f, which is the width of the critical section governing the bending shear design. Since the Standard does not cater for corner column connections, some semi-empirical formulas may be used to assess the punching shear strength of the footing (e.g. Loo and Falamaki 1992).

12.3.6 Reinforcement requirements

For bending design, column footings may be taken as slabs supported on columns. As such, the critical bending section is located at $0.7(D_c/2)$ or $0.7(b_c/2)$ (depending on the case) from the column centre line as shown in Figure 9.4(2). This assumption also leads to a conservative but logical design outcome. To ensure minimum capacity, Equation 3.5(5) may be followed.

$$p_{t.min} = 0.20(D/d)^2 f'_{ct.f}/f_{sy} \qquad \text{Equation 12.3(18)}$$

12.3.7 Design example

Problem

Given a rectangular column with $D_c = 500$ mm and $b_c = 300$ mm. The design actions $N^* = 1400$ kN and $M^* = 200$ kNm; the soil bearing capacity $q_f = 300$ kPa; $f'_c = 20$ MPa; and exposure classification A1 applies.

Design an asymmetrical (pad) footing in such a way that the subsoil pressure is uniformly distributed. Use N20 bars for bending in one layer each way and provide shear reinforcement as required.

Solution

The design specifications are illustrated in Figure 12.3(6). The eccentricity $e = M^*/N^* = 142.9$ mm as per Equation 12.3(8). Note that, as cautioned in Section 12.2.5, the concrete cover must be increased unless the base of the footing is well levelled and compacted.

Using Equation 12.3(12) and assuming $\rho_w = 24$ kN/m³

$$D_f = \left(\frac{0.5 \times 1400}{0.3(300 - 1.2 \times 24 \times D)}\right)^{\frac{1}{2}} = \left(\frac{700}{90 - 8.64D}\right)^{\frac{1}{2}} \qquad \text{Equation (a)}$$

and

Equation 12.3(10): $b_f = 0.6 D_f$ \qquad Equation (b)

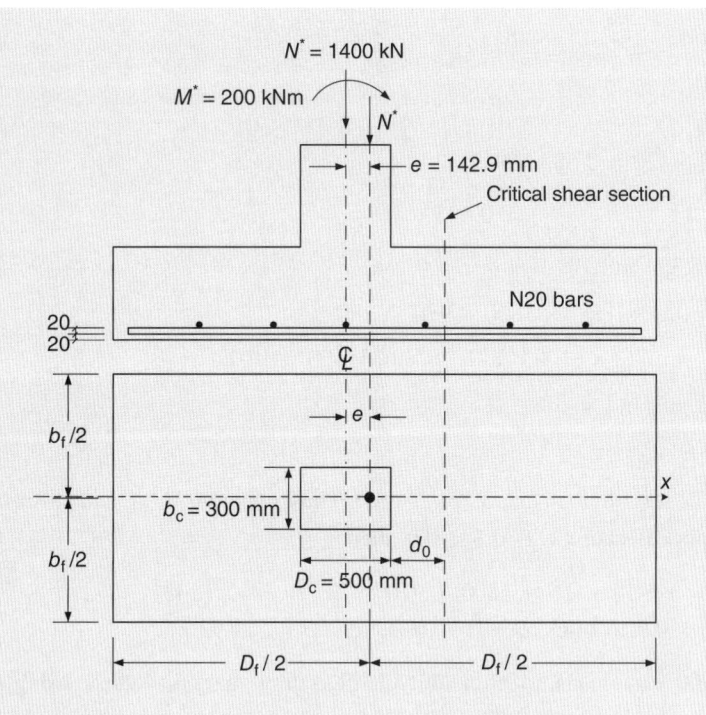

Figure 12.3(6) Design specifications for the example asymmetrical pad footing

With Equations (a) and (b) in hand, follow the relevant steps given in Section 12.2.2, which would lead to the required design.

Trial 1

Step 1
Assume $D = 1000$ mm and:
 Equation (a): $D_f = 2.933$ m (rounded to 3.0 m)
 Equation (b): $b_f = 1.8$ m
$d_o = 1000 - 20 - 20/2 = 970$ mm

Step 2
The design shear force at the critical shear section – see Figure 12.3(6) – is

$$V^* = b_f(D_f/2 + e - D_c/2 - d_o)(q_f - 0.9\rho_w D) \qquad \text{Equation (c)}$$

Therefore

$$V^* = 1.8(1.5 + 0.1429 - 0.5/2 - 0.97)(300 - 0.9 \times 24 \times 1) = 211.92 \text{ kN}$$

Step 3
As the section is deep enough to be reinforceable, this step is not required to be checked.

Step 4
For a single layer of steel,
$d = d_o = 970$ mm

Equation 2.2(2): $f'_{ct.f} = 0.6\sqrt{20} = 2.683$ MPa

Equation 6.3(5)b: $\beta_1 = 1.1\left(1.6 - \dfrac{970}{1000}\right) = 0.693 < 0.8$

Thus, $\beta_1 = 0.8$, Equations 6.3(6) and (8): $\beta_2 = \beta_3 = 1.0$, and

Equation 12.3(13):
$$V_{uc} = 0.8 \times 1 \times 1 \times 1800 \times 970 \times \sqrt[3]{20}$$
$$\times \left[0.20 \times (1000/970)^2 \times 2.683/500\right]^{\frac{1}{3}} \times 10^{-3}$$
$$= 396.15 \text{ kN}$$

Step 5
Check Cases A, B, C and D. It is obvious that $0.5\phi V_{uc} < V^*$. Thus, compute $V_{u.min}$ and check Case D (since $D > 750$ mm). Because $\sqrt{f'_c}\, 0.1 < 0.6$, then:

Equation 6.3(15): $V_{u.min} = 396.15 + 0.6 \times 1800 \times 970 \times 10^{-3} = 1443.75$ kN

Equation 6.3(13): $\phi V_{u.min} = 0.7 \times 1443.75 = 1010.63$ kN $> V^* = 211.92$ kN

The assumed $D = 1000$ mm leads to an unduly large safety margin. Revise using $D = 750$ mm, the minimum depth for which Case D still governs.

Trial 2

Step 1
Assume $D = 750$ mm, then:

Equation (a): $D_f = 2.895$ m (rounded to 3.0 m)

Equation (b): $b_f = 1.8$ m

$d_o = 750 - 20 - 20/2 = 720$ mm

Step 2
Equation (c): $V^* = 1.8\,(1.5 + 0.1429 - 0.5/2 - 0.72)\,(300 - 0.9 \times 24 \times 0.72)$
$= 344.53$ kN

Step 3
Not required as explained for trial 1.

Step 4
For the assumed total depth of 750 mm:
$d = d_o = 720$ mm

Equation 6.3(5)b: $\beta_1 = 1.1\left(1.6 - \dfrac{720}{1000}\right) = 0.968$

Equation 12.3(13):
$$V_{uc} = 0.968 \times 1 \times 1 \times 1800 \times 720 \times \sqrt[3]{20} \times \left[0.20 \times (750/720)^2 \times 2.683/500\right]^{\frac{1}{3}} \times 10^{-3}$$
$$= 358.26 \text{ kN}$$

Equation 6.3(15):
$$V_{u.min} = 358.26 + 0.6 \times 1800 \times 720 \times 10^{-3} = 1135.86 \text{ kN}$$

Step 5

Check Case D – Equation 6.3(13) prevails since

$$0.5\phi V_{uc} = 125.39 \text{ kN} < V^* < \phi V_{u.min} = 795.1 \text{ kN}$$

Consequently, the spacing (s) for the shear reinforcement is the lesser of $0.75D$ and 500 mm, or $s = 500$ mm. Equation 6.3(12) gives $A_{sv.min} = 630$ mm^2.

As can be found in Table 2.3(1), six N12 gives $A_{sv} = 678$ mm^2. This is acceptable.

Step 6

Not required for this example.

Step 7

Not required for this example.

Step 8

The layout of A_{sv} is shown in Figure 12.3(7). Note that $s = 400$ mm is used to neatly fit in the required A_{sv} across the critical sections on the left and right sides of the column. Also note that the six N12 shear reinforcement over the width b_f is made up of three ties as shown in Figure 12.3(7)b.

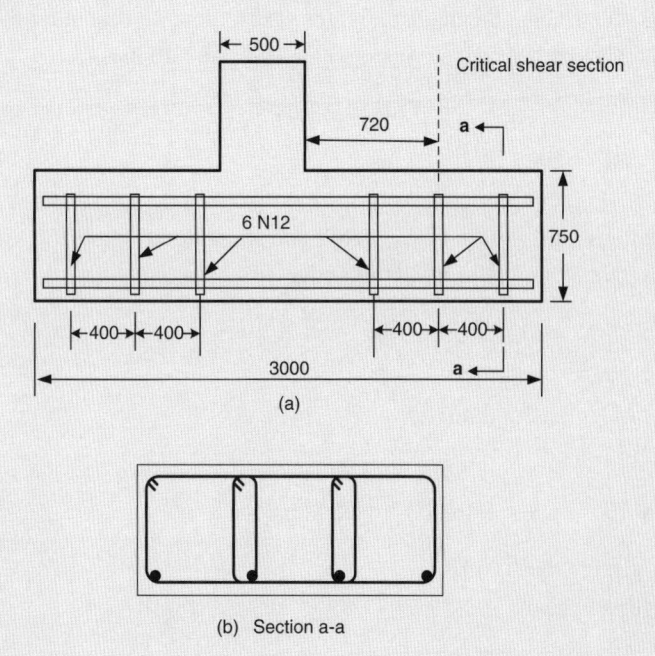

Figure 12.3(7) Shear reinforcement details

Note: all dimensions are in mm.

Punching shear check

The footing can be treated as an edge column with overhang on the short (left or heel) side of the footing. To be on the conservative side, the overhang may be ignored in computing the critical shear perimeter. Noting that $d_{om} = 720 - 10 = 710$ mm and:

Equation 9.6(4): $u = 2(500 + 710/2) + (300 + 710) = 2720$ mm

Equation 9.6(6): $f_{cv} = 0.17\left(1 + \dfrac{2}{500/300}\right)\sqrt{f'_c} = 0.374\sqrt{f'_c} \leq 0.34\sqrt{f'_c}$

Thus $f_{cv} = 0.34\sqrt{20} = 1.521$ MPa and:

Equation 9.6(3): $V_{uo} = 2720 \times 710 \times 1.521 \times 10^{-3} = 2937.4$ kN

and as shown in Figure 12.3(6), $M_v^* = 200$ kNm

Equation 9.6(9): $V_u = \dfrac{2937.4}{1 + \dfrac{2720 \times 200 \times 10^6}{8 \times 1400 \times 10^3 \times 1210 \times 710}} = 2780.2$ kN

Equation 9.6(10): $\phi V_u = 0.7 \times 2780.2 = 1946.1 > N^* = 1400$ kN

Therefore, the punching shear strength is more than adequate.

Steps 9 and 10

Bending design covers the moments about the major and minor axes. First, A_{st} design for the moment about the major axis must be determined. For a single layer of N20 bars:

$d = d_o = 720$ mm
Equation 3.3(2)a: $\alpha_2 = 0.85$ (for $f'_c = 20$ MPa)
Equation 3.3(2)b: $\gamma = 0.85$

Following the recommendations illustrated in Figure 9.4(2) and referring to Figure 12.2(6), the critical bending section is located at L_b, from the right edge of the footing where

$$L_b = \dfrac{D_f}{2} + e - \dfrac{0.7 D_c}{2}$$

That is, $L_b = 3/2 + 0.1429 - 0.7 \times 0.5/2 = 1.4679$ m
Thus

$$M^* = (300 - 0.9 \times 24 \times 0.75) \times 1.8 \times 1.4679^2/2 = 550.36 \text{ kNm}$$

and Equation 3.5(7): $\xi = 0.85 \times 20/500 = 0.034$.
Assuming $\phi = 0.8$, then Equation 3.5(6):

$$p_t = 0.034 - \sqrt{0.034^2 - \dfrac{2 \times 0.034 \times 550.36 \times 10^6}{0.8 \times 1800 \times 720^2 \times 500}} = 0.001508$$

Check to ensure that the assumed value of $\phi = 0.8$ is valid. From Equation 3.4(8)

$$k_u = \frac{0.001508 \times 500}{0.85 \times 0.85 \times 20} = 0.0522$$

Since $d = d_o$, then $k_{uo} = k_u = 0.0522$ and Equations 3.4(20)a with b yield $\phi = 0.8$.
This is acceptable.

Equation 12.3(18): $p_{t.min} = 0.20 \times (750/720)^2 \times 0.6 \times \sqrt{20}/500 = 0.001165$
< 0.001508

Therefore, $A_{st} = 0.001508 \times 1800 \times 720 = 1955$ mm^2 and from Table 2.3(1), seven N20 gives $A_{st} = 2198$ mm^2. This is acceptable. Note that since $p_t > p_{t.min}$, $V_{uc} = 358.26$ kN as computed in step 4 is conservative and safe.

Once the A_{st} design for moment about the major axis is completed as above, A_{st} design for the moment about the minor axis is to be carried out. For a single layer of N20 bars, $d = d_{major} - 20 = 700$ mm. For the minor direction

$$L_b = b_f/2 - 0.7b_c/2 = 1.8/2 - 0.7 \times 0.3/2 = 0.795 \text{ m}$$

Thus

$$M^* = (300 - 0.9 \times 24 \times 0.75) \times 3 \times 0.795^2/2 = 269.05 \text{ kNm}$$

If we assume $\phi = 0.8$, then Equation 3.5(6):

$$p_t = 0.034 - \sqrt{0.034^2 - \frac{2 \times 0.034 \times 269.05 \times 10^6}{0.8 \times 3000 \times 700^2 \times 500}} = 0.0004607$$

Check for ϕ as follows.
From Equation 3.4(8)

$$k_u = \frac{0.0004607 \times 500}{0.85 \times 0.85 \times 20} = 0.0159$$

and since $d_o = 700$ mm

$$k_{uo} = \frac{k_u d}{d_o} = \frac{0.0159 \times 720}{700} = 0.01635$$

Equations 3.4(20)a with b yield $\phi = 0.8$. This is acceptable.
Now, Equation 12.3(18):

$$p_{t.min} = 0.20 \times (750/700)^2 \times 0.6 \times \sqrt{20}/500 = 0.001232 > p_t$$

Therefore, $A_{st} = 0.001232 \times 3000 \times 700 = 2588$ mm^2. From Table 2.3(1), nine N20 gives $A_{st} = 2826$ mm^2. This is acceptable.

The layout of the reinforcing bars is illustrated in Figure 12.3(8). In practice, Figures 12.3(7) and 12.3(8) should be combined into one working drawing.

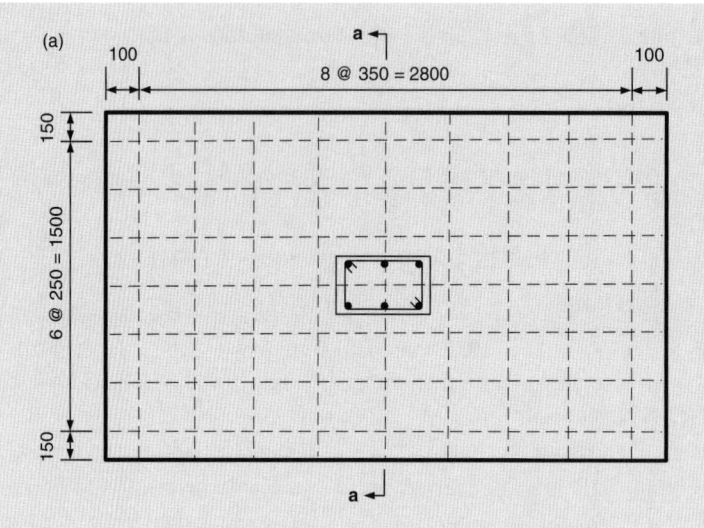

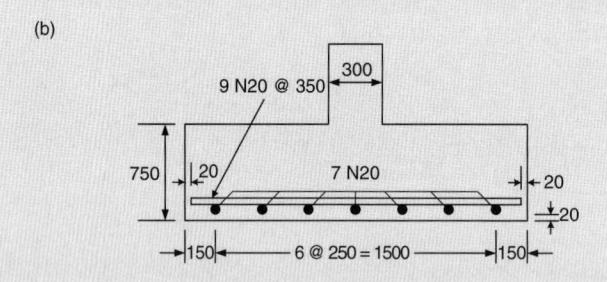

Figure 12.3(8) Bending reinforcement: **(a)** plan and **(b)** section a-a

Note: all dimensions are in mm.

Step 11
Check the bond of the major moment direction. For Equation 8.2(1):

$k_1 = 1.3$
$k_2 = (132 - 20)/100 = 1.12$
$k_3 = 1 - 0.15(20 - 20)/20 = 1$

Therefore, Equation 8.2(1):

$$L_{\text{sy.tb}} = 0.5 \times 1.3 \times 1 \times 500 \times 20 / \left(1.12 \times \sqrt{20}\right) = 1298 \text{ mm } (> 300 \text{ mm})$$

Since the available

$$L_{\text{sy.t}} < (L_b - \text{cover}) = 1468 - 20 = 1448 \text{ mm}$$

and

$$L_{\text{sy.t}} < (D_f - L_b - \text{cover}) = 3000 - 1468 - 20 = 1512 \text{ mm}$$

then straight bars of length 3000 − 2 × 20 = 2960 mm have adequate development length at the ultimate state. Depending on the quality of excavation, a shorter bar length should be used resulting in more concrete cover.

Check the bond of the minor moment direction. Similar to the major moment direction calculations, Equation 8.2(1):

$$L_{sy.tb} = 1298 \text{ mm}$$

At either end of the N20 bar, the available length for stress development is

$$\frac{b_f}{2} - \text{cover} - 0.7\frac{b_c}{2} = 900 - 20 - 0.7 \times 300/2 = 775 \text{ mm} < L_{sy.tb} = 1298 \text{ mm}$$

which appears inadequate. However, Equation 8.2(7) indicates that the required stress development length is $L_{st} = M^* L_{sy.tb}/M_{u.min} = 269.05\, L_{sy.tb}/M_{u.min}$ where, as per Equation 3.4(10), the minor moment capacity

$$M_{u.min} = 2826 \times 500 \times 700 \times \left(1 - \frac{2826 \times 500}{2 \times 0.85 \times 3000 \times 700 \times 20}\right) \times 10^{-6}$$

$$= 969.5 \text{ kNm}$$

or

$$L_{st} = 269.05 \times 1298/969.5 = 360.2 \text{ mm} < 775 \text{ mm}$$

Thus straight bars of 1800 − 2 × 20 = 1760 mm have adequate bond strength in the minor bending direction. A shorter bar length should be used to provide more concrete cover as necessary.

Step 12

Now the shrinkage and temperature steel must be determined. The A_{st} in the major bending direction is greater than the $A_{st.min}$ specified in Equation 12.3(18), whereas in the minor bending direction $A_{st} = A_{st.min}$. Hence, no additional shrinkage and temperature steel is required.

12.4 PILE CAPS

The analysis of forces acting on pile caps and their design are similar to those of column footings. The main difference lies in the reaction to the column loading. For footings, the reaction is provided by the subsoil in the form of pressure, uniform or otherwise; for pile caps, the reaction comprises the axial forces in the piles.

12.4.1 Concentric column loading

For the symmetrically loaded square-shaped pile cap shown in Figure 12.4(1), the axial force in each of the five piles

$$P_p = (N^* + G^*)/5 \quad \quad \text{Equation 12.4(1)}$$

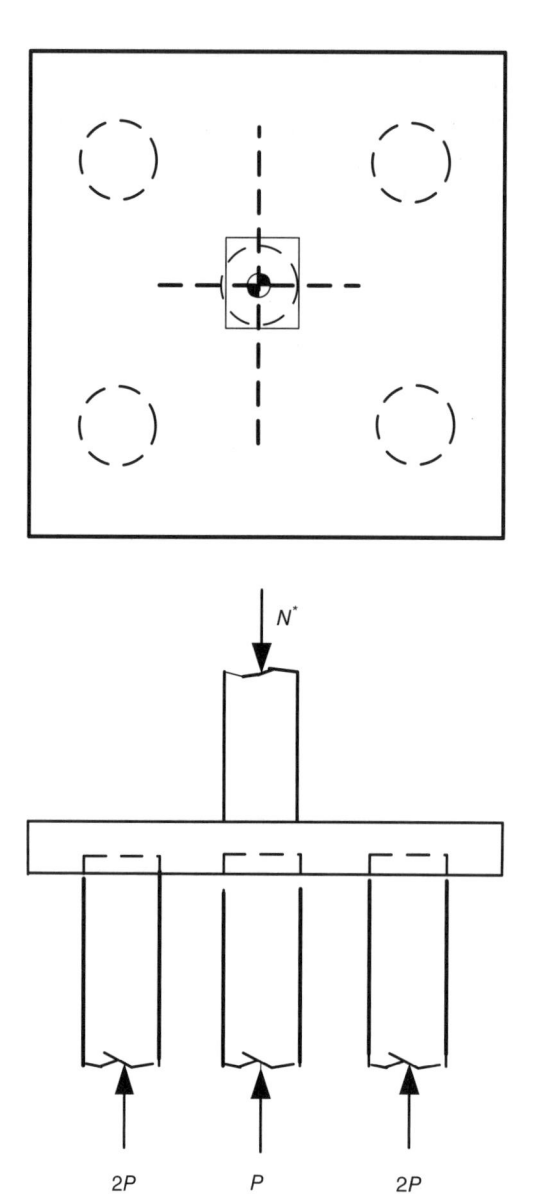

Figure 12.4(1) Symmetrically loaded pile cap

Because of symmetry, the consideration of transverse shear on any of the four critical shear sections next to the column faces yields the depth (D) of the pile cap. On the other hand, punching shear consideration around the column leads to an alternative D. The design D is the larger of the two values. The example below illustrates the procedure involved in the pile-cap design.

Illustrative example

Problem

Details of a centrally loaded square column to be supported on four circular concrete piles are shown in Figure 12.4(2). Given $N^* = 2000$ kN and $f'_c = 25$ MPa, obtain the overall depth (D) of the pile cap.

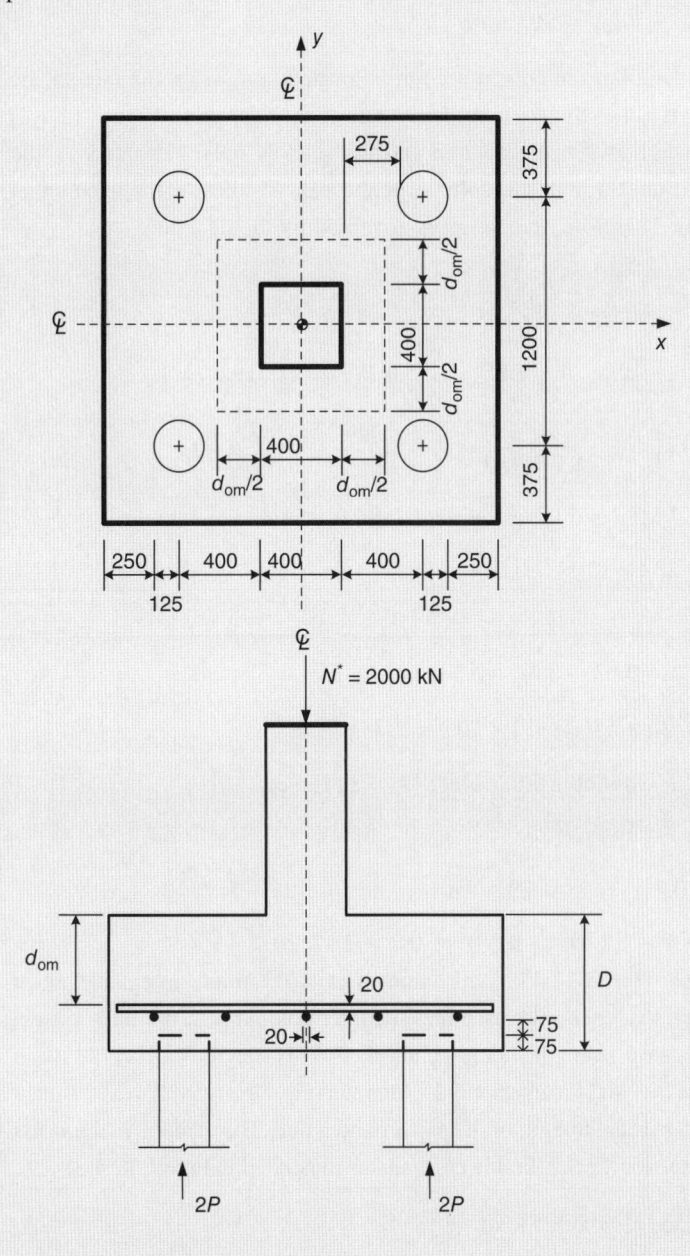

Figure 12.4(2) A centrally loaded square column supported on four circular concrete piles

Note: all dimensions in mm.

Solution
The depth of the section is to be determined by shear strength consideration. The pile load

$$P_p^* = (N^* + 1.2\rho_w b_f D_f D)/4 \qquad \text{Equation (a)}$$

Assuming $D = 700$ mm and using N20 bars for reinforcement,

$$d_o = 700 - 75 - 75 - 20 - 20/2 = 520 \text{ mm}.$$

For bending shear consideration, the critical section is located at d_o from the face of the support, which in this case is either the column or any pair of the four piles. Figure 12.4(2) shows that the clear distance between the column and the piles is only 275 mm (i.e. the critical shear section would cut through the column or the pair of piles). This means that bending shear is not critical. Therefore, the design is governed by the punching shear.

Punching shear design
For punching shear design, we have:

Equation (a): $P_p^* = (2000 + 1.2 \times 24 \times 1.95 \times 1.95 \times 0.7)/4 = 519.16$ kN

Equation 9.6(6): $f_{cv} = 0.17\left(1 + \dfrac{2}{1.95/1.95}\right)\sqrt{f_c'} = 0.51\sqrt{f_c'} > 0.34\sqrt{f_c'}$

Adopt $f_{cv} = 0.34\sqrt{25} = 1.7$ MPa and the mean value of

$$d_{om} = (700 - 2 \times 75 - 20) \times 10^{-3} = 0.53 \text{ m}$$

Then:

Equation 9.6(5): $u = 2(0.4 + 0.53 + 0.49 + 0.53) = 3.72$ m

Equation 9.6(3): $V_{uo} = 3720 \times 530 \times 1.7 \times 10^{-3} = 3351.72$ kN

Since the square pile cap is symmetrically loaded, $M^* = 0$. Hence:

Equation 9.6(9): $V_u = V_{uo} = 3351.72$ kN

Equation 9.6(10): $\phi V_u = 0.7 V_u = 2346.2$ kN

In Figure 12.4(2), $N^* = 2000 + 1.2 (0.4 + 0.53)^2 \times 0.7 \times 24 = 2017.4$ kN

Since, $\phi V_u = 2346.2$ kN $> N^* = 2017.4$ kN, the assumed $D = 700$ mm is acceptable, albeit conservative. Note, however, that each pile resembles a corner column in a flat plate system around which the punching shear strength should be checked. This may be carried out using semi-empirical formulas (see Loo and Falamaki 1992). Alternatively, a conservative assessment may be made of the punching shear strength of the cap around a pile. This is done in a process similar to that described in Section 12.3.7. In this case, and as illustrated in Figure 12.4(3)

$$M^*_v < P^*_p(0.125 + 0.53/2) = 202.5 \text{ kNm}$$

Let $M^*_v = 202.5$ kNm.

Conservatively, and referring to Figure 12.4(3), the circular pile section may be converted into a square with $b_c = D_c = \sqrt{\pi \times 125^2} = 221.6$ mm.

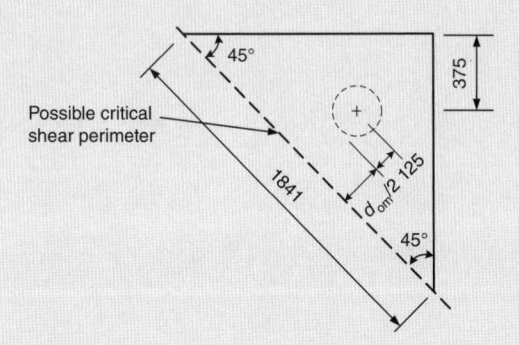

Figure 12.4(3) Determination of critical shear perimeter for the punching shear check of the pile cap

Note: all dimensions are in mm.

Then, the critical shear perimeter is obtained using Equation 9.6(4) or

$$u = 2a + a_2$$

where again conservatively,

$$a = 221.6 + d_{om}/2 = 486.6 \text{ mm}$$

and

$$a_2 = 221.6 + d_{om} = 751.6 \text{ mm}$$

Thus $u = 1724.8$ mm.

Note that this u is shorter and hence more critical than another possible shear perimeter – see Figure 12.4(3). Then

Equation 9.6(3) : $V_{uo} = 1724.8 \times 530 \times 1.7 \times 10^{-3} = 1554.0$ kN

Equation 9.6(9) : $V_u = \dfrac{1554.0}{1 + \dfrac{1.7248 \times 202.5}{8 \times 519.16 \times 0.4866 \times 0.53}} = 1171.9$ kN

Equation 9.6(10): $\phi V_u = 820.3$ kN $> P^*_p = 519.16$ kN

Thus, punching shear strength around the pile appears to be adequate.

To complete the pile-cap design, proceed to compute the bending steel A_{st} in each of the x and y directions – see Figure 12.4(2). The process is similar to that used in Sections 12.2.5

and 12.3.7 for wall and column footings, respectively. Note, however, that the bending span in either direction is

$$L_b = 400 + 0.3 \times \frac{400}{2} = 460 \text{ mm}$$
$$M^*{}_x = M^*{}_y = 2 \times 519.16 \times 0.460 - 24 \times 1.95 \times 0.7 \times 0.460^2/2 = 474.2 \text{ kNm}$$
$$d_x = 700 - 75 - 75 - 20 - 20/2 = 520 \text{ mm and } d_y = d_x + 20 = 540 \text{ mm}$$

After computing the bending reinforcement in both directions, ensure that the bars have adequate stress development length. If they don't, use hooks or cogs.

12.4.2 Biaxial bending

Figure 12.4(4)a depicts a column subjected to biaxial bending, or a combination of N^*, $M^*{}_x$ and $M^*{}_y$. The location of the resultant ($4P$) of the four equally loaded piles may be determined by solving

$$\sum M_{aboutx} = 0 \text{ for } e_y \qquad \text{Equation 12.4(2)}$$

and

$$\sum M_{abouty} = 0 \text{ for } e_x \qquad \text{Equation 12.4(3)}$$

With e_x and e_y calculated, the plan area of the pile cap can readily be determined by simple proportion. The values of b_f and D_f are governed by the load bearing capacity of the pile (P_p) where

$$P_p \geq P^* = \frac{N^* + G^*}{4} \qquad \text{Equation 12.4(4)}$$

The procedure involved in the design of pile cap supporting a column under biaxial bending loads is given below. Note that for multiple columns the design can be carried out in a similar manner.

For a typical pile cap as shown in Figure 12.4(4), the longer side of the rectangular cap is designated D_f and the shorter side b_f. The critical shear section ($d_o b_f$) may be taken as located at a distance d_o from the face of the column, which is normally bigger in size. This being the case, the pile cap can be designed in a process similar to that under a concentric load (e.g. Section 12.4.1). Following the design for transverse shear, it is necessary to check for punching shear strength as for the case of a corner-column connection.

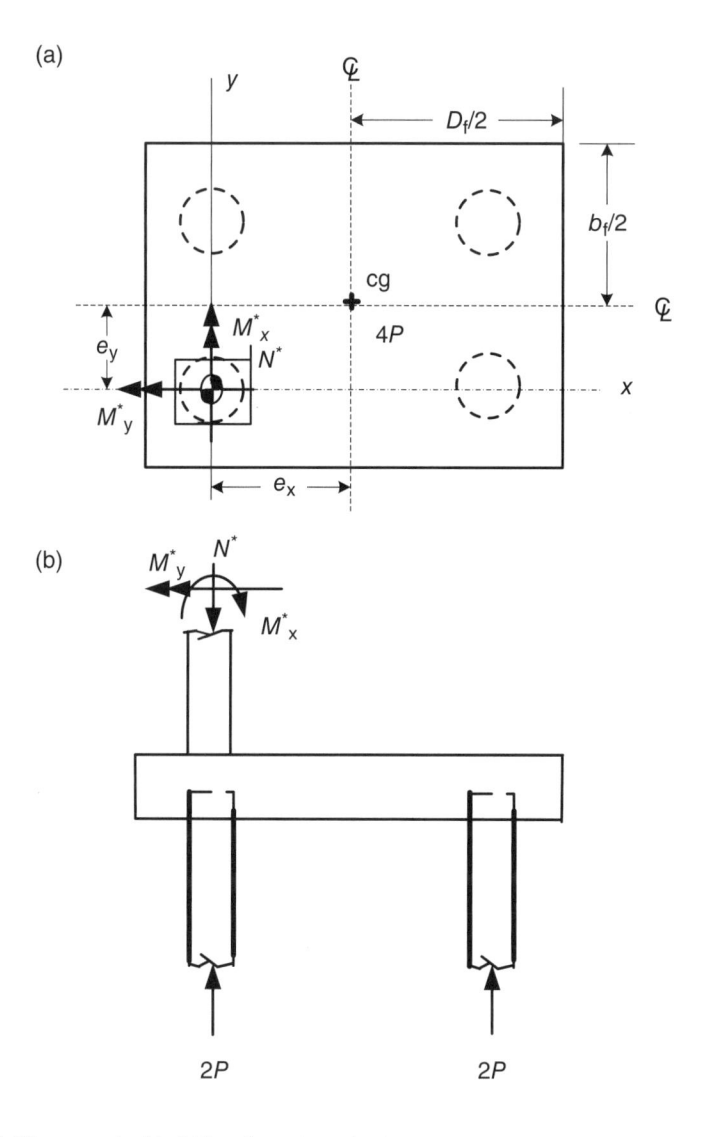

Figure 12.4(4) Pile cap under biaxial bending column loads

12.5 RETAINING WALLS

12.5.1 General remarks

Geotechnical engineering structures designed to retain horizontal earth or backfill pressure are referred to as retaining walls. Redrawn from a major and comprehensive book on foundation engineering (Bowles 1996), Figure 12.5(1) illustrates the types of reinforced concrete retaining walls usually encountered in practice.

These include:
- gravity walls, which may be made of stone masonry, brick or plain concrete construction – Figure 12.5(1)a
- semigravity walls of concrete construction, requiring minimum amount of reinforcement since much of the lateral earth pressure is resisted by the weight of the wall – Figure 12.5(1)c
- bridge abutments, which are normally of reinforced concrete construction – Figure 12.5(1)d – built integrally with the 'wing walls' as illustrated in Figure 12.5(2) to resist earth pressure in the direction other than those resisted by the abutment
- counterforts, which are of reinforced concrete construction, but the earth pressure is resisted mainly by the props or back-stays (covered under the backfill) in tension, noting that retaining walls with the props in compression on the other side of the backfill or the front face of the wall are referred to as buttresses – Figure 12.5(1)e
- crib walls with individual reinforced concrete components assembled to perform the earth-retaining function – Figure 12.5(1)f
- reinforced concrete retaining walls, which retain the earth or backfill pressure and act as a cantilever structure – Figure 12.5(1)b.

This section presents only the analysis and design of the cantilever retaining walls illustrated in Figure 12.5(1)b, as the remaining wall types are beyond the scope of this work.

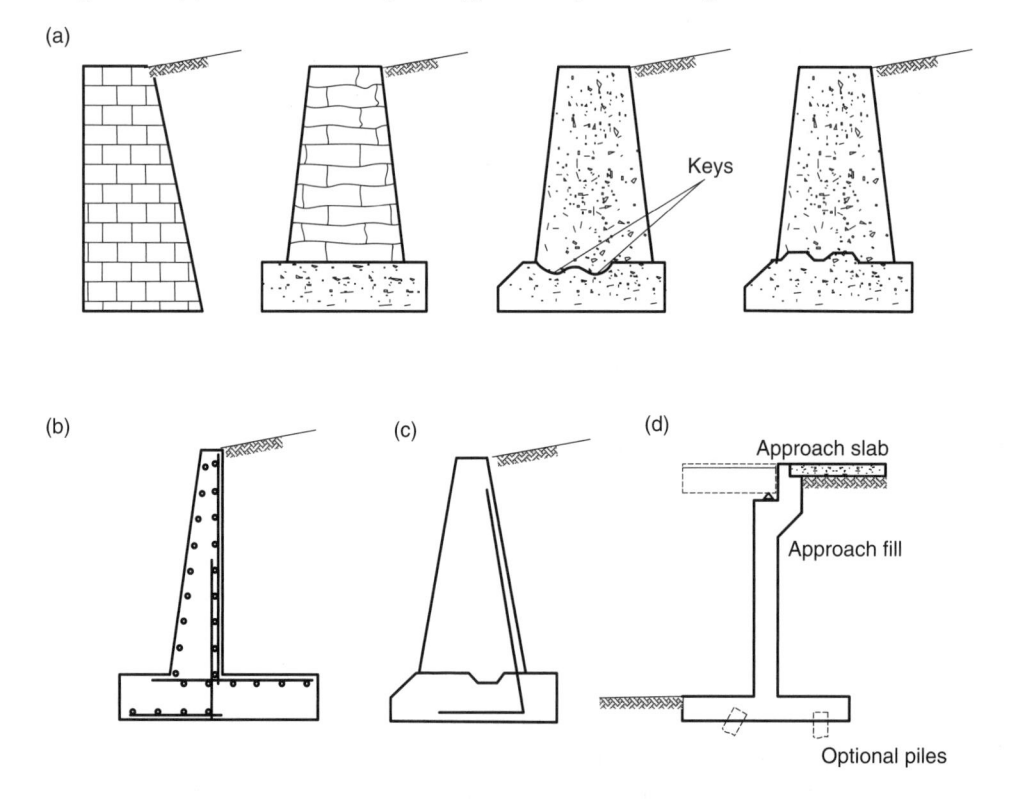

Figure 12.5(1) Types of retaining structures: **(a)** gravity wall; **(b)** cantilever wall; **(c)** semigravity wall; **(d)** bridge abutment; **(e)** counterfort; and **(f)** crib wall

Source: redrawn from Bowles (1996).

(e)

(f)

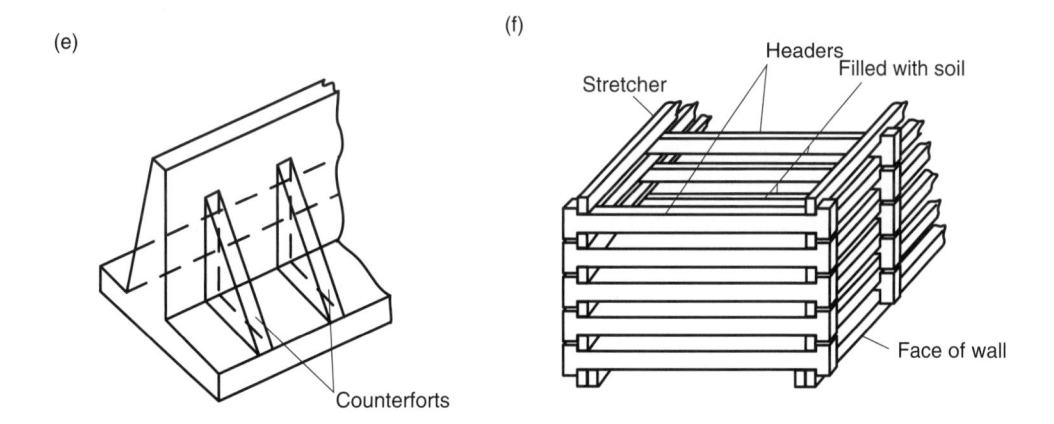

Stretcher
Headers
Filled with soil
Face of wall
Counterforts

Figure 12.5(1) (*cont.*)

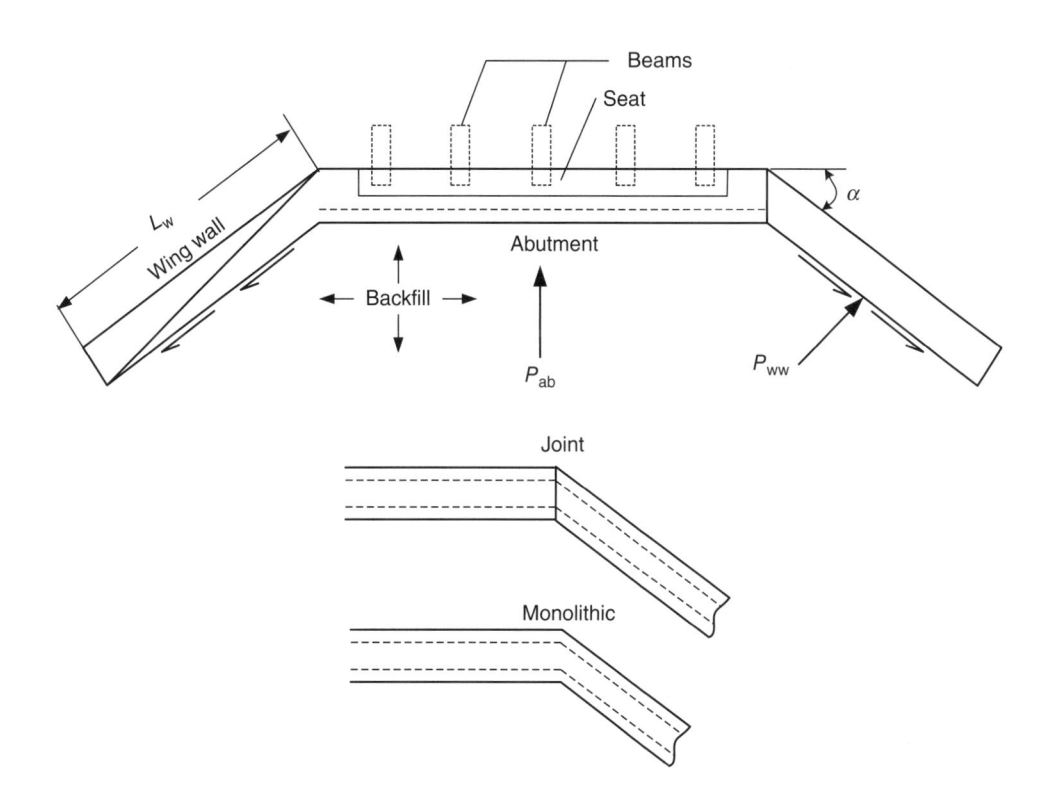

Beams
Seat
L_w
Wing wall
Abutment
Backfill
P_{ab}
P_{ww}
α

Joint

Monolithic

Figure 12.5(2) Bridge abutment and wing walls

12.5.2 Stability considerations

Figure 12.5(3) shows details of a typical retaining wall with definitions of various components and aspects of the structure. For such cantilever structures, Bowles (1996) suggests some preliminary design dimensions as illustrated in Figure 12.5(4).

For overall wall stability against overturning and sliding, the design earth pressures are detailed in Figure 12.5(5). The relevant earth pressure coefficients and parameters are listed below:

- β, slope angle of the backfill
- $\beta' = \beta$ or ϕ
- ϕ, angle of internal friction
- ρ_w, weight density of soil
- K_a, active earth pressure coefficient
- K_p, passive earth pressure coefficient
- δ, angle of friction between soil and wall or soil and soil
- P_{ah}, horizontal component of the Rankine or Coulomb lateral earth pressure against the vertical line *ab* (virtual back) in Figure 12.5(5)
- P_{av}, vertical shear resistance on virtual back that develops as the wall tends to turn over
- F_r, friction between base and soil.

Under the force system as depicted in Figure 12.5(5), the overall stability of the retaining wall must be ensured (Bowles 1996). This includes guarding against:

- sliding at the base, which is produced by earth pressure on vertical plane *ab*
- overturning about the toe at point *O*
- rotational shear failure as illustrated in Figure 12.5(6)
- bearing-capacity failure resulting in excessive settlements and wall tilt.

As the overall design of the retaining system is essentially a geotechnical engineering problem, the reader should refer to the text by Bowles (1996) for an extensive discussion.

For the reinforced concrete design of the cantilever wall, the earth pressures and resulting forces are detailed in Figure 12.5(7). The keys to computing the design forces are K_a, the active earth pressure coefficient of the soil, and δ, the angle of friction between the soil and the back face of the wall.

Note that stability design is also dealt with in AS 4678–2002 *Earth-Retaining Structures*, which identifies, for cantilever walls, the following five limit modes for ultimate limit state design:

1. *Limit Mode U1* – sliding within or at the base of the wall
2. *Limit Mode U2* – rotation of the wall
3. *Limit Mode U3* – rupture of component and connections
4. *Limit Mode U5* – global failure mechanisms
5. *Limit Mode U6* – bearing failure.

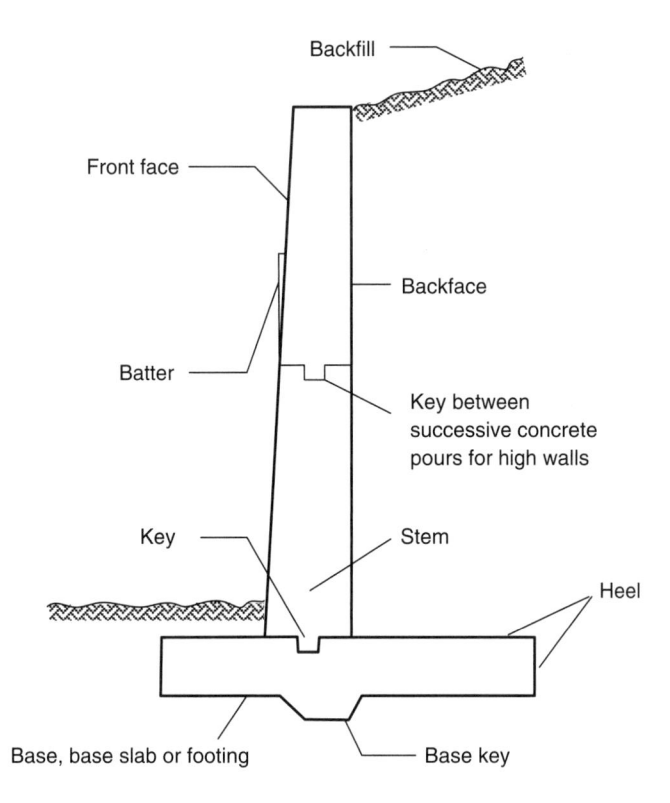

Figure 12.5(3) Details of cantilever retaining wall

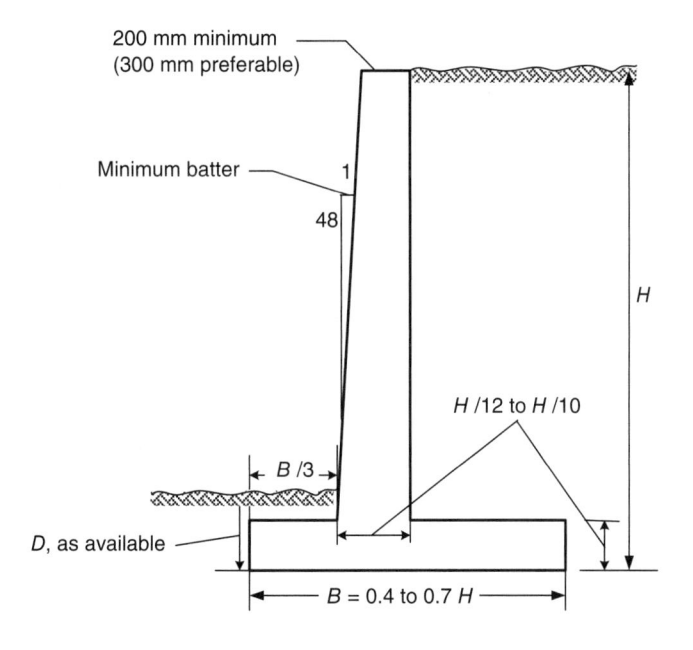

Figure 12.5(4) Preliminary design dimensions

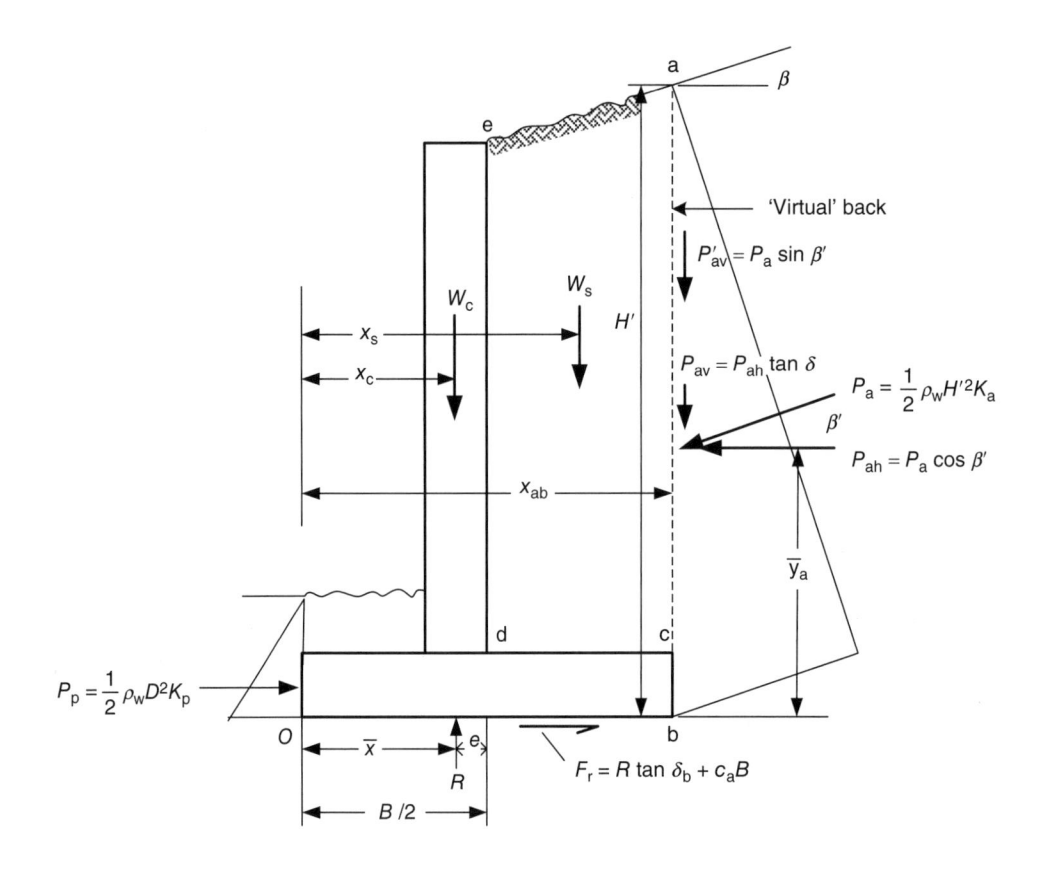

Figure 12.5(5) Design earth pressures for wall stability consideration

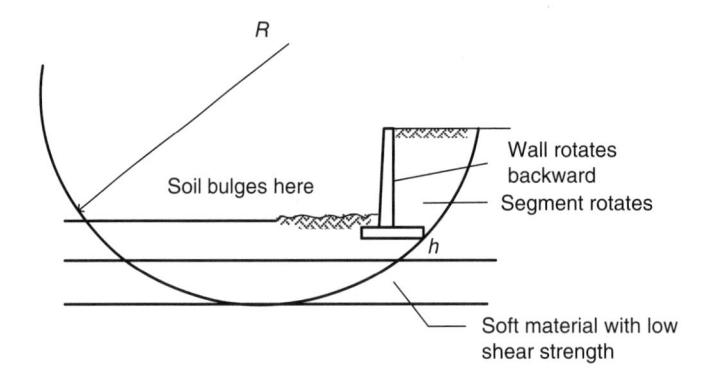

Figure 12.5(6) Rotational failure

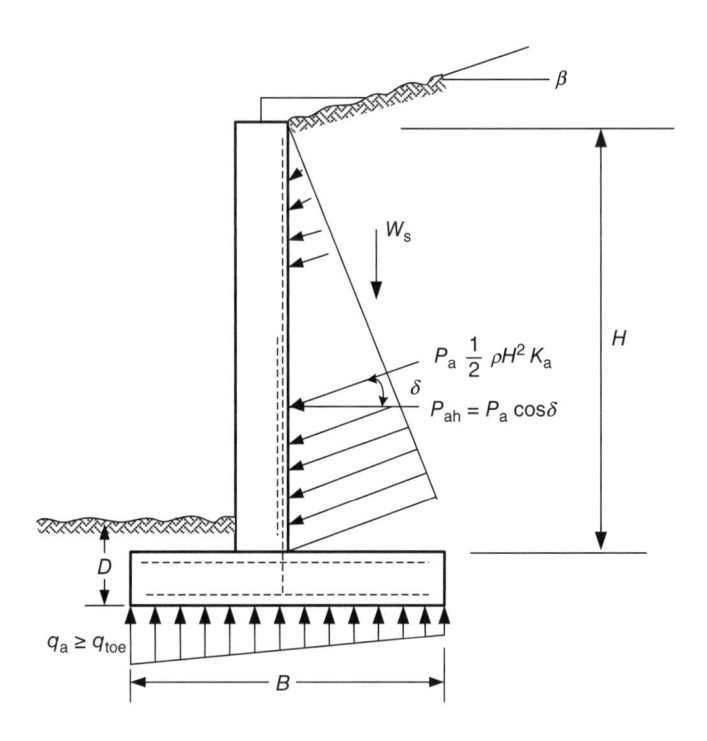

Figure 12.5(7) Earth pressures and resulting forces

These failure modes are illustrated in Figures 12.5(8)a to e.

In addition, checks on serviceability limit states must be carried out on the following three limit modes:
1. *Limit Mode S1* – rotation of the structure
2. *Limit Mode S2* – translation or bulging of the retaining structure
3. *Limit Mode S3* – settlement of the structure.

These are depicted in Figure 12.5(9).

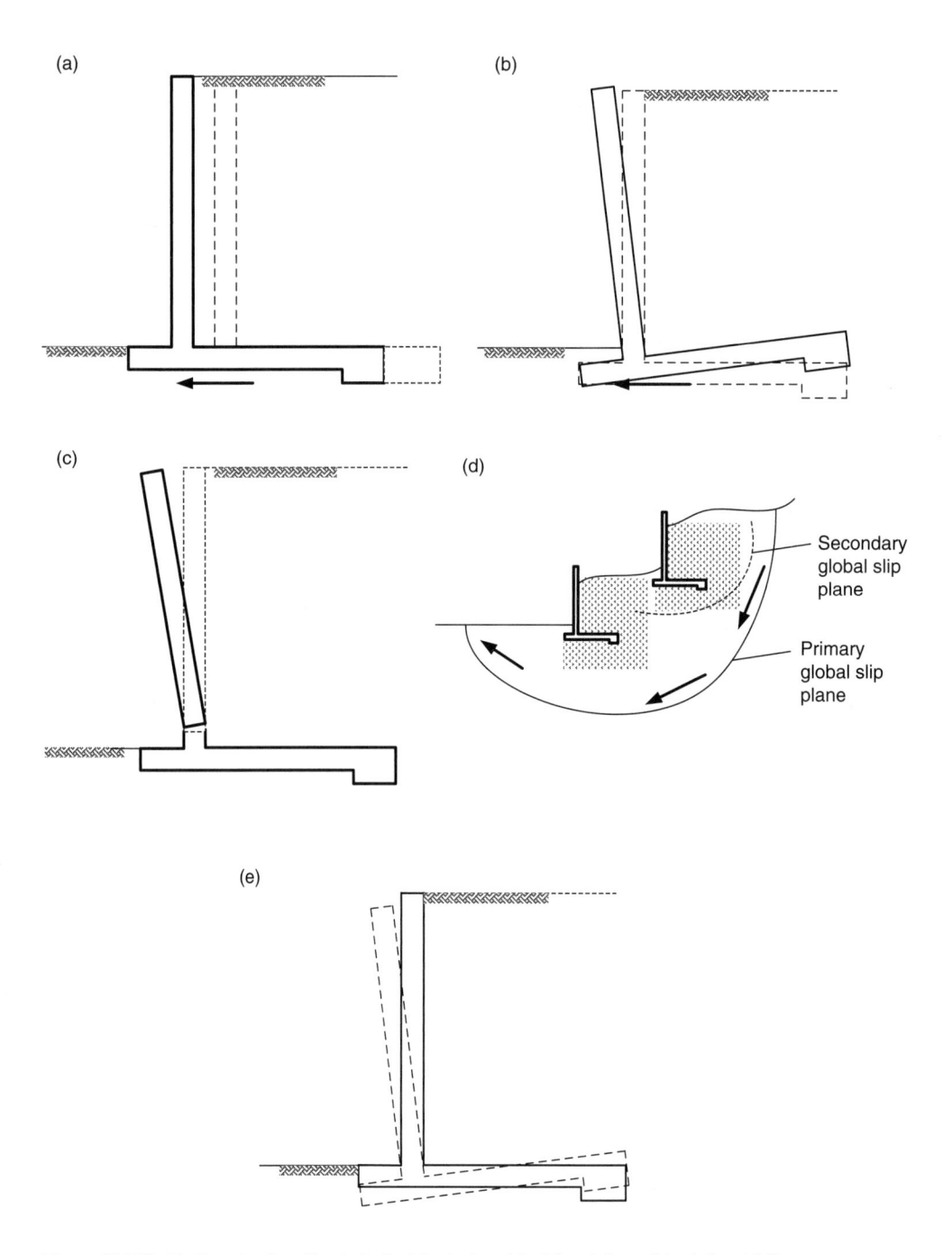

Figure 12.5(8) Limit modes for ultimate limit state design: **(a)** sliding failure; **(b)** rotational failure; **(c)** component failure; **(d)** global failure; and **(e)** rotational and translational failure

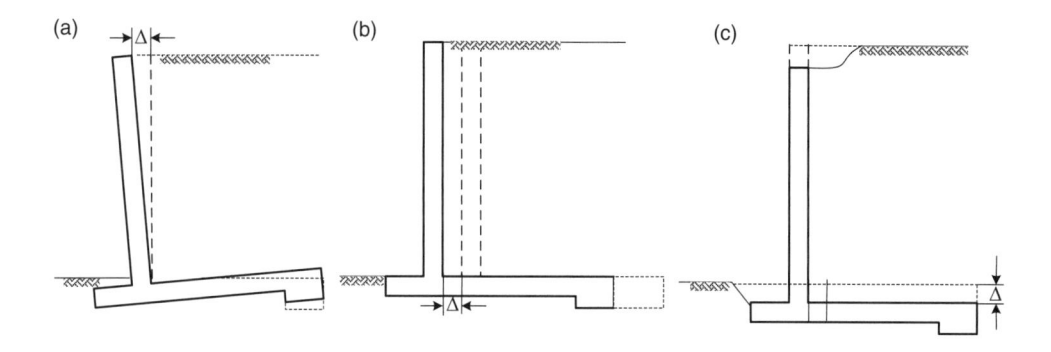

Figure 12.5(9) Limit modes for serviceability limit state checks: **(a)** rotation; **(b)** translation; and **(c)** settlement

12.5.3 Active earth pressure

The value of the active earth pressure (p_a) behind a retaining wall is a function of the backfill properties, which in turn depend upon the type of soil used as backfill. Clause D2.2.5 of AS 4678–2002 states that 'whilst a wide range of fills may be used as backfill behind retaining walls, selected cohesionless granular fill placed in a controlled manner behind the wall is usually the most desirable'. It should also be noted that AS 4678–2002 provides 'no recommendations for other fill' or cohesive fill materials.

As per Clause E1 of AS 4678–2002, the values of p_a for granular soils may be computed based on the Rankine–Bell design model. Figure 12.5(10) illustrates the distribution of p_a for two cases of granular backfill:
1. with (sloping) surcharge having a surface inclination (β)
2. without sloping surcharge (i.e. $\beta = 0$).

The general equation for the lateral pressure is

$$p_a = K_a \rho_w z \qquad \text{Equation 12.5(1)}$$

where

$$K_a = \frac{\cos\beta - \sqrt{\cos^2\beta - \cos^2\phi}}{\cos\beta + \sqrt{\cos^2\beta - \cos^2\phi}} \qquad \text{Equation 12.5(2)}$$

For the case of granular backfill without sloping surcharge, i.e. with $\beta = 0$, Equation 12.5(2) becomes

$$K_a = \tan^2(45° - \phi/2) \qquad \text{Equation 12.5(3)}$$

In Equations 12.5(1) to (3), ρ_w is the unit weight of the backfill, z is the depth of the pressure profile as defined in Figure 12.5(10) and ϕ is the characteristic effective internal friction angle of the backfill.

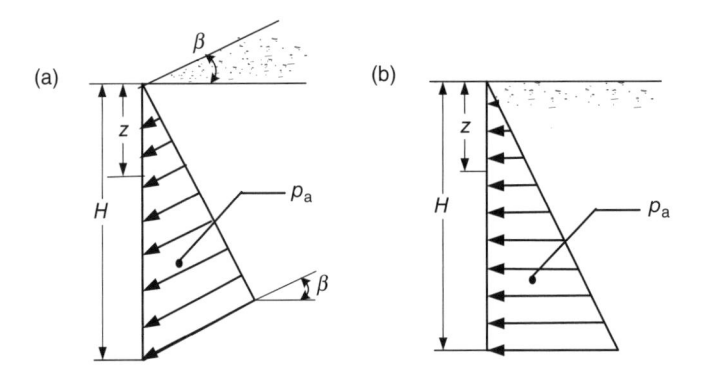

Figure 12.5(10) Typical earth or backfill pressure profiles

Note that for level backfill, $\beta = 0$, and Equation 12.5(2) reduces to Equation 12.5(3). For selected granular backfill, the values of ρ_w are given in Table D1 of AS 4678–2002. This table is reproduced here as Table 12.5(1). Clause D2.2.3, AS 4678–2002 recommends that an estimated ϕ may be taken as

$$\phi = 30° + k_A + k_B + k_C \quad \text{Equation 12.5(4)}$$

where k_A, k_B and k_C are functions of the angularity, grading and density of the backfill particles, respectively. Some recommended values of k_A, k_B and k_C may be found in Table 12.5(2), which is a reproduction of Table D2 in AS 4678–2002.

Table 12.5(1) Unit weights of granular backfills

Material	ρ_m – moist bulk weight (kN/m³) Loose	Dense	ρ_s – saturated bulk weight (kN/m³) Loose	Dense
Gravel	16.0	18.0	20.0	21.0
Well-graded sand and gravel	19.0	21.0	21.5	23.0
Coarse or medium sand	16.5	18.5	20.0	21.5
Well-graded sand	18.0	21.0	20.5	22.5
Fine or silty sand	17.0	19.0	20.0	21.5
Rock fill	15.0	17.5	19.5	21.0
Brick hardcore	13.0	17.5	16.5	19.0
Slag fill	12.0	15.0	18.0	20.0
Ash fill	6.5	10.0	13.0	15.0

Source: Standards Australia (2002). AS 4678-2002 *Earth-Retaining Structures*. Standards Australia, Sydney, NSW. © Standards Australia Limited. Copied by Cambridge University Press with the permission of Standards Australia and Standards New Zealand under Licence 1712-c038.

Table 12.5(2) Values of k_A, k_B and k_C (for computing φ for siliceous sands and gravels) – intermediate values of k_A, k_B and k_C may be obtained by interpolation

Angularity[1]	k_A
Rounded	0
Sub-angular	2
Angular	4
Grading of soil[2, 3]	k_B
Uniform	0
Moderate grading	2
Well graded	4
N' (below 300 mm)[4]	k_C
<10	0
20	2
40	6
60	9

Notes:

1. Angularity is estimated from visual description of soil.
2. Grading may be determined from grading curve by the use of coefficient of uniformity = D_{60}/D_{10} where D_{10} and D_{60} are particle sizes such that, in the sample, 10% of the material is finer than D_{10} and D_{60}, and 60% is finer than D_{60}.

Grading	Coefficient of uniformity
Uniform	<2
Moderate grading	2–6
Well graded	6

3. A step-graded soil should be treated as uniform or moderately graded soil according to the grading of the finer fraction.
4. N' from results of standard penetration test modified where necessary.

Source: Standards Australia (2002). AS 4678–2002 *Earth-Retaining Structures*. Standards Australia, Sydney, NSW. © Standards Australia Limited. Copied by Cambridge University Press with the permission of Standards Australia and Standards New Zealand under Licence 1712-c038.

12.5.4 Design subsoil pressures

Following a satisfactory global stability analysis and geotechnical design, a typical cantilever wall of a 1-metre run is shown in Figure 12.5(11). In addition to the superimposed live load (W_{SL}), the vertical forces include the weights of the backfill, surcharge and front surcharge, designated as W_{BF}, W_{SL} and W_{FS}, respectively. Also, the vertical forces include the weight of the retaining wall components (i.e. the wall and its base W_{W1}, W_{W2} and W_{W3}). Note that AS 4678–2002 recommends only cohesionless backfill. Thus, the vertical friction force along the

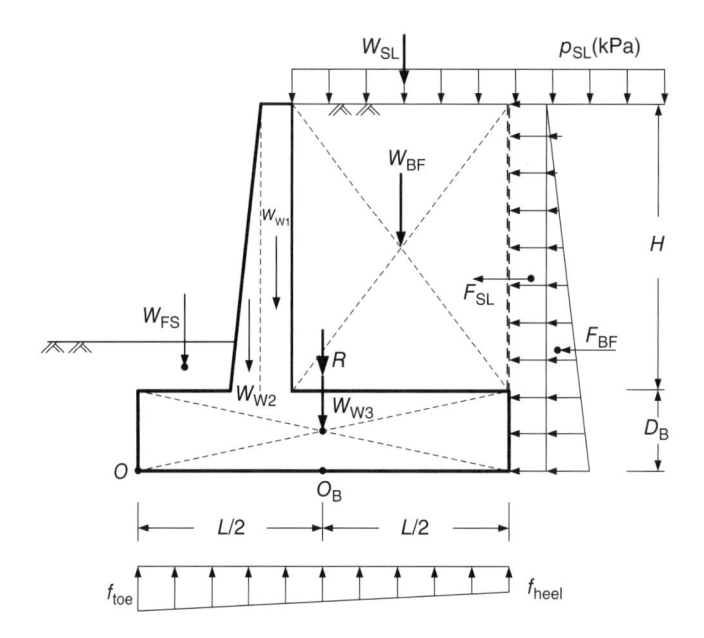

Figure 12.5(11) Vertical and horizontal actions on a typical retaining wall

vertical line a–c–b in Figure 12.5(5) is negligible. The horizontal forces, on the other hand, are the pressure resultants due to the backfill and superimposed live load, F_{BF} and F_{SL}, respectively. Also, the horizontal forces are the passive pressure resultant due to the front surcharge, which is not shown in this sliding stability analysis.

The resultant vertical force at the centre of the wall base, O_B, is

$$R = \beta_W(W_{W1} + W_{W2} + W_{W3}) + \beta_{SL}W_{SL} + \beta_{BF}W_{BF} + \beta_{FS}W_{FS} \qquad \text{Equation 12.5(5)}$$

where the various subscripted β are the relevant load combination factors (see Section 1.3).

Assuming that anticlockwise direction is positive, the resulting moment due to the vertical and horizontal forces at O_B is given as

$$M_R = F_{SL}L_{SL} + F_{BF}L_{BF} + \beta_W W_{W1}L_{W1} + \beta_W W_{W2}L_{W2} + \beta_{FS}W_{FS}L_{FS} \\ - \beta_{SL}W_{SL}L_{SL} - \beta_{BF}W_{BF}L_{BF} \qquad \text{Equation 12.5(6)}$$

where each subscripted L represents the lever arm of the associated force with respect to O_B, and the horizontal forces

$$F_{SL} = \beta_{SL}K_a p_{SL}(H + D_B) \qquad \text{Equation 12.5(7)}$$

and

$$F_{BF} = \frac{1}{2}\beta_{BF}K_a \rho_{w.BF}(H + D_B)^2 \qquad \text{Equation 12.5(8)}$$

in which p_{SL} is the superimposed distributed live load, H is the height of the wall, D_B is the depth of the base and $\rho_{w.BF}$ is the unit weight of the backfill.

Making use of Equations 12.5(5) and (6), the subsoil pressures at the toe and heel, as shown in Figure 12.5(11), are obtained as

$$f_{toe} = R/L + 6M_R/L^2 \qquad \text{Equation 12.5(9)}$$

and

$$f_{heel} = R/L - 6M_R/L^2 \qquad \text{Equation 12.5(10)}$$

respectively, where L is the width of the base.

In design, f_{toe} and f_{heel}, under the most severe load combination, must be less than the soil bearing capacity (q_f). It is also desirable that $f_{heel} \geq 0$, in which case the stability against overturning of the retaining wall with respect to the toe, or point O in Figure 12.5(11), is assured. No separate check is necessary. However, if part of the base sustains no soil pressure at the heel end and beyond (towards the toe), ensure that

$$RL/2 - M_R \geq F_{SL}(H + D_B)/2 + F_{BF}(H + D_B)/3 \qquad \text{Equation 12.5(11)}$$

where M_R is given in Equation 12.5(6).

12.5.5 Design moments and shear forces

The design actions for the wall, toe and heel can be obtained using simple statics. At the root of the wall, the design moment with direction as shown in Figure 12.5(12) is

$$M^*_{O1} = \beta_{SL}K_a p_{SL} H l_{b1} + \frac{1}{2}\beta_{BF}K_a \rho_{w.BF} H^2 l_{b2}/2 \qquad \text{Equation 12.5(12)}$$

where in accordance with the recommendation given in Figure 9.4(2) the cantilever span is

$$l_{b1} = \frac{H}{2} + 0.15 D_B \qquad \text{Equation 12.5(13)}$$

and

$$l_{b2} = H/3 + 0.15 D_B \qquad \text{Equation 12.5(14)}$$

Similarly, the heel moment

$$M^*_{O2} = (\beta_{SL} W_{SL} + \beta_{BF} W_{BF})(L_{heel}/2 + 0.15 D_w) + \beta_w W_{heel}(L_{heel} + 0.15 D_w)/2 \\ - P_{1.heel}(L_{heel} + 0.15 D_w)/2 - P_{2.heel}(L_{heel} + 0.15 D_w)/3 \qquad \text{Equation 12.5(15)}$$

where W_{heel} is the weight of the heel over ($L_{heel} + 0.15 D_w$), and $P_{1.heel}$ and $P_{2.heel}$ are the resultants of the subsoil pressure block along the heel as shown in Figure 12.5(13).

Finally, the toe moment

$$M^*_{O3} = P_{1.toe}(L_{toe} + 0.15 D_w)/2 + 2P_{2.toe}(L_{toe} + 0.15 D_w)/3 \\ - \beta_{FS} W_{FS}(L_{toe}/2 + 0.15 D_w) - \beta_w W_{toe}(L_{toe} + 0.15 D_w)/2 \qquad \text{Equation 12.5(16)}$$

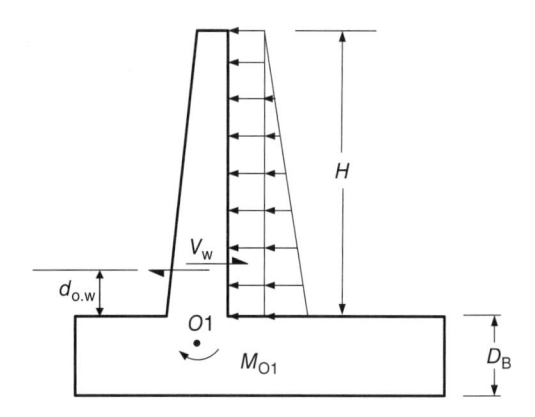

Figure 12.5(12) Position and direction of the design moment and shear at the root of the wall

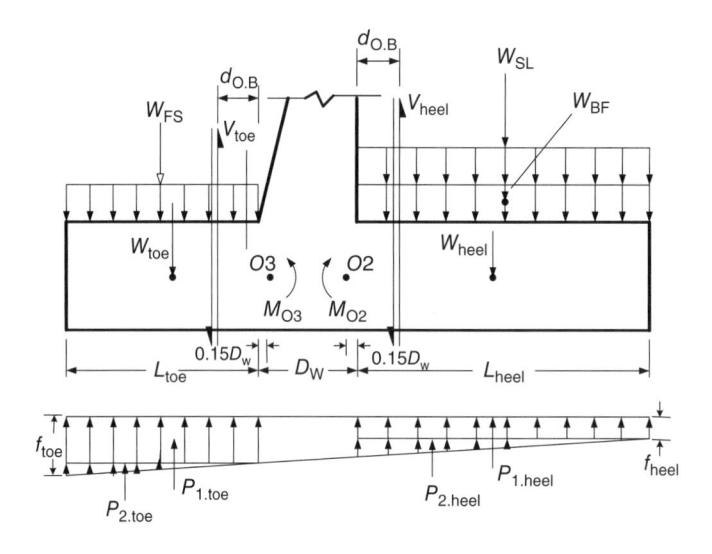

Figure 12.5(13) Subsoil pressure distribution, design moments and shear forces

Note that ΣM^* may not be equal to zero because M^*_{O1}, M^*_{O2} and M^*_{O3} are not necessarily due to the same load combination. Even for the same load case, the three moments would not sum up to zero, simply because they are with respect to three different points – $O1$, $O2$ and $O3$.

The design shear force (V_w) for the wall is shown in Figure 12.5(12) acting at the critical shear section $d_{o.w}$ from the top of the base, where the shear effective depth ($d_{o.w}$) is as defined in Figure 6.3(1). For the base, the design shear forces (V_{toe} and V_{heel}) are shown in Figure 12.5(13); they occur at $d_{o.toe}$ and $d_{o.heel}$ from the corresponding faces of the wall, respectively. Similarly, $d_{o.toe}$ and $d_{o.heel}$ are defined in Figure 6.3(1).

The values of V^*_w, V^*_{toe} and V^*_{heel} can be computed following the usual process (i.e. by considering the equilibrium of forces at each of the critical shear sections concerned).

12.5.6 Load combinations

In Equations 12.5(5) and (6), each of the load factors associated with the various vertical forces must be so assigned to produce the most severe outcomes. In this case, it is the subsoil pressures f_{toe} and f_{heel} as given in Equations 12.5(9) and (10), respectively. Consequently, for the superimposed live load W_{SL}, β_{SL} may be taken as 1.5 or ψ_c – see Table 1.3(1). For the self-weights of the wall components and the unit weights of the surcharge and backfill, the corresponding β values may be 1.2 or 0.9, as appropriate. The process of determining the design values for f_{toe}, f_{heel} and the various moments and shear forces are elaborated in the following section.

12.5.7 Illustrative example

Problem

Figure 12.5(14) shows details of a reinforced concrete cantilever retaining wall. The superimposed live load, $p = 15$ kPa, and the wall and base dimensions are the outcomes of a global geotechnical analysis, including sliding stability check, followed by a preliminary design exercise.

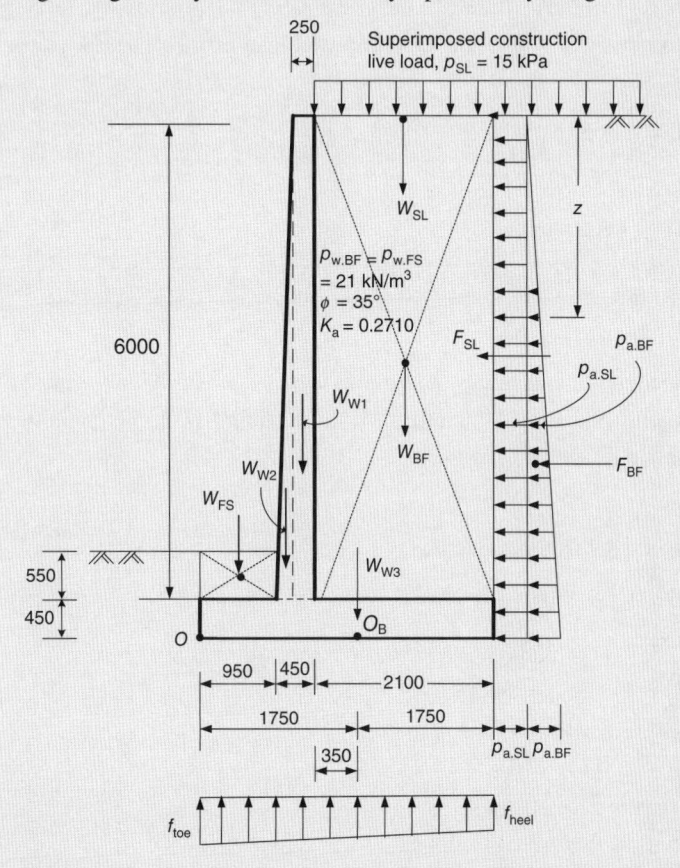

Figure 12.5(14) Details of the example reinforced concrete cantilever retaining wall

Note: all dimensions are in mm.

Given the unit weights of the cohesionless backfill and the front surcharge $\rho_{w.BF} = \rho_{w.FS} = 21$ kN/m^3, with a characteristic effective internal friction angle $\phi = 35°$, subsoil bearing capacity $q_f = 250$ kPa and $f'_c = 25$ MPa. Use D500N bars only.

Compute the subsoil pressures at the toe and heel, f_{toe} and f_{heel}, respectively, and check the overturning stability. Then carry out a full reinforced concrete design of the retaining structure. The passive earth pressure due to the front surcharge may be ignored.

Solution

The design steps required are enumerated below.

Step 1

Compute K_a and the lateral active earth pressures. For the levelled backfill, $\beta = 0$, and Equation 12.5(2) gives

$$K_a = \frac{1 - \sqrt{1 - \cos^2 35°}}{1 + \sqrt{1 - \cos^2 35°}} = 0.2710$$

At a given depth z, the earth pressures due to the superimposed load and the backfill as per Equation 12.5(1) are

$$p_{a.SL} = 0.2710 \times 15 \times 1 = 4.065 \text{ kPa}$$

and

$$p_{a.BF} = 0.2710 \times 21z \times 1 = 5.691z \text{ kPa}$$

respectively.

Step 2

Now compute various vertical and lateral forces. Assuming $\rho_{w.concrete} = 25$ kN/m^3 with the given $\rho_{w.BF} = \rho_{w.FS} = 21$ kN/m^3 and for a 1-metre run of the retaining wall, the quantities required in Equations 12.5(5) and (6) are as follows.

Forces (kN):

(a) $\beta_W W_{W1} = (0.25 \times 6 \times 1) \times 25\beta_W = 37.5\beta_W$
(b) $\beta_W W_{W2} = \left(\frac{0.2}{2} \times 6 \times 1\right) \times 25\beta_W = 15.0\beta_W$
(c) $\beta_W W_{W3} = (0.45 \times 3.5 \times 1) \times 25\beta_W = 39.375\beta_W$
(d) $\beta_{SL} W_{SL} = (15 \times 2.1 \times 1) \times \beta_{SL} = 31.5\beta_{SL}$
(e) $\beta_{BF} W_{BF} = (6 \times 2.1 \times 1) \times 21\beta_{BF} = 264.6\beta_{BF}$
(f) $\beta_{FS} W_{FS} = (0.55 \times 0.95 \times 1) \times 21\beta_{FS} = 10.973\beta_{FS}$
(g) $F_{SL} = 4.065\beta_{SL} \times (6 + 0.45) = 26.22\beta_{SL}$
(h) $F_{BF} = 5.691 \times 6.45\beta_{BF} \times 6.45/2 = 118.38\beta_{BF}$

Lever arms with respect to O_B (m):

(i) $L_{W1} = \dfrac{0.25}{2} + 2.1 - 1.75 = 0.475$

(j) $L_{W2} = \dfrac{0.2}{3} + 0.25 + 2.1 - 1.75 = 0.70$

(k) $L_{FS} = \dfrac{0.95}{2} + 0.45 + 2.1 - 1.75 = 1.275$

(l) $L_{SL} = L_{BF} = 1.75 - \dfrac{2.1}{2} = 0.70$

(m) $L_{SL.H} = (6 + 0.45)/2 = 3.225$

(n) $L_{BF.H} = (6 + 0.45)/3 = 2.15$

Step 3

Now calculate the load combinations (LC). Considering the aggravating and reversal effects of each of the various vertical and induced lateral forces, three load combination cases are required to produce the most critical outcomes. The details are presented in Table 12.5(3).

Table 12.5(3) Load factors for critical load combination cases

Load combination	β_W	β_{FS}	β_{SL}	β_{BF}
LC1	1.2	1.2	1.5	1.2
LC2	1.2	1.2	0.4	0.9
LC3	0.9	0.9	1.5	1.2

Step 4

Compute R and M_R at O_B for LC1, 2 and 3. With values obtained in step 2, the resultant vertical force as per **Equation 12.5(5)** is

$$R = (37.5 + 15.0 + 39.375)\beta_W + 31.5\beta_{SL} + 264.6\beta_{BF} + 10.973\beta_{FS}$$
$$= 91.875\beta_W + 31.5\beta_{SL} + 264.6\beta_{BF} + 10.973\beta_{FS} \quad \text{Equation (a)}$$

Substituting the corresponding load factors, which are the subscripted β values for LC1 in Table 12.5(3)

$R = 488.19$ kN

where

$\beta_W = 1.2; \beta_{SL} = 1.5; \beta_{BF} = 1.2;$ and $\beta_{FS} = 1.2$

The moment about O_B (positive anticlockwise) as per **Equation 12.5(6)** and with reference to Table 12.5(3), is

$$M_R = 26.22\beta_{SL} \times 3.225 + 118.38\beta_{BF} \times 2.15 + 37.5\beta_W \times 0.475 + 15.0\beta_W$$
$$\times 0.70 + 10.973\beta_{FS} \times 1.275 - 31.5\beta_{SL} \times 0.70 - 264.6\beta_{BF} \times 0.70$$

which is similar to

$$M_R = 84.560\beta_{SL} + 254.517\beta_{BF} + 28.313\beta_W + 13.991\beta_{FS} - 22.050\beta_{SL} - 185.220\beta_{BF}$$ Equation (b)

from which for LC1, $M_R = 227.686$ kNm

Following the same process, Equation (a) gives

$R = 374.16$ kN for LC2

and

$R = 457.33$ kN for LC3

Then, Equation (b) leads to

$M_R = 138.13$ kNm for LC2

and

$M_R = 214.99$ kNm for LC3

Step 5

Compute f_{toe} and f_{heel} and check overturning stability. The subsoil pressures at the toe and heel are obtained through Equations 12.5(9) and (10), respectively, as

$$f_{toe} = R/3.5 + 6M_R/3.5^2$$ Equation (c)

and

$$f_{heel} = R/3.5 - 6M_R/3.5^2$$ Equation (d)

The outcomes for R and M_R due to the three critical load combinations are summarised below:

LC1: $R = 488.19$ kN; $M_R = 227.69$ kNm
LC2: $R = 374.16$ kN; $M_R = 138.13$ kNm
LC3: $R = 457.33$ kN; $M_R = 214.99$ kNm

Now the subsoil pressures for the three load combination cases are computed using Equations (c) and (d):

LC1: $f_{toe} = 251.00$ kPa; $f_{heel} = 27.96$ kPa
LC2: $f_{toe} = 174.56$ kPa; $f_{heel} = 39.25$ kPa
LC3: $f_{toe} = 235.97$ kPa; $f_{heel} = 25.36$ kPa

These results show that none of the subsoil pressures exceed the given bearing capacity $q_f = 250$ kPa. Note, however, that for LC1 it is slightly larger but can be taken as 250 kPa. This means that the retaining wall dimensions are acceptable.

Since all the pressure values are positive (compressive), the retaining wall is stable against overturning.

Step 6

Compute the design moments (M^*). The root moment of the wall for LC1 is given by Equation 12.5(12) with $\beta_{SL} = 1.5$ and $\beta_{BF} = 1.2$ as

$$M^*_{O1} = 1.5 \times 4.065 \times 6 \times l_{b1} + 0.5 \times 1.2 \times 0.271 \times 21 \times 6^2 \times l_{b2}$$
$$= 36.585 l_{b1} + 122.926 l_{b2}$$

where from Equation 12.5(13)

$$l_{b1} = 6/2 + 0.15 \times 0.45 = 3.068 \text{ m}$$

and Equation 12.5(14) gives

$$l_{b2} = 6/3 + 0.15 \times 0.45 = 2.068 \text{ m}$$

or

$$M^*_{O1} = 366.45 \text{ kNm}$$

For LC2, $\beta_{SL} = 0.4$ and $\beta_{BF} = 0.9$, which gives

$$M^*_{O1} = 9.756 l_{b1} + 92.194 l_{b2}$$

from which and with $l_{b1} = 3.068$ m and $l_{b2} = 2.068$ m

$$M^*_{O1} = 220.59 \text{ kNm}$$

For LC3, with $\beta_{SL} = 1.5$ and $\beta_{BF} = 1.2$, the moment is the same as for LC1, which is

$$M^*_{O1} = 366.45 \text{ kNm}$$

For the heel moment, Equation 12.5(15) yields

$$M^*_{O2} = (31.5\beta_{SL} + 264.6\beta_{BF})(2.1/2 + 0.15 \times 0.45) + \beta_W(0.45 \times 2.1 \\
\times 1 \times 25)(2.1 + 0.15 \times 0.45)/2 \\
- P_{1.heel}(2.1 + 0.15 \times 0.45)/2 \\
- P_{2.heel}(2.1 + 0.15 \times 0.45)/3 \\
= 35.201\beta_{SL} + 295.691\beta_{BF} + 25.604\beta_W - 1.084 P_{1.heel} \\
- 0.723 P_{2.heel}$$ Equation (e)

For LC1, $\beta_{SL} = 1.5, \beta_{BF} = 1.2, \beta_w = 1.2$,

$$P_{1.heel} = 27.97 \times (2.1 + 0.15 \times 0.45) = 27.97 \times 2.1675 = 60.625 \text{ kN}$$

and

$$P_{2.heel} = \left[(250) - 27.97 \times \frac{2.1675}{3.5}\right] \times 2.1675/2 = 149.02 \text{ kN}$$

based on which Equation (e) gives

$$M^*_{O2} = 264.90 \text{ kNm}$$

For LC2, $\beta_{SL} = 0.4, \beta_{BF} = 0.9, \beta_W = 1.2$,

$$P_{1.\text{heel}} = 39.25 \times (2.1 + 0.15 \times 0.45) = 39.25 \times 2.1675 = 85.074 \text{ kN}$$

and

$$P_{2.\text{heel}} = \left[(174.56 - 39.25) \times \frac{2.1675}{3.5}\right] \times 2.1675/2 = 90.81 \text{ kN}$$

Then from Equation (e)

$$M^*_{O2} = 153.05 \text{ kNm}$$

For LC3, $\beta_{SL} = 1.5, \beta_{BF} = 1.2, \beta_W = 0.9$,

$$P_{1.\text{heel}} = 25.36 \times (2.1 + 0.15 \times 0.45) = 25.36 \times 2.1675 = 54.968 \text{ kN}$$

and

$$P_{2.\text{heel}} = \left[(235.97 - 25.36) \times \frac{2.1675}{3.5}\right] \times 2.1675/2 = 141.35 \text{ kN}$$

Then from Equation (e)

$$M^*_{O2} = 268.89 \text{ kNm}$$

Similarly, the toe moment is given by Equation 12.5(16) as

$$\begin{aligned} M^*_{O3} &= P_{1.\text{toe}}(0.95 + 0.15 \times 0.45)/2 + 2P_{2.\text{toe}} \times 1.0175/3 - \beta_{FS} \times 10.973 \times \\ & \quad (0.95/2 + 0.15 \times 0.45) - \beta_W \times (0.95 + 0.15 \times 0.45) \times 0.45 \times 25 \\ & \quad \times 1.0175/2 \\ &= 0.5088 P_{1.\text{toe}} + 0.6783 P_{2.\text{toe}} - 5.953 \beta_{FS} - 5.824 \beta_W \end{aligned} \qquad \text{Equation (f)}$$

For LC1, $\beta_{FS} = \beta_W = 1.2$

$$P_{1.\text{toe}} = \left[27.97 + (250 - 27.97) \times \frac{3.5 - 1.0175}{3.5}\right] \times 1.0175 = 188.695 \text{ kN}$$

and

$$P_{2.\text{toe}} = \frac{1}{2} \times (250 - 185.45) \times 1.0175 = 32.84 \text{ kN}$$

Equation (f) thus gives

$$M^*_{O3} = 104.15 \text{ kNm}$$

For LC2, $\beta_{FS} = \beta_W = 1.2$

$$P_{1.\text{toe}} = \left[39.25 + (174.56 - 39.25) \times \frac{3.5 - 1.0175}{3.5}\right] \times 1.0175 = 137.60 \text{ kN}$$

and

$$P_{2.toe} = \frac{1}{2} \times (174.56 - 135.223) \times 1.0175 = 20.013 \text{ kN}$$

from which Equation (f) gives

$$M^*_{O3} = 69.45 \text{ kNm}$$

For LC3, $\beta_{FS} = \beta_W = 0.9$

$$P_{1.toe} = \left[25.36 + (235.97 - 25.36) \times \frac{3.5 - 1.0175}{3.5}\right] \times 1.0175 = 177.80 \text{ kN}$$

and

$$P_{2.toe} = \frac{1}{2} \times (235.97 - 174.743) \times 1.0175 = 31.15 \text{ kN}$$

Hence, Equation (f) gives

$$M^*_{O3} = 100.99 \text{ kNm}$$

Finally, the root, heel and toe moments for the three load combination cases are shown in Table 12.5(4).

Table 12.5(4) Root, heel and toe moments for the three load combination cases

Load combination case	Root moment, M^*_{O1} (kNm)	Heel moment, M^*_{O2} (kNm)	Toe moment, M^*_{O3} (kNm)
LC1	366.45	264.90	104.15
LC2	220.59	153.05	69.45
LC3	366.45	268.89	100.99

From Table 12.5(4), the design moments are obtained as

- root moment, $M^*_{O1} = 366.45$ kNm
- heel moment, $M^*_{O2} = 268.89$ kNm
- toe moment, $M^*_{O3} = 104.15$ kNm.

Step 7

Now determine the design bending reinforcement, A_{st}.

For the root section of the **wall**,

$M^*_{O1} = 366.45$ kNm, $d = 410$ mm and Equation 3.5(7) gives
$\xi = 0.85 \times 25/500 = 0.0425$

Assuming $\phi = 0.8$, Equation 3.5(6) gives

$$p_t = 0.0425 - \sqrt{0.0425^2 - \frac{2 \times 0.0425 \times 366.45 \times 10^6}{0.8 \times 1000 \times 410^2 \times 500}} = 0.005853$$

Equation 3.4(8) gives

$$k_u = \frac{0.005853 \times 500}{0.85 \times 0.85 \times 25} = 0.1620$$

Since $k_{uo} = k_u = 0.1620$, $\phi = 0.8$ is confirmed as per Equations 3.4(20)a with b. Equation 12.3(18) gives

$$p_{t.min} = 0.20 \times \left(\frac{450}{410}\right)^2 \times 0.6\sqrt{25}/500 = 0.001446 < p_t = 0.005853$$

Therefore,

$$A_{st} = 0.005853 \times 1000 \times 410 = 2400 \text{ mm}^2/\text{m}$$

When Table 2.3(2) is consulted, N20 @ 125 mm gives $A_{st} = 2512$ mm^2/m, which is larger than 2400 mm^2/m by only 4.67%. This is acceptable.

This amount of steel is required only at the root of the cantilever wall. As desired, for a more economical design, some of the bars may be curtailed at levels towards the top of the wall where the moment diminishes rapidly, remembering that the bars must be extended beyond the curtailment level to provide adequate stress development for the curtailed bars.

In a practical design, a deflection check should be performed. However, in the present case of an inwardly tapered wall, even excessive deflection would not be apparent to a layperson.

For the **heel** section

$$M^*_{O2} = 268.89 \text{ kN/m}, d = 410 \text{ mm}$$

and Equation 3.5(6) gives

$$p_t = 0.0425 - \sqrt{0.0425^2 - \frac{2 \times 0.0425 \times 268.89 \times 10^6}{0.8 \times 1000 \times 410^2 \times 500}} = 0.004207 > p_{t.min}$$

Therefore,

$$A_{st} = 0.004207 \times 1000 \times 410 = 1725 \text{ mm}^2/\text{m}$$

From Table 2.3(2), N20 @ 175 mm gives $A_{st} = 1794$ mm^2/m $>$ 1725 mm^2/m. This is acceptable.

For the **toe** section,

$$M^*_{O3} = 104.15 \text{ kN/m}, d = 410 \text{ mm}$$

and Equation 3.5(6) gives

$$p_t = 0.0425 - \sqrt{0.0425^2 - \frac{2 \times 0.0425 \times 104.15 \times 10^6}{0.8 \times 1000 \times 410^2 \times 500}} = 0.001578 > p_{t.\,min}$$

Therefore,

$$A_{st} = 0.001578 \times 1000 \times 410 = 647 \text{ mm}^2/\text{m}$$

From Table 2.3(2), N20 @ 300 mm gives $A_{st} = 1047$ mm^2/m > 647 mm^2/m. This is acceptable.

Also, for crack control purposes, as shown in Table 1.4(5) (slabs), $s < 300$ mm. Therefore, use N20 @ 300 mm.

For a more economical solution, use N16 @ 300 mm, which gives $A_{st} = 670$ mm^2/m. This is acceptable.

Step 8

Compute design shear V^* for the **wall**. LC1 governs as per Table 12.5(3). With reference to Figure 12.5(15)

$$d_o \approx 410 - \frac{410}{6000} \times (450 - 250) = 396 \text{ mm}$$

$$p_{a.SL} = 4.065 \text{ kPa}$$

$$p_{a.BF} = 5.691 \times 5.604 = 31.892 \text{ kPa}$$

For a 1-metre run

$$V^*_{root} = 1.5 \times 4.065 \times 5.604 \times 1 + 1.2 \times 31.892 \times 5.604/2 \times 1 = 141.404 \text{ kN}$$

Now calculate design shear V^* for the **heel**. LC1 governs here.

In Figure 12.5(16)

$d_{o.heel} = 410$ mm
$W'_{SL} = 1.5 \times 15 \times 1.69 \times 1 = 38.025$ kN
$W'_{BF} = 1.2 \times 21 \times 6 \times 1.69 \times 1 = 255.528$ kN
$W'_{W3} = 1.2 \times (0.45 \times 1.69 \times 1) \times 25 = 22.815$ kN

Earth pressure resultant is

$$W'_{heel} = 27.97 \times 1.69 \times 1 + \frac{1.69}{3.5} \times (251 - 27.97) \times \frac{1.69}{2} = 47.269 + 90.999$$

$$= 138.268 \text{ kN}$$

$$V^*_{heel} = 38.025 + 255.528 + 22.815 - 138.268 = 178.10 \text{ kN}$$

Compute design shear V^* for the **toe**. LC3 governs here.

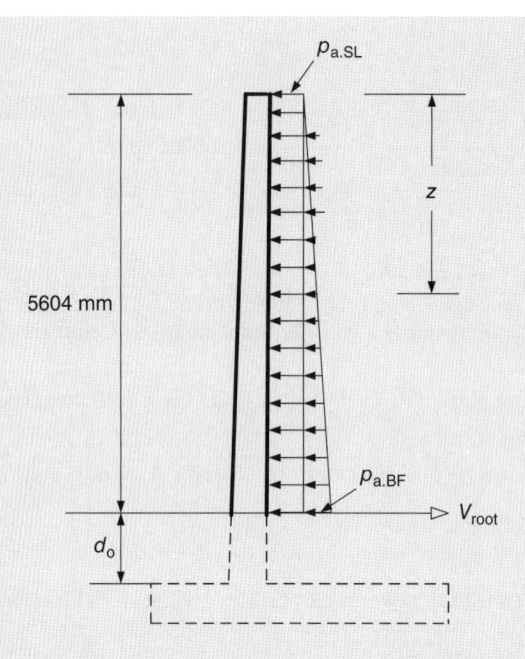

Figure 12.5(15) Location and determination of critical shear force for the wall

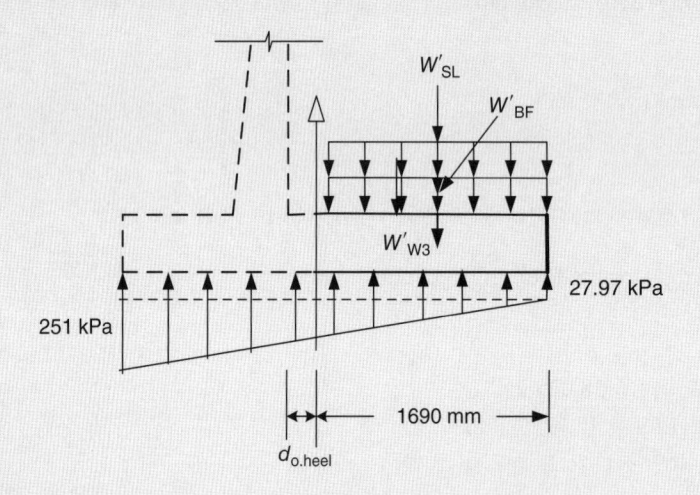

Figure 12.5(16) Location and determination of critical shear force for the heel

In Figure 12.5(17),

$d_{o.toe} = 410$ mm

Earth pressure resultant is

$$0.54 \times 235.97 - \frac{0.54}{3.5} \times (235.97 - 25.36) \times \frac{0.54}{2} = 118.65 \text{ kN}$$

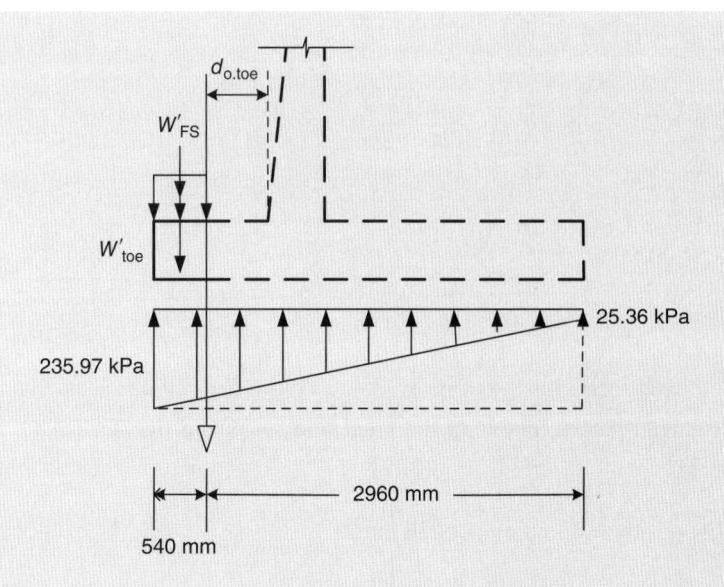

Figure 12.5(17) Location and determination of critical shear force for the toe

$$W'_{FS} = 0.9 \times 0.55 \times 0.54 \times 21 = 5.6133 \text{ kN}$$
$$W'_{toe} = 0.9 \times 0.45 \times 0.54 \times 25 = 5.4675 \text{ kN}$$
$$V^*_{toe} = 118.650 - 5.613 - 5.468 = 107.569 \text{ kN}$$

Step 9

Check the shear capacity of the **wall**.

$$d_o \approx 396 \text{ mm}$$

Equation 6.3(5)b gives

$$\beta_1 = 1.1 \left(1.6 - \frac{396}{1000} \right) = 1.3244 > 0.8$$

Equation 6.3(6) gives

$$\beta_2 = 1.0$$

Equation 6.3(8) gives

$$\beta_3 = 1.0$$

Equation 6.3(4) gives

$$V_{uc} = 1.3244 \times 1 \times 1 \times 1000 \times 396 \times f_{cv} \sqrt[3]{\frac{2512}{1000 \times 396}}$$

where Equation 6.3(4)a is

$$f_{cv} = \sqrt[3]{25} = 2.9240 < 4.00 \text{ MPa}$$

This is acceptable. Therefore,

$$V_{uc} = 283.9 \text{ kN}$$

Now perform the shear strength checks following the procedure set out in Section 6.3.5. Case A is not applicable since

$$D \approx 450 - \frac{396}{6000} \times (450 - 250) = 436.8 \text{ mm} > 250 \text{ mm}$$

Case B is not acceptable because Equation 6.3(10) gives

$$0.5 \times 0.7 \times 283.9 = 99.4 \text{ kN} < V^*_{root} = 141.404 \text{ kN}$$

Case C is not applicable since $D < 750$ mm.

Since none of Case A, B or C prevails, there are two approaches to proceed to satisfy the shear strength provisions in the Standard:

(a) increase the overall depth D and/or f'_c of the wall, such that Case B or C prevails, and no shear reinforcement is required
(b) satisfy the stipulation in Case D and provide the minimum shear reinforcement required.

From the Case B check, above, it is apparent that V_{uc} needs to be increased to approximately (141.404/0.35) = 404 kN to satisfy Equation 6.3(10). In view of Equation 6.3(4), the wall depth at its root would need to be substantially increased to avoid using shear reinforcement. Also, the increase in wall depth would render it into Case C where $D > 750$ mm, which would also mandate the use of shear reinforcement. In the present example, D would be much larger than 750 mm! In view of the dilemma and for illustrative purposes, approach (b) is followed here. To help explain this, see the comments given at the end of this (wall shear) section.

For Case D, Equation 6.3(15) gives

$$V_{u.min} = V_{uc} + 0.1\sqrt{25}b_w d_o = V_{uc} + 0.5 b_w d_o < V_{uc} + 0.6 b_w d$$

Therefore,

$$V_{u.min} = V_{uc} + 0.6 b_w d_o = 283.9 + 0.6 \times 1000 \times 396 \times 10^{-3} = 521.5 \text{ kN}$$

Since $\phi V_{u.min} > V^*_{root}$, Equation 6.3(13) is satisfied. Now proceed to design the shear reinforcement. Using Equation 6.3(12)

$$A_{sv.min} = 0.06\sqrt{25}b_w s/f_{sy.f} < 0.35 b_w s/f_{sy.f}$$

and adopt

$$A_{sv.min} = 0.35 b_w s/f_{sy.f}$$

in which the spacing

$$s \leq 0.75D \approx 0.75 \times 436.8 = 327.6 \text{ mm}$$

Use $s = 320$ mm and Equation 6.3(12) to get

$$A_{sv.min} = 0.35 \times 1000 \times 320/500 = 224 \text{ mm}^2$$

Use N12 ties @ 320 mm ($A_{sv} = 226$ mm^2) starting from the base of the wall, and terminate at $(z - D) \approx (z - 396)$; z is the distance from the top of the wall to a level where Equation 6.3(10) prevails or the Case B check applies. In which case, and with reference to Figure 12.5(15), z is computed by locating the level where the sum of the active pressure resultant equals $0.5\phi V_{uc}$ or

$$4.065z + 5.691z \times z/2 = 99.4$$

from which, $z = 5.239$ m.

Thus, the level where shear reinforcement is no longer needed is at $(5.239 - 0.396) = 4.843$ m. This distance can be taken as 4500 mm from the top of the wall.

Comments

It should be noted here that the wall height (6000 mm) is much greater than the thickness of the wall (250 mm at the top and 450 mm at the base). Thus, the slender wall (unlike the wall footing, column footing or pile cap discussed in earlier sections of this chapter) may be taken as a slab. If this is the case, the designer may invoke Clause 8.2.5(b) – the shear strength of the wall is adequate if $\phi V_{uc} \geq V^*$. In our case, we have

$$\phi V_{uc} = 0.7 \times 283.9 = 198.73 \text{ kN} > V^*_{root} = 141.404 \text{ kN}$$

Thus, shear reinforcement may not be required for the wall. However, the same does not apply to the cases of the heel and the toe in the discussion below.

Now check the shear capacity of the **heel**. With $V^*_{heel} = 171.104$ kN and $d_o = 410$ mm, shear reinforcement is required as per the provisions in Case D (i.e. Equation 6.3(13) prevails). Use N12 ties @ 320 mm.

Finally, check the shear capacity of the **toe**. With $V^*_{toe} = 107.569$ kN and $d_o = 410$ mm, relevant computations confirm that the design also falls within the purview of Case D. Thus, use N12 @ 320 mm.

Step 10
Provide shrinkage and temperature steel using Equation 9.4(14) which gives

$$A_{s.min} = 1.75 \times 1000 \times 410 \times 10^{-3} = 717.5 \text{ mm}^2/\text{m}$$

Use two layers of N12 bars @ 300 mm, which gives $A_s = 754$ mm^2/m. This is acceptable.

Step 11
Check the stress development, that is, the adequacy of the bond strength at the base section of the wall, and the root sections of the toe and heel. The process is similar to that used in Section 12.2.5 – step 11.

Step 12
The final design for the reinforced concrete retaining wall structure is shown in Figure 12.5(18). Note that the shear reinforcement is still provided for the wall, which may not be necessary in view of the comments given earlier in step 9 with respect to the shear design of the wall.

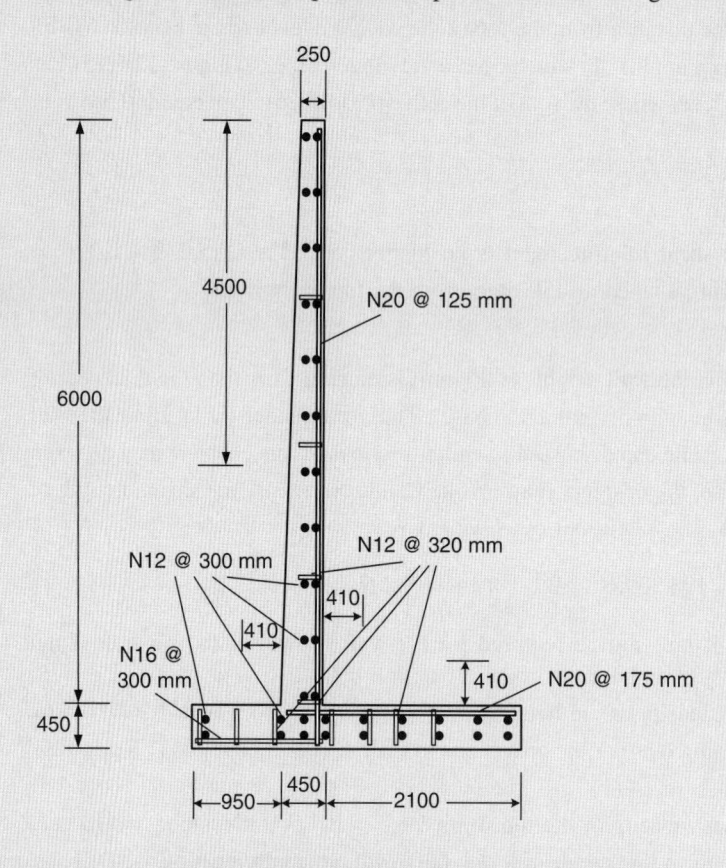

Figure 12.5(18) Reinforcement layout for the reinforced concrete retaining wall structure

Note: all dimensions are in mm; clear cover for bending steel 30 mm; for temperature steel not less than 30 mm

12.6 SUMMARY

The analysis of the forces acting above and below typical wall footings and their design are presented. The treatments for footings supporting single and multiple columns are given and pile caps are dealt with. Illustrative and design examples are given to highlight the application of the analysis and design procedures.

The horizontal soil pressure analysis and the design of retaining walls are discussed in some detail which also includes illustrative and design examples.

12.7 PROBLEMS

1. Given a reinforced concrete wall, 300 mm thick and subjected to a dead load $DL = 250$ kN/m and a live load $LL = 60$ kN/m each applied concentrically along the centre plane of the wall. Take $f'_c = 40$ MPa, $\rho_w = 24$ kN/m³, the soil bearing capacity $q_f = 200$ kPa and that A2 exposure classification applies. Design the symmetrical wall (strip) footing in such a way that the subsoil pressure is uniformly distributed. Use N16 or N20 bars in one layer and use no shear reinforcement.

2. Design a square column footing for a 400 mm square tied interior column that supports a dead load $DL = 900$ kN and a live load $LL = 750$ kN. The soil bearing capacity $q_f = 250$ kPa, $f'_c = 20$ MPa, $\rho_w = 24$ kN/m³ and exposure classification A1 applies. Use N24 bars for bending in one layer each way and use no shear reinforcement.

3. Rederive Equations 12.5(5), (6) and (11) for a retaining wall with all the same details as illustrated in Figure 12.5(11), but with a sloping surcharge at an angle β replacing the live load (p_{SL}).

4. The trial section of a semigravity plain concrete retaining wall is shown in Figure 12.7(1). Check the safety of the wall against overturning, sliding, and bearing pressure under the footing.

The given unit weight of the backfill $\rho_{w.BF} = 17.3$ kN/m³, with a characteristic effective internal friction angle $\phi = 35°$, coefficient of friction between concrete and soil $\mu = 0.5$, subsoil bearing capacity $q_f = 150$ kPa, $\rho_w = 24$ kN/m³ and $f'_c = 20$ MPa. The passive earth pressure due to the front surcharge may be ignored.

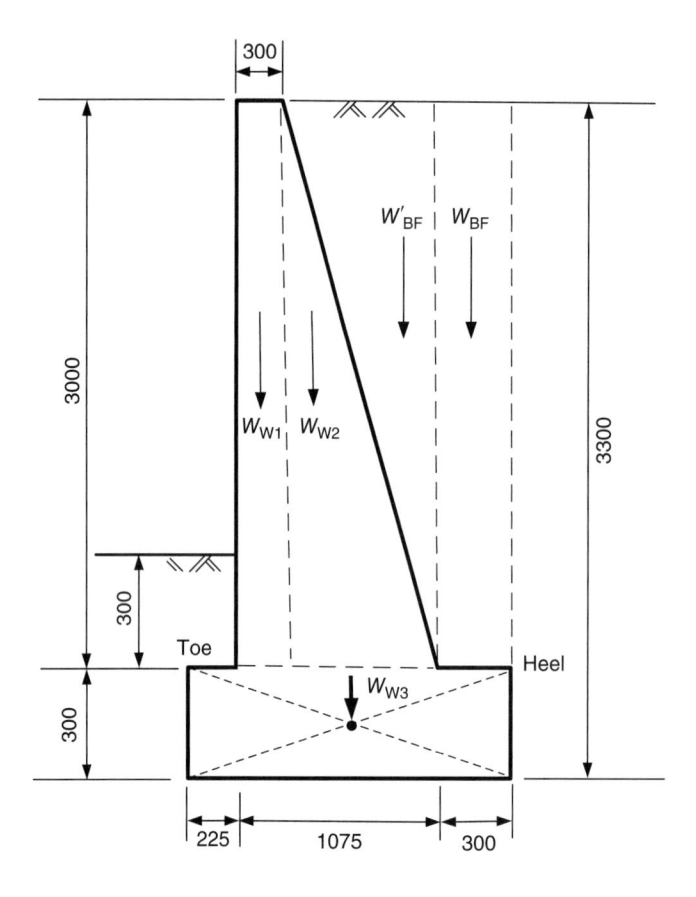

Figure 12.7(1) Details of the plain concrete semigravity retaining wall

Note: all dimensions are in mm.

5. A semigravity retaining wall of plain concrete with $\rho_w = 24$ kN/m^3 is shown in Figure 12.7(2). The bank of supported earth is assumed to weigh 17.3 kN/m^3, to have a ϕ of 30° and a coefficient of friction against sliding on soil of 0.5. Determine the safety factors against overturning and sliding, as well as the bearing pressure underneath the toe of the footing. You may ignore the passive earth pressure due to the front surcharge.

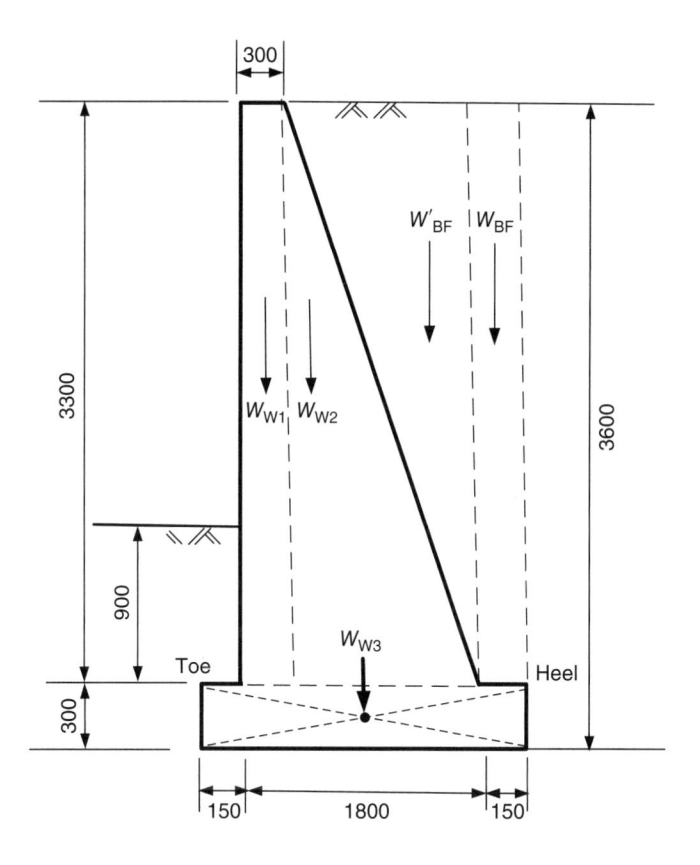

Figure 12.7(2) Details of the semigravity retaining wall

Note: all dimensions are in mm.

6. Repeat the illustrative example given in Section 12.5.7 for a wall similar to that shown in Figure 12.5(14), but tapered behind the wall (rather than in front). Use the same wall dimensions (i.e. 250 mm at the top and 450 mm at root).

376 PART 1 REINFORCED CONCRETE

7. Figure 12.7(3) shows details of a reinforced concrete cantilever retaining wall. The superimposed live load $p_{SL} = 15$ kPa, and the wall and base dimensions are the outcomes of a global geotechnical analysis, including sliding stability check, followed by a preliminary design exercise.

Given the unit weight of the cohesionless backfill $\rho_{w.BF} = 12.5$ kN/m³, with a characteristic effective internal friction angle $\phi = 30°$, subsoil bearing capacity $q_f = 200$ kPa, $\rho_w = 25$ kN/m³ and $f'_c = 20$ MPa. Use D500N bars only.

Compute the subsoil pressures at the toe and heel, f_{toe} and f_{heel}, respectively, and check the overturning stability. Then carry out a full reinforced concrete design of the retaining structure.

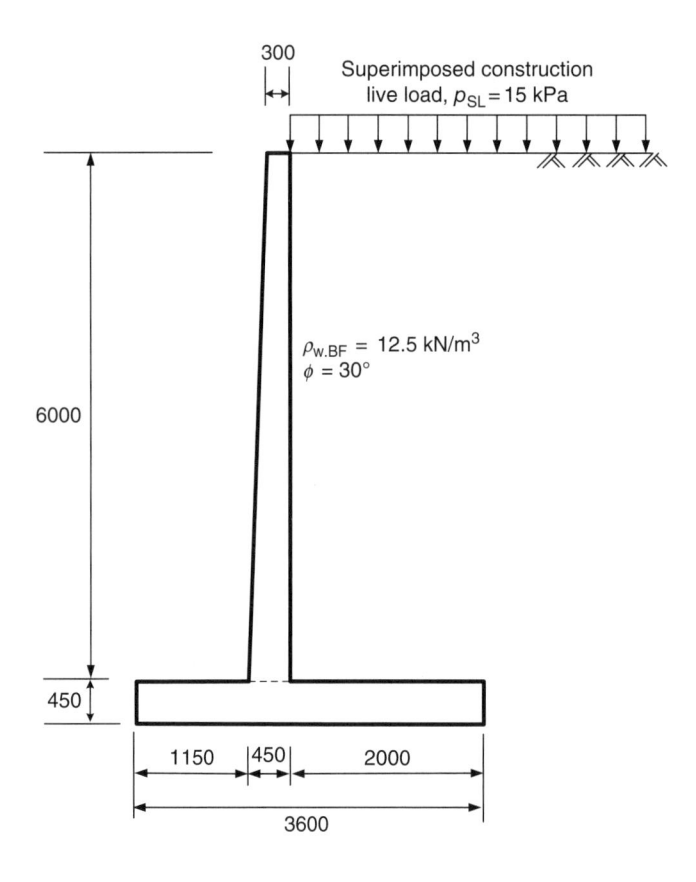

Figure 12.7(3) Details of the reinforced concrete cantilever retaining wall

Note: all dimensions are in mm.

STRUT-AND-TIE MODELLING OF CONCRETE STRUCTURES

13

13.1 INTRODUCTION

'Strut-and-tie modelling' is described in Section 7 of AS 3600–2009 (the Standard). Since the original issue of AS 3600 (1988), the current Standard is the first update in which a separate section is devoted to this type of modelling. This chapter serves to complement Section 7 of the Standard.

Figure 13.1(1) shows some typical reinforced and prestressed concrete structures and elements. In terms of stress distribution characteristics in response to external loads, each

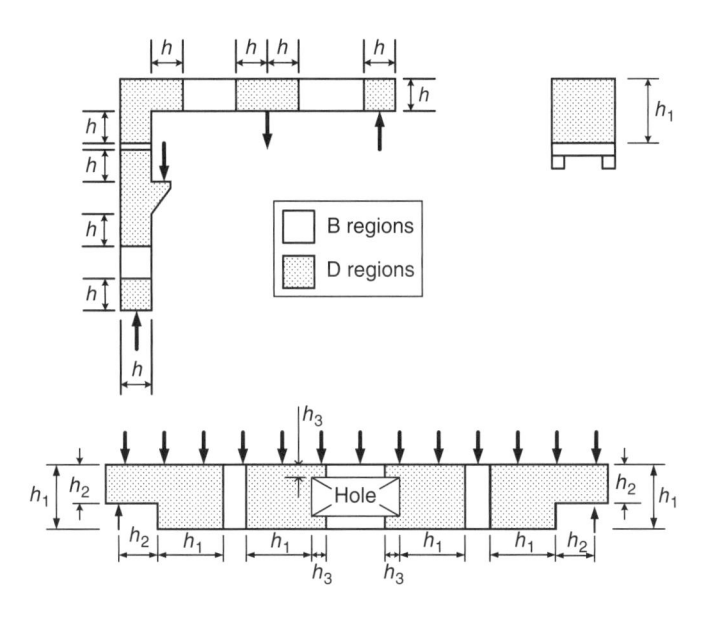

Figure 13.1(1) Typical B and D regions of concrete structures and elements

Source: Redrawn from Schlaich, Schafer and Jennewein (1987)

structure or element can be divided into the so-called 'B' and 'D' regions (Schlaich, Schafer and Jennewein 1987; Schlaich and Schafer 1991). In general, B regions are dominated by bending and D regions (the 'distributed' or discontinuous regions) are dominated by non-flexural stresses.

The analysis and design of various B regions have been extensively explored in most chapters of this book. That the strength behaviour of B (bending) regions can be accurately determined or designed for, using explicit formulas or well-prescribed analytical procedures, is beyond doubt. However, the same cannot be said of the behaviour of the D regions. For these, the analysis or design in general requires rather crude empirical formulas or, alternatively, the aid of sophisticated computer-based numerical procedures, such as the finite element method. The strut-and-tie modelling technique, on the other hand, can provide a direct design process for many types of D region.

The fundamental concepts of the strut-and-tie model (STM) and some recent developments of the technique are discussed in this chapter, typical models taking the form of statically determinate truss systems are illustrated, and the Standard's specifications for various types of struts, ties and 'nodes' are also described.

13.2 FUNDAMENTALS

A typical deep beam under a single concentrated load (P) at midspan is illustrated in Figure 13.2(l)a. The zones between the external load and the two reactions may be taken as a D region. Under the load P and the reactions $P/2$, the compressive and tensile stress distributions, as represented by the stress trajectories, are presented schematically in Figure 13.2(l)a. It is obvious that zones a–d–b–a and a–e–c–a are dominated by compressive stresses, whereas zone f–g–h is dominated by tensile stresses. As the span–depth ratio L/h becomes smaller, the compressive and tensile stress fields (zones) would become more and more distinctive. This observation leads to the conclusion that for some geometric configurations the prominent stress fields may be acceptably simulated by individual stress resultants. Figure 13.2(l)b represents one such 'discretisation' of the continuous stress plot for the deep beam in Figure 13.2(l)a. Two more elaborate discretised models are given in Figures 13.2(l)c and d, respectively.

Previously, application of similar discretisation concepts in bridge analysis led to the simulation or approximation of such continuums as the orthotropic plate and box girder by (discrete or skeletal) grillage and space-frame systems, respectively (Cusens and Pama 1975). The extreme but by far the most accurate and versatile form of simulation is now the all too well-known 'discrete' or finite element technique. For some complicated D regions,

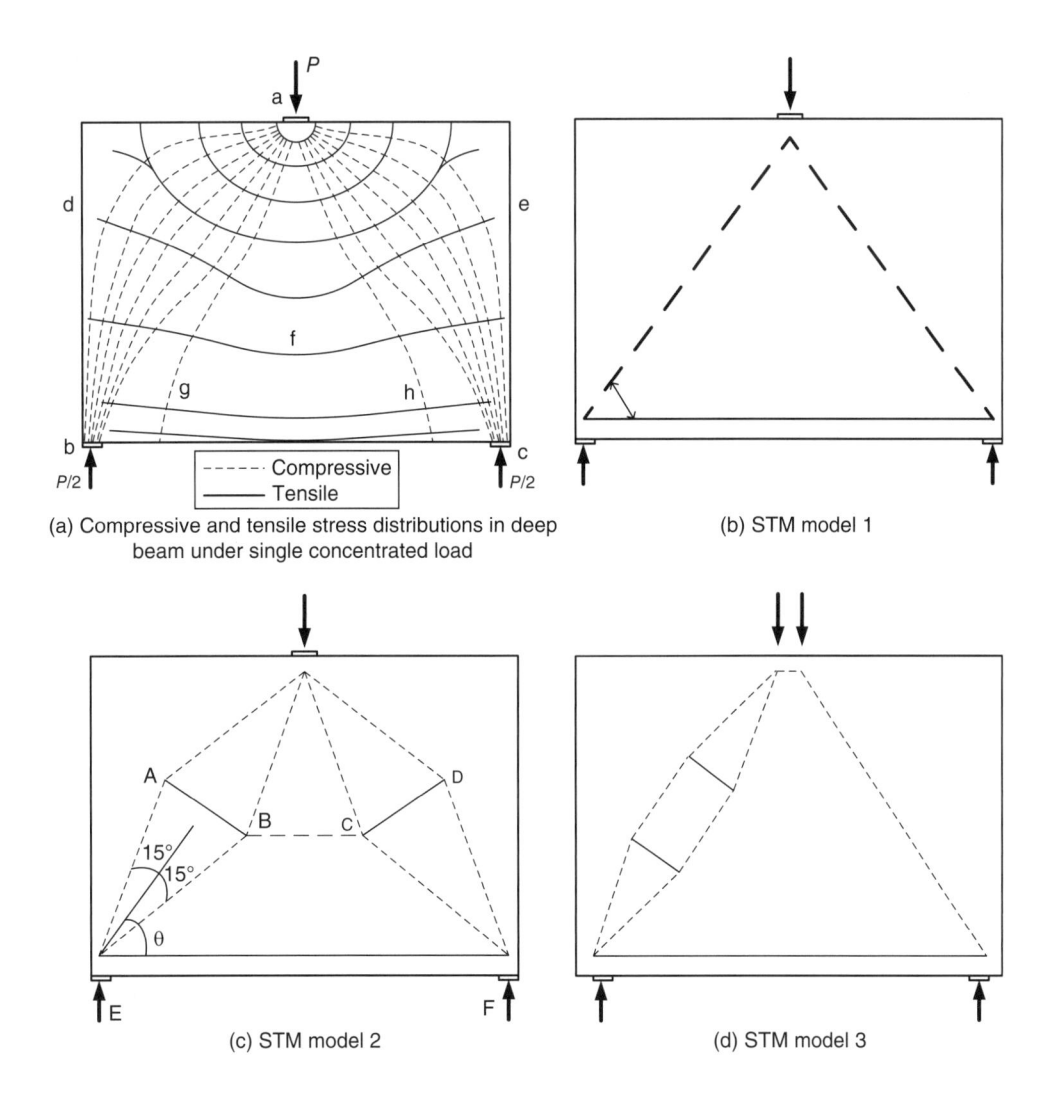

Figure 13.2(1) **(a)** Compressive and tensile stress distributions in deep beam under single concentrated load; **(b)**, **(c)** and **(d)** 'discretised' stress distribution in three deep-beam and feasible strut-and-tie (STM) models

the use of this computer-based approach is believed to be more effective than strut-and-tie modelling.

Note that in Figures 13.2(1)b, c and d, the self-equilibrating stress resultants are the forces in the relevant trusses. Each truss comprises compression members – 'struts' – (represented by the dotted lines) and tension members – 'ties' – (solid lines). Here lies the genesis of the 'strut-and-tie' technique for modelling concrete structures and elements with prominent D regions such as deep beams, walls, corbels, deep-pile caps and other similar continuums.

Also, in each of Figures 13.2(1)b, c and d, the struts together simulate the compressive stress fields, whereas the ties approximate the tensile fields. By nature of the simulation process, the struts and ties are connected by 'nodes' – just like in a truss.

13.3 STRUTS, TIES AND NODES

Figures 13.2(1)b, c and d depict some STMs, in idealised forms, for the simple deep beam shown in Figure 13.2(1)a. In practical models, the struts and ties are connected by concrete 'hydrostatic' nodes, each of which sustains a uniform stress on all its surfaces. As an example, the practical model, which corresponds to Figure 13.2(1)b, is detailed in Figure 13.3(1).

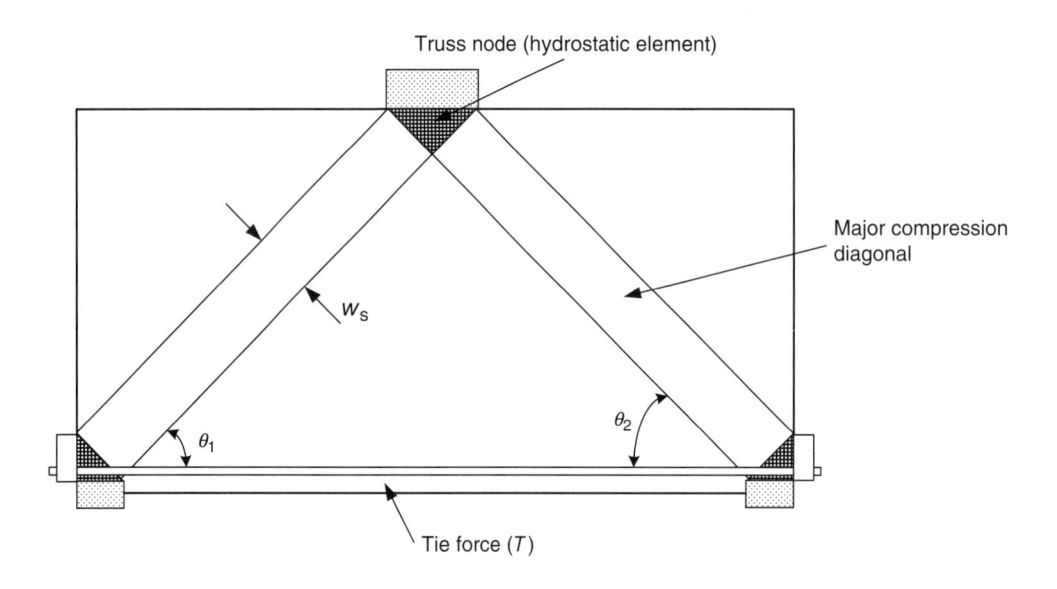

Figure 13.3(1) Practical strut-and-tie model for symmetrically loaded deep beam

The classical theory of plasticity dictates that, for a lower bound or conservative design (Fu 2001), the crushing of any of the concrete struts or nodes (or both) must not occur prior to the yielding of the steel ties. It is also important to note that adequate stress development is provided to prevent premature bond failure of steel reinforcement, which replaces the ties in reality. Likewise, proper tendon anchorage must be available in prestressed structures. Such conditions can be ensured in the structural design process by specifying the proper concrete strength properties and by the correct bond and anchorage design of the reinforcement and tendons. Two examples of STMs for simple and continuous deep beams are reproduced in Figure 13.3(2), which also defines the node types. This includes CCC nodes, where only struts meet, and CCT nodes, where two (or more) struts and a tie meet. In addition, a CTT node (not depicted in the figure) is where one strut meets with two (or more) ties.

(a) Simply supported

(b) Two-span continuous

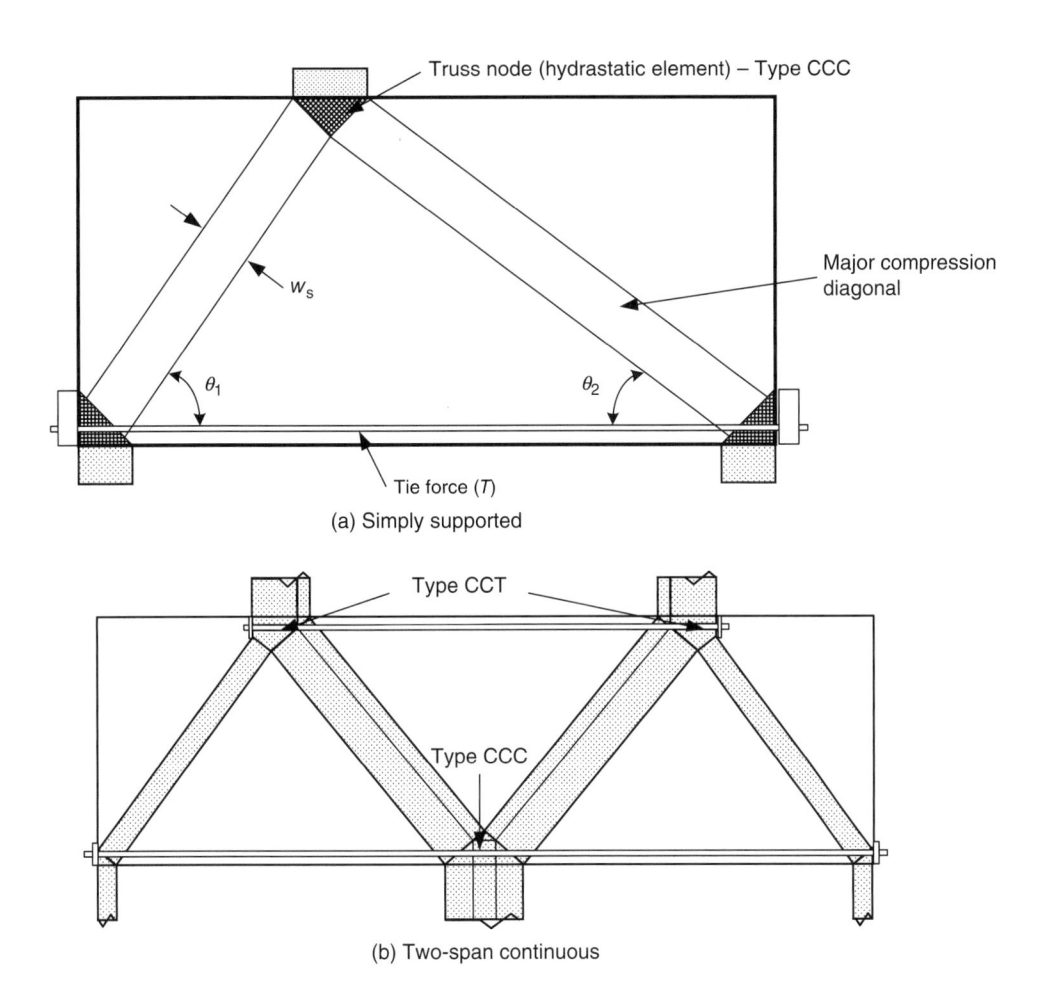

Figure 13.3(2) Strut-and-tie models for **(a)** a simply supported beam and **(b)** a continuous deep beam

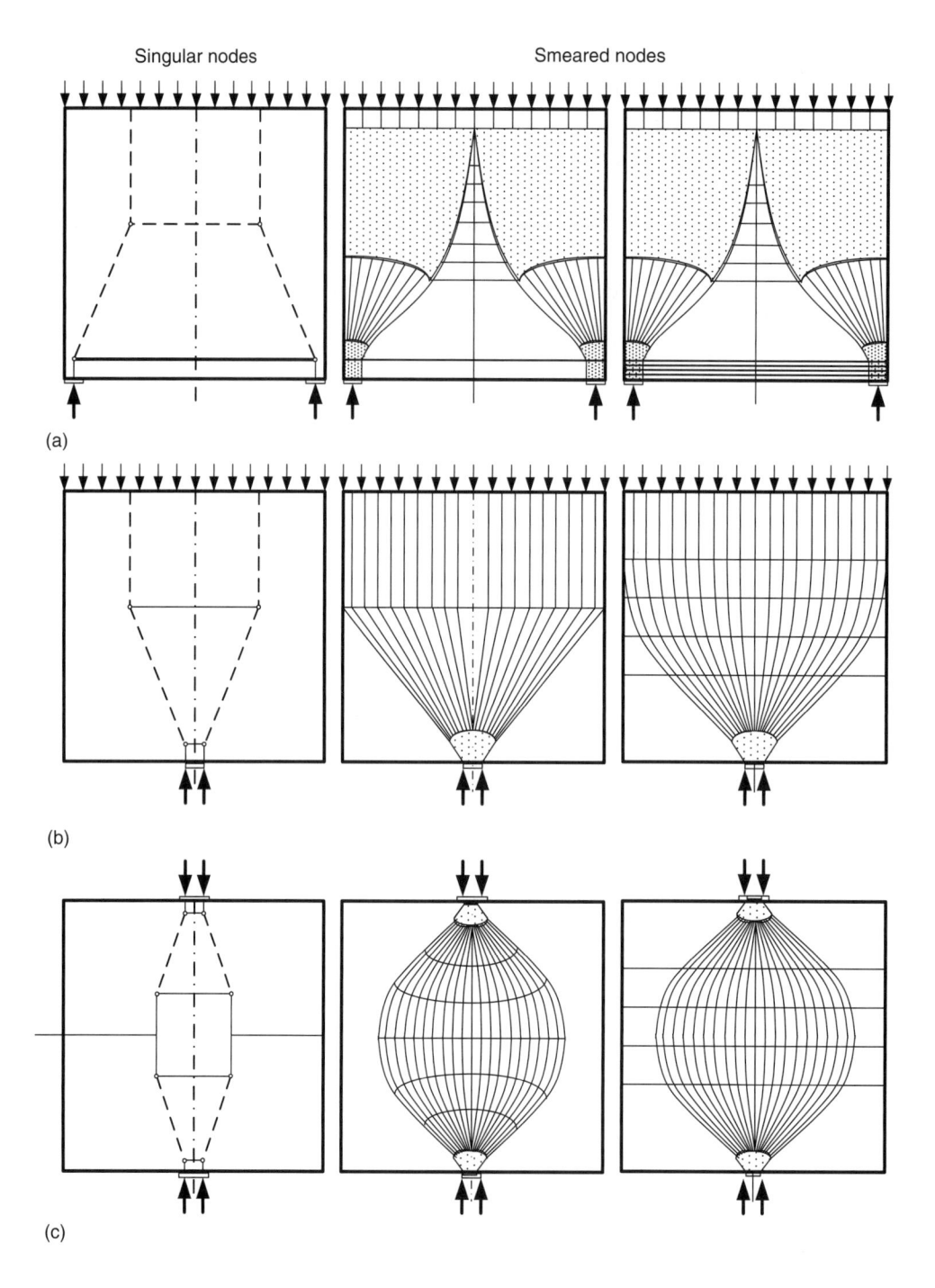

Figure 13.3(3) Some simple and elaborate strut-and-tie models: **(a)** uniform load on a deep beam; **(b)** uniform load on a deep beam with single central support; and **(c)** opposing concentrated loads on a deep member

For illustrative purposes, more deep beams and related structures under various loading and support conditions, and their corresponding simple and elaborate STMs, are given in Figures 13.3(3)a, b and c. For the simple models the struts and ties are connected by 'singular' nodes whereas 'smeared' nodes are required in the elaborate models.

A survey of some elaborate STMs indicates that there are three types of struts as characterised by their shapes (prismatic, fan-shape, half bottle and full bottle) – see Figure 13.3(4). The fan and bottle-shaped struts, which entail more complicated calculations in general, provide better simulations of the original continuum.

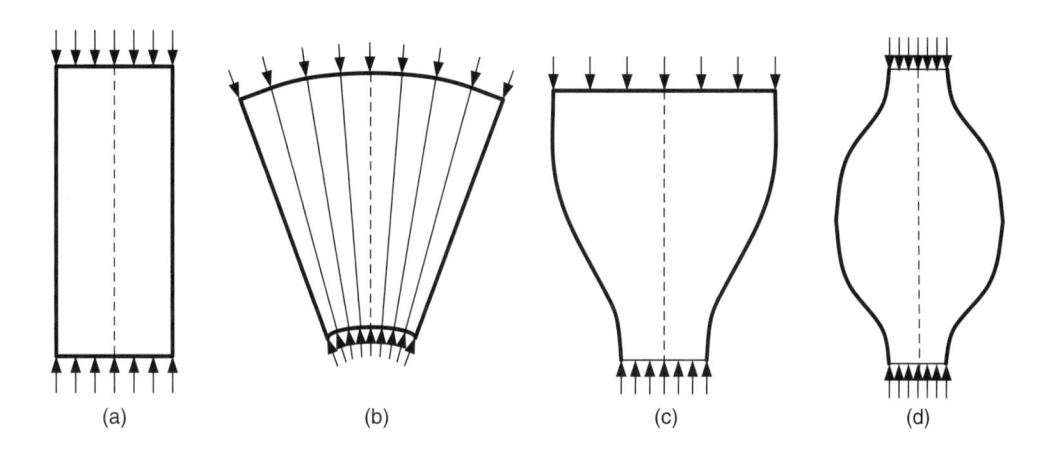

Figure 13.3(4) Prismatic and elaborately-shaped struts: **(a)** prism; **(b)** fan; **(c)** bottle (half); and **(d)** bottle (full)

13.4 COMMON TYPES OF STRUT-AND-TIE MODELS

Additional selected STMs suitable for simulating various commonly encountered nonflexural concrete structures and components are reproduced in Figure 13.4(1) from published literature to make the reader aware of them. These include deep beam, pier cap, post-tensioned member, corbel, rigid-frame joint and stepped beam. Interested readers should refer to the excellent works of Schlaich, Schafer and Jennewein (1987) and Fu (2001) for an in-depth exposition.

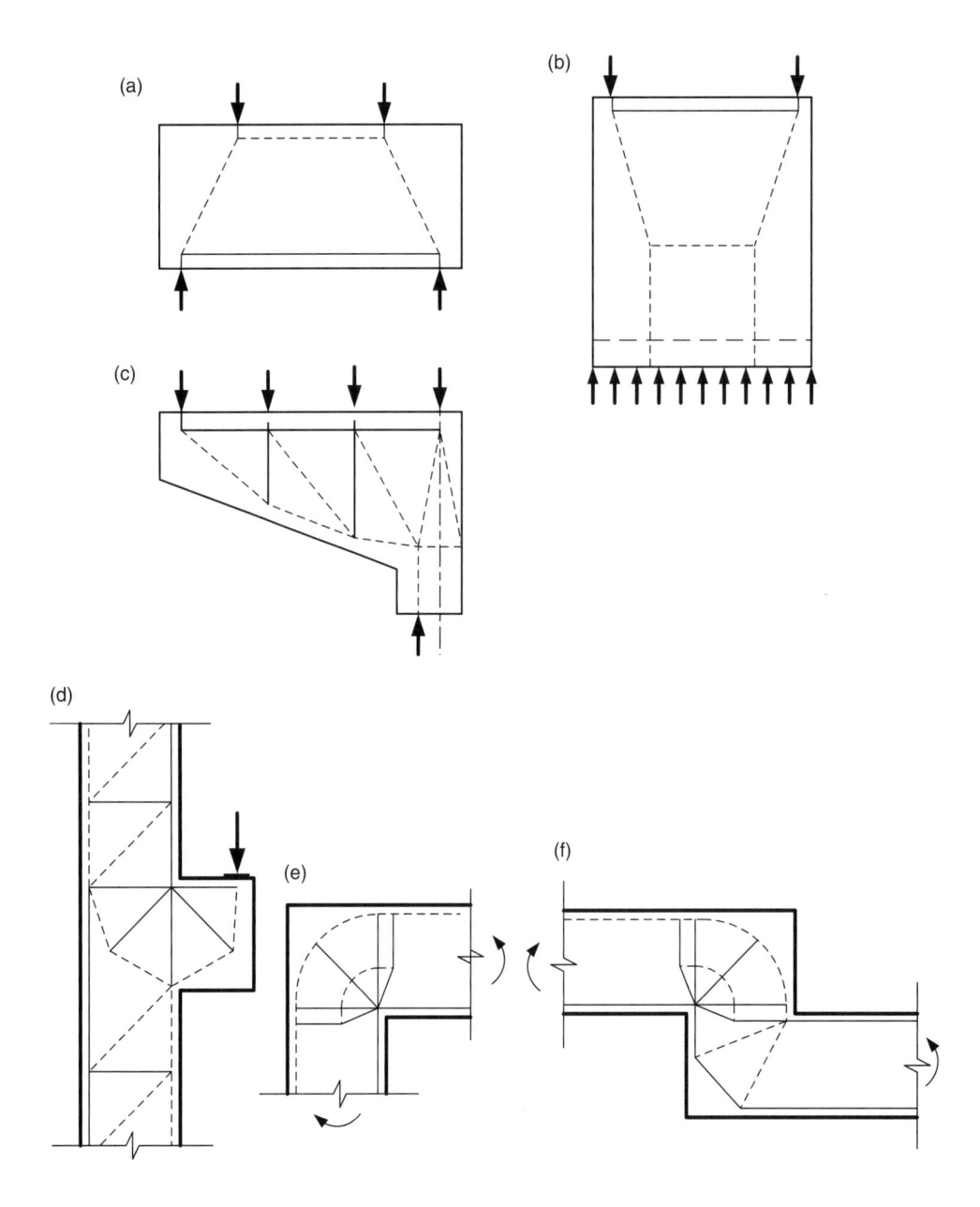

Figure 13.4(1) Strut-and-tie models for various structures and components: **(a)** deep beam; **(b)** post-tensioned end block; **(c)** pier cap; **(d)** corbel; **(e)** frame joint; and **(f)** stepped beam

13.5 DEVELOPMENTS

Since its first introduction in the 1980s, the strut-and-tie modelling technique has undergone very extensive development by concrete engineering researchers around the

globe. In addition to the work of Schlaich, Schafer and Jennewein (1987) and the notes of Fu (2001), the published literature contains numerous analytical and laboratory-based papers on:
- the cyclic hysteretic response of reinforced concrete structures utilising an idealised uniaxial fibre model (To, Ingham and Sritharan 2000)
- solid reinforced and prestressed deep beams (e.g. Yun and Lee 2005; Quintero-Febres, Parra-Montesinos and Weight 2006; Park and Kuchma 2007; Zhang and Tan 2007)
- deep beams with openings (e.g. Maxwell and Breen 2000; Brown and Bayrak 2007; Guan and Doh 2007; Ley, Riding, Widiento and Breen 2007)
- deep beams under monotonic and reversed cyclic loading (Alcocer and Uribe 2008)
- application of microtruss and strut-and-tie models to predict the nonlinear response of reinforced concrete structures (Nagarajan, Jayadeep and Pillai 2009).

Strut-and-tie models have also been used as a tool to:
- investigate the strength behaviour of deep beams (e.g. Tang and Tan 2004; Russo, Venir and Pauletta 2005; Arabzadeh, Rahaie and Aghayari 2009) and that of headed bars in CCT nodes (see Section 13.3), taking into consideration the bearing at bar heads plus the bond between the steel and concrete (Chun and Hong 2009)
- calculate the rotation capacities in a collapse load analysis for plastic hinges (Lopes and do Carmo 2006)
- develop headed steel bars to improve the efficiency of beam-column joints (Hong, Chun, Lee and Oh 2007)
- analyse the shear strength of prestressed concrete deep beams in which concrete softening effects are also taken into consideration (Wang and Meng 2008)
- design stocky pile caps while underscoring the inadequacy of existing design code provisions (Souza, Kuchma, Park and Bittencourt 2009)
- model the behaviour of dapped-end beams, which are encountered from time to time in practice (Mattock 2012).

It is axiomatic that, for a given structure, an infinite number of possible STMs exist. Thus, the traditional trial and error approach could be a tedious way of configuring an efficient model. To address this problem, the topology optimisation technique, the evolutionary structural optimisation (ESO) process and genetic algorithms have been used to search for the required or optimal strut-and-tie configurations for structures of various complicated shapes (Ali and White 2001; Liang, Xie and Steven 2000, 2001; Elia, Palmisano, Vitone and Vitone 2002; Guan et al. 2003; Guan, Eurviriyanukul and Loo 2006; Kwak and Noh 2006; Nagarajan and Pillai 2008; Bruggi 2009; Perera, Vique, Arteaga and Diego 2009; Perera and Vique 2009). The ESO method has also been extended to handle three-dimensional structures (Leu, Huang, Chen and Liao 2006). A computer-based interactive process is also available for the analysis and design of structural concrete based on the strut-and-tie modelling technique (Tjhin and Kuchma 2007).

Efforts by researchers to refine the traditional approach have led to the development of the 'softened' STMs (Hwang, Yu and Lee 2000; Bakir and Boduroglu 2005). The 'softened' models approach, originated by Hsu (1988, 1993), takes into consideration the equilibrium of stress resultants, the compatibility of strains and the constitutive laws of cracked reinforced concrete. The adjective 'softened' refers to the strain-softening characteristics of reinforced concrete in compression and in tension, which are not accounted for in the traditional STMs. The technique to determine the minimum tension reinforcement for the traditional bottle-shaped struts has been published by Brown and Bayrak (2006). In an attempt to enhance the efficiency of STMs, Sahoo, Singh and Bhargava (2009) analytically and experimentally studied the compressive stress dispersion in bottle-shaped struts, and were able to quantify the resulting transverse tensile forces. The laboratory-based investigation of Kuchma et al. (2008) reveals potential concerns over the applicability of strut-and-tie modelling for the design of complex D regions. In particular, a poor selection for the shape of the model can lead to unacceptable levels of cracking and damage under service loading.

Brown and Bayrak (2008a) studied the published literature and collated the test results of 596 deep beams subjected to single and third-point loads, and with shear span-depth ratios of less than two. Based on these results, they examined the strain energy stored in various feasible strut-and-tie configurations used to model the test beams, and found that a single, direct strut between the load and one of the reaction points (i.e. a tie-arch model – see Figure 13.2(l)b) was the preferred arrangement. Note that the best model is one that entails the minimum strain energy. They also indicated that, for beams with heavy web reinforcement or having larger shear span-depth ratios (i.e. more than two), more complicated truss or 'two-panel' models would be appropriate (e.g. see Figure 13.2(l)c). They then evaluated the performances of the ACI 318–2005 and AASHTO LRFD (1998) recommendations on strut-and-tie modelling. This led to the conclusion that the two national codes could produce unsafe design outcomes. In a companion article, Brown and Bayrak (2008b) proposed new design provisions that account for the effects of strut inclination, concrete strength and the amount of reinforcement used. The new method raises safety to acceptable levels – only 5% of the test beams would have failed at a load lower than the predicted strengths. In a similar comparative study based on a database of 868 deep-beam test results assembled from previous research and 37 of their own fabricated and tested beams, Tuchscherer, Birrcher and Bayrak (2011) found the design provisions of the ACI 318–2008 and AASHTO LRFD (2008) to be inefficient and overly conservative. To improve the efficiency and accuracy of the strut-and-tie design provisions, the writers proposed a new and simple strut-and-tie modelling procedure for the strength design of deep beams. From a different perspective, Skibbe (2010) compared dimension requirements, concrete quantities, steel quantities and constructability of the two methods – strut-and-tie modelling (STM) and the deep-beam method (DBM), through the design of three different deep beams. Note that, for the design of deep beams, ACI 318–2008 recommends either the STM or DBM. According to Skibbe (2010), the main advantage of the STM lies in its ability to have decreased member depth without decreasing shear reinforcement spacing. Thus, if the member depth is not a concern in the design, the preferred method is DBM.

Readers interested in further studying the strut-and-tie modelling technique should refer to the papers referenced in this chapter; however, this list is by no means exhaustive, as it covers only a selection of easily accessible works.

13.6 SPECIFICATIONS IN AS 3600

The concrete engineering profession has fully accepted the strut-and-tie modelling technique, and it has therefore been included in the Canadian Standard CSA A23.3 *Design of Concrete Structures* (1994), the *AASHTO LRFD Bridge Design Specifications* (1998) and the American Concrete Institute ACI 318 (2002) – see Wight and Parra-Montesinos (2003). Now a separate chapter of AS 3600–2009 (Section 7) is devoted to strut-and-tie-modelling, whereas in earlier issues the technique was included as a design tool for nonflexural elements.

Section 7 of the Standard recognises the merits and limitations of the STMs as a basis both for strength analysis and design of nonflexural members, and nonflexural or D regions of concrete structures. In selecting a model, Clause 7.1 specifies the following:

(a) Loads shall be applied at nodes, and the struts and ties shall be subjected only to axial force.
(b) The model shall provide load paths to carry the loads and other actions to the supports or into adjacent regions.
(c) The model shall be in equilibrium with the applied loads and the reactions.
(d) In determining the geometry of the model, the dimensions of the struts, ties, and nodal zones shall be taken into account.
(e) Ties shall be permitted to cross struts.
(f) Struts shall cross or intersect only at nodes.
(g) For reinforced concrete elements at a node point, the angle between the axes of any strut and any tie shall be not less than 30°.
(h) For prestressed concrete members at a node point, the angle between the axes of any strut and any tie with a tendon acting as the reinforcement shall be not less than 20°.

Note that specifications (a) to (c), as well as (e) and (f) above, are assumptions used in the static analysis of trusses. However, some strut-and-tie components and indeed models are considered unstable but nevertheless acceptable only because of symmetry. Some such examples may be found in Figures 13.2(l)d, 13.3(3)a, b and c (i.e. the singular-node models), as well as Figure 13.4(l) (i.e. the deep-beam and end-block models). The restrictions on the angular distance between any strut and tie meeting at a node are to ensure the effectiveness of the model.

13.6.1 Concrete struts

Three types of struts are specified in Clause 7.2 – prismatic, fan-shaped and bottle-shaped. For use in determining the design strength of prismatic struts, the strut efficiency factor $\beta_s = 1$. For fan and bottle-shaped stress fields (see Figure 13.3(4)) that are not confined

$$\beta_s = \frac{1}{1.0 + 0.66 \cot^2\theta} \quad \text{but } 0.3 \leq \beta_s \leq 1.0 \qquad \text{Equation 13.6(1)}$$

The angle θ is measured between the axis of the strut and the axis of a tie passing through a common node (see Figure 13.6(1)). Where more than one tie passes through a node, or where the angle θ is different for nodes at each end of a strut, the smallest value of θ is to be used in determining β_s.

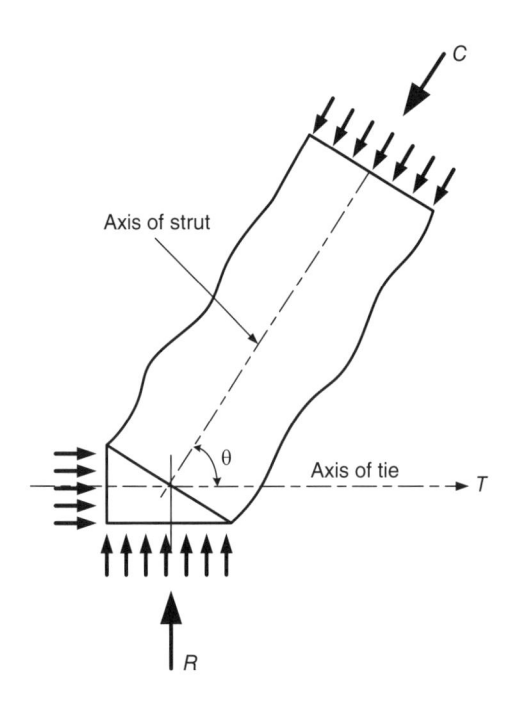

Figure 13.6(1) Definition of θ

Source: Standards Australia (2009). AS 3600–2009 *Concrete Structures*. Standards Australia, Sydney, NSW.
© Standards Australia Limited. Copied by Cambridge University Press with the permission of Standards Australia and Standards New Zealand under Licence 1712-c038.

The design strength of a concrete strut

$$f_{\text{strut}} = \phi \beta_s 0.9 f'_c A_c \qquad \text{Equation 13.6(2)}$$

where $\phi = 0.6$ as per Table 2.2.4 of the Standard and:
- A_c = the smallest cross-sectional area of the concrete strut at any point along its length and measured normal to the line of action of the strut
- $\beta_s = 1$ for prismatic struts and is defined in Equation 13.6(1), for fan- and bottle-shaped struts.

Note that compression reinforcement may be used to increase the strength of a strut, in which case it should be designed as a prismatic pin-ended short column. The reinforcement must be laterally restrained by ties or spirals as defined in Figure 10.10(1).

For bottle-shaped struts, to control cracking at service load and prevent splitting failure at ultimate state, the bursting force (T_j) for determining the transverse reinforcement is taken as the greater of the two values determined as follows:

1. The force is calculated such that the bursting strength of the strut after cracking is not less than that before cracking. The bursting tension across the strut at cracking may be taken as

$$T_{b.cr} = 0.7 b l_b f'_{ct} \qquad \text{Equation 13.6(3)}$$

and

2. The force is calculated using an equilibrium model, consistent with the bottle shapes as shown in Figure 13.6(2). To ensure adequate crack control, splitting forces shall be assessed at both service and ultimate limit states. The divergence angle (α) for the bottle strut – see Figure 13.6(2) – must be assessed for each situation but must not be less than:
   - $\tan \alpha = 1/2$ for serviceability
   - $\tan \alpha = 1/5$ for strength.

   Bursting reinforcement must be evenly distributed throughout the bursting zone (l_b) – Figure 13.6(2) – where

$$l_b = \sqrt{z^2 + a^2} - d_c \qquad \text{Equation 13.6(4)}$$

and a and z are the shear span and the component normal to the shear span, respectively – see Figure 13.6(2).

The bursting reinforcement must be uniformly distributed throughout the bursting zone (l_b) and be placed in either two orthogonal directions at angles γ_1 and γ_2 to the axis of the strut, or in one direction at an angle γ to the axis of the strut. If the reinforcement is in only one direction, γ is not to be less than 40°.

To provide adequate strength, the quantity of reinforcement must be such that

$$\phi \sum A_{si} f_{sy} \sin \gamma_i \geq T_b^* \qquad \text{Equation 13.6(5)}$$

and to provide adequate crack control, the quantity of reinforcement must be such that

$$\phi \sum A_{si} f_{si} \sin \gamma_i \geq T_{b.s}^* \qquad \text{Equation 13.6(6)}$$

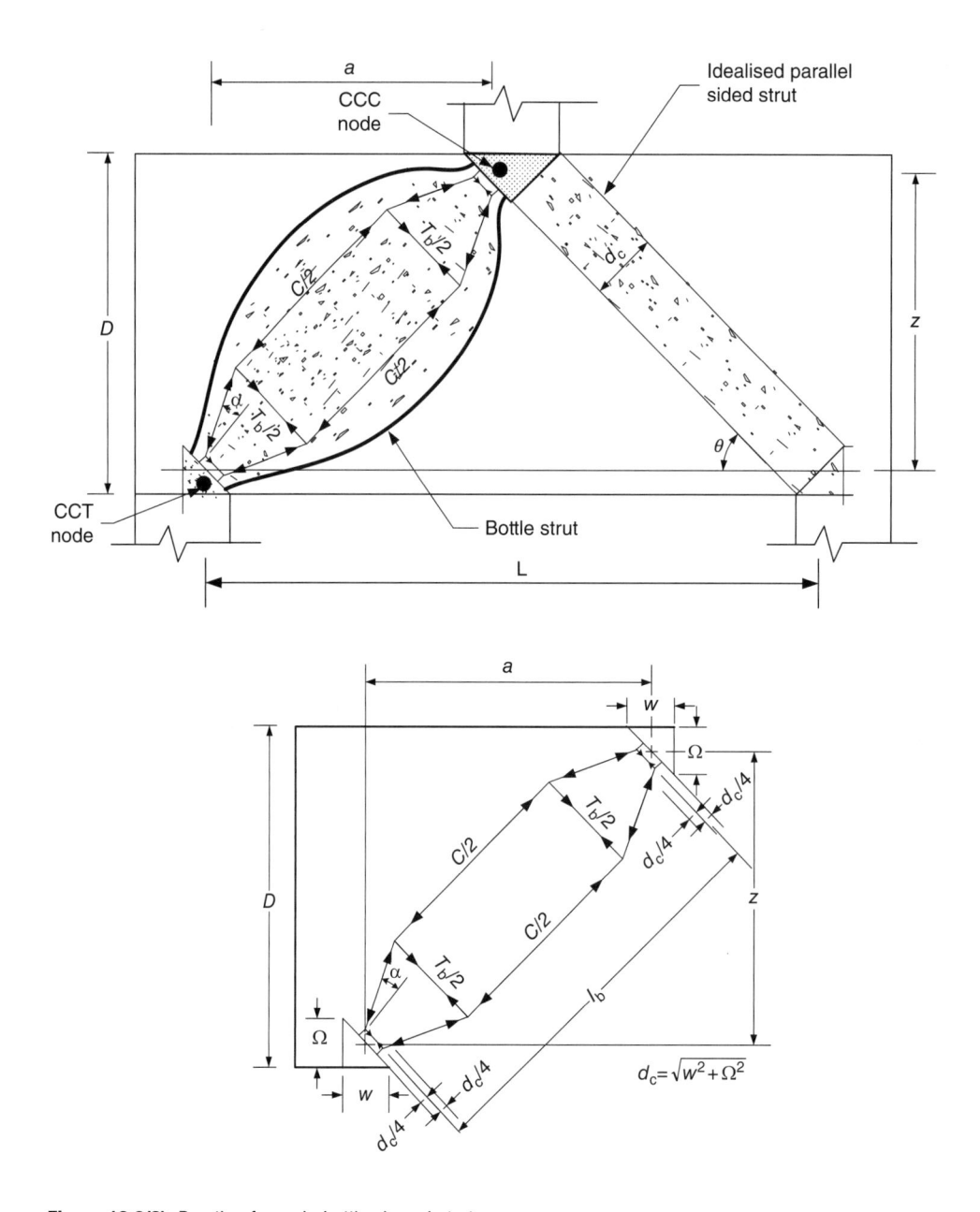

Figure 13.6(2) Bursting forces in bottle-shaped struts

Source: Standards Australia (2009). AS 3600–2009 *Concrete Structures.* Standards Australia, Sydney, NSW. © Standards Australia Limited. Copied by Cambridge University Press with the permission of Standards Australia and Standards New Zealand under Licence 1712-c038.

In the above expressions T_b^* and $T_{b.s}^*$ are bursting forces calculated using the design load for strength and serviceability, respectively; A_{si} is the area of reinforcement in directions 1 and 2 crossing a strut at an angle γ_i to the axis of the strut – see Figure 13.6(3); $\phi = 0.6$; and f_{si} is the serviceability limit stress in the reinforcement but not greater than the following:
- 250 MPa where a minor degree of control over cracking is required
- 200 MPa where a moderate degree of control over cracking is required
- 150 MPa where a strong degree of control over cracking is required.

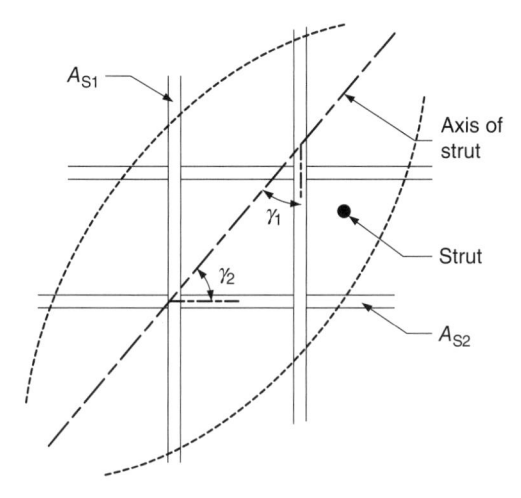

Figure 13.6(3) Reinforcement against bursting

Source: Standards Australia (2009). AS 3600–2009 *Concrete Structures*. Standards Australia, Sydney, NSW. © Standards Australia Limited. Copied by Cambridge University Press with the permission of Standards Australia and Standards New Zealand under Licence 1712-c038.

Also for prestressed concrete, the change in stress in the tendons after the point of decompression is not to exceed the limits given above, as appropriate.

13.6.2 Steel ties

Ties in a selected model must consist of reinforcing bars or prestressing tendons (or both). The reinforcement or tendons must be evenly distributed across the nodal regions at each end of the tie and arranged in such a way that the resultant tensile force coincides with the axis of the tie in the STM.

The design strength of a tie shall be taken as $\phi[A_{st}f_{sy} + A_p(\sigma_{p.ef} + \Delta\sigma_p)]$ where $(\sigma_{p.ef} + \Delta\sigma_p)$ must not exceed f_{py} and $\phi = 0.8$ as per Table 2.2.4 of the Standard.

To provide adequate anchorage at each end of the tie, the reinforcement or tendon must be extended beyond the node to achieve the design strength of the tie at the node and anchored in accordance with that given in Section 8.2 of this book.

13.6.3 Nodes

Clause 7.4 specifies three types of node as distinguished by the arrangement of the entering struts and ties, and the confinement thus provided, as follows:
- CCC – there are only struts (C) entering the node
- CCT – there are two or more struts and a single tension tie (T) entering the node
- CTT – there are two or more tension ties entering the node. Some examples of these node types are shown in Figure 13.3(2).

Where confinement is not provided to the nodal region, the design strength of the node shall be such that the principal compressive stress on any nodal face, determined from the normal shear stresses on that face, is not greater than $\phi\beta_n 0.9 f'_c$ where $\phi = 0.6$ and:
- $\beta_n = 1.0$ for CCC nodes
- $\beta_n = 0.8$ for CCT nodes
- $\beta_n = 0.6$ for CTT nodes.

Further, where confinement is provided to the nodal region, the design strength of the node may be determined by tests or calculations, considering the confinement, but must not exceed a value corresponding to a maximum compressive principal stress on any face of $1.8\,\phi f'_c$.

13.6.4 Additional specifications

When analysing an STM to determine the internal forces in the struts and ties, the requirements of Clause 6.1.1 must be satisfied, and Clauses 6.1.2 and 6.8.2 must be complied with.

When strut-and-tie modelling is used for strength design, the requirements of Clause 2.2.4 must be satisfied. Note that Clause 2.2.4 stipulates a list of six strength checks for design adequacy.

When design for strength is based on strut-and-tie modelling, separate checks must be undertaken to ensure that the design requirements for serviceability are satisfied.

13.6.5 Illustrative example

Problem

Using the strut-and-tie modelling (STM) approach of the Standard, design a 600-mm-wide transfer girder spanning 5 m with a column at midspan with ultimate factored load of 5340 kN. The girder is supported by 600 mm square columns. The overall depth of the girder is 2.25 m as shown in Figure 13.6(4). The ultimate shear diagram is shown in Figure 13.6(5). Use $f'_c = 50$ MPa and N bars only for reinforcement.

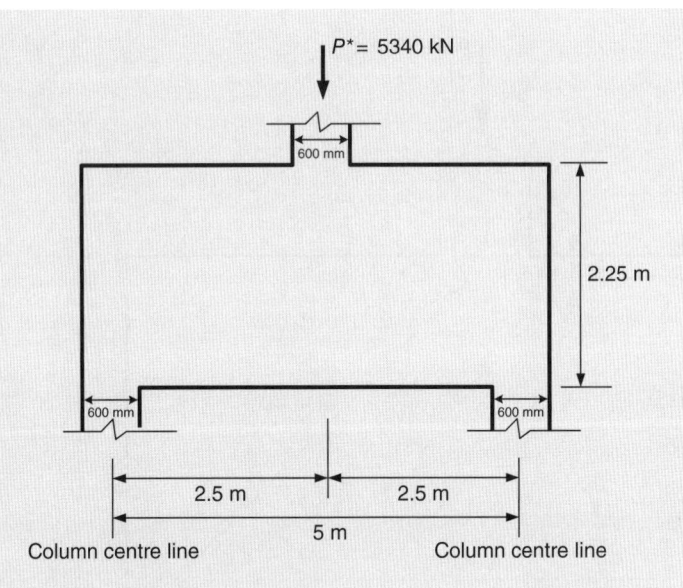

Figure 13.6(4) STM design example – transfer girder

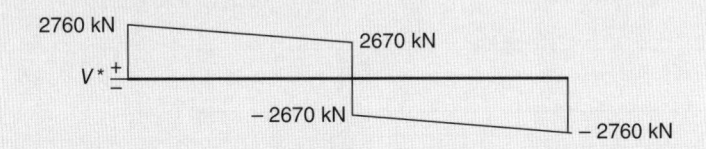

Figure 13.6(5) STM design example – shear diagram

Solution

Select and establish strut-and-tie model and node locations

Assume that the loads are carried by a strut-and-tie system consisting of two direct struts running from the top loading column to the supporting columns and a tie connecting the struts horizontally. The geometry of the assumed STM is shown in Figure 13.6(6).

Because of the heavy loads applied on the structure and the minimum allowable height being used, much deeper node locations are required. After multiple iterations, the node at location C at the loading point is determined to be 300 mm from the top of the girder, and the node location at the supports as 350 mm from the bottom of the girder as shown in Figure 13.6(6). Note that after the design, if the final nodal locations show a difference of roughly 50 mm or less, the original locations are deemed acceptable because the forces in the strut may increase only by 1–2%, which should not change the final design.

Figure 13.6(6) STM design example – assumed STM and node locations

Based on the assumed nodal locations as above (see Figure 13.6(6)), the angle between the struts and the tie = $\tan^{-1}\left(\frac{2250-300-350}{2500}\right)$ = 32.6° > 30°. As per Clause 7.1 of AS 3600–2009, it is acceptable.

Determine forces in struts and ties

From the geometry of the girder with reference to Figure 13.6(6),

$$\text{length of strut CA} = \sqrt{(2250-300-350)^2 + (2500)^2} = 2968.2 \text{ mm}$$

$$\text{length of strut CB} = \sqrt{(2250-300-350)^2 + (2500)^2} = 2968.2 \text{ mm}$$

Thus,

$$\text{force in strut CA} = 2760 \times \frac{2968.2}{2250-300-350} = 5120.1 \text{ kN}$$

force in strut CB = $2760 \times \dfrac{2968.2}{2250 - 300 - 350} = 5120.1$ kN

force in tie AB = $2760 \times \dfrac{2500}{2250 - 300 - 350} = 4312.5$ kN

Determine effective concrete strength in nodes and struts

Enough space exists within the girder for bottle-shaped struts to be formed in struts CA and CB. Also, bursting reinforcement will be provided to resist cracking.
Thus, Equation 13.6(1):

$$\beta_s = \dfrac{1}{1.0 + 0.66 \cot^2 32.6^0} = 0.383 > 0.3 \text{ which is acceptable}$$

Making use of Equation 13.6(2), the effective concrete strength is:

$$\phi \beta_s 0.9 f'_c = 0.6 \times 0.383 \times 0.9 \times 50 = 10.34 \text{ MPa}$$

The struts within the columns do not have enough space for a bottle-shaped strut to form. Thus $\beta_s = 1.0$ and the effective concrete strength:

$$\phi \beta_s 0.9 f'_c = 0.6 \times 1 \times 0.9 \times 50 = 27.0 \text{ MPa}$$

For the nodal region at C, a CCC situation prevails. Thus, $\beta_n = 1.0$ and the principal compressive stress on the nodal face is:

$$\phi \beta_n 0.9 f'_c = 0.6 \times 1 \times 0.9 \times 50 = 27.0 \text{ MPa}$$

For the nodal region at A and B, a CCT situation prevails. Thus, $\beta_n = 0.80$ and the principal compressive stress on the nodal face:

$$\phi \beta_n 0.9 f'_c = 0.6 \times 0.8 \times 0.9 \times 50 = 21.6 \text{ MPa}$$

Determine STM geometry

Hydrostatic nodal regions are used herein. Hence, the stresses on each face of the region must be identical and the faces are perpendicular to the axis of the struts. Extended nodal zones may be used, but hydrostatic nodal regions are easy to use for this type of loading and also add some conservatism in the design by requiring a larger nodal zone. As hydrostatic nodal zones are used, the minimum value for the effective concrete strength (i.e. 10.34 MPa) must be used in calculating the widths of the struts and the height of the tie to ensure a static situation.

Using Equation 13.6(2) with $A_c = d_c t$, where t is the thickness of the strut (which is 600 mm in this example), the strut width

$$d_c = \dfrac{f_{\text{strut}}}{\phi \beta_s 0.9 f'_c t}$$

Thus,

$$\text{width of strut CA, } d_{c.CA} = \frac{5120.1 \times 10^3}{10.34 \times 600} = 825.3 \text{ mm}$$

$$\text{width of strut CB, } d_{c.CB} = \frac{5120.1 \times 10^3}{10.34 \times 600} = 825.3 \text{ mm}$$

$$\text{width of strut A, } d_{c.A} = \frac{2760 \times 10^3}{10.34 \times 600} = 444.9 \text{ mm}$$

$$\text{width of strut B, } d_{c.B} = \frac{2760 \times 10^3}{10.34 \times 600} = 444.9 \text{ mm}$$

$$\text{width of strut C1, } d_{c.C1} = \frac{2670 \times 10^3}{10.34 \times 600} = 430.4 \text{ mm}$$

$$\text{width of strut C2, } d_{c.C2} = \frac{2670 \times 10^3}{10.34 \times 600} = 430.4 \text{ mm}$$

$$\text{height of tie AB, } d_{c.AB} = \frac{4312.5 \times 10^3}{10.34 \times 600} = 695.1 \text{ mm}$$

For the extra compression strut width required ($\geq$ (430.4 + 430.4) = 860.8 mm) within the column applying the loads, a 875 mm × 600 mm column is required. All other dimensions fit within the girder and supporting columns and follow the STM guidelines stipulated in Clause 7.1 of the Standard, as shown in Figure 13.6(7).

The actual node locations are determined using geometry as shown in Figure 13.6(7). The node at C is 699.8/2 = 349.9 mm from the top of the girder, which is within 50 mm of the initial 300 mm used for the design, and the nodes at A and B are 695.1/2 = 347.55 mm from the bottom of the girder, which is also very close to the 350 mm initially selected.

If these nodes were much further apart, new initial node locations would have to be selected and every quantity recalculated until the differences were appropriate.

Determine steel in the tie

For the force in tie AB, $F^*_{AB} = 4312.5$ kN, the area of steel should be

$$A_{st} \geq \frac{F^*_{AB}}{\phi f_{sy}} = \frac{4312.5 \times 10^3}{0.8 \times 500} = 10\,781.3 \text{ mm}^2$$

Use four rows of four N32 bars in each row or A_{st} = 12 864 mm^2 > 10 781.3 mm^2, which is acceptable.

The tension tie reinforcement of four rows of four N32 bars in each row spaced at 64 mm centre to centre is shown in Figure 13.6(8).

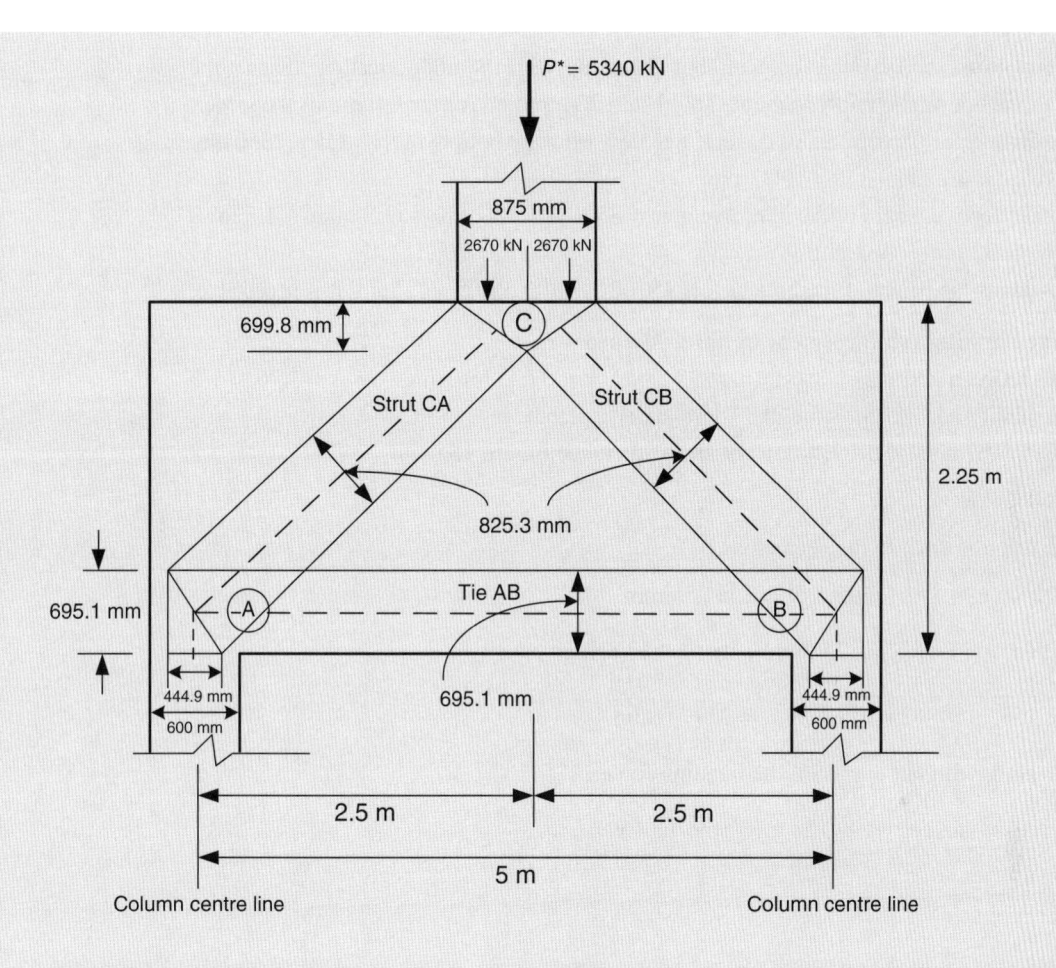

Figure 13.6(7) STM design example – STM and node geometry

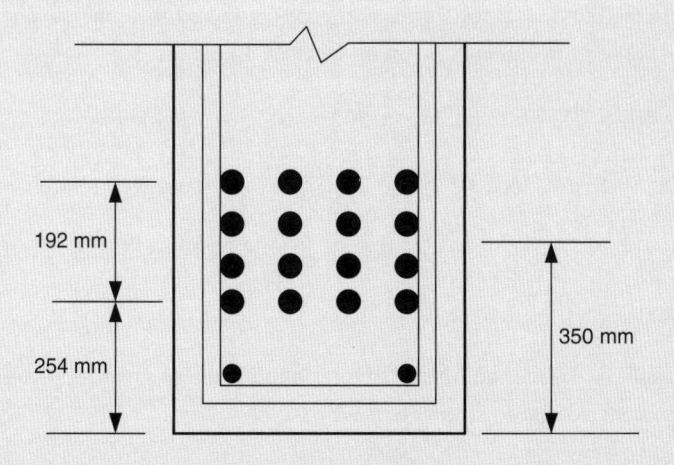

Figure 13.6(8) STM design example – tension tie reinforcement

The centroid of the tie reinforcement should line up with the node location; that is, the centroid of the bottom tie reinforcement should start 254 mm above the bottom of the girder. Thus, the distance of the centroid of tie reinforcement from the bottom of the girder = 350 mm and as a result $d = 2250 - 350 = 1900$ mm.

The total effective height of the reinforcement = 350 + 2 rows of bars @ 32 mm + 1.5 rows of spacing @ 32 mm = 462 mm.

Check against the height of the tie which is $350 \times 2 = 700$ mm > 462 mm. It is acceptable.

Determine bursting reinforcement in bottle-shaped struts

The angle between stirrups and struts, $\alpha = 90° - 32.6° = 57.4°$ for which $\tan \alpha = 1.76 > 1/2$ for serviceability and 1/5 for strength. These are acceptable.

Because of symmetry, the left or right shear span (see Figure 13.6(7))

$a = 2500$ mm

The component normal to the shear span

$z = 2250 - 699.8/2 - 695.1/2 = 1552.55$ mm

Thus, with $d_{c.CA} = d_{c.CB} = 825.3$ mm, Equation 13.6(4) gives

$$l_b = \sqrt{1552.55^2 + 2500^2} - 825.3 = 2117.6 \text{ mm}$$

For strength, with $\tan \alpha = 1.76$, the bursting force

$$T_b^* = f_{strut} \times \tan \alpha = 5120.1 \times 1.76 = 9011.4 \text{ kN}$$

and the area of steel

$$A_s \geq T_b^*/(\phi f_{sy}) = 9011.4 \times 10^3/(0.8 \times 500) = 2253 \text{ mm}^2$$

At the loading point where the bursting crack forms, the force carried by the concrete and to be transferred to the bursting reinforcement is given by Equation 13.6(3) as

$$T_{b.cr} = 0.7 b l_b f'_{ct} = 0.7 \times 600 \times 2117.6 \times \left(0.36 \times \sqrt{50}\right) \times 10^{-3} = 2264 \text{ kN}$$

As per Clause 7.2.4 of the Standard, since $T_b^* > T_{b.cr}$, only the minimum web reinforcement is required or

$$A_{s.min} = \frac{T_{b.cr}}{\phi f_{sy}} = \frac{2264 \times 10^3}{0.8 \times 500} = 5660 \text{ mm}^2$$

Comparing the above A_s and $A_{s.min}$, the minimum bursting reinforcement governs. Thus, the area of bursting reinforcement normal to strut CA or CB is:

$A_{st} = 5660$ mm^2

Thus the reinforcement ratio $p_t = A_s/l_b t = 5660/(2117.6 \times 600) = 0.00445$.

The force across the bursting plane is maintained if orthogonal reinforcement is placed parallel and normal to the axis of the member such that

$p_{th} = p_t \sin\theta = 0.00445 \times \sin 32.6° = 0.00240$

$p_{tv} = p_t \cos\theta = 0.00445 \times \cos 32.6° = 0.00375$

The vertical web reinforcement in each shear span, considering bursting reinforcement, is p_{tv} = 0.00375. Adopting two layers (one layer at each face) of 16 mm diameter bars, with a total bar area across the 600 mm thick section of 402 mm^2, gives a bar spacing requirement of

$s = 402/(0.00375 \times 600) = 178.7$ mm

Use N16 stirrups at 175 mm spacing for the vertical reinforcement for the total span.

For the horizontal reinforcement, p_{th} = 0.00240. Using two layers of N16 bars (A_s = 402 mm^2) gives a bar spacing requirement of

$s = 402/(0.00240 \times 600) = 279.2$ mm

Use two layers of N16 bars at 275 mm spacing as the longitudinal reinforcement for the total span.

Final design

The completed design of the girder is shown in Figure 13.6(9), with dimensions and the reinforcement.

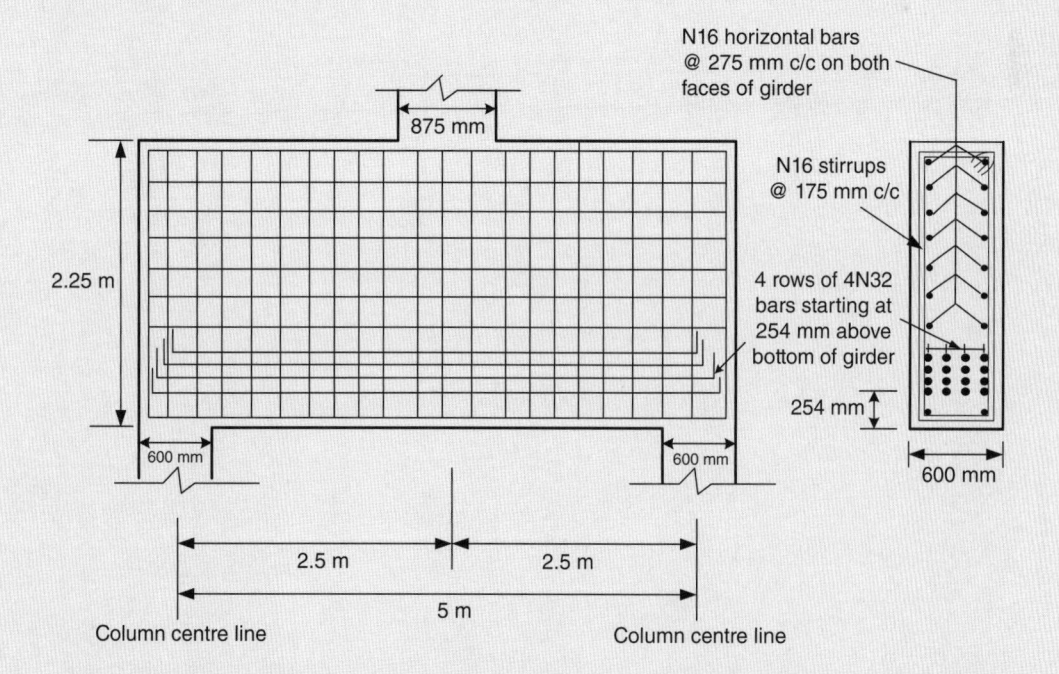

Figure 13.6(9) STM design example – final design section

13.7 SUMMARY

The fundamental concepts of the strut-and-tie model (STM) and some recent developments of the technique are discussed with an extensive literature review. Typical models taking the form of some statically determinate trusses are illustrated. Also described in detail are the Standard specifications for various types of struts, ties and 'nodes'. To demonstrate the application of the STM technique, the design of a transfer girder is presented in full.

PART 2

PRESTRESSED CONCRETE

INTRODUCTION TO PRESTRESSED CONCRETE 14

14.1 INTRODUCTION

Prestressing may be seen as an elaborate and active way to reinforce concrete when its tensile capacity is insufficient. Whereas the traditional reinforcement becomes active mainly after the concrete has exceeded its cracking strength, the purpose of prestressing is to prevent cracking from occurring. This is done by introducing compressive stress in the concrete to neutralise the anticipated tensile stress developed under load.

In a traditional reinforced concrete design, the safety margin can always be increased by providing more reinforcement. The same may not be true in prestressed concrete, as over-prestressing can cause cracking or perhaps failure before even any external loading is applied. As a result, prestressed concrete analysis and design are more complicated and mechanics-based than for reinforced concrete, which relies more on empirical formulas. In practice, prestressed concrete also requires a higher level of technology in its construction.

By nature, prestressing is more efficient than the traditional reinforcement in that the stress in concrete, either tensile or compressive (caused by self-weight or other forms of dead load), can be neutralised before any additional (live) loading is applied. Consequently, for a given design, the maximum permissible prestressed concrete span can be considerably larger than a reinforced one. Following some fundamentals given in this chapter, Chapter 15 presents the bending theory of fully prestressed concrete beams based on the critical stress state criteria (which ensures that no cracking or overstressing in tension or compression would ever occur throughout the life of the beam under service load). The design of beams in bending using the critical stress state approach is given in Chapter 16. To comply with the Australian Standard (AS 3600–2009 (the Standard), in practice, all prestressed beams must have the required strength. The bending strength analysis of fully and partially prestressed concrete beams is discussed in Chapter 17. The last chapter, Chapter 18, reviews the analysis and design of end blocks for prestress anchorage.

14.2 NON-ENGINEERING EXAMPLES OF PRESTRESSING

14.2.1 Wooden barrel

Figure 14.2(1) shows an assembly of curved barrel staves with tapered ends that are held together by the metal hoops forced from both ends towards the centre. The forced expansion of the hoops produces the compressive stress between the staves. To enhance efficiency, the hoops may be heated up before installation. It is well known that a well-made wooden barrel cannot only hold itself together, but is also water tight.

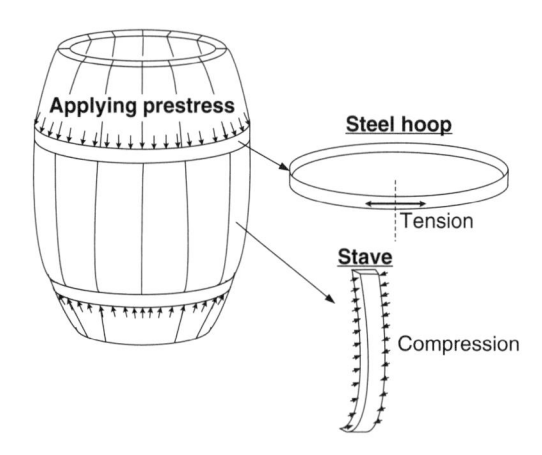

Figure 14.2(1) Internal stresses in wooden barrel assembly

14.2.2 Stack of books

Despite the collective weight of a stack of books, if pressed together by two hands hard enough they can be held together, as shown in Figure 14.2(2), and be moved as a single elongated object from one location to another. Experience would suggest that the object would collapse if the lateral pressure is inadequate. This is the principle of prestressing at work: the stack of books may be seen as a 'beam'; the weight of the books constitutes the uniformly distributed dead load; and the two self-equilibrating compressive actions of the hands constitute the 'prestressing' force. Other weights (live loads) may be added on top of the books if the hand pressure is high enough. In fact, this simple principle may also be relied upon to understand the segmental construction of a post-tensioned multispan concrete box girder bridge – see Figure 14.2(3). The stressing of the box segments together to form the bridge girder incrementally is designed to carry the traffic and other loads in addition to the self-weight of the bridge.

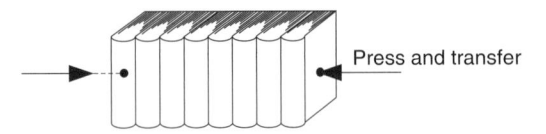

Figure 14.2(2) Moving of a stack of books with two hands

Figure 14.2(3) Segmental construction of a box bridge

Source: Courtesy of Brisbane City Council <https://commons.wikimedia.org/wiki/File:Go_Between_Bridge_Construction_(5160262570).jpg>

14.3 PRINCIPLE OF SUPERPOSITION

Figure 14.3(1)a illustrates a beam subjected to a pair of axial concentric and self-equilibrating forces (H), the lateral concentrated loads (P_1 and P_2) and the uniformly distributed loading (w). The axial force H produces, at a given beam cross-section, the uniform stress H/A as in Figure 14.3(1)c. The lateral loads lead to the linearly distributed stress (LDS) over the section with the maximum compressive and tensile stresses located at the top and bottom fibres, respectively – see Figure 14.3(1)d.

By the principle of superposition for linear and elastic structures, the two stress distributions due to H and the lateral loads, which are obtained separately, may be combined or superposed to produce the final stress diagram. Depending on the magnitude of H and the intensities of the lateral loads, the final stress distribution may take one of two forms: the entire section is under compression, or under compressive and tensile stresses. The two stress-distribution patterns, respectively, are shown in Figures 14.3(1)e and f. Note that this process of superposition is valid for any linear–elastic system. Therefore, the beam may be constructed of steel, timber and concrete, as long as the concrete sustains no cracking. This means the beam is considered linear and elastic.

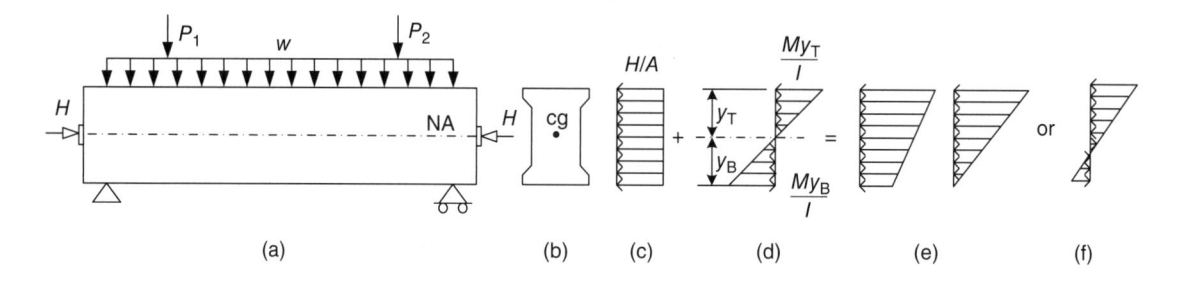

Figure 14.3(1) Stress profiles under lateral loads and concentric axial forces

To further understand the principle of prestressed concrete, consider the beam in Figure 14.3(2)a subjected to the same lateral forces and a pair of self-equilibrating forces H are now applied with an eccentricity (e) measured downwards from the centre of gravity – see Figure 14.3(2)a. The eccentric H may be transformed into a concentric H plus a clockwise moment He as depicted in Figure 14.3(2)b.

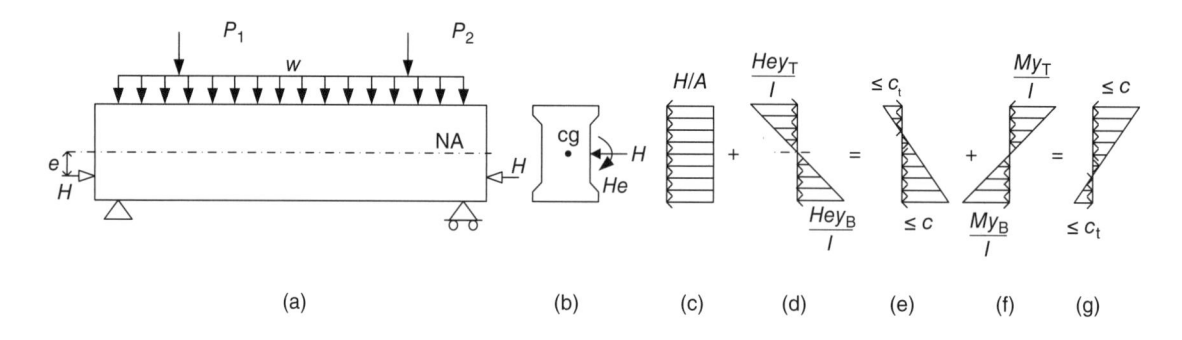

Figure 14.3(2) Stress profiles under lateral loads and eccentric axial forces

As for the case in Figure 14.3(1), the now concentric H produces the uniform stress H/A, but the clockwise moment He produces the LDS profile with maximum tensile and compressive stresses at the top and bottom fibres, respectively. Superimposing the stresses in Figures 14.3(2)c and d leads to the LDS in Figure 14.3(2)e. The concrete does not crack as long as the top fibre stress is less than or equal to the permissible concrete tensile stress (c_t). By the same token, the section is not overstressed in compression as long as the extreme bottom fibre stress is below or equal to the permissible concrete compressive stress (c).

The LDS in Figure 14.3(2)f is due to the lateral loads P_1, P_2 and w. Superimposing the stresses in Figures 14.3(2)e and f produces the final LDS in Figure 14.3(2)g. Again, the concrete beam is not overstressed if, now, the top and bottom fibre stresses are lower than

c and c_t, respectively. Note that in a prestressed concrete beam, H is the prestress force. For better efficiency, the eccentricity (e) should be so chosen or designed so that the maximum tensile and compressive stresses in both Figures 14.3(2)e and g should be as close as possible to the corresponding permissible stresses. The practical aspects of producing the prestress force H in pre- and post-tensioned systems are discussed in Section 14.4. The permissible stresses c and c_t as per the Standard are given in Section 15.4.

The theory of prestressed concrete discussed in Part 2 is based on linear–elastic or permissible stress principles. This is in contrast to Part 1 of the book on reinforced concrete, which is based on inelastic and nonlinear, ultimate strength principles. Linear–elastic principles are more analytical or implemented through fundamental and generic formulas, independent of specific construction materials as long as these materials, concrete included, can be assumed to behave in a linear and elastic manner. Ultimate strength theory, as discussed in Part 1, is mainly applied via empirical formulas or equations calibrated using laboratory test results. Such formulas are applicable specifically to the given material – in our case, reinforced concrete and in specific cross-sectional shapes. Section 4.2 explains that the analysis and design formulas for rectangular sections are different from those for T-sections and other flanged sections. For irregular sections, including circular ones, numerical and iterative procedures may be needed for analysis and design (Section 4.3). By contrast, the critical stress state (CSS) theory of prestressed concrete discussed here is implemented using only six equations for analysis (see Chapter 15) and three for design (see Chapter 16). They are applicable to beams with any given cross-sectional shape. The prestressed concrete beam so designed will not crack throughout its life under service load; therefore, the beam can be satisfactorily assumed to behave in a linear–elastic manner.

Why are there only nine equations in total that are capable of covering all cross-sectional shapes? It is simply because the required geometric properties of a given beam section can be represented by just four parameters: the cross-sectional area (A), location of the centroid, elastic section modulus (Z) and moment of inertia (I). The proficiency in applying the prestressed concrete theory is considered a manifestation of how well one can apply and manipulate the six equations in analysis, and the three equations in design. This is different from reinforced concrete analysis and design, which entails the correct application of the appropriate set of mostly empirical formulas.

14.4 TYPES OF PRESTRESSING

Depending on the design, a prestressed beam may be pretensioned or post-tensioned. In general, precast beams to be simply supported are pretensioned, although large single-span box beams for bridge construction are often post-tensioned. Continuous multispan beams are post-tensioned without exception.

14.4.1 Pretensioning

The steps for pretensioning beams in a precast form as depicted in Figure 14.4(1)a are as follows:

1. Pull the tendons fixed at the dead end, one at a time, at predetermined locations from the stressed (live) end. Each tendon is anchored at the dead-end buttress using a barrel-and-wedge anchorage device and each is stressed at the live end with an appropriate prestressing jack using another anchorage device – see Figure 14.4(1)b. After the required tension is attained, the anchorage device now butting against the live-end buttress also will serve to maintain the tension in the tendon when and after the jack pulling the tendon is deactivated. Depending on the length of the prestress bed – see Figure 14.4(1)a – more than one beam can be manufactured in a single operation.
2. Cast concrete into the formwork following the proper compaction procedure.
3. After the concrete has hardened to the required strength, sever the tendons (in between the beam ends, and between the beam ends and the buttresses) to 'transfer' the tension in the tendons to the concrete. This completes the pretensioning process.

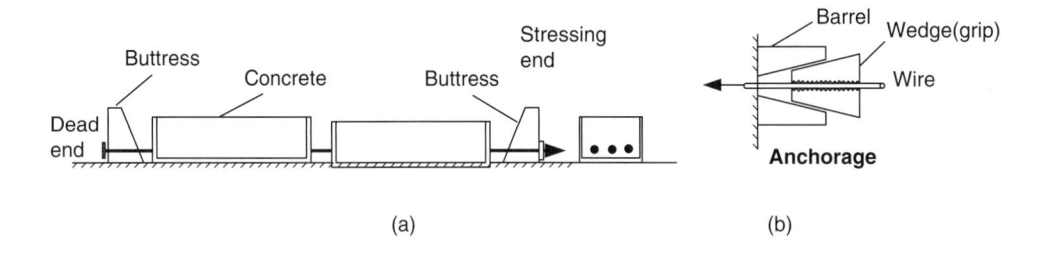

Figure 14.4(1) Pretensioned prestressing system: **(a)** prestressing bed and **(b)** anchorage system

14.4.2 Post-tensioning

The process for manufacturing a post-tensioned beam follows the steps below and is shown schematically in Figure 14.4(2).

1. Locate the required strands of tendons or prestress cables, in the desired or predetermined profile. The cables are enclosed in a water-tight sheath; the wedge-barrel anchorage systems at the live and dead ends are proprietary products in most cases. Examples of these are shown in Figures 14.4(3)a, b and c, respectively.
2. Cast and properly compact the concrete in the formwork with the cables encased in the water-tight sheath unstressed.
3. When the required concrete strength is attained, commence the stressing or post-tensioning process to produce the required tension in the group of cables. Again, the stressing operation and the equipment needed are generally patented. An example is shown in Figure 14.4(4). The operation is self-explanatory.

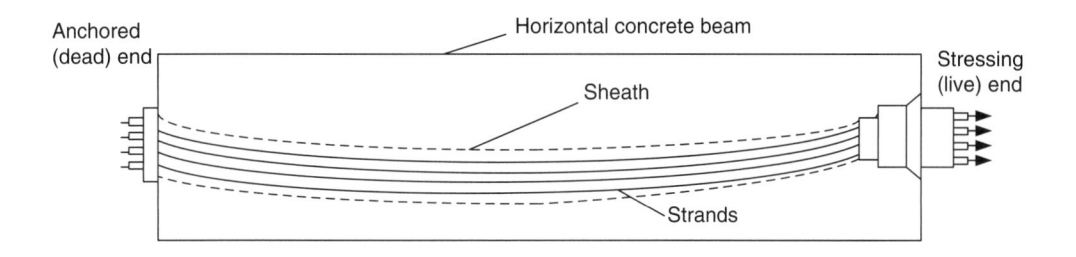

Figure 14.4(2) Details of a post-tensioned beam

In Figure 14.4(4), all the cables in the sheath are stressed simultaneously to the required tension. The same post-tensioning process is applicable to a beam with more than one group of cables, again one at a time. Note that the helical reinforcement around the cable group at the beam end serves to prevent failure due the 'bursting stress' (see Section 18.3.4).

14.5 PARTIAL PRESTRESSING

Applying to both pretensioning and post-tensioning, partial prestressing and partially prestressed members have been variously defined as follows:

- 'When a member is designed so that ... some tensile stresses will be produced in the member under working load, then it is termed partially prestressed.' (Lin and Burns 1981)
- '... members are referred to as being partially prestressed [if] the permissible tensile stresses are kept sufficiently low so that no visible cracking occurs, and [particularly if] the tensile stresses are restricted such that crack widths do not exceed 0.1 mm for very severe environments and 0.2 mm for other conditions.' (Kong and Evans 1987)
- 'A member is usually said to be partially prestressed if the precompression is not sufficient to prevent cracks from forming under the full service load.' (Warner and Faulkes 1988)
- 'Members that are designed to crack at the full service load are often called *partially prestressed*.' (Gilbert, Mickleborough and Ranzi 2016)
- '... visible or invisible cracks can occur if calculated concrete tensile stresses occur under design loads. If these cracks open and close under fluctuating service loads, the member is considered to be partially prestressed.' (Post-Tensioning Institute 2006)
- partial prestressing is the level of prestressing at which 'flexural [concrete] tensile stress and some limited cracking [can occur] under full service load.' (Darwin, Dolan and Nilson 2016)

While 'partial prestressing is not widely used in buildings and bridge structures' (Post-Tensioning Institute 2006), the application of lower levels of prestressing may be necessary in practice – for example, to control cambers and/or for economic reasons.

Some aspects of the analysis and design of partially prestressed beams are covered in Sections 15.1 and 16.1, respectively.

(a) Stressing anchorage VSL Type Gc

(b) Dead end anchorage VSL Type P

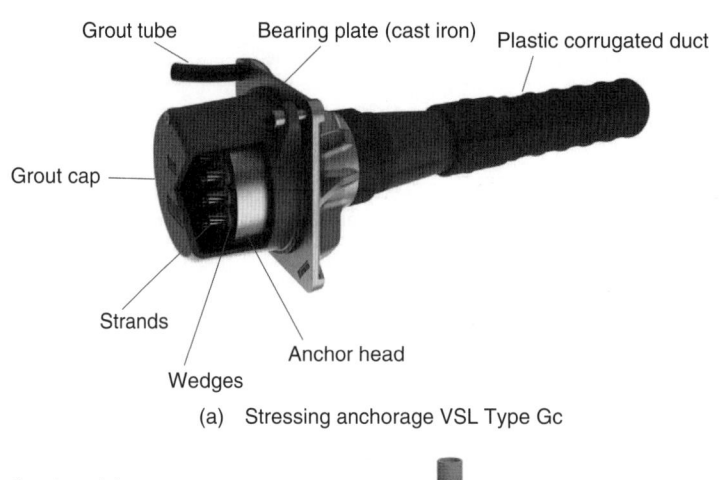

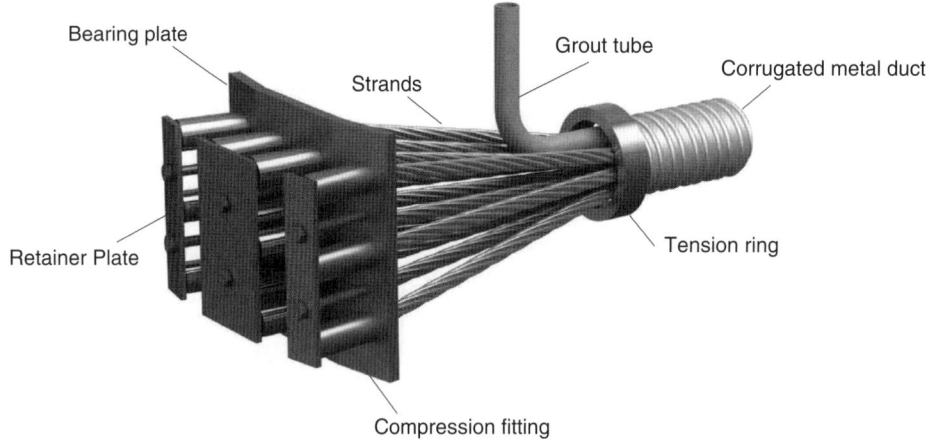

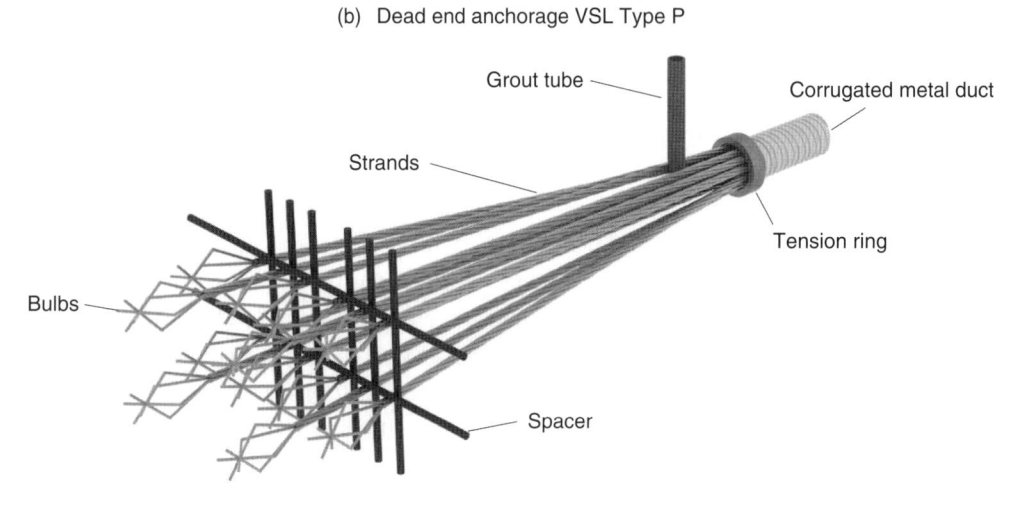

(c) Dead end anchorage VSL Type H

Figure 14.4(3) Examples of strands of tendons or prestress cables

Source: <http://www.vsl.com>

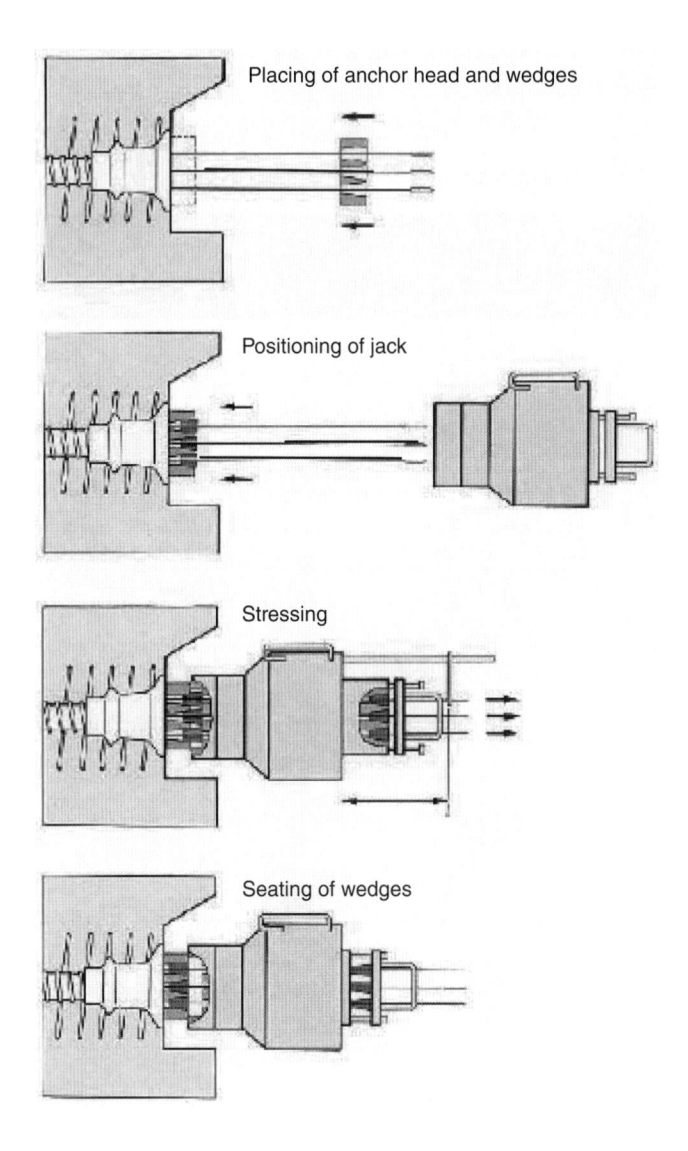

Figure 14.4(4) Stressing operation

Source: <http://www.vsl.com>

14.6 TENSILE STRENGTH OF TENDONS AND CABLES

Prestressing tendons and cables are made of high-tensile steel – the strength of which is more than three times that of the Class L and N reinforcing bars. Such a high strength is mandatory for an effective prestress or for the tendons to sustain adequate elongations, thereby providing the required compression in the beam. The compression is there to carry the dead and live loads, as well as to absorb the loss in prestress caused by several factors, including the elastic and long-term shortening of the concrete that makes up the beam.

Table 14.6(1) details the various grades of prestressing steel and the corresponding tensile strengths as specified in AS/NZS 4672.1–2007.

Table 14.6(1) Different grades of prestressing steel

Product	Type	Nominal wire/ strand diameter (mm)	Nominal cross-sectional area (mm^2)	Nominal tensile strength (MPa)
As-drawn wire	Common wire sizes	5.0	19.6	1770
		7.0	38.5	1670
Stress-relieved wire	Common wire sizes	7.0	38.5	1670
		5.03	19.9	1700
Quenched and tempered wire	Plain	6.0	28.3	1570
		7.0	38.3	
		8.0	50.3	
		10.0	78.3	
		12.2	117.0	
		14.0	154.0	
		16.0	201.0	
	Ribbed	6.2	30.2	1570
		7.2	40.7	
		8.0	50.3	
		10.0	78.5	
		12.0	113.0	
		14.0	154.0	
		16.0	201.0	
	Grooved or indented	7.1	40.0	1570
		9.0	64.0	
		10.7	90.0	
		12.6	125.0	
Strands	7-wire ordinary	9.5	55.0	1850
		12.7	98.6	1870
		15.2	143.0	1750
		15.2	143.0	1830
	7-wire compacted	15.2	165.0	1820
		18.0	223.0	1700
	19-wire	21.8	113.0	1810

Source: Standards Australia (2007). AS/NZS 4672.1–2007 *Steel Prestressing Materials Part 1: General Requirements*, Standards Australia, Sydney, NSW. © Standards Australia Limited. Copied by Cambridge University Press with the permission of Standards Australia and Standards New Zealand under Licence 1712-c038.

14.7 AUSTRALIAN STANDARD PRECAST PRESTRESSED CONCRETE BRIDGE GIRDER SECTIONS

Standard sections for precast prestressed concrete girders have been adopted in AS 5100.5–2017 for bridge design. Details of such standard sections are given in Appendix C, where Figure C(1) relates to I-girders and Figure C(2) covers the 'super T-girders'. For bridge beams, concrete cover in excess of 25 mm may be required for durability and other purposes. In this case, thicker webs than those of the standard sections are necessary.

14.8 SUMMARY

Some fundamentals of prestressed concrete are covered in this chapter. The standard types of prestressing as well as the processes for pretensioning and post-tensioning beams are discussed in some detail. Partial prestressing is also described. Various grades of prestressing steel and the corresponding tensile strengths (as specified in Australian Standards) are presented in a ready-to-use form. For completeness, details of Australian standard sections for precast prestressed concrete bridge girders are included in Appendix C.

15 CRITICAL STRESS STATE ANALYSIS OF BEAMS

15.1 INTRODUCTION

By definition, a fully prestressed beam sustains neither tensile cracking nor overstress in compression, under any given service load. Achieving these no-crack and no-overstress conditions throughout the working life of a beam – when prestress losses occur instantaneously and continuously – is a complicated problem. The critical stress state (CSS) approach presented in this chapter provides a fool-proof solution to this otherwise intractable problem. It is a linear–elastic method and is valid subject to the following assumptions:

- The plane section remains plane after bending.
- The material behaves elastically.
- The beam section is homogenous and uncracked.
- The principle of superposition holds.

Note that the CSS approach is suitable for partially prestressed beams sustaining tensile stresses below the concrete cracking strength (see Section 14.5).

15.2 NOTATION

Figure 15.2(1)a illustrates a typical section of a prestressed I-shaped bridge beam. It may be idealised as shown in Figure 15.2(1)b in which the resultant H of the individual prestressing forces is located at the effective centre of prestress, or with an eccentricity (e_B) from the neutral axis (NA). The effective centre is the centre of gravity (action) of the individual prestressing forces. Its location, or the value of e_B, can be determined by simple statics.

To develop the CSS equations in a systematic manner, specific notation needs to be followed. As shown in Figure 15.2(1)b, the eccentricity (e) is positive upwards and negative downwards. Consequently, the moment with respect to NA created by H is $-He_B$. The sign convention for the beam moment is given in Figure 15.2(2): positive (+) where it produces compression above NA, and negative (–) for tension below it.

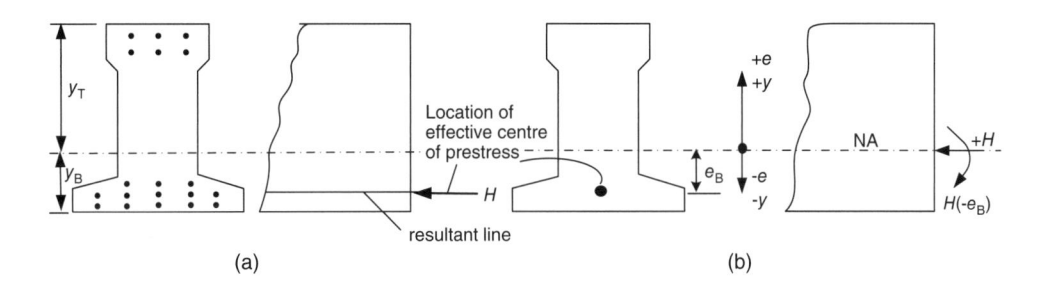

Figure 15.2(1) Prestressed I-beam: (a) typical section and (b) idealised section

Figure 15.2(2) Moment sign convention

The concrete stress f_c is positive (+) for compression and negative (−) for tension. Note that this convention may be opposite to that adopted in some books on the mechanics of materials.

For the section shown in Figure 15.2(1), under a working load moment (M_w), we can write, by virtue of the principle of superposition, the following equations:
for the extreme top-fibre stress

$$f_{cT} = \frac{H}{A} + \frac{H(-e_B)y_T}{I} + \frac{M_w y_T}{I} = \frac{H}{A} - \frac{He_B y_T}{I} + \frac{M_w y_T}{I} \quad \text{Equation 15.2(1)}$$

for the extreme bottom-fibre stress

$$f_{cB} = \frac{H}{A} + \frac{H(-e_B)(-y_B)}{I} + \frac{M_w(-y_B)}{I} = \frac{H}{A} + \frac{He_B y_B}{I} - \frac{M_w y_B}{I} \quad \text{Equation 15.2(2)}$$

Equations 15.2(1) and (2) may be presented graphically – see Figure 15.2(3) – recognising the sign conventions to be adopted in Section 15.7 for the CSS equations. If it were not

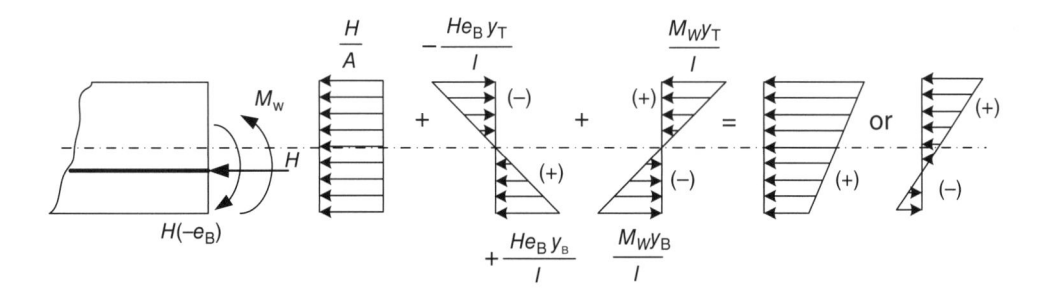

Figure 15.2(3) Stress distributions in prestressed beam under working load moment (M_w)

for the unavoidable loss of prestress in prestressed beams, it could be said that these two equations are all that are needed for prestressed-concrete analysis.

However, prestress loss is unavoidable, and the analysis becomes more complicated as a result.

15.3 LOSS OF PRESTRESS

15.3.1 Standard provisions

Provisions for loss of prestress in tendons are detailed in Clause 3.4 of AS 3600–2009 (the Standard). There are two categories of losses: immediate (Clause 3.4.2) and time dependent (Clause 3.4.3).

In general, immediate losses include those that are caused by the elastic shortening of concrete, losses in prestressing jacks and anchorages during the stressing process, and friction along post-tensioned cables. Time-dependent or deferred losses comprise shrinkage, creep in concrete and relaxation in steel. Clause 3.4 recommends empirical formulas for all except loss due to elastic shortening of concrete. The reader is referred to the Standard for the application of the empirical formulas. The following example demonstrates a process for calculating the elastic loss.

Problem

The given properties for the beam in Figure 15.3(1) are:

- the initial modulus (E_c) and cross-sectional area (A) for concrete
- Young's modulus (E_s) for steel and a prestress force (H).

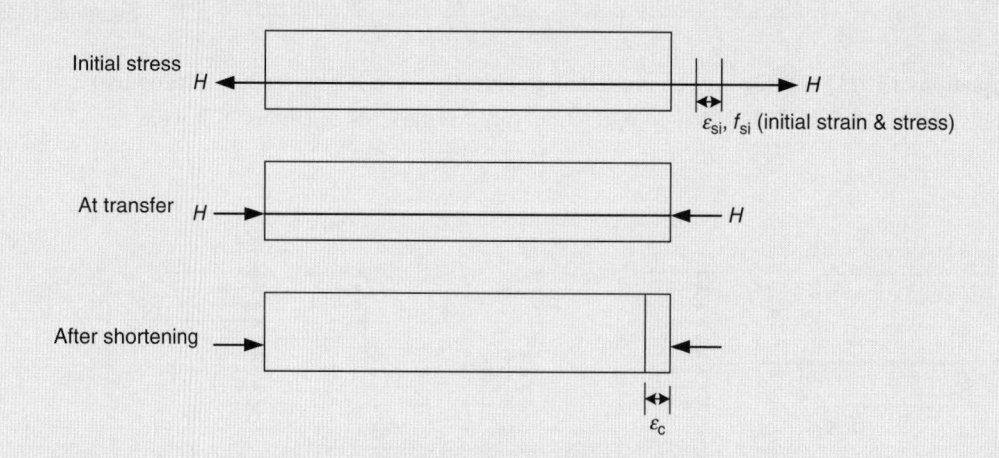

Figure 15.3(1) Loss of prestress in a beam due to the elastic shortening of concrete

Solution

After elastic loss

$$f_s = f_{si} - \text{elastic loss} = f_{si} - E_s \varepsilon_c = f_{si} - E_s \frac{H}{AE_c} \qquad \text{Equation 15.3(1)}$$

and hence the loss due to the elastic shortening of concrete $= \dfrac{E_s}{E_c} \dfrac{H}{A}$.

Note that in Equation 15.3(1) the gross concrete area (A) is used without accounting for the steel area. For more accurate calculations, the effects of the steel area should be included or the transformed section should be used instead (see Example 1 in Section 15.3.2).

15.3.2 Examples of prestress loss due to elastic shortening of concrete

To further illustrate the treatment of various immediate losses of prestress resulting from the instantaneous shortening of concrete at transfer, more discussion and additional examples are given below for both pretensioned and post-tensioned beams.

Example 1: Axially pretensioned member

Problem

In Figure 15.3(2), the straight pretensioned concrete member (12 m long with a cross-section of 380 mm × 380 mm) is concentrically prestressed with 750 mm² of steel tendons, which are anchored to the buttress walls with a stress of 1035 MPa. If $E_c = 34500$ MPa and $E_s = 205000$ MPa, compute the loss of prestress at transfer.

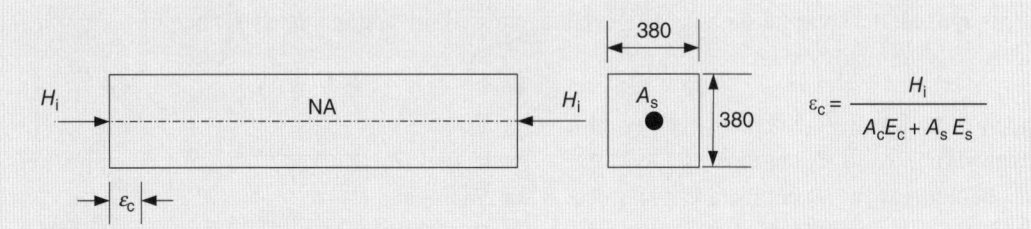

Figure 15.3(2) Axially pretensioned member for Example 1

Solution

Under the initial prestress force (H_i), the concrete beam sustains a compressive strain (ε_c). Hence, the loss in tendon stress

$$\Delta f_s = E_s \varepsilon_c = \frac{E_s H_i}{A_c E_c + A_s E_s} = \frac{\dfrac{E_s}{E_c} H_i}{A_c + \dfrac{E_s}{E_c} A_s} \approx \frac{E_s}{E_c} \frac{H_i}{A_g} \qquad \text{Equation 15.3(2)}$$

and by substituting the various given quantities we have

$$\Delta f_s = \frac{\dfrac{205\,000}{34\,500} \times 1035 \times 750}{(380^2 - 750) + \dfrac{205\,000}{34\,500} \times 750} = 31.1 \text{ MPa}$$

Example 2: Post-tensioned member

A beam post-tensioned in succession using five groups of tendons or cables is shown in Figure 15.3(3). Note that the total prestress $H = H_1 + H_2 + H_3 + H_4 + H_5$.

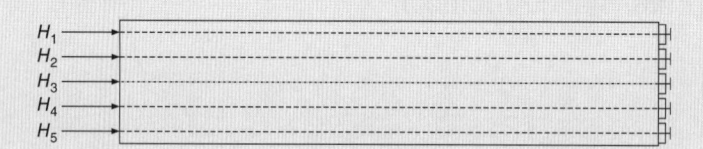

Figure 15.3(3) The example post-tensioned beam

When H_1 is applied, the concrete shortens as it is being stressed (by approximately ε_1) but H_1 sustains no loss. Similarly, when:

- H_2 is applied, no loss to H_2 but there is a loss of $E_s \varepsilon_2$ for H_1
- H_3 is applied, no loss to H_3 but there is a loss of $E_s \varepsilon_3$ for H_1 and H_2
- H_4 is applied, no loss to H_4 but there is a loss of $E_s \varepsilon_4$ for H_1, H_2 and H_3
- H_5 is applied, no loss to H_5 but there is a loss of $E_s \varepsilon_5$ for H_1, H_2, H_3 and H_4.

If H_2 to H_5 are identical, then the loss to H_1 is $4E_s \varepsilon$ and to:

- H_2 is $3E_s \varepsilon$
- H_3 is $2E_s \varepsilon$
- H_4 is $E_s \varepsilon$.

Example 3: Post-tension loss and compensation

Problem

Calculate the loss of prestress at transfer if given a beam similar to that in Example 2 above, but with four groups of post-tensioning tendons (each consisting of 187.5 mm^2).

Solution

Loss to:

- $H_1 = \dfrac{E_s}{E_c} \dfrac{3H}{A_g} = \dfrac{205\,000 \times (3 \times (187.5 \times 1035))}{34\,500 \times 380^2} = 24$ MPa
- $H_2 = \dfrac{E_s}{E_c} \dfrac{2H}{A_g} = \dfrac{205\,000 \times (2 \times (187.5 \times 1035))}{34\,500 \times 380^2} = 16$ MPa
- $H_3 = \dfrac{E_s}{E_c} \dfrac{H}{A_g} = \dfrac{205\,000 \times (187.5 \times 1035)}{34\,500 \times 380^2} = 8$ MPa
- $H_4 = 0$ MPa

Therefore, the total loss = 48 MPa, averaging 12 MPa per tendon group.
To compensate for the losses, we should:

- for H_1, stress to (1035 + 24) = 1059 MPa
- for H_2, stress to (1035 + 16) = 1051 MPa
- for H_3, stress to (1035 + 8) = 1043 MPa
- for H_4, stress to 1035 MPa.

Alternatively, we can take one of the two following options:

- Stress all of the tendons to the original level of 1035 MPa and allow for the average loss of 12 MPa in the design (i.e. in evaluating the effective prestress coefficient, η – see Section 15.3.3).
- Stress all of the tendons to 1035 + 12 = 1047 MPa and assume the beam sustains no elastic loss.

Example 4: Loss due to bending of the member

Figure 15.3(4) illustrates a prestressed beam bent by the internal moment due to H applied at e_B from the NA.

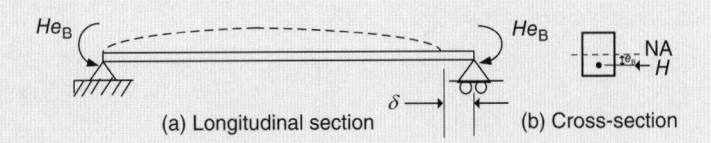

Figure 15.3(4) Bending of a prestressed beam due to internal moment

For pretensioned beams, since the steel is bonded to the concrete forming an integral part, shortening of the tendon (by δ) due to bending of the member is not regarded as loss and it does not incur any immediate loss.

For beams that are post-tensioned and stressed in stages, where transfer occurs before grouting and for unbonded beams, the amount of shortening δ should be computed. By ignoring the second order 'P–Δ' effect and assuming that the deflected shape in Figure 15.3(4) takes the form of a single sine wave, we obtain:

$$\delta = \frac{L^3}{256}\left(\frac{\pi H e_B}{E_c I}\right)^2 \qquad \text{Equation 15.3(3)}$$

where L is the length of the beam; I is the gross moment of inertia; and H, e_B and E_c are as defined previously.

15.3.3 Effective prestress coefficient

To develop the CSS equations, it is necessary to introduce a coefficient to quantify the effective prestress after all the losses – immediate and time dependent – have occurred. The effective prestress coefficient

$$\eta = (H - \text{all losses})/H \qquad \text{Equation 15.3(4)}$$

in which the losses are expressed in terms of force. The value of η in most cases is between 0.6 and 0.9.

Note that in all the ensuing chapters and sections, the value of η is given or assumed (where necessary). In practice, the total loss may be computed using relevant formulas recommended in Clause 3.4 and in Section 15.3.2.

15.3.4 Stress equations at transfer and after loss

The transfer of prestress occurs at the instant when all of the tendons are severed for a pretensioned beam (Section 14.4.1). For a post-tensioned beam (Section 14.4.2), the transfer becomes effective at the instant the required prestress is attained and the elastic shortening has occurred.

The equations for bending analysis are at transfer:

$$f_{cT} = \frac{H}{A} - \frac{H e_B y_T}{I} + \frac{M_w y_T}{I} \qquad \text{Equation 15.3(5)}$$

$$f_{cB} = \frac{H}{A} + \frac{H e_B y_B}{I} - \frac{M_w y_B}{I} \qquad \text{Equation 15.3(6)}$$

After loss:

$$f_{cT} = \eta\left(\frac{H}{A} - \frac{He_B y_T}{I}\right) + \frac{M_w y_T}{I} \qquad \text{Equation 15.3(7)}$$

$$f_{cB} = \eta\left(\frac{H}{A} + \frac{He_B y_B}{I}\right) - \frac{M_w y_B}{I} \qquad \text{Equation 15.3(8)}$$

In Equations 15.3(7) and (8), η is the effective prestress coefficient as given in Equation 15.3(4).

15.4 PERMISSIBLE STRESSES c AND c_t

As discussed previously, the CSS method of analysis and design is based on the permissible stress concept. A prestressed beam is not expected to crack or be overstressed in compression under any service load at any stage of its working life. This includes at transfer and after all losses have occurred. So that the beam will not crack, the maximum tensile stress that will ever occur must be less than the permissible tensile stress, c_t. Similarly, the maximum-ever compressive stress must be less than the permissible compressive stress, c. The values of c and c_t at transfer and after loss, as recommended in the permissible stress-based prestressed concrete Standard AS 1481–1978, are summarised in Table 15.4(1).

Table 15.4(1) Permissible compressive and tensile stresses

Stress	At transfer	After loss
Compressive	$c = \alpha_1 f'_{cp}$	$c = \beta_1 f'_c$
Tensile	$c_t = \alpha_2 \sqrt{f'_{cp}}$	$c_t = \beta_2 \sqrt{f'_c}$

Notes: f'_c is the characteristic strength of concrete; f'_{cp} is the minimum compressive strength at transfer; $\alpha_1 = 0.5 - 0.6$, $\alpha_2 = 0.0 - 0.5$, $\beta_1 = 0.3 - 0.45$ and $\beta_2 = 0.0 - 0.63$ are the safety factors

It may be recalled that since 1988, the Standard AS 3600 has been based on the limit-state concept. The newly revised version of 2009 is no exception. However, implicitly Clause 8.1.6.2 of the Standard recommends

$$c = 0.6 f_{cp} \qquad \text{Equation 15.4(1)}$$

where f_{cp} is the concrete compressive strength at transfer.

For unreinforced sections, Clause 8.6.2 indicates that

$$c_t = 0.25 \sqrt{f'_c} \qquad \text{Equation 15.4(2)}$$

and for sections with reinforcement or bonded tendons

$$c_t = 0.6\sqrt{f'_c} \qquad \text{Equation 15.4(3)}$$

The values of c and c_t are important criteria in the CSS analysis and design of prestressed beams under service load.

15.5 MAXIMUM AND MINIMUM EXTERNAL MOMENTS

The moment experienced by a prestressed beam under external load at service level – including self-weight and other dead loads – varies with time. Note that this does not apply to prestress that is considered internal to the beam. For example, in the case of a bridge beam, the variation is cyclical, reflecting the traffic flow patterns on a daily basis. In addition, there is the occasional or one-off loading that can produce extreme maximum or minimum moment to the beam. To ensure that the beam will not be stressed beyond the permissible level in tension or compression, it is necessary to determine the minimum and maximum moments that can possibly occur in a given beam immediately after the transfer of prestress, and throughout its working life.

The two extreme moments that can envelop the moment variation in a prestressed beam are the:

1. minimum moment, M_1, which may be positive (+) or negative (–) according to the sign convention defined in Figure 15.2(2)
2. maximum moment, M_2, which can only be positive (+), again as defined in Figure 15.2(2).

For a beam cast in situ on simple supports under a dead-load moment (M_g) and a maximum live-load moment (M_q)

$$M_1 = M_g \qquad \text{Equation 15.5(1)}$$

and

$$M_2 = M_g + M_q \qquad \text{Equation 15.5(2)}$$

Note that the in situ beam specification is important, since M_1 and M_2 are different for a precast beam, which needs to be transported from the casting site and placed onto the supports by a crane. They are different because during the transport and construction process extra loads or different patterns of loading would be imposed on the beam. For example, the lifting force on the beam during placement could change M_1 from being positive to negative (see Example 3 below).

To elaborate on the definition of M_1 and M_2, three examples are given below.

Problem

A simply supported beam is loaded as shown in Figure 15.5(1). Compute M_1 and M_2 at the midspan section.

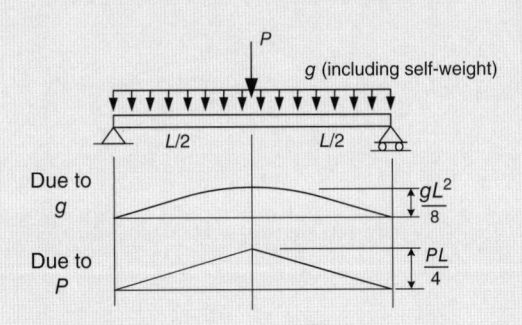

Figure 15.5(1) A simply supported beam – loading and moment diagrams

Solution

From the bending moment diagrams in Figure 15.5(1), we obtain

$$M_1 = \frac{gL^2}{8} \quad M_2 = \frac{PL}{4} + \frac{gL^2}{8}$$

Problem

A simply supported beam with overhang is shown in Figure 15.5(2). It is subjected to a moving load P in addition to the dead load. Determine (qualitatively) the M_1 and M_2 at sections B and C of the beam.

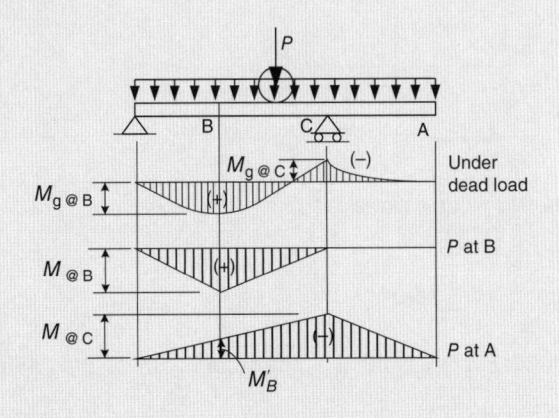

Figure 15.5(2) A simple beam with overhang – loading and moment diagrams

Solution

For the given loading, the diagrams for the dead-load moment, the live-load moment with P at any given point B between the two supports, and that with P at the tip of the overhang can be drawn (Figure 15.5(2)). Using these diagrams, M_1 and M_2 can be determined.

For section B

$$M_1 = M_{g@B} - M'_B \text{ (which may be } + \text{ or } -\text{)}$$
$$M_2 = M_{g@B} + M_{@B} \text{ (which is always+)}$$

For section C

$$M_1 = -M_{g@C}$$
$$M_2 = -\left(M_{g@C} + M_{@C}\right)$$

However, by definition, M_2 cannot be negative. But for computational purposes, both M_1 and M_2 may be taken as positive. Or

$$M_1 = M_{g@C}$$

and

$$M_2 = M_{g@C} + M_{@C}$$

This has the effect of hypothetically inverting the section to comply with the notation adopted in Figure 15.2(2), which corresponds to the required prestress force H being located below the NA. After computing H and the eccentricity (e), the section has to be reverted to its real position. This means H has to be acting above the NA as shown in Figure 15.5(3).

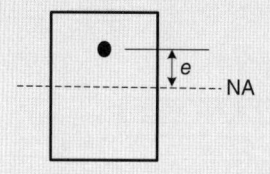

Figure 15.5(3) Actual location of prestress for negative moment at C

Problem

If during construction, the beam in Example 1 is lifted as shown in Figure 15.5(4). What are the new M_1 and M_2?

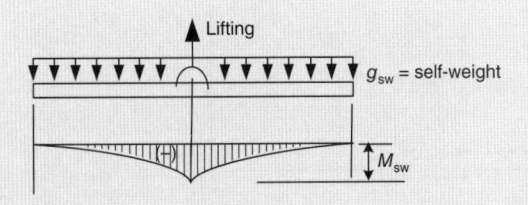

Figure 15.5(4) A beam being lifted for placement

Solution

Since the beam is lifted during the placement process at midspan, the minimum moment M_1 now becomes the negative moment caused by the lifting force. Or

$$M_1 = M_{sw} = -\frac{g_{sw}L^2}{8}$$

The maximum moment remains unchanged. Or

$$M_2 = \frac{gL^2}{8} + \frac{PL}{4}$$

15.6 CASE A AND CASE B PRESTRESSING

15.6.1 Fundamentals

It is mentioned in Section 15.5 that prestress can be used to neutralise the dead-load moment. More than that, the effects due to the minimum moment M_1, be it positive or negative, can also be neutralised in the same manner. Obviously, for a beam with a positive M_1, which produces compressive or positive stress in the top fibre, H should be so located to produce a tensile or negative stress at the same fibre. By the same token, H ought to be so located to produce a compressive stress at the top fibre to counter the tensile stress created by a negative M_1.

Case A prestressing is one that is required for a beam with a positive M_1; Case B prestressing is required for a negative M_1. These are discussed further in Section 15.6.2.

Due to prestress alone, the top-fibre stress (f_{cT}) in the beam section shown in Figure 15.6(1) is given as

$$f_{cT} = \frac{H}{A} - \frac{He_B y_T}{I} = \frac{H}{A}\left(1 - \frac{e_B y_T}{I/A}\right) \qquad \text{Equation 15.6(1)}$$

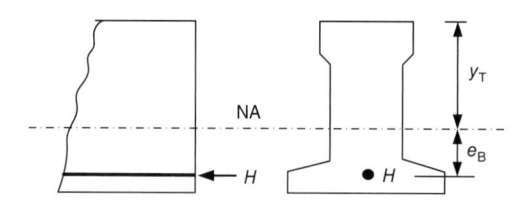

Figure 15.6(1) A typical prestressed beam section

But $\frac{I}{A} = k^2$ where $k = \sqrt{\frac{I}{A}}$ is the radius of gyration. Thus,

$$f_{cT} = \frac{H}{A}\left(1 - \frac{e_B y_T}{k^2}\right) = \frac{H}{A}\left(1 - \frac{e_B}{k^2/y_T}\right) \quad \text{Equation 15.6(2)}$$

In Equation 15.6(2), if $e_B > \frac{k^2}{y_T}$, f_{cT} is negative (i.e. tensile) – it requires Case A prestressing. Alternatively, if $e_B < \frac{k^2}{y_T}$, f_{cT} is positive (i.e. compressive) – it requires Case B prestressing.

Note that, for a rectangular beam with an overall depth D,

$$\frac{k^2}{y_T} = \frac{D}{6} \quad \text{Equation 15.6(3)}$$

which helps define the 'middle third' of a short column under uniaxial bending and the 'kern' under biaxial bending. It is well known that an axial load applied within the middle third or kern of a rectangular column produces no tensile stress anywhere within the column cross-section.

15.6.2 Applying Case A and Case B

Let's consider the case of a beam with positive M_1 and positive M_2. The top-fibre stress f_{cT} is compressive or positive under M_1 as shown in Figure 15.6(2). For an economical design, this positive f_{cT} due to M_1 should be neutralised by appropriate prestressing. Therefore, the f_{cT} to be produced by prestress alone should be negative. This requires Case A prestressing.

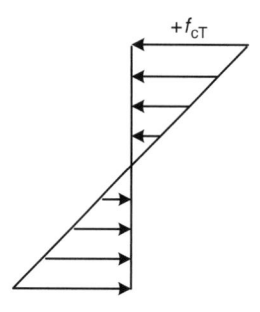

Figure 15.6(2) Stress distribution for a beam under positive M_1

For a beam with negative M_1 and positive M_2, the f_{cT} is tensile or negative under M_1 – see Figure 15.6(3). Again, to have an economical design, the f_{cT} produced by prestress alone should be positive. This requires Case B prestressing.

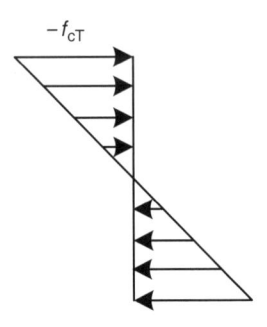

Figure 15.6(3) Stress distribution for a beam under negative M_1

15.7 CRITICAL STRESS STATE (CSS) EQUATIONS

We are ready to develop the CSS equations for analysis of beams with Case A and Case B prestressing based on the:
- assumptions in Section 15.1
- notation in Section 15.2, and incorporating the
  – effective prestress coefficient η (defined in Section 15.3)
  – permissible stresses c and c_t (defined in Section 15.4)
  – minimum (M_1) and maximum (M_2) moments (defined in Section 15.5).

15.7.1 Case A prestressing

Case A prestressing is required when $e_B > \dfrac{k^2}{y_T}$.

For a beam section **under M_1**, the top-fibre stress is

$$f_{cT} = \left(\frac{H}{A} - \frac{He_B y_T}{I}\right) + \frac{M_1 y_T}{I} \geq -c_t \qquad \text{Equation (A-1)}$$

and the bottom-fibre stress is

$$f_{cB} = \left(\frac{H}{A} + \frac{He_B y_B}{I}\right) - \frac{M_1 y_B}{I} \leq c \qquad \text{Equation (A-2)}$$

Equations A-1 and A-2 may be presented graphically as the solid lines in Figure 15.7(1). Note that there is a change of equation numbering system in this and some subsequent sections, where 'A' stands for Case A and 'B', Case B. This selective change is for ease of local reference and identification.

After the loss of prestress has occurred, the solid lines in Figure 15.7(1) will be replaced by the dotted lines. By comparing the solid and dotted lines, it is obvious that both f_{cT} and f_{cB} are lower due to the loss of prestress. In other words, the top- and bottom-fibre stresses are more critical before the loss has occurred. Therefore, Equations A-1 and A-2, which do not have η applied to H, are the CSS equations under M_1.

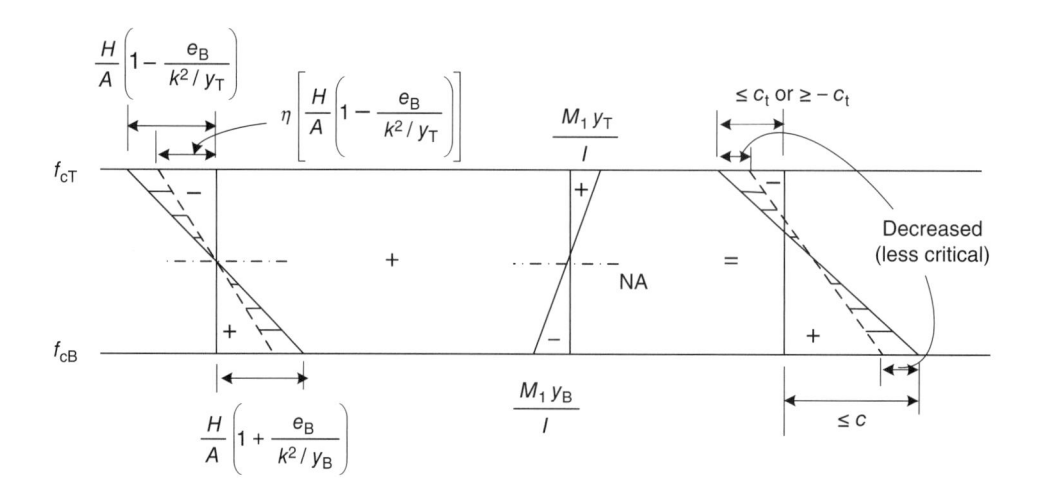

Figure 15.7(1) Graphical presentation of arguments for Equations A-1 and A-2

For the same beam section **under M_2**, the top- and bottom-fibre stresses before and after the loss of prestress has occurred, respectively, are represented by the solid and dotted lines in Figure 15.7(2).

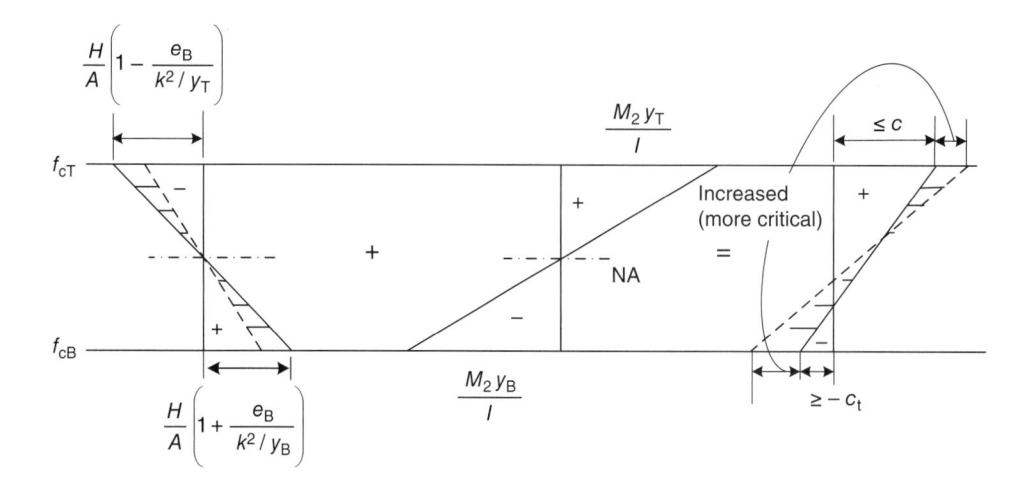

Figure 15.7(2) Graphical presentation of arguments for Equations A-3 and A-4

By assuming losses to have occurred, both f_{cT} and f_{cB} become more critical – see Figure 15.7(2). Thus, η needs to be incorporated, yielding the following equations:

$$\text{Top-fibre stress, } f_{cT} = \eta\left(\frac{H}{A} - \frac{H e_B y_T}{I}\right) + \frac{M_2 y_T}{I} \leq c \qquad \text{Equation (A-3)}$$

$$\text{Bottom-fibre stress, } f_{cB} = \eta\left(\frac{H}{A} + \frac{H e_B y_B}{I}\right) - \frac{M_2 y_B}{I} \geq -c_t \qquad \text{Equation (A-4)}$$

Equations A-3 and A-4 are the CSS equations for the beam under M_2.

15.7.2 Case B prestressing

Case B prestressing is required when $e_B < \dfrac{k^2}{y_T}$.

Using the same deductive process as used for Case A, and with reference to Figure 15.7(3), the CSS equations for Case B under M_1 are given below. Note that in these equations, η is only applied to f_{cT} and not to f_{cB}, because the bottom-fibre stress is less critical after the loss of prestress has occurred.

$$f_{cT} = \eta\left(\frac{H}{A} - \frac{H e_B y_T}{I}\right) + \frac{M_1 y_T}{I} \geq -c_t \qquad \text{Equation (B-1)}$$

$$f_{cB} = \left(\frac{H}{A} + \frac{H e_B y_B}{I}\right) - \frac{M_1 y_B}{I} \leq c \qquad \text{Equation (B-2)}$$

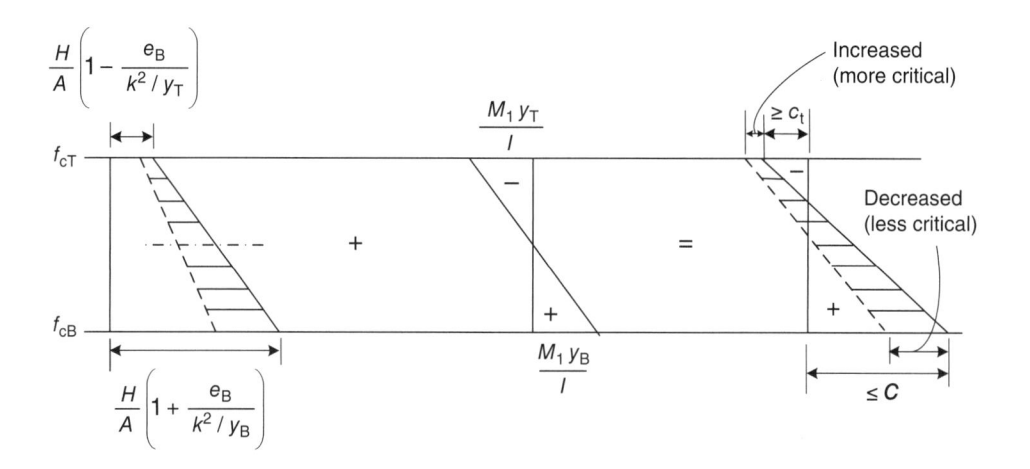

Figure 15.7(3) Graphical presentation of arguments for Equations B-1 and B-2

Finally, following the same argument with reference to Figure 15.7(4), the CSS equations for Case B prestressing **under M_2** for the top- (Equation B-3) and bottom-fibre stresses (Equation B-4) are:

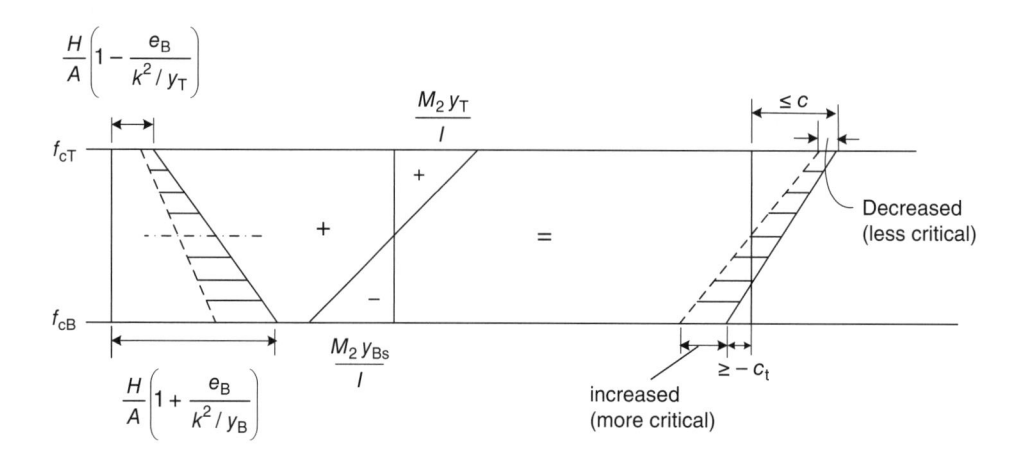

$$f_{cT} = \left(\frac{H}{A} - \frac{He_B y_T}{I}\right) + \frac{M_2 y_T}{I} \leq c \qquad \text{Equation (B-3)}$$

$$f_{cB} = \eta\left(\frac{H}{A} + \frac{He_B y_B}{I}\right) - \frac{M_2 y_B}{I} \geq -c_t \qquad \text{Equation (B-4)}$$

Figure 15.7(4) Graphical presentation of arguments for Equations B-3 and B-4

15.7.3 Summary of Case A and Case B equations

For ease of reference, the eight CSS equations for Case A and Case B prestressing may be collated as below.

Case A $\left(\text{for positive } M_1;\ e_B > \dfrac{k^2}{y_T}\right)$

Under M_1

$$f_{cT} = \left(\frac{H}{A} - \frac{He_B y_T}{I}\right) + \frac{M_1 y_T}{I} \geq -c_t \qquad \text{Equation (A-1)}$$

$$f_{cB} = \left(\frac{H}{A} + \frac{He_B y_B}{I}\right) - \frac{M_1 y_B}{I} \leq c \qquad \text{Equation (A-2)}$$

Under M_2

$$f_{cT} = \eta\left(\frac{H}{A} - \frac{He_B y_T}{I}\right) + \frac{M_2 y_T}{I} \leq c \qquad \text{Equation (A-3)}$$

$$f_{cB} = \eta\left(\frac{H}{A} + \frac{He_B y_B}{I}\right) - \frac{M_2 y_B}{I} \geq -c_t \qquad \text{Equation (A-4)}$$

Case B $\left(\text{for negative } M_1; e_B < \dfrac{k^2}{y_T}\right)$

Under M_1

$$f_{cT} = \eta\left(\frac{H}{A} - \frac{He_B y_T}{I}\right) + \frac{M_1 y_T}{I} \geq -c_t \qquad \text{Equation (B-1)}$$

$$f_{cB} = \left(\frac{H}{A} + \frac{He_B y_B}{I}\right) - \frac{M_1 y_B}{I} \leq c \qquad \text{Equation (B-2)/(A-2)}$$

Under M_2

$$f_{cT} = \left(\frac{H}{A} - \frac{He_B y_T}{I}\right) + \frac{M_2 y_T}{I} \leq c \qquad \text{Equation (B-3)}$$

$$f_{cB} = \eta\left(\frac{H}{A} + \frac{He_B y_B}{I}\right) - \frac{M_2 y_B}{I} \geq -c_t \qquad \text{Equation (B-4)/(A-4)}$$

Note that in the above list Equations B-2 and A-2 are identical, as are Equations B-4 and A-4. Thus, there are only six independent equations for the CSS analysis of all fully prestressed beams under service loads. In addition, no cracking or overstressing would occur throughout the life of the beam from the instant after the transfer of prestress. Therefore, prestressed beam analysis is merely a matter of suitably and correctly applying the six equations to provide the required solution for a given problem.

15.8 APPLICATION OF CSS EQUATIONS

To demonstrate the application of the CSS equations, an example is given below.

Problem

A post-tensioned concrete beam has a 500 mm deep section, which is prestressed symmetrically with respect to its centroid ($I = 20.68 \times 10^8$ mm^4 and $A = 72.5 \times 10^3$ mm^2). If the initial prestress force is $H = 667.5$ kN, determine the:

- maximum uniformly distributed live load that may be applied on a simply supported span of 15.25 m
- required eccentricity of the prestress force at midspan.

Assume $c = 17.25$ MPa, $c_t = 0$, $\eta = 0.85$ and $\rho_w = 24$ kN/m^3.

Solution

The beam, because it is simply supported and under normal dead and live loads (i.e. has positive M_1 and positive M_2), requires Case A prestressing.

Thus

$$M_1 = M_d; M_2 = M_d + M_1$$

and

$$y_T = y_B = y = 250 \text{ mm}$$

Then

Equation (A-1): $\dfrac{H}{A} - \dfrac{He_B y}{I} + \dfrac{M_d y}{I} \geq 0$ \hfill Equation (1)

Equation (A-2): $\dfrac{H}{A} + \dfrac{He_B y}{I} - \dfrac{M_d y}{I} \leq c$ \hfill Equation (2)

Equation (A-3): $\eta \left(\dfrac{H}{A} - \dfrac{He_B y}{I} \right) + \dfrac{(M_d + M_1)y}{I} \leq c$ \hfill Equation (3)

Equation (A-4): $\eta \left(\dfrac{H}{A} + \dfrac{He_B y}{I} \right) - \dfrac{(M_d + M_1)y}{I} \geq 0$ \hfill Equation (4)

The unknowns in the above four equations are e_B and M_1.
Equations (3) and (4) may be rewritten as (5) and (6), respectively.

$$M_1 \leq \dfrac{I}{y} \left[c - \dfrac{M_d y}{I} - \eta \left(\dfrac{H}{A} - \dfrac{He_B y}{I} \right) \right] \quad \text{Equation (5)}$$

$$M_1 \leq \dfrac{I}{y} \left[0 - \dfrac{M_d y}{I} + \eta \left(\dfrac{H}{A} + \dfrac{He_B y}{I} \right) \right] \quad \text{Equation (6)}$$

Then

Equation (1): $e_B \leq \dfrac{I}{Hy} \left[\dfrac{H}{A} + \dfrac{M_d y}{I} \right]$ \hfill Equation (7)

Equation (2): $e_B \leq \dfrac{I}{Hy} \left[c - \dfrac{H}{A} + \dfrac{M_d y}{I} \right]$ \hfill Equation (8)

But

$$M_d = \dfrac{wL^2}{8} = (24 \times 10^{-6} \times 72.5 \times 10^3) \times \dfrac{(15.25 \times 10^3)^2}{8}$$
$$= 50.582 \times 10^6 \text{ Nmm.}$$

From Equation (7), we have

$$e_B \leq \frac{20.68 \times 10^8}{667\,500 \times 250} \left[\frac{667\,500}{72\,500} + \frac{50.582 \times 10^6 \times 250}{20.68 \times 10^8}\right] \leq 12.393(9.207 + 6.115)$$
$$\leq 189.9 \text{ mm}$$

From Equation (8), we have

$$e_B \leq 12.393\,(17.25 - 9.207 + 6.115) \leq 175.5 \text{ mm}$$

Thus, we have for the required eccentricity, $e_B = 175.5$ mm.
Then

Equation (5): $M_1 \leq \dfrac{20.68 \times 10^8}{250}$

$$\left[17.25 - 6.115 - 0.85 \times 9.207 + 0.85 \times \frac{667\,500 \times 175.5 \times 250}{20.68 \times 10^8}\right]$$
$$\leq 8.272 \times 10^6 (17.25 - 6.115 - 7.826 + 12.038)$$
$$\leq 126.92 \text{ kNm}$$

Similarly,

Equation (6): $M_1 \leq 8.272 \times 10^6 (-6.115 + 7.826 + 12.038)$
$$\leq 113.73 \text{ kNm}$$

Thus, we have $M_1 \leq 113.73$ kNm.

Since $M_1 = \dfrac{w_1 L^2}{8}$

we have for the maximum live load $w_1 = \dfrac{M_1 8}{L^2} = \dfrac{8 \times 113.73}{15.25^2} = 3.91$ kN/m

15.9 SUMMARY

Based on the fundamentals of critical stress state method and by incorporating the Standard recommendations on permissible stresses, the formulas for analysing prestressed concrete beams for Case A and Case B prestressing are developed. Applications of these formulas are illustrated using fully worked examples.

15.10 PROBLEMS

1. Rederive Equation 15.3(3), assuming that the deflected shape in Figure 15.3(4) takes the form of a parabola.
2. A rectangular concrete beam ($b = 250$ mm and $D = 500$ mm) is prestressed with thirty 5.08 mm-diameter high-tensile wires, initially to 1200 MPa at an eccentricity of 175 mm. Assuming $\eta = 0.80$, $c = 15$ MPa and $c_t = 1.5$ MPa, determine the maximum live-load bending moment that the beam is capable of carrying. Also determine the simple span over which the beam must be supported if this live-load moment is to be carried. Use $\rho_w = 24$ kN/m^3.
3. A simply supported rectangular beam with a span of 7 m is 100 mm wide and 250 mm deep (overall). The eccentricity of the prestressing force at midspan is 70 mm. Considering only the midspan section, calculate:
   (a) the minimum required prestressing force H
   (b) the maximum uniformly distributed superimposed load that the beam can carry.
   Assume $c = 15$ MPa, $c_t = 0$, $\eta = 0.85$ and $\rho_w = 24$ kN/m^3.
4. A simply supported prestressed concrete floor beam with a 6.1 m span is prestressed by 65 mm^2 of tendons 25 mm from the top, and 80 mm^2 of tendons 25 mm from the bottom of an I-beam section. The I-beam section is 200 mm × 200 mm overall with web and flanges each 50 mm thick. The ultimate steel strength $f_p = 2400$ MPa and the initial stress in the wires (after elastic losses) is $0.6 f_p$. Take $c = 14$ MPa, $c_t = 1$ MPa and $\eta = 0.80$.

    Determine the superimposed uniform service load that the beam can carry.
5. A simple beam with overhang under two (stationary) loading cases (LCs) – LC1 and LC2 – is shown in Figure 15.10(1) with its cross-section. LC1 and LC2 can act simultaneously or independently in service.

    Compute the minimum moment M_1 and maximum moment M_2 (as defined in the CSS approach) that the beam will experience at each of sections B and C.

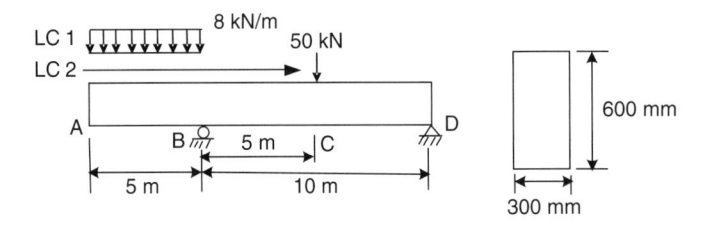

Figure 15.10(1) Details of a simply supported beam with overhang

6. A simply supported prestressed concrete floor beam with a 10 m span is stressed by 1100 mm² of tendons 120 mm from the bottom fibre and 800 mm² of tendons 100 mm from the top fibre of a rectangular section as shown in Figure 15.10(2). The ultimate steel strength $f_p = 1770$ MPa and the initial stress in the wires is $0.6f_p$. Take $c = 14$ MPa, $c_t = 1$ MPa and $\eta = 0.80$.

   Determine the (working) superimposed uniform load that the beam can carry.

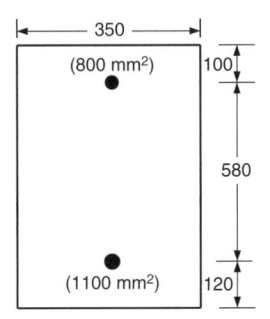

Figure 15.10(2) Details of a simply supported rectangular prestressed concrete beam

Note: all dimensions are in mm.

7. A simply supported prestressed concrete floor beam with a 12 m span is stressed by 1200 mm² and 900 mm² of tendons 100 mm respectively from the bottom and top fibres of an I-beam section, as illustrated in Figure 15.10(3). The ultimate steel strength $f_p = 1770$ MPa and the initial stress in the tendons is $0.6f_p$. Take $c = 16$ MPa, $c_t = 1$ MPa, $\eta = 0.85$ and $\rho_w = 24$ kN/m³.

   Using the critical stress state (CSS) approach, determine the (working) superimposed uniform load that the beam can carry.

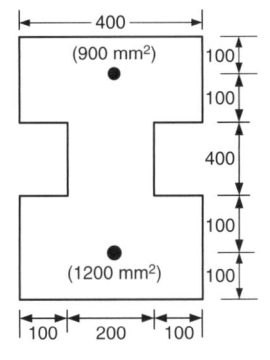

Figure 15.10(3) Details of a simply supported prestressed concrete I-beam

Note: all dimensions are in mm.

8. A simply supported rectangular prestressed concrete beam with a 12 m span is shown in Figure 15.10(4). It is stressed by 1200 mm² and 900 mm² of tendons, 100 mm respectively from the bottom and top fibres of the section. The ultimate steel strength f_p = 1770 MPa and the initial stress in the wires is $0.6f_p$. Assume c = 16 MPa, c_t = 1 MPa, η = 0.85 and ρ_w = 24 kN/m³.

    Using the critical stress state (CSS) approach, determine the (working) superimposed uniform load that the beam can carry.

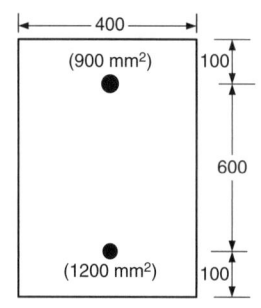

Figure 15.10(4) Details of a simply supported prestressed concrete beam

Note: all dimensions are in mm.

9. A cantilever prestressed concrete beam with a 5 m span is stressed by 1100 mm² of tendons 120 mm from the top fibre and 800 mm² of tendons 100 mm from the bottom fibre of a precast section. The dimensional details of the rectangular beam are as shown in Figure 15.10(5). The ultimate steel strength f_p = 1770 MPa and the initial stress in the wires is $0.6f_p$ Assuming c = 14 MPa, c_t = 1 MPa and η = 0.80, determine the (working) superimposed uniform load which the beam can carry.

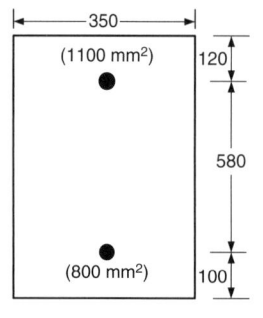

Figure 15.10(5) Details of a cantilever prestressed concrete beam

Note: all dimensions are in mm.

CRITICAL STRESS STATE DESIGN OF BEAMS 16

16.1 INTRODUCTION

In the bending design of prestressed members in general, and of beams in particular, the process below should be followed:

- Critical stress state (CSS) criteria must be satisfied at all stages of the life of the beam (i.e. at transfer, during handling, during construction, under service load conditions and after losses have occurred).
- If applicable, shear and torsion must be designed for and the CSS design modified if necessary (see Clauses 8.2–8.4 of AS 3600–2009 [the Standard]).
- The design must be checked for adequacy under ultimate load conditions (see Clause 8.1 of the Standard).
- End-block stresses must be estimated, and reinforcement provided (Clause 12.5 of the Standard).
- Deflections must be assessed and kept within acceptable limits (see Clause 8.5 of the Standard).

There are other general design requirements that have to be met, including durability (Section 4 of the Standard), fire resistance (Section 5), material properties (Section 3), and other serviceability considerations including crack control (for partially prestressed beams only) and vibration (Clause 9.5). For cracked partially prestressed beams, Equations 5.7(1) and (2) respectively may be used to estimate the average and maximum crack widths.

This chapter mainly presents the CSS approach to bending design or how the first item in the above bulleted list is satisfied. Chapter 17 covers, in some detail, the ultimate strength check for fully and partially prestressed beams (the third bullet point above). Chapter 18 presents the end-block design for prestressing anchorages. The reader is referred to the Standard for details of the other design considerations listed above. It is worth noting that design topics such as shear, torsion, durability, material properties and crack control for reinforced beams have been discussed in detail in Part 1. The design procedures are similar, if not identical, for prestressed beams. Equations for shear and torsion design of reinforced beams, as well as deflection calculations, are truncated from the unified formulas given in the Standard for reinforced and prestressed beams. As durability and fire resistance requirements affect mainly detailing including concrete cover for steel, the Standard provisions are the same for the two types of beams.

16.2 FORMULAS AND PROCEDURES – CASE A

The four CSS equations for Case A analysis from Section 15.7.3 are again shown below:

$$\frac{H}{A} - \frac{H e_B y_T}{I} + \frac{M_1 y_T}{I} \geq -c_t \qquad \text{Equation (A-1)}$$

$$\frac{H}{A} + \frac{H e_B y_B}{I} - \frac{M_1 y_B}{I} \leq c \qquad \text{Equation (A-2)}$$

$$\eta \left(\frac{H}{A} - \frac{H e_B y_T}{I} \right) + \frac{M_2 y_T}{I} \leq c \qquad \text{Equation (A-3)}$$

$$\eta \left(\frac{H}{A} + \frac{H e_B y_B}{I} \right) - \frac{M_2 y_B}{I} \geq -c_t \qquad \text{Equation (A-4)}$$

The main design objectives are to obtain the required cross-sectional properties and prestress details of the beam. Cross-sectional properties to be determined are the two elastic section moduli Z_T and Z_B, with respect to the top and bottom stresses. The minimum prestressing force (H) and the maximum eccentricity (e_B) constitute the required prestress details. Section 16.2.1 discusses the procedure for determining the section moduli, and Section 16.2.2 presents the use of Magnel's plot to obtain the minimum H and the associated eccentricity e_B. The design steps for Case A beams are given in Section 16.2.3.

16.2.1 Elastic section moduli

In Equations A-1 to A-4, I/y_T and I/y_B are Z_T and Z_B, respectively. Hence, the objective now is to work on the four inequalities – that is, solving Equations A-1 and A-3 for Z_T, and Equations A-2 and A-4 for Z_B.

The process to obtain Z_T and Z_B is shown below.

$$\text{(A-1)} \times (-\eta): \quad -\eta \left(\frac{H}{A} - \frac{H e_B y_T}{I} \right) - \eta \frac{M_1 y_T}{I} \leq \eta c_t$$

$$\text{(A-3)}: \quad \eta \left(\frac{H}{A} - \frac{H e_B y_T}{I} \right) + \frac{M_2 y_T}{I} \leq c$$

Combining the two above equations gives

$$(M_2 - \eta M_1) \frac{y_T}{I} \leq c + \eta c_t$$

from which the section modulus

$$\frac{I}{y_T} = Z_T \geq \frac{M_2 - \eta M_1}{c + \eta c_t}$$

Equation (A1.3)

Similarly, (A-2) × η − (A-4) leads to

$$\frac{I}{y_B} = Z_B \geq \frac{M_2 - \eta M_1}{\eta c + c_t}$$

Equation (A2.4)

16.2.2 Magnel's plot for Case A

If we choose a section using Equations (A1.3) and (A2.4), it will lead to a solution for H and e_B, which will satisfy Equations (A-1), (A-2), (A-3) and (A-4). The solution, however, may not be a practical one. For example, e_B may be such that the centre of gravity of the prestressing force or the location of H is required to be outside of the section!

How do we determine the minimum value of H and the corresponding value of e_B? Magnel (1954) considered this question and came up with the following elegant yet simple approach that started by plotting e_B against $1/H$.

From Equation (A-1), we can write

$$\left(1 - \frac{e_B y_T}{k^2}\right) \geq \left(-c_t - \frac{M_1 y_T}{I}\right) \frac{A}{H}$$

Or

$$e_B \leq \frac{k^2}{y_T} + \frac{1}{H}\left(M_1 + \frac{c_t I}{y_T}\right)$$

Equation (1)

Similarly, from Equation (A-2), we have

$$e_B \leq -\frac{k^2}{y_B} + \frac{1}{H}\left(M_1 + \frac{cI}{y_B}\right)$$

Equation (2)

from Equation (A-3)

$$e_B \geq \frac{k^2}{y_T} + \frac{1}{H}\left(M_2 - \frac{cI}{y_T}\right)\frac{1}{\eta}$$

Equation (3)

and finally from Equation (A-4),

$$e_B \geq -\frac{k^2}{y_B} + \frac{1}{H}\left(M_2 - \frac{c_t I}{y_B}\right)\frac{1}{\eta}$$

Equation (4)

Plotting e_B against $1/H$ based on Equations (1) to (4) above leads to the four lines as drawn in Figure 16.2(1). The diagram is referred to as a Magnel plot.

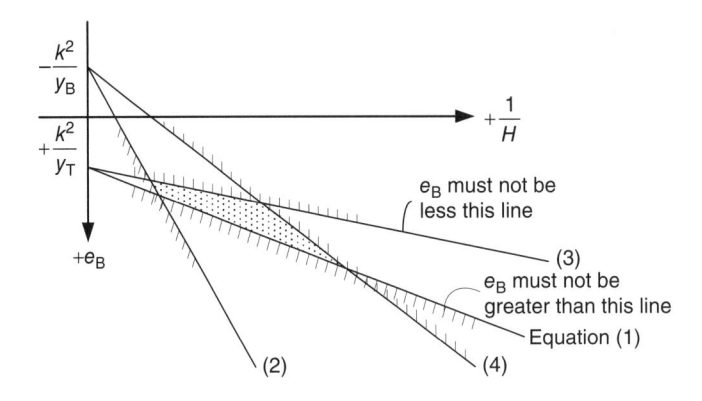

Figure 16.2(1) Magnel's plot for Case A beams

The significance and interpretation of the Magnel plot are:
- The four lines enclose a safe zone within which a design satisfies the inequalities stipulated in Equations A-1 to A-4.
- An ideal design would be for the four lines to meet at one point (i.e. the shaded area converges to a single point).
- To determine the minimum H (i.e. maximum $1/H$), solve Equations (1) and (4) for H and e_B.

16.2.3 Design steps

Using the interplay among the four CSS inequalities, as revealed in the Magnel plot, the steps to be followed in the design of a Case A beam may be summarised as follows:
1. Assume a beam section and calculate M_1 and M_2.
2. Obtain the required section using Equations (A1.3) and (A2.4) from Section 16.2.1.
3. If the required section is the same as the assumed one (with an appropriate tolerance), accept the section. Otherwise, repeat the process until they are the same, and then calculate M_1 and M_2 of the accepted section.
4. Solve Equations (1) and (4) from Section 16.2.2 as identities for the maximum $1/H$ (i.e. minimum H) and maximum e_B. If e_B permits the required concrete cover for the prestressing tendons, then this solution should be acceptable. However, if e_B does not permit the required concrete cover, then
   (a) choose an alternative section providing greater depth and solve Equations (1) and (4) as identities again or
   (b) keep the section, reduce e_B to the maximum allowed by the section and solve Equation (4) for the minimum H. Also, check Equation (3) to ensure that the lower limit of e_B is not violated as required in Magnel's plot (Figure 16.2(1)).

16.3 FORMULAS AND PROCEDURES – CASE B

16.3.1 Elastic section moduli

The analysis formulas for Case B are again shown below (from Section 15.7.3).

$$\eta\left(\frac{H}{A} - \frac{H e_B y_T}{I}\right) + \frac{M_1 y_T}{I} \geq -c_t \quad \text{Equation (B-1)}$$

$$\frac{H}{A} + \frac{H e_B y_B}{I} - \frac{M_1 y_B}{I} \leq c \quad \text{Equation (B-2)/(A-2)}$$

$$\left(\frac{H}{A} - \frac{H e_B y_T}{I}\right) + \frac{M_2 y_T}{I} \leq c \quad \text{Equation (B-3)}$$

$$\eta\left(\frac{H}{A} + \frac{H e_B y_B}{I}\right) - \frac{M_2 y_B}{I} \geq -c_t \quad \text{Equation (B-4)/(A-4)}$$

Parallel to the procedure used in Section 16.2 for Case A design, the Z_T and Z_B for a Case B beam can be obtained by solving Equations B-1 and B-3, and by Equations B-2 and B-4 (or more exactly A-2 and A-4), respectively.

The processes are demonstrated below.

$$(B\text{-}3) \times (-\eta): \quad -\eta\left(\frac{H}{A} - \frac{H e_B y_T}{I}\right) - \eta\frac{M_2 y_T}{I} \geq -\eta c$$

$$(B\text{-}1): \quad \eta\left(\frac{H}{A} - \frac{H e_B y_T}{I}\right) + \frac{M_1 y_T}{I} \geq -c_t$$

Combining the two above equations yields

$$(M_1 - \eta M_2)\frac{y_T}{I} \geq -(\eta c + c_t)$$

From this inequality, we obtain the section modulus

$$\frac{I}{y_T} = Z_T \geq \frac{(\eta M_2 - M_1)}{\eta c + c_t} \quad \text{Equation (B1.3)}$$

Similarly, $\eta \times$ (B-2) – (B-4), leads to

$$\frac{I}{y_B} = Z_B \geq \frac{M_2 - \eta M_1}{\eta c + c_t} \quad \text{Equation (B2.4) = (A2.4)}$$

16.3.2 Magnel's plot for Case B

The Magnel plot for Case B is presented in Figure 16.3(1), which is constructed using a procedure parallel to that given in Section 16.2.2. The significance, interpretation and application of Case B mirror those of Case A.

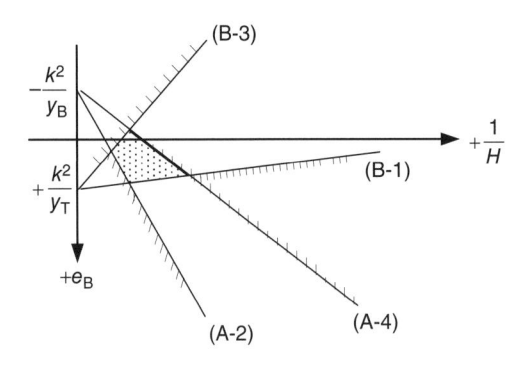

Figure 16.3(1) Magnel's plot for Case B beams

16.3.3 Design steps

Recognising the interplay among the four CSS inequalities, as revealed in the Magnel plot given in Figure 16.3(1), the steps to follow to design a Case B beam are:

1. Choose a beam section that satisfies Equations (B1.3) and (B2.4) from Section 16.3.1. A trial and error process may be needed to reach the required section.
2. Solve Equations (B-l) and (B-4) as identities for H and e_B. If the value of e_B is not acceptable, either
   (a) choose an alternative section
   or
   (b) retain the section and adopt a maximum e_B – then calculate H from Equation (B-4) and check to ensure that Equation (B-3) is not violated.

16.4 DESIGN EXAMPLES

Three numerical examples are given in Sections 16.4.1–16.4.3 to demonstrate the application of the CSS design procedures for Case A and Case B beams.

16.4.1 Simply supported beam

Problem

A rectangular beam, 300 mm wide, has a simply supported span of 21.35 m. The live-load moment at the midspan section is 335 kNm. Dead load is acting once prestress is applied.

Assume $c = 10.5$ MPa, $c_t = 0$ and $\eta = 0.85$, and the initial stress in the prestressing wire immediately after transfer is 825 MPa. For concrete $\rho_w = 24$ kN/m³, use 5.08 mm diameter tendons. Design a suitable midspan section.

Solution

The sectional properties as given in Figure 16.4(1) is $y_T = y_B = y$.

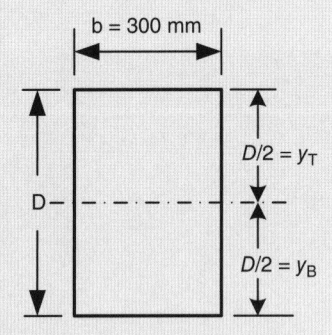

Figure 16.4(1) Cross-sectional details of the example beam

$$M_1 = M_{d.L} (+)$$
$$M_2 = M_{d.L} + M_{L.L} (+)$$

Therefore, Case A applies.

Equation (A1.3): $\dfrac{I}{y_T} \geq \dfrac{M_2 - \eta M_1}{c + \eta c_t}$

Equation (A2.4): $\dfrac{I}{y_B} \geq \dfrac{M_2 - \eta M_1}{\eta c + c_t}$

The latter is the governing equation. In our case, $c_t = 0$ and

$$\frac{I}{y} = \frac{bD^2}{6} \geq \frac{M_{L.L} + M_{d.L} - \eta M_{d.L}}{\eta c}$$

That is,

$$\frac{bD^2}{6} \geq \frac{335 \times 10^6 + 0.15 \times M_{d.L}}{8.925} \qquad \text{Equation (i)}$$

It is obvious from **Equation (i)** that prestressed concrete beams are more efficient in carrying dead weight as the section is required to carry only 0.15 or $(1 - \eta)$ times the dead-load moment. Therefore, they can have a longer unsupported span than reinforced concrete beams.

Trial 1
Assume $D = 900$ mm. We then have

$$w_d = \frac{300 \times 900}{10^6} \times 24 = 6.48 \text{ kN/m}$$

That is, $M_{d.L} = \dfrac{w_d L^2}{8} = \dfrac{6.48 \times 21.35^2}{8} = 369.2 \text{ kNm}$

Equation (i): $\dfrac{300 D^2}{6} \geq \dfrac{(355 + 0.15 \times 369.2) \times 10^6}{8.925} \geq 43.74 \times 10^6$

That is, $D \geq \sqrt{\dfrac{6 \times 43.74 \times 10^6}{300}}$ or $D \geq 935$ mm

This is not acceptable.

Trial 2
Assume $D = 950$ mm and we have $w_d = 6.84$ kN/m (i.e. $M_{d.L} = 389.7$ kNm) and

Equation (i) yields $D \geq 939$ mm

This is acceptable, and we can accept $D = 950$ mm.

We can now obtain the minimum H and the corresponding (maximum) e_B by solving Equations A-1 and A-4 as identities:

Equation (A-I): $\dfrac{H}{A} - \dfrac{He_B y}{I} + \dfrac{M_{d.L} y}{I} = 0$

Equation (A-4): $\eta\left(\dfrac{H}{A} + \dfrac{He_B y}{I}\right) - \dfrac{(M_{d.L} + M_{L.L}) y}{I} = 0$

where

$$\frac{y}{I} = \frac{6}{bD^2} = \frac{6}{0.3 \times 0.95^2} = 22.16 \text{ m}^{-3}$$

$$M_{d.L} \times \frac{y}{I} = 389.7 \times 22.16 = 8635.8 \text{ kN/m}^2$$

$$(M_{d.L} + M_{L.L}) \times \frac{y}{I} = (389.7 + 335) \times 22.16 = 16\,059.4 \text{ kN/m}^2$$

and

$$(M_{d.L} + M_{L.L}) \times \frac{y}{\eta I} = 18\,893.4 \text{ kN/m}^2$$

Thus

Equation (A-1) becomes $\dfrac{H}{A} - \dfrac{He_B y}{I} + 8635.8 = 0$ .... Equation (ii)

Equation (A-4) becomes $\dfrac{H}{A} + \dfrac{He_B y}{I} - 18\,893.4 = 0$ .... Equation (iii)

Combining Equations (ii) and (iii) yields

$$\frac{2H}{A} = 10\,257.6$$

or

$$H/A = 5128.8 \text{ kN/m}^2$$

From Equation (ii), we have

$$5128.8 - (5128.8 \times 0.3 \times 0.95) \times 22.16 e_B + 8635.8 = 0$$

from which

$$e_B = \frac{5128.8 + 8635.8}{5128.8 \times 0.3 \times 0.95 \times 22.16} = 0.425 \text{ m}$$

and $H = 1462$ kN.

Therefore, the steel area $A_s = \dfrac{H}{f_s} = \dfrac{1462 \times 10^3}{825} = 1772 \text{ mm}^2$

The number of 5.08 mm diameter wires $= \dfrac{1772}{20.25} = 88$

Now check the acceptability of the section as shown in Figure 16.4(2).

It is obvious that with 50 mm cover, the section is inadequate to accommodate 88 wires.

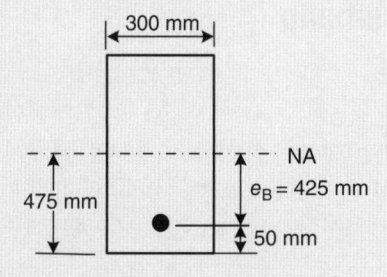

Figure 16.4(2) Location of prestressing steel for the calculated e_B

Thus, we either (1) choose an alternative section, or (2) keep the section and revise e_B and H. If we follow procedure (2) – since the section is relatively deep – this means we accept $b \times D = 300$ mm $\times 950$ mm.

After consideration for minimum limits for covers and spacings, it was decided that the maximum permissible $e_B = 0.35$ m. Then

Equation (iii): $H\left(\dfrac{1}{0.3 \times 0.95} + 0.35 \times 22.16\right) - 18\,893.4 = 0$

or

$$H = \dfrac{18\,893.4}{3.51 + 7.76} = 1677 \text{ kN}$$

We should now check Equation (A-3) to ensure that e_B is within the minimum limit

$$\eta\left(\dfrac{H}{A} - \dfrac{He_B y}{I}\right) + \dfrac{(M_{d.L} + M_{L.L})y}{I} \leq 10.5 \times 10^3 \text{ kN/m}^2$$

which gives

$$0.85\left(\dfrac{1677}{0.3 \times 0.95} - 1677 \times 0.35 \times 22.16\right) + 16\,059.4 = 10\,006 < 10\,500 \text{ kN/m}^2$$

Therefore, this is acceptable.

For

$$A_s = 1677 \times 10^3/825 = 2033 \text{ mm}^2$$

the number of 5.08 mm diameter wires = 2033/20.25 = 100.4. Therefore, the required number of wires is 100.

16.4.2 Simple beam with overhang

Problem

Design a simply supported beam with overhang as shown in Figure 16.4(3). The beam is 12 m long, including a cantilever span of 3 m, and carries a concentrated moving load of 100 kN. Assume $\rho_w = 24$ kN/m^3, $c = 14$ MPa, $c_t = 0$, $b \approx D/2$ and $\eta = 0.80$ for the design.

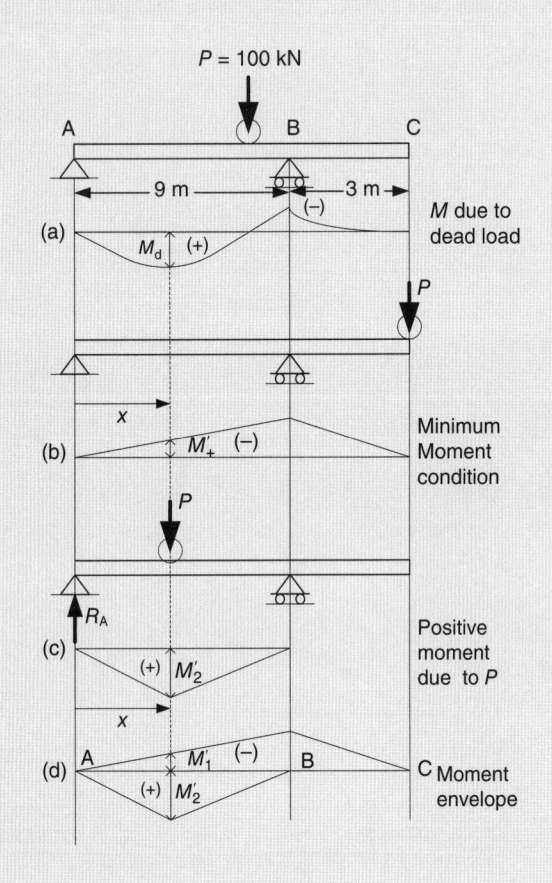

CHAPTER 16 CRITICAL STRESS STATE DESIGN OF BEAMS **447**

Figure 16.4(3) Loading configuration and moment diagrams for the example beam

Solution

The design process involves 10 steps as elaborated below.

1. In Equations (A1.3), (A2.4) and (B1.3), I/y depends on $(M_2 - \eta M_1)$ or $(\eta M_2 - M_1)$. Consider $\eta \approx 1$; then $I/y \propto (M_2 - M_1)$.
2. For a section at a distance x from A, we have from Figures 16.4(3)a and d,

$$M_1 = M_d - M'_1$$
$$M_2 = M_d + M'_2$$

$M_2 - M_1 = M'_1 + M'_2$, which is the live-load moment variation – see Figure 16.4(3)d.

3. For a beam with uniform cross-section, I/y should be chosen for the maximum value of $(M_2 - M_1)$. At the section x from the left support,

$$M'_1 = \frac{x}{9}(3P)$$

$$M'_2 = R_A x = \left(\frac{9-x}{9}\right)Px$$

Therefore

$$\overline{M} = M_2 - M_1 = M'_1 + M'_2 = P\left[\frac{x}{3} + \frac{x(9-x)}{9}\right]$$

To maximise the function, let

$$\frac{d\overline{M}}{dx} = P\left(\frac{1}{3} + 1 - \frac{2x}{9}\right) = 0$$

from which $x = 6$ m.

4. At $x = 6$ m

$$\overline{M} = \overline{M}_{\max} = P\left(\frac{6}{3} + \frac{3}{9} \times 6\right) = 4P$$

$$M_1 = M_d - \frac{6P}{3} = (M_d - 200) \text{ kNm} \qquad \text{Equation (i)}$$

$$M_2 = M_d + \frac{(9-6) \times 6}{9} P = (M_d + 200) \text{ kNm} \qquad \text{Equation (ii)}$$

5. Trial 1
Assume $b \times D = 300$ mm $\times$ 600 mm.
This gives $w_d = 4.32$ kN/m.

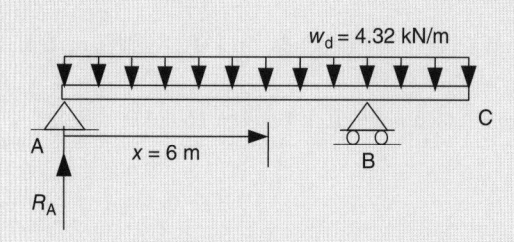

Figure 16.4(4) First trial beam subjected to assumed self-weight

With reference to Figure 16.4(4), we have

$$R_A = (4.32 \times 12 \times 3)/9 = 17.28 \text{ kN}$$

and

$$M_d = 17.28 \times 6 - (4.32 \times 6^2)/2 = 25.92 \text{ kNm}$$

Then

Equation (i): $M_1 = -174.08$ kNm

Equation (ii): $M_2 = 225.92$ kNm

Thus, we have Case B prestressing.

Equation (B1.3): $\dfrac{300D^2}{6} \geq \dfrac{\eta M_2 - M_1}{\eta c}$

from which $D \geq 796$ mm. This is not acceptable.

6. Further trials

After some more trials, the final results were found to be

$$b \times D = 380 \text{ mm} \times 760 \text{ mm}$$

$$H = 2100 \text{ kN and } e_B = 13.7 \text{ mm}$$

Note that, for the final trial, w_d = 6.93 kN/m, M_1 = –158.42 kNm and M_2 = 241.58 kNm.

7. Check stresses in the section at x = 6 m.

For the final selected section, we have

$$\frac{H}{A} = \frac{2100 \times 10^3}{380 \times 760} = 7.271$$

$$\frac{I}{y} = \frac{380 \times 760^2}{6} = 36.58 \times 10^6 \text{ mm}^3$$

$$\frac{He_B y}{I} = 0.786$$

$$\frac{M_1 y}{I} = -4.331 \text{ MPa}$$

$$\frac{M_2 y}{I} = 6.604 \text{ MPa}$$

Then

Equation (B-1): $0.8\,(7.271 - 0.786) - 4.331 = 2.115\ (> 0)$

Equation (B-2): $7.271 + 0.786 + 4.331 = 12.39\ (< 14 \text{ MPa})$

Equation (B-3): $7.271 - 0.786 + 6.604 = 13.09\ (< 14 \text{ MPa})$

Equation (B-4): $0.8(7.271 + 0.786) - 6.604 = -0.158\ (\approx 0)$

Therefore, the design is acceptable. Now we must determine the limiting zone within which the post-tensioned cable must be located.

8. The limiting values of e_B can be determined as follows.
   At $x = 6$ m – see Figure 16.4(4) – we have from step 6

   $$e_B = +13.7 \text{ mm}$$

   At sections A and C, $M_1 = M_2 = 0$.
   Then Equation (A-l) becomes

   $$\frac{H}{A} - \frac{He_B y_T}{I} \geq -c_t (\text{i.e.} \geq 0)$$

   from which

   $$e_B = +126.7 \text{ mm}$$

   Similarly, from Equation (A-4)

   $$e_B = -126.7 \text{ mm}$$

9. At section B, we need to know the moments for determining the limiting values of e_B.
   The dead- and live-load moments at section B are

   $$M_d = -\frac{W_d \times 3^2}{2} = -\frac{6.93 \times 9}{2} = -31.19 \text{ kNm}$$

   and

   $$M_l = -P \times 3 = -100 \times 3 = -300 \text{ kNm}$$

   from which the minimum and maximum moments are computed as

   minimum: $M_1 = -300 - 31.19 = -331.19$ kNm

   maximum: $M_2 = -31.19$ kNm

   Or, by 'inverting' the section (as discussed in Section 15.5 – Example 2), we have

   $$M_1 = 31.19 \text{ kNm } (> 0)$$

   and

   $$M_2 = 331.19 \text{ kNm}$$

   Now, Case A prestressing applies.
   We then have

   $$\frac{M_1 y}{I} = 0.853 \text{ MPa} \quad \text{and} \quad \frac{M_2 y}{I} = 9.054 \text{ MPa}$$

Then

Equation (A-I): $7.271 - 0.0574\,e_B + 0.853 \geq 0$ (i.e. $e_B \leq 142$ mm)

Equation (A-2): $7.271 + 0.0574\,e_B - 0.853 \leq 14$ (i.e. $e_B \leq 132$ mm)

Equation (A-3): $0.8(7.271 - 0.0574e_B) + 9.054 \leq 14$ (i.e. $e_B \geq 19$ mm)

Equation (A-4): $0.8(7.271 + 0.0574e_B) - 9.054 > 0$ (i.e. $e_B \geq 70$ mm)

Thus, $70 \leq e_B \leq 132$ mm.

10. The limiting zone for the post-tensioned cable can be drawn based on the vertical limits for e_B obtained in steps 8 and 9. This is given in Figure 16.4(5).
If more than one cable is used, they must be symmetrical with respect to the centre line of the limiting zone.

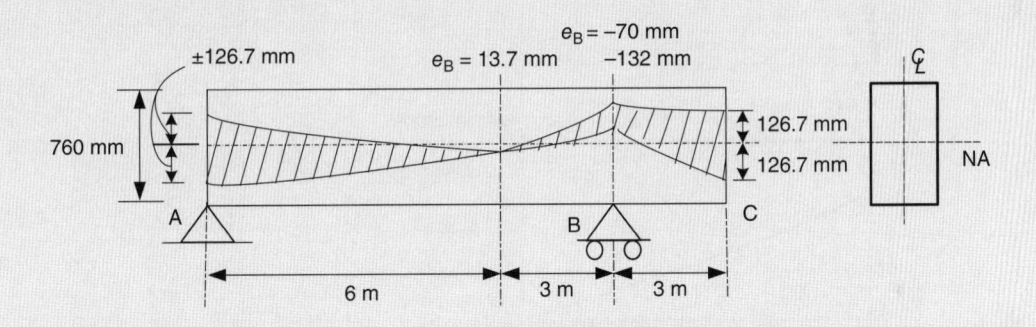

Figure 16.4(5) Layout of the limiting zone for the post-tensioned cable

16.4.3 Cantilever beam

Problem

Determine the minimum prestressing force and the cable limiting zone of a post-tensioned cantilever beam, 5 m long, to carry in addition to its own weight a concentrated moving load of 250 kN as illustrated in Figure 16.4(6). Take $\rho_w = 24$ kN/m^3, $c = 20$ MPa, $c_t = 2$ MPa and $\eta = 0.80$. For the design, select an appropriate Australian Standard (AS) I-girder section as shown in Figure C(1) in Appendix C.

452 PART 2 PRESTRESSED CONCRETE

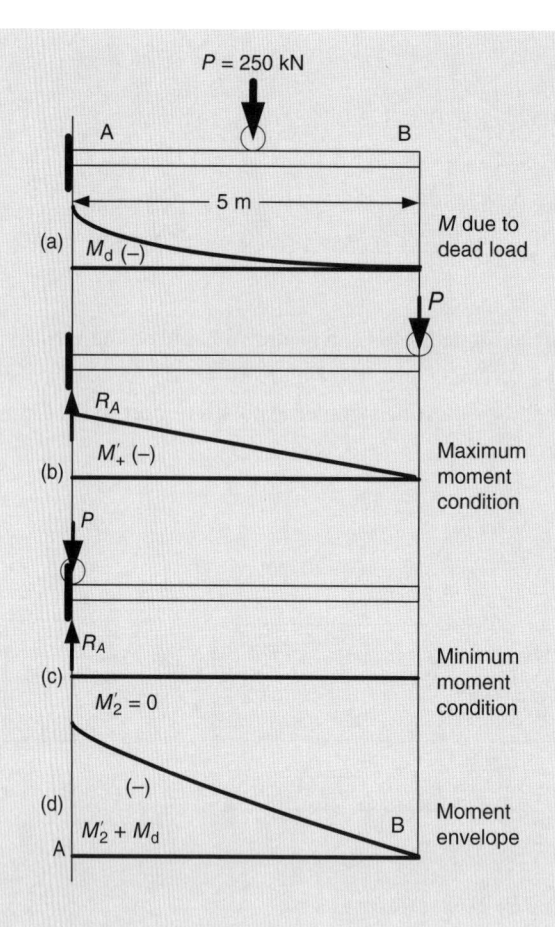

Figure 16.4(6) Example post-tensioned cantilever beam
Notes: section properties: A_g = 317 000 mm^2, Z_T = 82 900 000 mm^3, Z_B = 91 100 000 mm^3,
I = 49 900 000 000 mm^4, y_T = 602 mm, y_B = 548 mm

Solution

1. By 'inverting' the section, we have

   $$M_1 = M_d + M_2' > 0;$$
   $$M_2 = M_d + M_1'$$

   Therefore, this is a Case A prestressing.

   $$M_1' = P \times L = 250 \times 5 = 1250 \text{ kNm}$$

2. Trial 1
   Use a Type 3 section as detailed in Figure 16.4(7) and use the sectional properties given in Figure C(1) in Appendix C for the section.

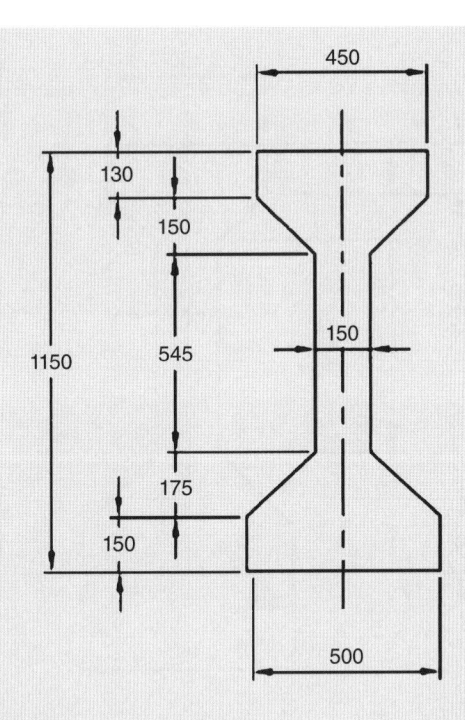

Figure 16.4(7) I-girder Type 3

The self-weight: $w_d = 317\,000 \times 10^{-6} \times 24 = 7.608$ kN/m
The dead-load moment: $M_d = w_d \times L^2/2 = 7.608 \times 5^2/2 = 95.1$ kNm
The minimum moment: $M_1 = M_d = 95.1$ kNm
The maximum moment: $M_2 = M_d + M_1' = 95.1 + 1250 = 1345.1$ kNm

Equation (A1.3): $Z_T \geq \dfrac{M_2 - \eta M_1}{c + nc_t} = 58.75 \times 10^6 < 82.9 \times 10^6$

Equation (A2.4): $Z_B \geq \dfrac{M_2 - \eta M_1}{nc + c_t} = 70.501 \times 10^6 < 91.1 \times 10^6$

The above equations indicate that both Z_T and Z_B for the selected section are acceptable. Thus, Type 3 section is acceptable though slightly conservative or oversized. To double-check appropriateness, try the next lower size beam.

3. Trial 2
   Use a Type 2 section.
   Following the process used in the first trial and with the sectional properties given in Figure 16.4(8), we obtain

$$w_d = 218\,000 \times 10^{-6} \times 24 = 5.232 \text{ kN/m} = 5232 \text{ N/m}$$
$$M_d = w_d \times L^2/2 = 5.232 \times 5^2/2 \times 10^{-3} = 65.4 \text{ kNm}$$
$$M_1 = M_d = 65.4 \text{ kNm}$$
$$M_2 = M_d + M_1' = 65.4 + 1250 = 1315.4 \text{ kNm}$$

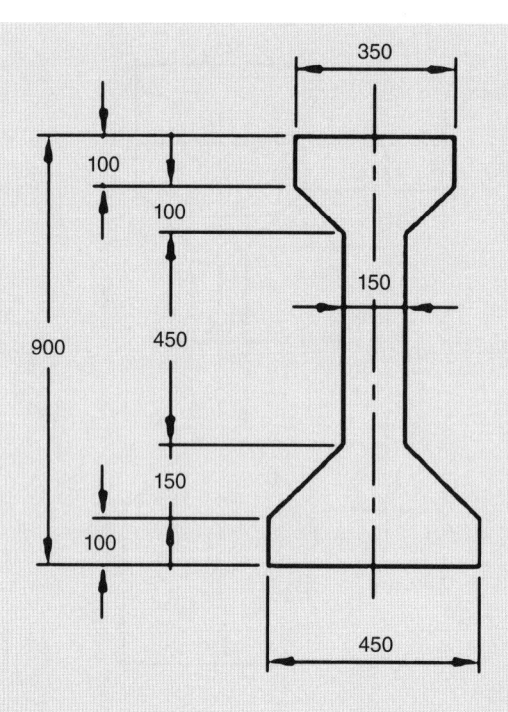

Figure 16.4(8) I-girder Type 2
Notes: section properties: $A_g = 218\,000$ mm^2, $Z_T = 41\,100\,000$ mm^3, $Z_B = 48\,100\,000$ mm^3,
$I = 19\,950\,000\,000$ mm^4, $y_B = 415$ mm, $y_T = 485$ mm

Then

Equation (A1.3): $$Z_T \geq \frac{M_2 - \eta M_1}{c + \eta c_t} = \frac{1315.4 - 0.8 \times 65.4}{20 + 0.8 \times 2} \times 10^6$$
$$= 58.476 \times 10^6 > 41.1 \times 10^6$$

This is not acceptable.

Equation (A2.4): $$Z_B \geq \frac{M_2 - \eta M_1}{\eta c + c_t} = \frac{1315.4 - 0.8 \times 65.4}{0.8 \times 20 + 2} \times 10^6$$
$$= 70.171 \times 10^6 > 48.1 \times 10^6$$

This is also not acceptable.

It is obvious that a Type 2 section is inadequate, and a Type 3 section must be used.

4. Determine the minimum H.

With the Type 3 section, we have

Equation (A-I) $\times \eta \times Z_T$ + Equation (A-4) $\times Z_B$

which yields

$$\eta \times \frac{H}{A_g} \times (Z_T + Z_B) + (\eta \times M_1 - M_2) \geq -c_t(\eta \times Z_T + Z_B)$$

from which

$$H \geq \frac{([-2 \times (0.8 \times 82.9 + 91.1) \times 10^6 - (0.8 \times 95.1 - 1345.1) \times 10^6] \times 317000 \times 10^{-3})}{0.8 \times (82.9 + 91.1) \times 10^6}$$

$= 2173\,\text{kN}$

Then

Equation (A-1): $e_B = \left(\dfrac{2173}{317\,000} \times 10^3 + \dfrac{95.1 \times 10^6}{82.9 \times 10^6} + 2\right) \times \dfrac{82.9 \times 10^6}{2173 \times 10^3}$

$\qquad\qquad\qquad = 382\,\text{nm}$

Adopt $H = 2200\,\text{kN}$.

5. Using the adopted H, determine maximum e_B.

   We have

   $\dfrac{H}{A_g} = 6.94\,\text{MPa}, \quad \dfrac{H}{Z_T} = 26.54 \times 10^{-3}\,\text{N/mm}^3, \quad \dfrac{H}{Z_B} = 24.15 \times 10^{-3}\,\text{N/mm}^3$

   $\dfrac{M_1}{Z_T} = 1.15\,\text{MPa}, \quad \dfrac{M_2}{Z_B} = 14.77\,\text{MPa}$

   Then

   Equation (A-1): $6.94 - 26.54 \times 10^{-3} \times e_B + 1.15 \geq -2$

   from which $e_B \leq 380$ mm, and

   Equation (A-4): $0.8(6.94 + 24.15 \times 10^{-3} \times e_B) - 14.77 \geq -2$

   from which $e_B \geq 374$ mm.

   Adopt $e_B = 375$ mm.

6. Check the fibre stress at section A of the beam.

   We have

   $\dfrac{H \times e_B}{Z_T} = 9.95\,\text{MPa}, \quad \dfrac{H \times e_B}{Z_B} = 9.06\,\text{MPa}, \quad \dfrac{M_1}{Z_T} = 1.15\,\text{MPa},$

   $\dfrac{M_1}{Z_B} = 1.04\,\text{MPa}, \quad \dfrac{M_2}{Z_T} = 16.23\,\text{MPa}, \quad \dfrac{M_2}{Z_B} = 14.77\,\text{MPa}$

   Then

   Equation (A-1): $6.94 - 9.95 + 1.15 = -1.86 > -2$

   Equation (A-2): $6.94 + 9.06 - 1.04 = 14.96 < 20$

   Equation (A-3): $0.8(6.94 - 9.95) + 16.23 = 13.82 < 20$

   Equation (A-4): $0.8(6.94 + 9.06) - 14.77 = -1.97 > -2$

Thus, the design is acceptable.

7. Determine the vertical limits for e_B at section B.
   Since $M_d = 0$ and $M_1 = M_2 = 0$

   Equation (A-1): $6.94 - 0.02654 e_B \geq -2$ (i.e. $e_B \leq 337$ mm)

   Equation (A-2): $6.94 + 0.02415 e_B \leq 20$ (i.e. $e_B \leq 541$ mm)

   Equation (A-3): $0.8(6.94 - 0.02654 e_B) \leq 20$ (i.e. $e_B \geq -680$ mm)

   Equation (A-4): $0.8(6.94 + 0.02415 e_B) \geq -2$ (i.e. $e_B \geq -391$ mm)

   Thus, the governing eccentricities are +337 mm and −391 mm.

8. The limiting zone
   The limiting zone is sketched in Figure 16.4(9) with the adopted e_B in step 5 and the limiting eccentricities from step 7. Note that the prestressing force $H = 2200$ kN.

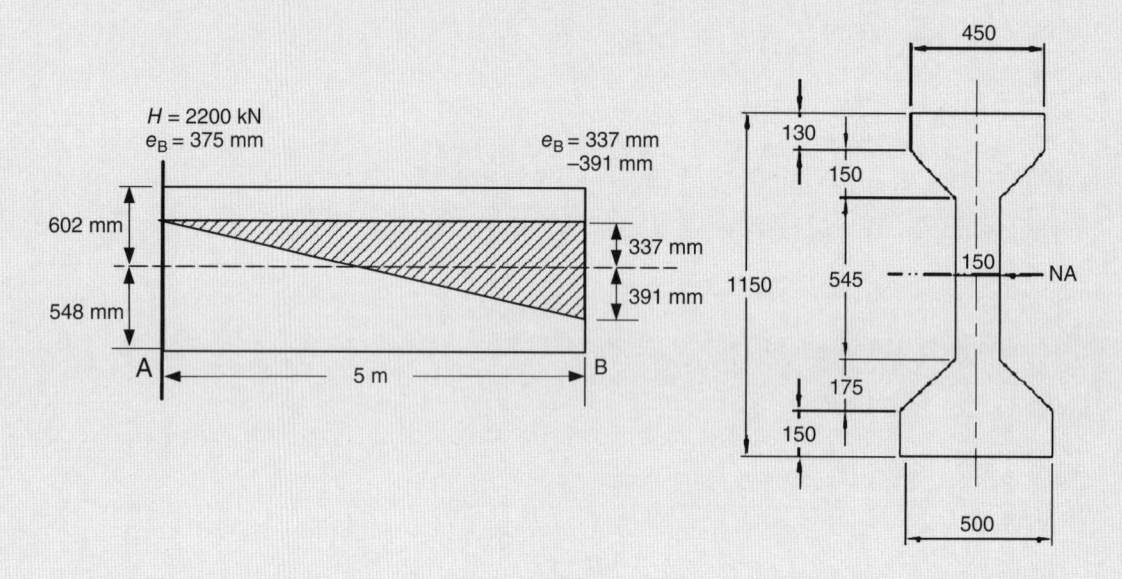

Figure 16.4(9) Layout of limiting zone for post-tensioned cable

16.5 SUMMARY

The critical stress state (CSS) method of bending design for both Case A and Case B prestressing is presented together with the design formulas and design procedures. The significance of Magnel's plots is explained and the design steps for both the prestressing cases are enumerated. Fully worked examples are given to illustrate the design of beam sections as well as the determination of post-tensioning forces and the cable limiting zones.

16.6 PROBLEMS

1. The beam in Section 16.4.1 is taken to be post-tensioned where transfer of prestress occurred before grouting. For the final design, compute the loss due to internal bending (He_B) caused by the prestress. You may take $E_c = 26000$ MPa.

2. A rectangular post-tensioned concrete beam is to be designed to carry a uniformly distributed live load of 7.5 kN/m over the entire simply supported span of 12 m. Assume a minimum cover of 100 mm, $\rho_w = 23$ kN/m³, $c = 15$ MPa, $c_t = 0$ and $\eta = 0.85$. Taking $b \approx D/2$, obtain suitable sectional dimensions, the minimum required prestressing force (H) and the e_B at midspan.

3. A prestressed post-tensioned rectangular beam is required to carry the uniformly distributed loads (a) and (b) positioned as shown in Figure 16.6(1), either simultaneously or independently. Obtain suitable sectional dimensions ($b \approx D/3$) and sketch the limiting zone for the prestressing cable giving the necessary values of the eccentricities. Assume $c = 14$ MPa, $c_t = 1.5$ MPa, $\eta = 0.80$ and $\rho_w = 23$ kN/m³.

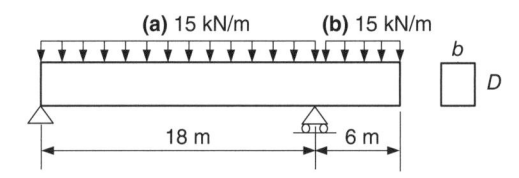

Figure 16.6(1) Load configuration and sectional details of a simple beam with overhang

4. A post-tensioned beam is to be designed to carry, in addition to its self-weight, a uniformly distributed live load of 13.5 kN/m over its entire simply supported span of 18 m. Assume a minimum cover of 100 mm (to H), $c = 15$ MPa, $c_t = 0$, $\eta = 0.85$, $\rho = 2400$ kg/m³ and use the critical stress state (CSS) approach to
   (a) obtain a suitable cross-section
   (b) compute the minimum required prestressing force H and the e_B at midspan.
   Notes:
   - You are required to use one of the Standard precast I-girder sections detailed in Figure C(1) in Appendix C.
   - A design with an oversized section is not acceptable.

5. A post-tensioned beam is to be designed to carry, in addition to its self-weight, a concentrated live load of 120 kN acting at the middle of a simply supported span of 18 m. Assuming a minimum cover of 100 mm (to H), $c = 20$ MPa, $c_t = 0$, $\eta = 0.85$, $\rho_w = 25$ kN/m³ and using the critical stress state (CSS) approach:
   (a) confirm via appropriate and correct calculations that the Type-2 I-girder as detailed in Figure C(1) in Appendix C is a satisfactory (i.e. not unduly conservative) section for this design; then
   (b) compute the minimum required prestressing force H and the e_B at midspan.
   Note in Appendix C that $Z_{y.sup} = Z_T$; $Z_{y.sub} = Z_B$; and $d_{y.sub} = y_B$.

17 ULTIMATE STRENGTH ANALYSIS OF BEAMS

17.1 INTRODUCTION

As stated in previous chapters, the critical stress state (CSS) design method is based on linear–elastic or permissible-stress principles. The beam design is considered to comply with AS 3600–2009 (the Standard) requirements only if its strength satisfies Equation 1.1(1) which stipulates for bending:

$$\phi M_u \geq M^* \quad \text{Equation 17.1(1)}$$

Before proceeding to develop the formulas for M_u, the following observations are made:
- The behaviour of prestressed beam at ultimate is similar to its reinforced counterpart.
- Assumptions made for reinforced beams in Section 3.3.1 are also valid for prestressed beams.
- The stress–strain curves and allied properties of prestressing and reinforcing steel are as given in Figure 17.1(1).
- The rectangular stress block as defined in Figure 3.3(3) is a valid equivalent to the actual stress distribution of concrete at ultimate.
- A prestressed section fails when $\varepsilon_c = 0.003$.

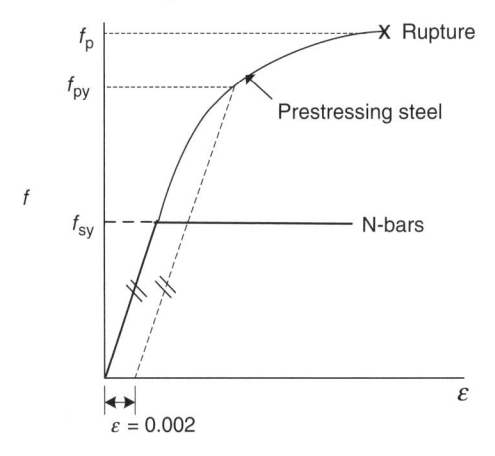

Figure 17.1(1) Stress–strain curves and allied properties of prestressing and reinforcing steel

Note: $E_s = 200000$ MPa
$E_p = 200000$ MPa for wires
  $= 195000$ MPa for strands

17.2 CRACKING MOMENT (M_{cr})

17.2.1 Formula

For the prestressed beam shown in Figure 17.2(1), the beam would crack if the bottom-fibre tensile stress reached the flexural tensile strength of concrete.

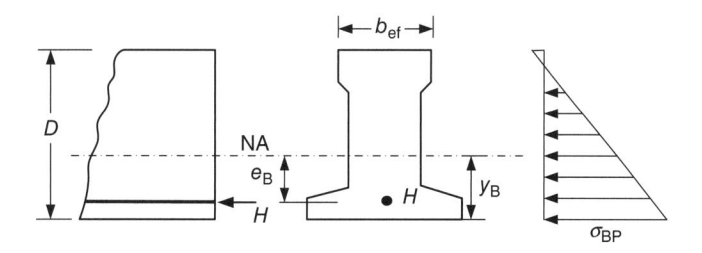

Figure 17.2(1) Stress distribution across the section of a typical prestressed beam

Based on Equation 15.3(8), and under the prestress force H at e_B from the neutral axis (NA), the compressive bottom-fibre stress after loss may be written as

$$\sigma_{BP} = \eta H \left(\frac{1}{A} + \frac{e_B y_B}{I} \right) = \frac{\eta H}{A} + \frac{\eta H e_B}{Z_B} \qquad \text{Equation 17.2(1)}$$

Thus, the cracking moment

$$M_{cr} = (f'_{ct.f} + \sigma_{BP}) \frac{I}{y_B} = Z_B \left(f'_{ct.f} + \eta \frac{H}{A} \right) + \eta H e_B \qquad \text{Equation 17.2(2)}$$

where $f'_{ct.f} = 0.6\sqrt{f'_c}$ as per Equation 2.2(2) and η is the effective prestress coefficient.

17.2.2 Illustrative example

To demonstrate the application of Equation 17.2(2), an example is given below.

Problem

Using the information given for the beam detailed in Section 15.8, what is M_{cr}, assuming $f'_c = 40$ MPa?

Solution

From Section 15.8, we know:

- $Z_B = 20.68 \times 10^8 / 250 = 8.272 \times 10^6$ mm^3
- $\eta H = 0.85 \times 667500 = 567375$ N

- $e_B = 175.5$ mm
- $A = 72.5 \times 10^3$ mm^2
- $f'_{ct.f} = 0.6\sqrt{f'_c} = 3.795$ MPa.

Substituting these values into Equation 17.2(2) yields

$$M_{cr} = \left[8.272 \times 10^6 \times (3.795 + 567375/72500) + 567375 \times 175.5\right] \times 10^{-6}$$
$$= 195.7 \text{ kNm} (> M_d + M_1 = 164.3 \text{ kNm})$$

Note that for the original design $c_t = 0$, which is conservative. Further, if $f'_{ct.f}$ is assumed to be zero, we have $M_{cr} = 164.3$ kNm, which is equal to $M_d + M_1$ (or M_2 in this case). This should be so, because Equation A-4 or B-4, as given in Section 15.7.3, prevails as an identity.

17.3 ULTIMATE MOMENT (M_u) FOR PARTIALLY PRESTRESSED SECTIONS

17.3.1 General equations

A partially prestressed beam may be defined as one that also includes traditional (nonprestressed) reinforcement for one reason or another – see the discussion in Section 17.4 for an example. A typical partially prestressed section is illustrated in Figure 17.3(1).

The procedure given below mirrors the one established in Section 3.4.3 for tension failure of reinforced beams with yielding A_{sc}.

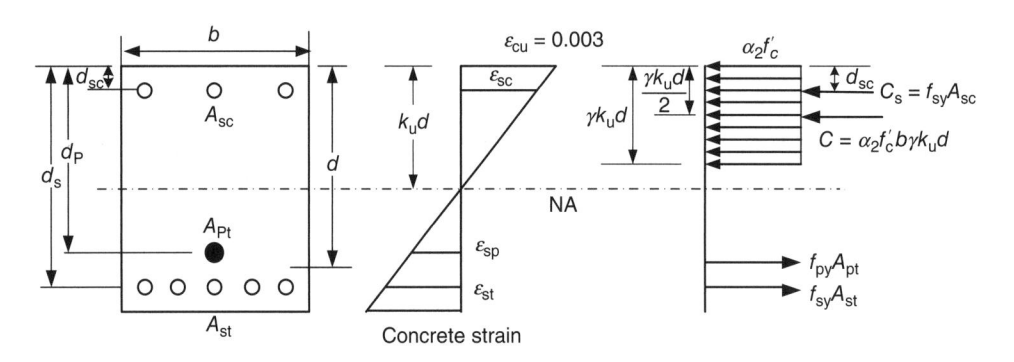

Figure 17.3(1) Stress and strain distribution across a typical partially prestressed beam section

Notes: A_{pt} is the total prestressing steel area in the tensile zone.
A_{st} and A_{sc} respectively are tensile and compressive reinforcement.
α_2 and γ are given in Equations 3.3(2)a and b, respectively.

By imposing $\Sigma F_x = 0$, we have

$$f_{py}A_{pt} + f_{sy}A_{st} = f_{sy}A_{sc} + \alpha_2 f'_c b\gamma k_u d$$

from which

$$k_u d = [f_{py}A_{pt} + f_{sy}(A_{st} - A_{sc})]/(\alpha_2 f'_c \gamma b) \qquad \text{Equation 17.3(1)}$$

where the mean effective depth is

$$d = (f_{py}A_{pt}d_p + f_{sy}A_{st}d_s)/(f_{py}A_{pt} + f_{sy}A_{st}) \qquad \text{Equation 17.3(2)}$$

Taking the moment about the top edge of the beam gives

$$M_u = f_{py}A_{pt}d_p + f_{sy}A_{st}d_s - f_{sy}A_{sc}d_{sc} - \alpha_2 f'_c b \frac{(\gamma k_u d)^2}{2} \qquad \text{Equation 17.3(3)}$$

Note that in Equation 17.3(3) the loss of prestress is assumed to have fully occurred; this which should be expected at the ultimate state, unless it is a case of premature failure or unanticipated fatal overloading.

17.3.2 Sections with bonded tendons

Equation 17.3(3) is a basic equation; it is valid only if A_{pt} can attain the yield stress at the ultimate state. The reader can recall from Figure 17.1(1) that prestressing steel does not have a yield plateau. Thus, the unlimited tensile strain in steel assumed in deriving Equation 17.3(3) (based on the second assumption in Section 3.3.1) is not valid. Consequently, the equation often gives erroneous ultimate moment results. To rectify this problem, some modifications are necessary.

Clause 8.1.7 of the Standard recommends that, for sections with bonded tendons (i.e. pretensioned and properly grouted post-tensioned beams), the tendon stress at the ultimate state or when failure occurs is

$$\sigma_{pu} = f_p(1 - k_1 k_2/\gamma) \qquad \text{Equation 17.3(4)}$$

where

$$\begin{aligned} k_1 &= 0.4 \quad \text{for} \quad f_{py}/f_p < 0.9 \\ &= 0.28 \quad \text{for} \quad f_{py}/f_p \geq 0.9 \end{aligned} \qquad \text{Equation 17.3(5)}$$

and

$$k_2 = [f_p A_{pt} + f_{sy}(A_{st} - A_{sc})]/(b_{ef}d_p f'_c) \qquad \text{Equation 17.3(6)}$$

If $d_{sc} > 0.15\, d_p$, then ignore A_{sc}, but ensure that

$$k_2 \geq 0.17 \quad \text{Equation 17.3(7)}$$

Replacing f_{py} by σ_{pu} in Equation 17.3(3) gives

$$M_u = \sigma_{pu} A_{pt} d_p + f_{sy} A_{st} d_s - f_{sy} A_{sc} d_{sc} - \alpha_2 f'_c b \frac{(\gamma k_u d)^2}{2} \quad \text{Equation 17.3(8)}$$

Note that Equation 17.3(4) is valid only if

$$\sigma_{p.ef} = \eta \sigma_{pi} \geq 0.5 f_p \quad \text{Equation 17.3(9)}$$

where σ_{pi} is the tendon stress at transfer and f_p is the rupture strength of the tendons – see Figure 17.1(1).

17.3.3 Sections with unbonded tendons

For unbonded tendons, which could lead to beam failure caused by a single or discrete fatal cracks as depicted in Figure 17.3(2), see Warner and Faulkes (1988). Clause 8.1.8 recommends the following equations for computing σ_{pu}:

- for span-to-overall depth ratio $L/D \leq 35$

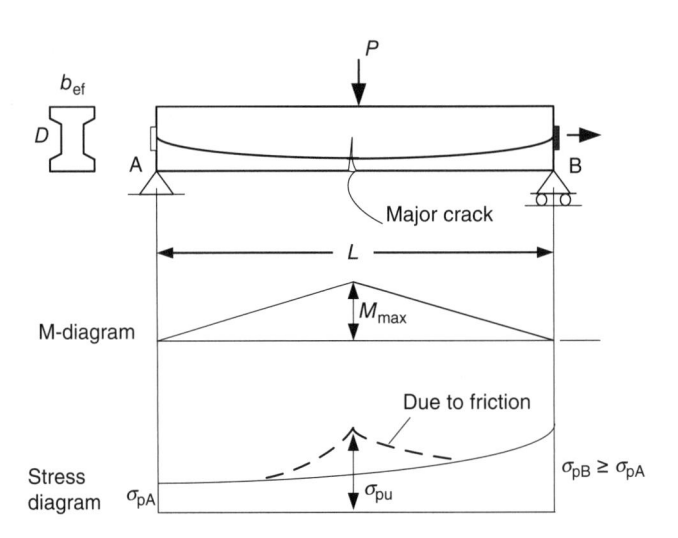

$$\sigma_{pu} = \sigma_{p.ef} + 70 + \frac{f'_c b_{ef} d_p}{100 A_{pt}} \leq \sigma_{p.ef} + 400 \quad \text{Equation 17.3(10)}$$

Figure 17.3(2) Ultimate tendon stress at failure for unbounded tendons

- for $L/D > 35$

$$\sigma_{pu} = \sigma_{p.ef} + 70 + \frac{f'_c b_{ef} d_p}{300 A_{pt}} \leq \sigma_{p.ef} + 200 \qquad \text{Equation 17.3(11)}$$

where:
- L and D are the span and depth of the beam – see Figure 17.3(2)
- $\sigma_{p.ef}$ is given in Equation 17.3(9)
- b_{ef} is defined in Equations 4.2(2) and (3)
- d_p and A_{pt} are described in Figure 17.3(1).

17.4 DUCTILITY REQUIREMENTS – REDUCED ULTIMATE MOMENT EQUATIONS

As discussed in Section 3.4.1, tension failure is desirable since it occurs only after the beam has sustained large deflection and widespread cracking. All this provides ample warning of an imminent disaster.

To ensure a tension failure, Clause 8.1.5 stipulates that

$$k_{uo} = \frac{k_u d}{d_o} \leq 0.36 \qquad \text{Equation 17.4(1)}$$

where, as defined in Section 3.4.5, the neutral axis (NA) parameter k_{uo} is with respect to d_o.

Note that to compute k_u Equation 17.3(1) should be used, but with f_{py} replaced by σ_{pu}. If, $k_{uo} > 0.36$, take $k_{uo} = 0.36$. This means that the effectiveness of some portion of A_{st} or A_{pt} is ignored. The effective tendon area can be obtained from Equation 17.3(1) as

$$A_{pt.ef} = \left[0.36\alpha_2 f'_c \gamma b d_o - f_{sy}(A_{st} - A_{sc})\right]/\sigma_{pu} \qquad \text{Equation 17.4(2)}$$

and the reduced ultimate moment as

$$M_{ud} = \sigma_{pu} A_{pt.ef} d_p + f_{sy} A_{st} d_s - f_{sy} A_{sc} d_{sc} - \alpha_2 f'_c b (\gamma 0.36 d_o)^2 / 2 \qquad \text{Equation 17.4(3)}$$

Note that for Equation 17.4(3) to be valid Clause 8.1.5(b) stipulates that

$$A_{sc} \geq 0.01 A'_c \qquad \text{Equation 17.4(4)}$$

where A'_c is the concrete area in compression or for a rectangular section,

$$A'_c = b\gamma k_u d \qquad \text{Equation 17.4(5)}$$

Clause 8.1.6.1 also specifies that

$$M_u \text{ or } M_{ud} \geq 1.2 M_{cr} \qquad \text{Equation 17.4(6)}$$

where the cracking moment M_{cr} is given in Equation 17.2(2).

17.5 DESIGN PROCEDURE

17.5.1 Recommended steps

The full design of a prestressed beam may take the following steps:
1. Design the beam using the CSS approach given in Chapter 15.
2. Check to ensure that $\phi M_u \geq M^*$ or $\phi M_{ud} \geq M^*$ as the case may be where M_u and M_{ud} are given in Equations 17.3(8) and 17.4(3), respectively. If M_u or M_{ud} is inadequate, add reinforcing steel ΔA_{st} where

$$\Delta A_{st} = \frac{\left[\frac{M^*}{\phi} - A_{pt}\sigma_{pu}(d_p - 0.15d_s)\right]}{(f_{sy}\alpha_2 d_s)} \qquad \text{Equation 17.5(1)}$$

Note that Equation 17.5(1) assumes the resultant of the compressive forces $(C + C_s)$ to be acting at the level $0.15d_s$ below the top beam fibre, where C and C_s are defined in Figure 17.3(1). This assumption entails some approximation. As necessary, a more accurate location of $(C + C_s)$ can be determined with reference to Figure 17.3(1) as

$$d_{cR} = (C_s d_{sc} + C\gamma k_u d/2)/(C + C_s) \qquad \text{Equation 17.5(2)}$$

Also note that the exact d_{cR} can be determined only by an iterative process because $k_u d$ is a function of the total A_{st} (i.e. the current A_{st} plus the ΔA_{st} to be computed).
3. Finally, check that $k_{uo} \leq 0.36$, and compute M_u or M_{ud} ensuring that ϕM_u or $\phi M_{ud} \geq M^*$. Resize the beam as necessary.

17.5.2 Illustrative example

Problem

For the section with unbonded tendons shown in Figure 17.5(1), compute the ultimate moment M_u using Equation 17.3(8).

Solution

Using Equation 17.3(10), determine that

$$\sigma_{pu} = 1100 + 70 + 40 \times 200 \times 625/(100 \times 405)$$
$$= 1293.5 \text{ MPa} < 1100 + 400 = 1500 \text{ MPa}$$

This is acceptable.
Now using Equation 17.3(1)

$$k_u d = (1293.5 \times 405 + 500 \times 1356)/(0.85 \times 40 \times 0.77 \times 200) = 229.5 \text{ mm}$$

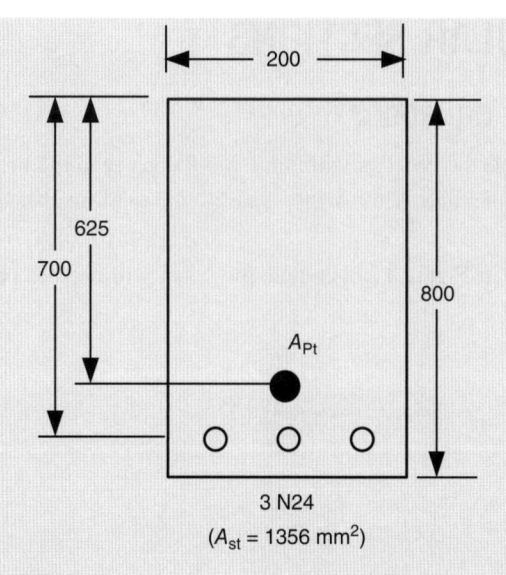

Figure 17.5(1) Details of a partially prestressed beam with unbonded tendons

Note: all dimensions are in mm.
$f'_c = 40$ MPa ($\alpha_2 = 0.85; \gamma = 0.77$); $A_{pt} = 405$ mm^2 unbonded
$f_{py} = 1710$ MPa; $f_p = 1850$ MP; $\sigma_{p.ef} = \eta\sigma_{pi} = 1100$ MPa
$L/D < 35$

and Equation 17.3(2) gives

$$d = (1293.5 \times 405 \times 625 + 500 \times 1356 \times 700)/(1293.5 \times 405 + 500 \times 1356)$$
$$= 667.3 \text{ mm}$$

However, $d_o = 700$ mm, and

$$k_{uo} = \frac{229.5}{700} = 0.328 < 0.36$$

which is acceptable.

The section is considered under-reinforced and complies with Equation 17.4(1). Finally, Equation 17.3(8) gives

$$M_u = \left[1293.5 \times 405 \times 625 + 500 \times 1356 \times 700 - 0.85 \times 40 \times 200 \times (0.77 \times 229.5)^2/2 \times 10^{-6}\right]$$
$$= 695.8 \text{ kNm}$$

17.6 NONRECTANGULAR SECTIONS

17.6.1 Ultimate moment equations

If given a nonrectangular section and the NA lies within the top flange, then the prestressed beam may be treated as a rectangular one. This is the same as for reinforced beams (see Section 4.2).

If this is not the case, as shown in Figure 17.6(1), then the beam will need to be treated differently.

Now we must use

$$C = \alpha_2 f'_c A'_c$$ Equation 17.6(1)

Then Equation 17.3(8) becomes

$$M_u = \sigma_{pu} A_{pt} d_p + f_{sy} A_{st} d_s - f_{sy} A_{sc} d_{sc} - \alpha_2 f'_c A'_c d'_c$$ Equation 17.6(2)

where

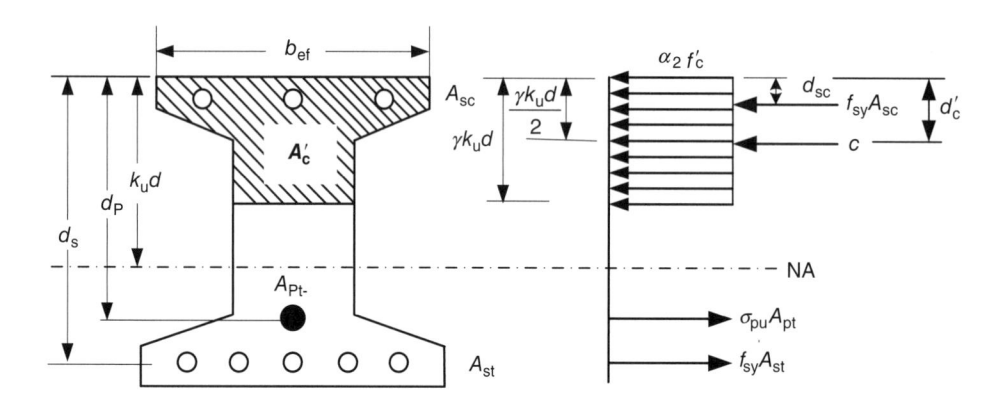

$$A'_c = \left[\sigma_{pu} A_{pt} + f_{sy}(A_{st} - A_{sc})\right]/(\alpha_2 f'_c)$$ Equation 17.6(3)

Figure 17.6(1) Nonrectangular partially prestressed section and internal forces

With A'_c in hand, the dimensions d'_c and $(\gamma k_u d)$ can be computed readily. Although d'_c is required in Equation 17.6(2), $(\gamma k_u d)$ is used in conjunction with Equation 17.3(2) to obtain k_u, which is needed in Equation 17.4(1) to ensure that the ductility requirement is satisfied.

Note that if k_u is greater than 0.4, then let $k_u = 0.4$ – see Figure 17.6(2) and determine A''_c accordingly. Equation 17.4(2) thus becomes

$$A_{pt.ef} = \left[\alpha_2 f'_c A''_c - f_{sy}(A_{st} - A_{sc})\right]/\sigma_{pu}$$ Equation 17.6(4)

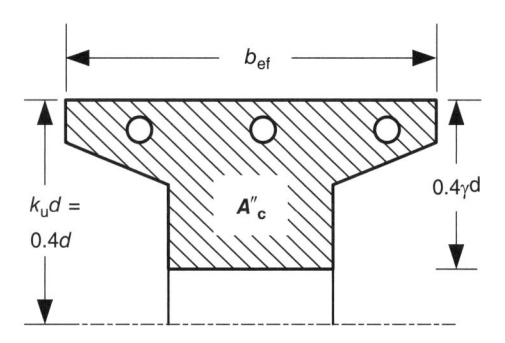

Figure 17.6(2) Concrete area in compression for the maximum permissible neutral axis depth

Now the reduced ultimate moment becomes

$$M_{ud} = \sigma_{pu}A_{pt.ef}d_p + f_{sy}(A_{st}d_s - A_{sc}d_{sc}) - \alpha_2 f'_c A''_c d''_c \qquad \text{Equation 17.6(5)}$$

where d'_c is the distance between C and the top fibre of the nonrectangular section.

17.6.2 Illustrative example

Problem

For the standard bridge-beam section as adopted in AS 5100.5–2017 and shown in Figure 17.6(3), compute the ultimate moment based on relevant recommendations from the Standard. All given values are identical to those prescribed for the rectangular beam in Section 17.5.2, except that $A_{pt} = 910$ mm^2 and the cable is bonded.

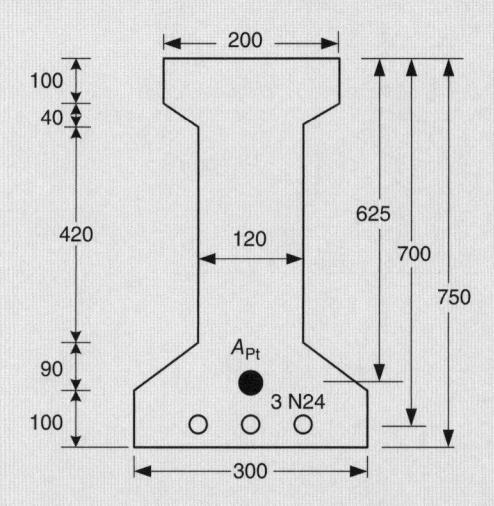

Figure 17.6(3) Standard bridge-beam section as per AS 5100.5–2017

Note: all dimensions are in mm.

Solution

From the following equations we have:

Equation 17.3(5): $k_1 = 0.28$

Equation 17.3(6): $k_2 = (1850 \times 910 + 500 \times 1356)/(200 \times 625 \times 40)$
$= 0.4723$

Equation 17.3(4): $\sigma_{pu} = 1850(1 - 0.28 \times 0.4723/0.77) = 1532.3\,\text{MPa}$

Equation 17.3(2): $d = (1532.3 \times 910 \times 625 + 500 \times 1356 \times 700)/$
$(1532.3 \times 910 + 500 \times 1356) = 649.5\,\text{mm}$

Equation 17.6(3): $A'_c = (1532.3 \times 910 + 500 \times 1356)/(0.85 \times 40)$
$= 60\,953\,\text{mm}^2$

Equating the above A'_c with the dimensions of the concrete area in compression, as shown in Figure 17.6(4), we have

$$60\,953 = 200 \times 100 + (120 + 200) \times 40/2 + (\gamma k_u d - 140) \times 120$$

from which $\gamma k_u d = 427.9$ mm. Thus

$$k_u = 427.9/(0.77 \times 649.5) = 0.86 \text{ and } k_{uo} = \frac{0.86 \times 649.5}{700} = 0.798 > 0.36$$

Imposing $k_{uo} = 0.36$ as required by Equation 17.4(1) leads to

$$\gamma k_u d = 0.77 \times 0.36 \times 700 = 194\,\text{mm}$$

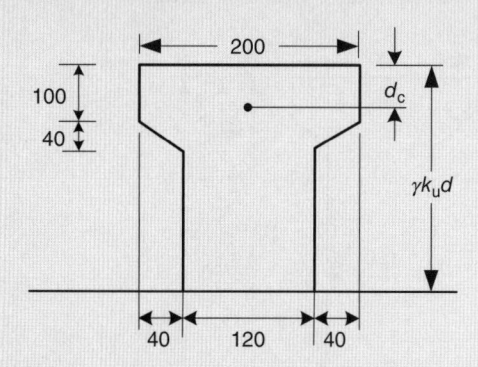

Figure 17.6(4) Concrete area in compression

Note: all dimensions are in mm.

With reference to Figure 17.6(4) we can obtain

$$A''_c = (194 - 140) \times 120 + (120 + 200) \times 40/2 + 200 \times 100 = 32\,880\,\text{mm}^2$$

By taking moment about the top edge of the section, the location of the compressive force C (see Figure 17.6(4)), can be computed as

$$d''_c = [200 \times 100 \times 50 + 120 \times 40 \times 120 + (2 \times 40 \times 40/2) \times (100 + 40/3)$$
$$+ 54 \times 120 \times (140 + 54/2)]/ = 86.4 A'_c \, \text{mm}$$

Hence

Equation 17.6(4): $A_{pt.ef} = (0.85 \times 40 \times 32\,800 - 500 \times 1356)/1532.3 = 287.1\,\text{mm}^2$
Equation 17.6(5): $M_{ud} = (1532.3 \times 287.1 \times 625 + 500 \times 1356 \times 700 - 0.85$
$$\times 40 \times 32\,880 \times 86.4) \times 10^{-6} = 653.0\,\text{kNm}$$

Finally, to comply with the ductility requirement, the required compression steel is given by Equation 17.4(4) as

$$A_{sc} = 0.01 A''_c = 329\,\text{mm}^2$$

Therefore, we can use three N12 or 339 mm².

17.7 SUMMARY

Formulas for analysing the ultimate strength moment capacity of partially prestressed concrete beams are developed. The Standard recommendations for sections with bonded and unbonded tendons are introduced. To comply with the Standard requirements for ductility, the reduced ultimate moment equations are derived and the calculation procedures summarised. The illustrative examples cover both rectangular and nonrectangular sections.

17.8 PROBLEMS

1. For the section in Figure 17.8(1), with $A_{pt} = 607\,\text{mm}^2$ (bonded), $f_p = 1900$ MPa, $f_{py} = 1700$ MPa, $\sigma_{p.ef} = 1200$ MPa, $A_{st} = 0$, and $f'_c = 40$ MPa:
   (a) Based on the requirement $\phi M_u = 1.2\, M_d + 1.5\, M_l$, compute M_l for a simply supported span of 15.25 m.
   (b) What is the required A_{sc} for ductility purposes?

470 PART 2 PRESTRESSED CONCRETE

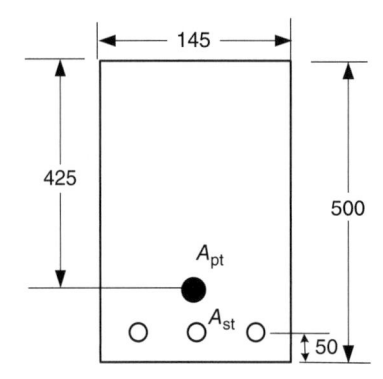

Figure 17.8(1) Cross-sectional details

Note: all dimensions are in mm.

2. The rectangular beam section in the example given in Section 17.5.2 is to develop an ultimate moment $M_u = 750$ kNm. Compute the additional amount of prestressing steel (ΔA_{pt}) required. Ensure that the Standard's ductility requirements are satisfied.
3. Compute the ultimate moment M_u for the section detailed in Figure 17.8(2). Given $f'_c = 40$ MPa; $A_{pt} = 910$ mm^2 bonded; $f_{py} = 1710$ MPa; $f_p = 1850$ MPa; $\sigma_{p.ef} = \eta\sigma_{pi} = 1100$ MPa; and $L/D < 35$.

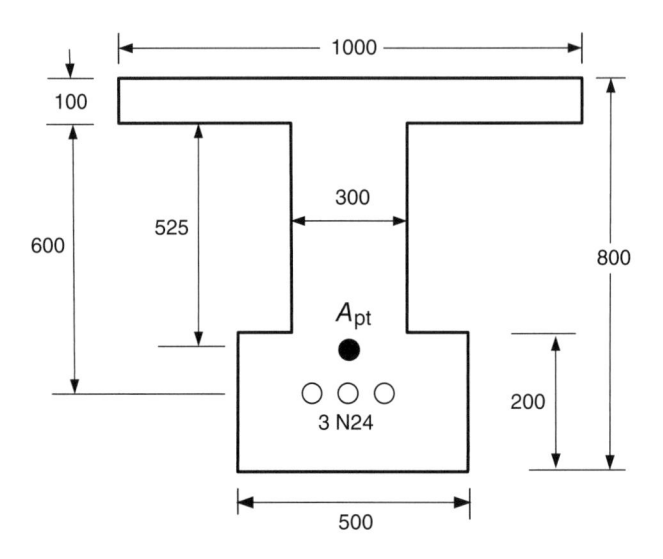

Figure 17.8(2) Cross-sectional details for Problem 3

Note: all dimensions are in mm.

4. Compute the ultimate moment M_u for the section detailed in Figure 17.8(3). Given $f'_c = 40$ MPa; $A_{pt} = 910$ mm^2 bonded; $f_{py} = 1710$ MPa; $f_p = 1850$ MPa; $\sigma_{p.ef} = \eta\sigma_{pi} = 1100$ MPa; and $L/D < 35$.

   You may assume that the beam satisfies the ductility requirements.

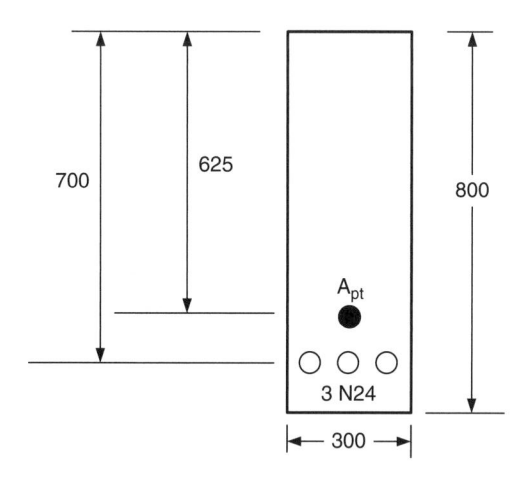

Figure 17.8(3) Cross-sectional details for Problem 4

Note: all dimensions are in mm.

5. The details of a partially prestressed concrete section are shown in Figure 17.8(4).
   (a) Given $f'_c = 40$ MPa, $A_{pt} = 910$ mm^2 bonded, $f_{py} = 1710$ MPa, $f_p = 1850$ MPa, $\sigma_{p.ef} = \eta\sigma_{pi} = 1100$ MPa and $L/D < 35$, compute the ultimate moment M_u. You may assume that the beam satisfied the ductility requirements.
   (b) Enumerate the relevant ductility requirements.

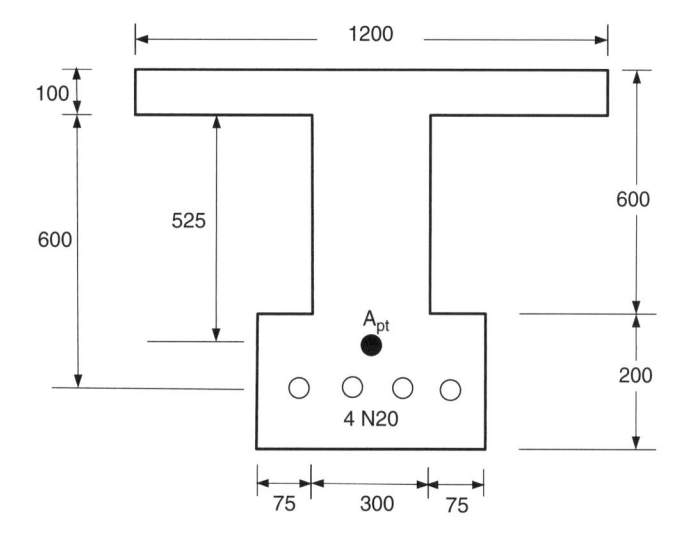

Figure 17.8(4) Details of the partially prestressed concrete section

Note: all dimensions are in mm.

6. Compute the ultimate moment M_u for the section detailed in Figure 17.8(5). Given $f'_c = 40$ MPa; $A_{pt} = 910$ mm² bonded; $f_{py} = 1720$ MPa; $f_p = 1900$ MPa; $\sigma_{p.ef} = \eta\sigma_{pi} = 1100$ MPa; and L/D < 35. You may assume that the beam satisfies the ductility requirements.

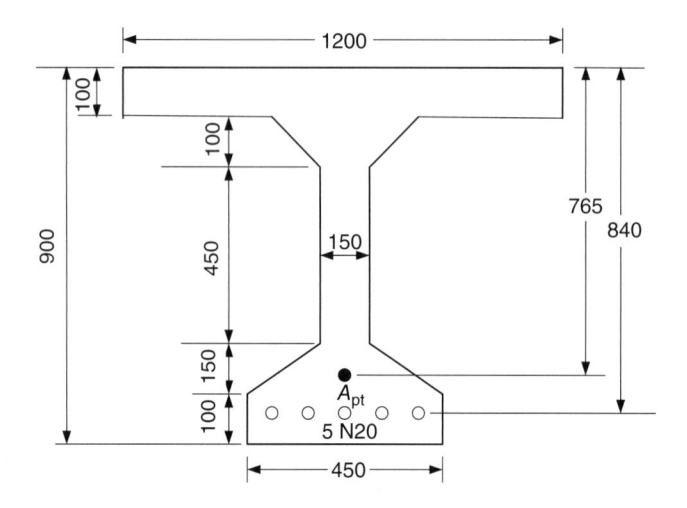

Figure 17.8(5) Cross-sectional details for Problem 6

Note: all dimensions are in mm.

END BLOCKS FOR PRESTRESSING ANCHORAGES 18

18.1 INTRODUCTION

It is recognised in Section 14.1 that prestressing tendons (either in the form of wire or strands of wire) reinforces the weaknesses of concrete in an active manner. Because of this, considerable concentrated forces are exerted at the extremities of a prestressed beam. At the end zones, these forces in pretensioned beams translate into intensive bond stresses in the steel concrete interface. In post-tensioned beams, they induce acute lateral tensile stresses and the anchor heads (see Figure 14.4(3)a) create high bearing stresses on the concrete ends.

These stresses need to be fully considered and carefully designed for, to prevent cracking and even premature failure in the end zones. A properly reinforced end zone is referred to as an end block.

The nature and distribution of the bond stress in the end zones of a pretensioned beam are given in Section 18.2, which also includes the design method recommended in AS 3600–2009 (the Standard). Section 18.3 identifies the three types of stresses induced by a post-tensioned anchorage system. Two of these are the bursting stress and the spalling stress, both of which are tensile, and orthogonal or transverse in direction to the axis of the post-tensioned tendon. There is also the bearing (compressive) stress on the concrete right behind the steel anchor head. The design for the bursting, spalling and bearing stresses is discussed in Section 18.4. Finally, the distribution and detailing of the end-block reinforcement are presented in Section 18.5.

18.2 PRETENSIONED BEAMS

For a pretensioning system, the transfer of the prestressing forces to concrete is achieved mainly by bond along the 'transmission length'. Figure 18.2(1)a shows a tendon in the end zone of a pretensioned beam before transfer. Note in Figure 18.2(1)b that, at and after transfer, the tendon diameter (R) at the free end is greater than the embedded diameter (r). This is a result of Poisson's effect.

The distribution of the resulting bond stress between tendon and concrete is illustrated in Figure 18.2(1)c. Figure 18.2(1)d depicts the gradually increasing tendon stress, which corresponds to the increase and decay of the bond stress.

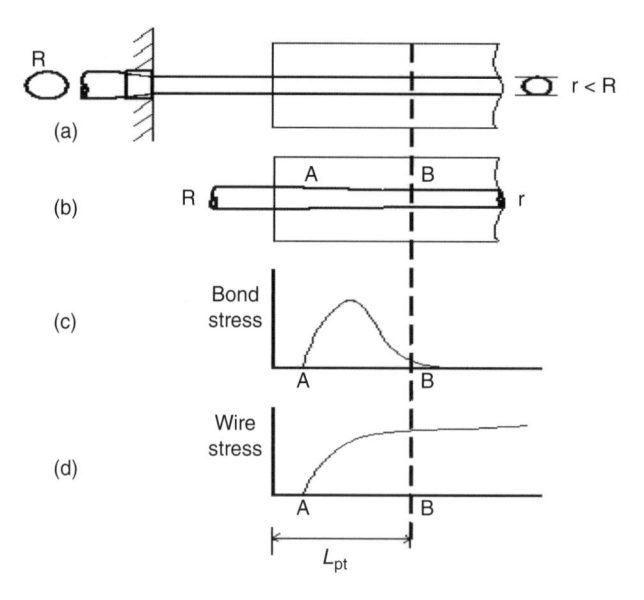

Figure 18.2(1) Pretensioning system

To account for these stress distribution characteristics, the Standard subscribes to the concept of the 'transmission length', which is represented by L_{pt} in Figure 18.2(1)d. The transmission length is parallel to the basic development length for reinforcing bars. The minimum values of L_{pt} for individual wire of diameter d_b, and strands are given in Table 13.3.2 of the Standard. The table is reproduced for convenient reference as Table 18.2(1) here. Note that these L_{pt} values are not affected by the loss of prestress, but they are valid only for a gradual rather than abrupt transfer of stress from tendon to concrete. A sudden release of stress to concrete causes a stress surge at the steel–concrete interface, which would need a transmission length longer than those given in Table 18.2(1). An abrupt transfer could cause premature bond failure.

Table 18.2(1) Minimum transmission length for pretensioned tendons

Type of tendon	L_{pt} for gradual release	
	$f_{cp} \geq 32$ MPa	$f_{cp} < 32$ MPa
Indented wire	100 d_b	175 d_b
Crimped wire	70 d_b	100 d_b
Ordinary and compact strand	60 d_b	60 d_b

Source: Standards Australia (2009). AS 3600–2009 *Concrete Structures*. Standards Australia, Sydney, NSW. © Standards Australia Limited. Copied by Cambridge University Press with the permission of Standards Australia and Standards New Zealand under Licence 1712-c038.

Figure 18.2(1)d shows that the stress-free point, A, over time is taken to be $0.1\,L_p$ from the end of the beam, while the bond-free point, B, is assumed unchanged.

In design, the stress development length L_p for pretensioned tendons in the form of indented or crimped wire is given as

$$L_p \geq 2.25\,L_{pt} \qquad \text{Equation 18.2(1)}$$

where L_{pt} is found in Table 18.2(1) for the given wire type.

For the seven-wire strands specified in AS/NZS 4672.1–2007, the design stress development length is given as

$$L_p = 0.145\,(\sigma_{pu} - 0.67\sigma_{p.ef})\,d_b \geq 60\,d_b \qquad \text{Equation 18.2(2)}$$

where σ_{pu} and $\sigma_{p.ef}$ in MPa are as defined in Sections 17.3.2 and 17.3.3 (noting that the expression within the parenthesis is used without units).

18.3 POST-TENSIONED BEAMS

The end zone of a post-tensioned beam with multiple tendons is shown in Figure 18.3(1). It is obvious that the prestress in the tendons is maintained by direct bearing of the end anchorages on the concrete. As such, each of the tendon forces acting through a steel anchorage block butting against the end face of the beam can be considered as an external force concentrated on the concrete whose line of action coincides with the axis of the tendon. This results in lateral bursting and spalling stresses, as well as the compressive bearing stress in the end-zone concrete. The nature, distribution and treatment of these stresses are discussed in the following sections.

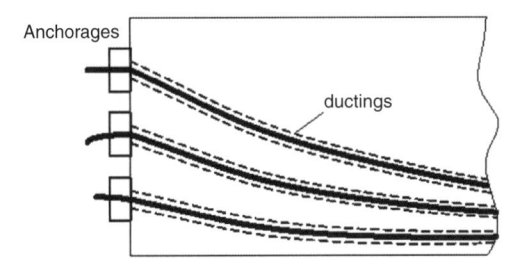

Figure 18.3(1) Post-tensioning system

18.3.1 Bursting stress

The two types of transverse tensile stress, bursting and spalling, are developed as a result of the post-tensioning and anchoring process. Figures 18.3(2)a and b show the end zone of a post-tensioned box beam with four tendons, which are symmetrical with respect to the centre of the beam section, C.

476 PART 2 PRESTRESSED CONCRETE

Under the action of the prestress (concentrated) force at anchorage A, the bursting stress is developed in the concrete zone immediately behind the anchorage. It radiates orthogonally to, and along the axis of, the tendon in planes containing the axis of the tendon – see Figure 18.3(2)c. The volume of the bursting stress is the stress block obtained by rotating the stress distribution curve – shown in Figure 18.3(2)c – 360° about the tendon axis. This indicates that the bursting stress has the tendency to blow the end-zone concrete in all of the directions orthogonal to and around the tendon axis.

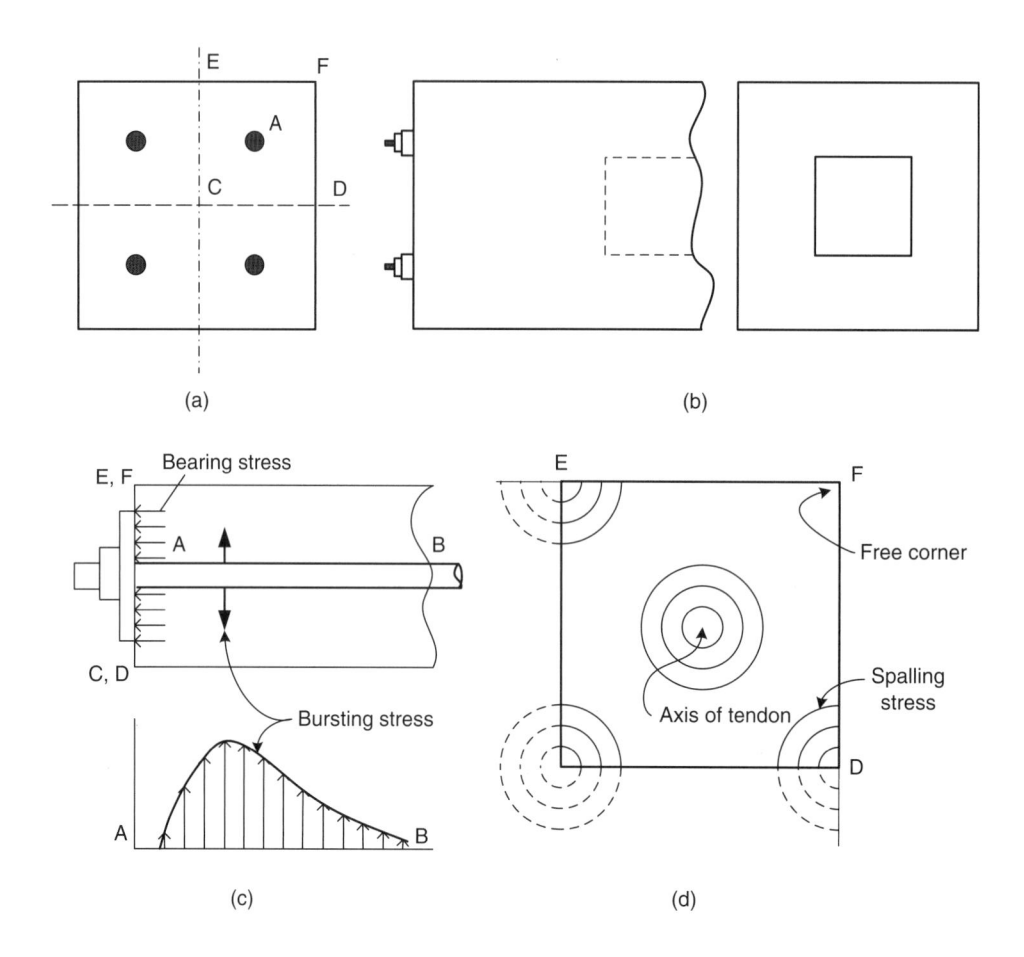

Figure 18.3(2) Nature and distribution of bursting, spalling and bearing stresses

18.3.2 Spalling stress

The spalling stress is developed around and between the anchorages. It acts transversely in planes parallel to, but not containing, the axis of the tendon. As shown in Figure 18.3(2)d, these stresses reach a maximum at the face of the anchorage and are especially important when the anchorages are very eccentric or widely spread.

18.3.3 Bearing stress

It is obvious in Figure 18.3(2)b that the compressive bearing stress occurs at the interface between each steel anchor and the end face of the concrete beam.

18.3.4 End blocks

In order to accommodate the anchorages, and to prevent cracking or premature failure caused by the bursting, spalling or bearing stresses (or all three), 'end blocks' are normally used. These require special design considerations. A portion of a typical end block with spiral reinforcement is seen in Figure 14.4(4). The spiral or helix is there to reinforce against the bursting stress.

18.4 END-BLOCK DESIGN

18.4.1 Geometry

Figure 18.4(1) shows the end faces of two typical post-tensioned beams, each with two anchorages. Note that the effects of these concentrated forces are critical during the stressing operation. In a symmetrical group of anchorages, only one typical anchorage force needs to be considered when determining the required reinforcement for the end block. In Figure 18.4(1), if the end block is designed using one of the two forces, then the required reinforcement may be used for the entire block. However, if the spacing is $< 0.3b$ or $s < 0.3D$, the two-in-one effects have to be accounted for. Note that b and D are the smaller dimensions in the two beams as shown in Figures 18.4(1)a and b, respectively.

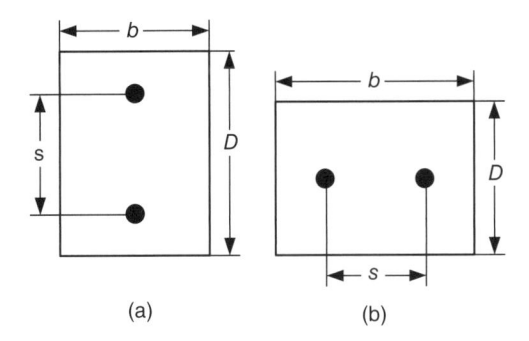

Figure 18.4(1) End faces of post-tensioned beams

18.4.2 Symmetrical prisms and design bursting forces

In the design of an end block for a given anchorage, the corresponding 'symmetrical prism' is defined in Figure 18.4(2)a. The volume of bursting stress, as discussed in Section 18.3.1, is replaced by two pairs of forces or stress resultants acting along the line of action of the tendon force (as per Clause 12.5.4). These forces are identified in Figures 18.4(2)b and c as T_1 acting horizontally and T_2 acting vertically. The values of these stress resultants are given as

$$T_1 = 0.25P(1 - k_{r1}) \, (\text{in kN})$$ Equation 18.4(1)

$$T_2 = 0.25P(1 - k_{r2}) \, (\text{in kN})$$ Equation 18.4(2)

where

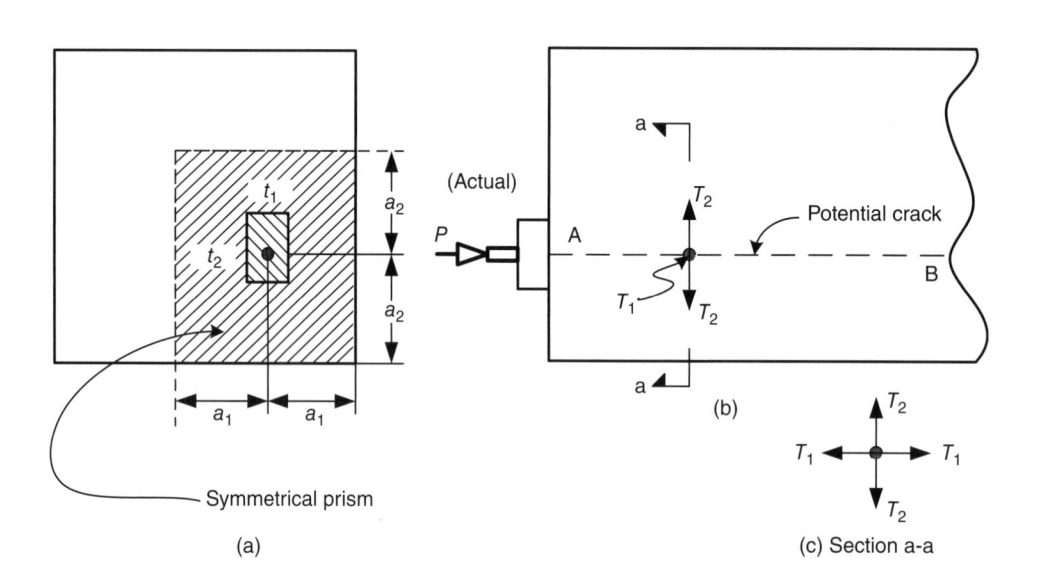

$$k_{r1} = t_1/2a_1 \quad \text{and} \quad k_{r2} = t_2/2a_2$$ Equation 18.4(3)

Figure 18.4(2) Symmetrical prism and transverse tensile forces

18.4.3 Design spalling force

Clause 12.5.5 recommends that the spalling-stress resultants due to an eccentric anchorage force as in Figure 18.4(3) may be represented by a pair of tensile forces, which are shown as T_3 in Figure 18.4(3)b with the locations of the potential cracks. Acting close to the end face of the beam, the design spalling force for a single eccentric anchorage is given as

$$T_3 = \frac{M_s}{0.5D}$$ Equation 18.4(4)

where the spalling moment M_s for an eccentricity of $e_B = D/4$ – see Figure 18.4(3) – is calculated as

$$M_s = 0.0495PD \qquad \text{Equation 18.4(5)}$$

in which P is the anchorage force and D is the overall depth of the beam. Note that Equation 18.4(5) is valid only for a D to t_2 ratio of 6 where t_2 – see Figure 18.4(2)a – is the bearing plate depth (Warner and Faulkes 1988).

For a pair of anchorages as shown in Figure 18.4(4)

$$T_3 = \frac{M_s}{0.6s} \qquad \text{Equation 18.4(6)}$$

where the spalling moment

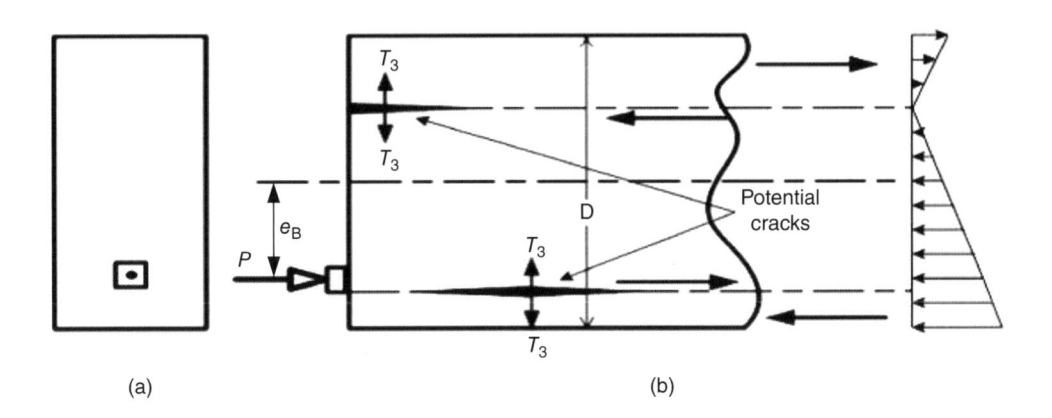

Figure 18.4(3) Beam with an eccentric anchorage

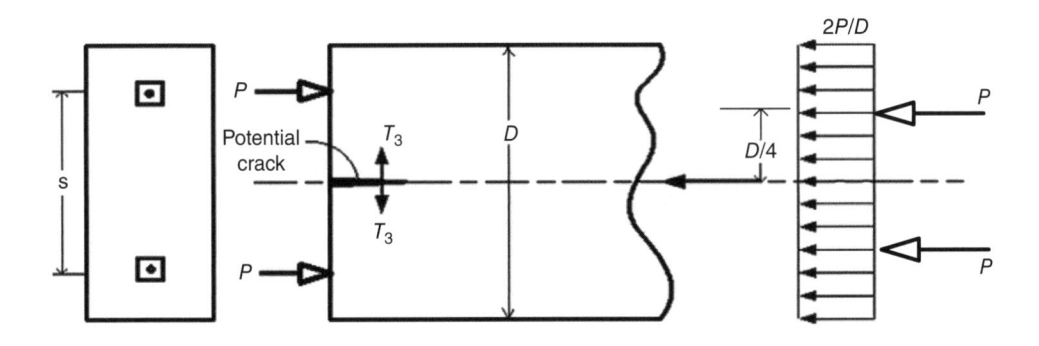

Figure 18.4(4) Beam with double symmetrical anchorages

$$M_s = P\left(\frac{s}{2} - \frac{D}{4}\right)$$ Equation 18.4(7)

in which P is the anchorage force, s is the anchorage spacing and D is the overall depth of the beam.

18.4.4 Design for bearing stress

Simple design guidelines and formulas for bearing stress in plain concrete or concrete without special confinement reinforcement are given in Clause 12.6. The bearing stress – see Figure 18.3(2) – is given as

$$\sigma_{be} \leq \phi 0.9 f'_c \sqrt{A_2/A_1} \quad \text{or} \quad \phi 1.8 f'_c$$ Equation 18.4(8)

whichever is the minimum, where A_1 is the bearing area, A_2 is the largest area of the supporting surface that is geometrically similar to and concentric with A_1, and $\phi = 0.6$ as per Clause 2.2.2(c)(ii). For examples of determining A_2, the reader is referred to the original source of Equation 18.4(8) in ACI 318–2005.

18.5 REINFORCEMENT AND DISTRIBUTION

With T_1, T_2 and T_3 now calculated, the corresponding reinforcement as per Clause 12.5.6 is determined as follows:

$$A_{s1} = T_1/150$$ Equation 18.5(1)

$$A_{s2} = T_2/150$$ Equation 18.5(2)

and

$$A_{s3} = T_3/150$$ Equation 18.5(3)

Note that in Equations 18.5(1), (2) and (3), T_1, T_2 and T_3 are in N, and A_{s1}, A_{s2} and A_{s3} are in mm^2.

The distributions of the reinforcement A_{s1} and A_{s2} for the bursting stress are detailed in Figure 18.5(1) as per Clause 12.5.6(a). The smaller reinforcement spacing specified near the extreme end of the beams is necessary due to the profile of the bond stress as shown in Figure 18.2(1)c.

The distribution of the spalling stress reinforcement is shown in Figure 18.5(2). Clause 12.5.6(b) recommends that A_{s3} should be provided as close to the loaded (end) face as possible.

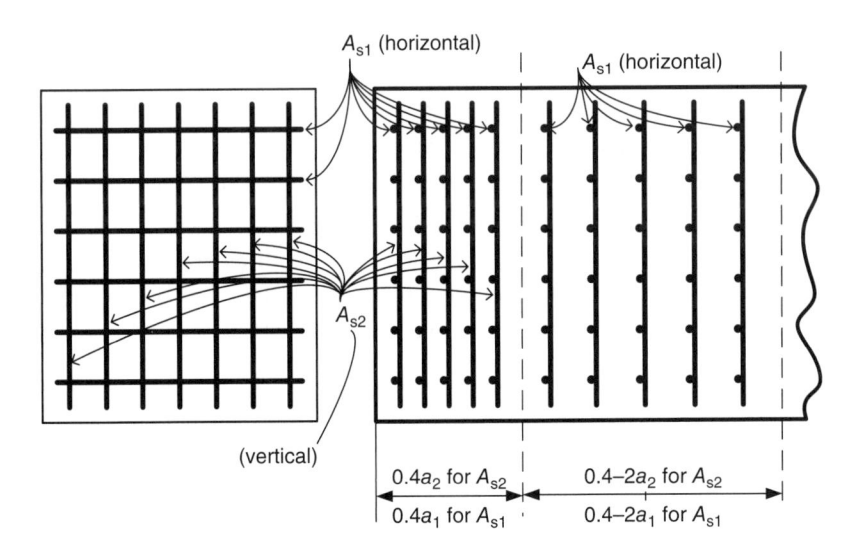

Figure 18.5(1) Distribution of bursting stress reinforcement

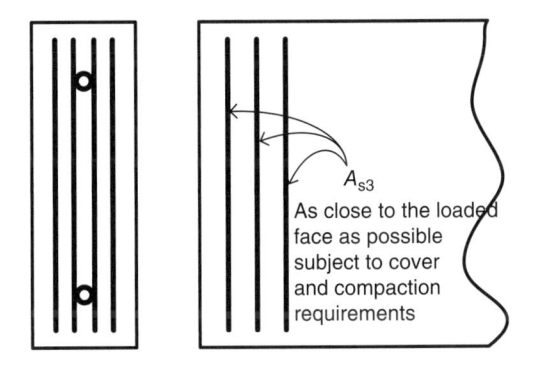

Figure 18.5(2) Distribution of spalling stress reinforcement

18.6 CRACK CONTROL

According to Clause 12.7 of the Standard, crack control requirements may be considered satisfied if the service stress in the reinforcement is not greater than the following:
- 250 MPa, where a minor degree of control over cracking is required
- 200 MPa, where a moderate degree of control over cracking is required
- 150 MPa, where a strong degree of control over cracking is required.

Note that in view of Equations 18.5(1), (2) and (3), where the steel stress is taken to be 150 MPa, the above requirements are automatically satisfied.

18.7 SUMMARY

The nature and distribution of the bond stress in the end zones of pretensioned beams are explained; the design method recommended in the Standard is introduced. Three types of stresses induced by a post-tensioned anchorage system are identified. Both the bursting stress and the spalling stress are tensile and orthogonal or transverse in direction to the axis of the post-tensioned tendon. The third type is the (compressive) bearing stress on the concrete right behind the steel anchor heads. Finally, the design for the bursting, spalling and bearing stresses is elaborated together with the distribution and detailing of the end-block reinforcement.

APPENDIX A   ELASTIC NEUTRAL AXIS

The elastic, or working stress method, has been dropped from the various design codes and Standards for many years. However, some of the analysis equations remain useful in computing deflections. In particular, neutral axis (NA) positions are needed in the assessments of the effective moments of inertia.

The procedure for determining the position of the NA of a fully cracked rectangular section is summarised in this appendix, accompanied by some useful formulas.

Considering the reinforced concrete beam in Figure A(1)b and taking moments of the areas about the elastic NA give for concrete

$$b(kd)^2/2 \qquad \text{Equation A(1)}$$

and for steel

$$nA_s(d-kd) \qquad \text{Equation A(2)}$$

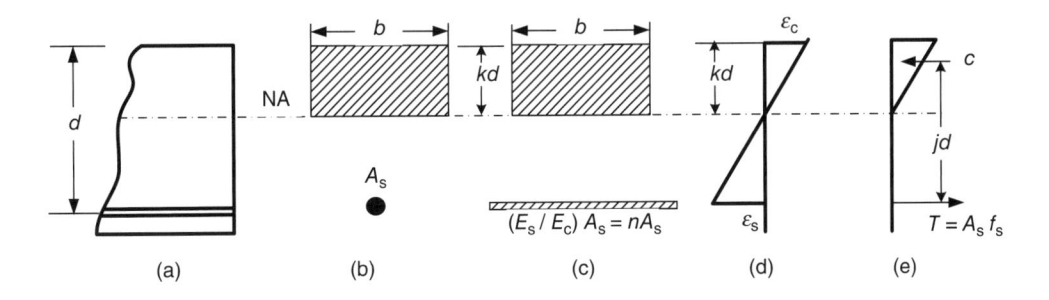

Figure A(1) Elastic neutral axis location for, and stress and strain distribution across, a rectangular reinforced concrete beam section

Imposing the requirement that Equation A(1) = Equation A(2) gives

$$b(kd)^2/2 = nA_s(d-kd) \qquad \text{Equation A(3)}$$

Now let the steel ratio be

$$p = \frac{A_s}{bd} \qquad \text{Equation A(4)}$$

Equation A(3) may then be written as

$$k^2 + 2pnk - 2pn = 0$$

APPENDIX A ELASTIC NEUTRAL AXIS

from which the neutral axis parameter

$$k = \sqrt{(pn)^2 + 2pn} - pn \qquad \text{Equation A(5)}$$

For the rectangular box section detailed in Figure A(2) and using the same procedure as above, the NA position can be expressed as

$$kd = \left\{ \sqrt{[np_1 + (\beta - 1)r]^2 - 2[np_1 + (\beta - 1)r^2/2]} - [np_1 + (\beta - 1)r] \right\} d \qquad \text{Equation A(6)}$$

where $n = E_s/E_c$; $p_1 = A_{st}/b_w d$; $\beta = b/b_w$; and $r = t/d$

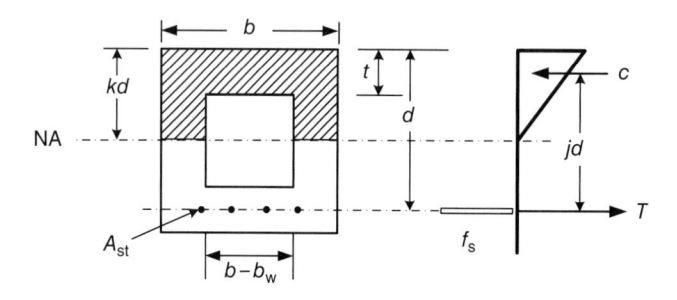

Figure A(2) Elastic neutral axis location for a rectangular box section

APPENDIX B CRITICAL SHEAR PERIMETER

The formulas for computing the values of the critical shear perimeter (u) for two commonly encountered column positions have been given in Section 9.6.3. Additional formulas are presented here for the corner and edge columns illustrated in Figure B(1).

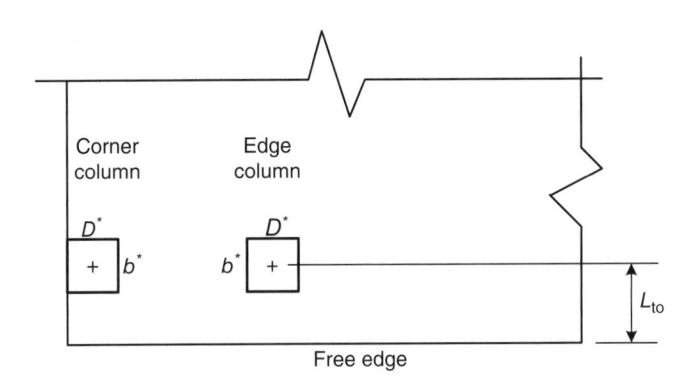

Figure B(1) Location of corner and edge columns

The following formulas are for corner columns:
- If $L_{to} = b^*/2$, then

$$u = b^* + D^* + d_{om} \qquad \text{Equation B(1)}$$

where b^* and D^* are the width and overall depth (respectively) of the column or drop panel as applicable; d_{om} is the mean shear effective depth of the slab.
- If $L_{to} \geq \frac{b^*}{2} + D^* + d_{om}$, then

$$u = b^* + 2(D^* + d_{om}) \qquad \text{Equation B(2)}$$

- If $L_{to} < D^* + d_{om} + \frac{b^*}{2}$, then

$$u = \frac{b^*}{2} + D^* + d_{om} + L_{to} \qquad \text{Equation B(3)}$$

For edge columns, the formulas below apply:
- If $L_{to} = b^*/2$, then

$$u = 2b^* + D^* + 2d_{om} \qquad \text{Equation B(4)}$$

- If $L_{to} \geq \frac{(b^* + D^*)}{2} + d_{om}$, then

$$u = 2(b^* + D^* + 2d_{om})$$ **Equation B(5)**

- If $L_{to} < \frac{(b^* + D^*)}{2} + d_{om}$, then

$$u = b^* + D^* + 2d_{om} + L_{to}$$ **Equation B(6)**

APPENDIX C AUSTRALIAN STANDARD PRECAST PRESTRESSED CONCRETE BRIDGE GIRDER SECTIONS

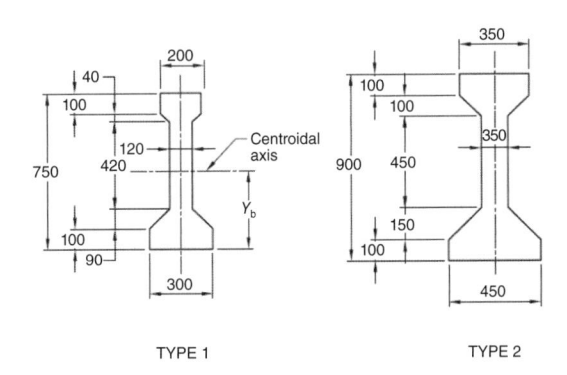

Girder type	A_g (mm^2)	$Z_{y.sup}$ (mm^3)	$Z_{y.sub}$ (mm^3)	I (mm^4)	$d_{y.sub}$ (mm)	Hypothetical thickness, t_h (girders only)(mm)
1	126 × 10^3	17.9 × 10^6	22.0 × 10^6	7400 × 10^6	337	120
2	218 × 10^3	41.1 × 10^6	48.1 × 10^6	19950 × 10^6	415	155
3	317 × 10^3	82.9 × 10^6	91.1 × 10^6	49900 × 10^6	548	180
4	443 × 10^3	135.9 × 10^6	168.6 × 10^6	105330 × 10^6	625	205

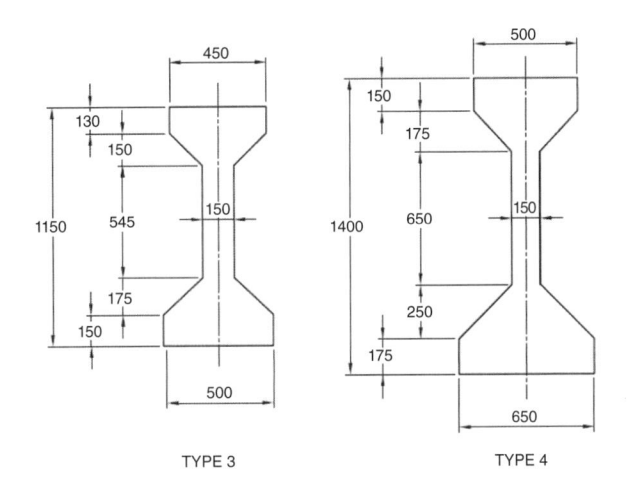

Dimensions in millimetres

Figure C(1) Standard precast prestressed concrete I-girder sections

Source: Standards Australia (2017). *AS 5100.5–2017 Bridge Design Part 5: Concrete*. Standards Australia Limited, Sydney, NSW. © Standards Australia Limited. Copied by Cambridge University Press with the permission of Standards Australia and Standards New Zealand under Licence 1712-c038.

Figure C(2) Standard precast prestressed concrete super T-girder sections

Source: Standards Australia (2017). *AS 5100.5–2017 Bridge Design Part 5: Concrete*. Standards Australia Limited, Sydney, NSW. © Standards Australia Limited. Copied by Cambridge University Press with the permission of Standards Australia and Standards New Zealand under Licence 1712-c038.

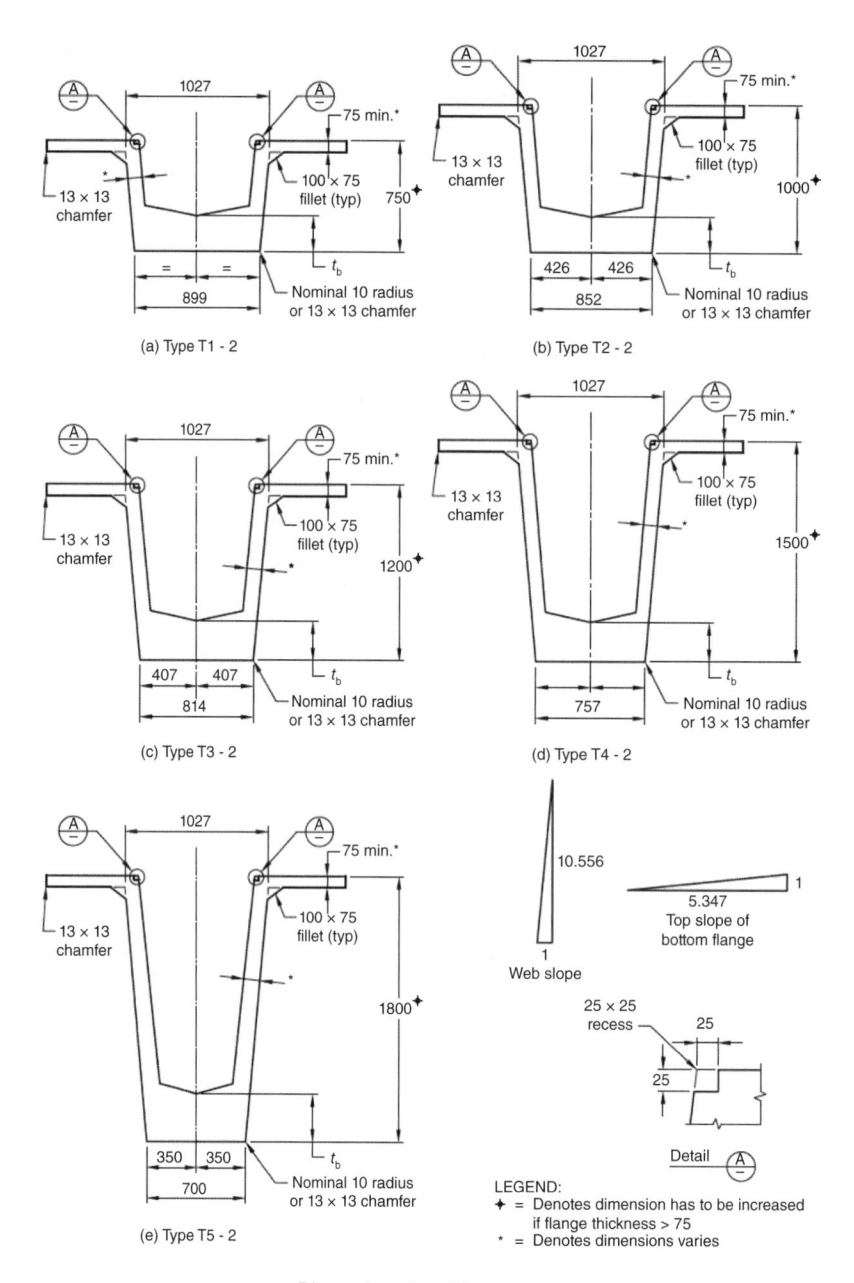

REFERENCES

AASHTO LRFD (1998). *AASHTO LRFD Bridge Design Specifications*. 2nd edition, American Association of State Highway and Transportation Officials, Washington DC.

AASHTO LRFD (2008). *AASHTO LRFD Bridge Design Specifications*. 4th edition – 2008 Interim Revisions, American Association of State Highway and Transportation Officials, Washington DC.

ACI 318–1995 (1995). *Building Code Requirements for Structural Concrete*. American Concrete Institute, Farmington Hills.

ACI 318–2002 (2002). *Building Code Requirements for Structural Concrete*. American Concrete Institute, Farmington Hills.

ACI 318–2005 (2005). *Building Code Requirements for Structural Concrete*. American Concrete Institute, Farmington Hills.

ACI 318–2008 (2008). *Building Code Requirements for Structural Concrete*. American Concrete Institute, Farmington Hills.

ACI 318–2014 (2014). *Building Code Requirements for Structural Concrete*. American Concrete Institute, Farmington Hills.

Alcocer SM and Uribe CM (2008). Monolithic and cyclic behaviour of deep beams designed using strut-and-tie models. *ACI Structural Journal* 105(3):327–337.

Ali MA and White RN (2001). Automatic generation of truss model for optimal design of reinforced concrete structures. *ACI Structural Journal* 98(4):431–442.

Arabzadeh A, Rahaie AR and Aghayari R (2009). A simple strut-and-tie model for prediction of ultimate shear strength of RC deep beams. *International Journal of Civil Engineering* 7(3):141–153.

AS 1012.9–2014 (2014). *Determination of the Compressive Strength of Concrete Specimens*. Standards Australia Limited, Sydney.

AS 1012.10–2000 (R2014) (2000, Reconfirmed 2014). *Determination of Indirect Tensile Strength of Concrete Cylinders (Brazil or Splitting Test)*. Standards Australia International Ltd, Strathfield.

AS 1012.11–2000 (R2014) (2000, Reconfirmed 2014). *Determination of the Modulus of Rupture*. Standards Australia International Ltd, Strathfield.

AS 1012.12.1–1998 (R2014) (1998, Reconfirmed 2014). *Determination of Mass per Unit Volume of Hardened Concrete – Rapid Measuring Method*. Standards Australia, North Sydney.

AS 1012.12.2–1998 (R2014) (1998, Reconfirmed 2014). *Determination of Mass per Unit Volume of Hardened Concrete – Water Displacement Method*. Standards Australia, North Sydney.

AS 1012.17–1997 (R2014) (1997, Recoinfirmed 2014). *Determination of the Static Chord Modulus of Elasticity and Poisson's Ratio of Concrete Specimens*. Standards Australia, North Sydney.

AS 1379–2007 (2007). *Specification and Supply of Concrete*. Standards Australia Limited, Sydney.

AS 1480–1982 (1982). *SAA Concrete Structures Code*. Standards Association of Australia, North Sydney.

AS 1481–1978 (1978). *SAA Prestressed Concrete Code*. Standards Association of Australia, North Sydney.

AS 3600–1988 (1988). *Concrete Structures*. Standards Association of Australia, North Sydney.

AS 3600–1994 (1994). *Concrete Structures*. Standards Australia, Homebush.

AS 3600–2001 (2001). *Concrete Structures*. Standards Australia International Ltd, Sydney.

AS 3600–2009 (2009). *Concrete Structures*. Standards Australia Limited, Sydney.

AS 3735–2001 (2001). *Concrete Structures for Retaining Liquids*. Standards Australia International Ltd, Sydney.

AS 4678–2002 (2002). *Earth-Retaining Structures*. Standards Australia, Sydney.

AS 5100.5–2017 (2017). *Bridge Design Part 5: Concrete*. Standards Australia Limited, Sydney.

AS/NZS 1170.0–2002 (2002). *Structural Design Actions Part 0: General Principles*. Standards Australia, Sydney and Standards New Zealand, Wellington.

AS/NZS 1170.1–2002 (R2016) (2002, Reconfirmed 2016). *Structural Design Actions Part 1: Permanent, Imposed and Other Actions*. Standards Australia, Sydney and Standards New Zealand, Wellington.

AS/NZS 4671–2001 (2001). *Steel Reinforcing Materials*. Standards Australia, Sydney and Standards New Zealand, Wellington.

AS/NZS 4672.1–2007 (2007). *Steel Prestressing Materials Part 1: General Requirements*. Standards Australia International Ltd, Sydney and Standards New Zealand, Wellington.

Bakir PG and Boduroglu HM (2005). Mechanical behaviour and non-linear analysis of short beams using softened truss and direct strut & tie models. *Engineering Structures* 27:639–651.

BCA (Building Code of Australia) (2011). *Building Code of Australia*, Australian Building Codes Board, Canberra.

Bowles JE (1996). *Foundation Analysis and Design*. 5th edition, McGraw-Hill, New York.

Bridge RO and Smith RG (1983). The assessment of the ultimate strain of concrete. *Proceedings, Symposium on Concrete 1983*. The Institution of Engineers, Australia: 75–82.

Brown MD and Bayrak O (2006). Minimum transverse reinforcement for bottle-shaped struts. *ACI Structural Journal* 103(6):813–821.

Brown MD and Bayrak O (2007). Investigation of deep beams with various load configurations. *ACI Structural Journal* 104(5):611–620.

Brown MD and Bayrak O (2008a). Design of deep beams using strut-and-tie models – Part I: Evaluating US provisions. *ACI Structural Journal* 105(4):395–404.

Brown MD and Bayrak O (2008b). Design of deep beams using strut-and-tie models – Part II: Design recommendations. *ACI Structural Journal* 105(4):405–413.

Bruggi M (2009). Generating strut-and-tie patterns for reinforced concrete structures using topology optimization. *Computers and Structures* 87:1483–1495.

BS 8007–1987 (1987). *Code of Practice for Design of Concrete Structures for Retaining Aqueous Liquids*. British Standards Institution, London.

BS 8110–1985 (1985). *Structural Use of Concrete Part 2. Code of Practice for Special Circumstances.* British Standards Institution, London.

BS 8110–1997 (1997). *Part 1: Structural Use of Concrete, Code of Practice for Design and Construction.* British Standards Institution, London.

CCAA (Cement Concrete & Aggregates Australia) (2010). *Fire Safety of Concrete Buildings.* Cement Concrete & Aggregates Australia, Sydney.

CEB (Comite Euro-International du Beton) (1978). *CEB-FIP Model Code for Concrete Structures.* English edition, Cement and Concrete Association, London.

Chi M and Kirstein AF (1958). Flexural cracks in reinforced concrete beams. *ACI Journal Proceedings* 54 (April):865–878.

Chowdhury SH and Loo YC (2001). A new formula for prediction of crack widths in reinforced and partially prestressed concrete beams. *Advances in Structural Engineering* 4(2):101–110.

Chowdhury SH and Loo YC (2002). Crack width predictions of concrete beams – New formulas and comparative study. In: *Proceedings of the 17th Australasian Conference on the Mechanics of Structures and Materials* (ACMSM 17), Gold Coast, 12–14 June 2002, 157–162.

Chowdhury SH and Loo YC (2006). Damping in high strength concrete beams – Prediction and comparison. In: *Proceedings of the 19th Australasian Conference on the Mechanics of Structures and Materials* (ACMSM 19), Christchurch, 29 November–1 December 2006, 631–636.

Chun S-C and Hong S-G (2009). Strut-and-tie models for headed bar development in C-C-T nodes. *ACI Structural Journal* 106(2):123–131.

CIA (Concrete Institute of Australia) (2010). *Reinforcement Detailing Handbook.* Concrete Institute of Australia, Rhodes.

Clark AP (1956). Cracking in reinforced concrete flexural members. *ACI Journal Proceedings* 52 (April):851–862.

CSA A23.3 (1994). *Design of Concrete Structures.* Canadian Standards Association, Rexdale.

Cusens AR and Pama RP (1975). *Bridge Deck Analysis.* John Wiley & Sons, London.

Darval PL and Brown HP (1976). *Fundamental of Reinforced Concrete Analysis and Design.* Macmillan, Melbourne.

Darwin D, Dolan CW and Nilson AH (2016). *Design of Concrete Structures.* 15th edition, McGraw-Hill, New York.

Doh JH and Fragomeni S (2004). Evaluation of experimental work on concrete walls in one- and two-way action. *Australian Journal of Structural Engineering* 6(1):37–52.

Elia G, Palmisano F, Vitone A and Vitone C (2002). An interactive procedure to design strut and tie models in reinforced concrete structures using the 'evolutionary structural optimisation' method. *Engineering Design Optimization. Product and Process Improvement*, P. Gosling (ed.), Newcastle, University of Newcastle-upon-Tyne: 60–66.

Eurocode 2 (EC2): ENV 1992–1–1 (1992). *Design of Concrete Structures, Part 1: General Rules and Rules for Buildings.* European Prestandard.

Eurocode 2 (EC2): BS EN 1992–1–1 (2004). *Design of Concrete Structures: General Rules and Rules for Buildings*. BSI, London.

Falamaki M and Loo YC (1992). Punching shear tests of half-scale reinforced concrete flat plate models with spandrel beams. *American Concrete Institute Structural Journal* 89(3):263–271.

Foster SJ, Kilpatrick AE and Warner RF (2010). *Reinforced Concrete Basics 2E – Analysis and Design of Reinforced Concrete Structures*. Pearson Australia, Frenchs Forest.

Fu CC (2001). *The strut-and-tie model of concrete structures*. The Bridge Engineering Software & Technology (BEST) Centre, University of Maryland, presented to the Maryland State Highway Administration, Baltimore, 21 August 2001.

Garda V, Guan H and Loo YC (2006). Prediction of punching shear failure behaviour of slab-edge column connections with varying size and location of openings. In: *Proceedings of the 19th Australasian Conference on the Mechanics of Structures and Materials* (ACMSM 19), Christchurch, 29 November–1 December 2006, 651–657.

Gilbert Rl (2008). Revisiting the tension stiffening effect in reinforced concrete slabs. *Australian Journal of Structural Engineering* 8(3):189–196.

Gilbert RI, Mickleborough NC and Ranzi G (2016). *Design of Prestressed Concrete to AS3600–2009*. 2nd edition, CRC Press, Boca Raton.

Guan H (2009). Prediction of punching shear failure behaviour of slab-edge column connections with varying opening and column parameters. *Advances in Structural Engineering* 12(1): 19–36.

Guan H, Chen YJ, Loo YC, Xie YM and Steven GP (2003). Bridge topology optimisation with stress, displacement and frequency constraints. *Computers and Structures* 81:131–145.

Guan H and Doh JH (2007). Development of strut-and-tie model in deep beams with web openings. *Advances in Structural Engineering* 10(6):697–711.

Guan H, Eurviriyanukul S and Loo YC (2006). Strut-and-tie models of deep beams influenced by size and location of web openings – An optimization approach. In: *Proceedings of the 10th East Asia-Pacific Conference on Structural Engineering and Construction* (EASEC-10), Bangkok, 3–5 August 2006, 143–148.

Guan H and Loo YC (1997). Layered finite element method in cracking and failure analysis of RC beams and beam-column-slab connections. *Structural Engineering and Mechanics* 5(1):645–662.

Guan H and Loo YC (2003). Failure analysis of column-slab connections with stud shear reinforcement. *Canadian Journal of Civil Engineering* 30:934–944.

HB71–2002 (2002). *Reinforced Concrete Design in Accordance with AS 3600–2001*. Standards Australia, Sydney.

HB71–2011 (2011). *Reinforced Concrete Design in Accordance with AS 3600–2009*. Standards Australia, Sydney.

Hognestad E (1962). High strength bars as concrete reinforcement – Part 2: Control of flexural cracking. *Journal of the PCA Research and Development Laboratories* 4(1):46–63.

Hong SG, Chun SC, Lee SH and Oh B (2007). Strut-and-tie model for development of headed bars in exterior beam-column joint. *ACI Structural Journal* 104(5):590–600.

Hsu TTC (1988). Softened truss model theory for shear and torsion. *ACI Structural Journal* 85(6):552–561.

Hsu TTC (1993). *Unified Theory of Reinforced Concrete*. CRC Press, Boca Raton, 1993.

Hwang SJ, Yu HW and Lee HJ (2000). Theory of interface shear capacity of reinforced concrete. *Journal of Structural Engineering* 126(6):700–707.

Kaar PH and Mattock AH (1963). High strength bars as concrete reinforcement – Part 4: Control of cracking. *Journal of the PCA Research and Development Laboratories* 5(1):15–38.

Kong FK and Evans RH (1987). *Reinforced and Prestressed Concrete*. 3rd edition, E & FN Spon, London.

Kuchma D, Yindeesuk S, Nagle T, Hart J and Lee HH (2008). Experimental validation of strut-and-tie method for complex regions. *ACI Structural Journal* 105(5):578–589.

Kwak H-G and Noh S-H (2006). Determination of strut-and-tie models using evolutionary structural optimization. *Engineering Structures* 28:1440–1449.

Lei IM (1983). *Size effects in model studies of concrete in torsion*. MEng thesis, Asian Institute of Technology, Bangkok.

Leu LJ, Huang CW, Chen CS and Liao YP (2006). Strut-and-tie methodology for three-dimensional reinforced concrete structure. *Journal of Structural Engineering* 132(6):929–938.

Ley MT, Riding KA, Widiento BS and Breen JE (2007). Experimental verification of strut-and-tie model design model. *ACI Structural Journal* 104(6):749–755.

Liang QQ, Xie YM and Steven GP (2000). Topology optimization of strut-and-tie models in reinforced concrete structures using an evolutionary procedure. *ACI Structural Journal* 97(2):322–332.

Liang QQ, Xie YM and Steven GP (2001). Generating optimal strut-and-tie models in prestressed concrete beams by performance-based optimization. *ACI Structural Journal* 98(2):226–232.

Lin TY and Burns NH (1981). *Design of Prestressed Concrete Structures*. 3rd edition, John Wiley & Sons, New York.

Loo YC (1988). *UNIDES-FP An Integrated Package for Design of Reinforced Concrete Fiat Plates on Personal Computer – User Manual*. Wollongong Uniadvice Ltd, Wollongong.

Loo YC and Chiang CL (1996). Punching shear strength analysis of post-tensioned flat plates with spandrel beams. *The Structural Engineer* 74(12):199–206.

Loo YC and Falamaki M (1992). Punching shear strength analysis of reinforced concrete flat plates with spandrel beams. *ACI Structural Journal* 89(4):375–383.

Loo YC and Sutandi TD (1986). Effective flange width formulas for T-beams. *Concrete International* 8(2):40–45.

Loo YC and Wong YW (1984). Deflection of reinforced box beams under repeated loadings. *Journal of the American Concrete Institute* 81(1):87–94.

Lopes SM and do Carmo RNF (2006). Deformable strut and tie model for the calculation of the plastic rotation capacity. *Computers and Structures* 84(31–32):2174–2183.

Lorell W (1968). *Advanced Reinforced Concrete Design – Lecture Notes*. Unpublished notes presented at the Asian Institute of Technology, Bangkok and recorded by Loo YC during the lecture, 3 January 1968.

Magnel G (1954). *Prestressed Concrete*. 3rd edition, McGraw-Hill, New York.

Mattock AH (2012). Strut-and-tie models for dapped-end beams. *Concrete International* February:35–40.

Maxwell BS and Breen JE (2000). Experimental evaluation of strut-and-tie applied to deep beam with opening. *ACI Structural Journal* 97(1):142–148.

Murdock LJ and Brook KM (1982). *Concrete Materials and Practice*. 5th edition, Edward Arnold, London.

Nagarajan P, Jayadeep UB and Pillai TMM (2009). Application of micro truss and strut and tie model for analysis and design of reinforced concrete structural elements. *Songklanakarin Journal of Science and Technology* 31(6):647–653.

Nagarajan P and Pillai TMM (2008). Development of strut and tie models for simply supported deep beams using topology optimization. *Songklanakarin Journal of Science and Technology* 30(5):641–647.

Nawy EG (1984). Flexural cracking of pre- and post-tensioned flanged beams. In: *Partial Prestressing, From Theory to Practice*, vol II, *Proceedings of the NATO Advanced Research Workshop*, Cohn MZ (ed.), Paris, 18–22 June 1984, 137–156.

Park JW and Kuchma D (2007). Strut-and-tie model analysis for strength prediction of deep beams. *ACI Structural Journal* 104(6):657–666.

Park R and Gamble WL (2000). *Reinforced Concrete Slabs*. 2nd edition, John Wiley & Sons, New York.

Park R and Paulay T (1975). *Reinforced Concrete Structures*. John Wiley & Sons, New York.

Perera R and Vique J (2009). Strut-and-tie modelling of reinforced concrete beams using genetic algorithms optimisation. *Construction and Building Materials* 23(8):2914–2925.

Perera R, Vique J, Arteaga A and Diego AD (2009). Shear capacity of reinforced concrete members strengthened in shear with FRP by using strut-and-tie models and genetic algorithms. *Composites Part B: Engineering* 40:714–726.

Post-Tensioning Institute (2006). *Post-Tensioning Manual*. 6th edition, Post-Tensioning Institute, Phoenix.

Quintero-Febres CG, Parra-Montesinos G and Weight JK (2006). Strength of struts in deep concrete members designed using strut-and-tie method. *ACI Structural Journal* 103(4):577–586.

Rangan BV (1993). Discussion. Punching shear tests of half-scale reinforced concrete flat plate models with spandrel beams, by Falamaki M and Loo YC 89-S25. *ACI Structural Journal* 90(2):219.

Russo G, Venir R and Pauletta M (2005). Shear strength model and design formula. *ACI Structural Journal* 102(3):429–437.

Ryan WG and Samarin A (1992). *Australian Concrete Technology*. Longman Cheshire, Melbourne.

Sahoo DK, Singh B and Bhargava P (2009). Investigation of dispersion of compression in bottle-shaped struts. *ACI Structural Journal* 106(2):178–186.

Salzmann A, Fragomeni S and Loo YC (2003). The damping analysis of experimental concrete beams under free-vibration. *Advances in Structural Engineering* 6(1):53–64.

Schlaich J and Schafer K (1991). Design and detailing of structural concrete using strut-and-tie models. *The Structural Engineer* 69(6):113–125.

Schlaich J, Schafer K and Jennewein M (1987). *Toward a Consistent Design of Structural Concrete*. Special report to the Comite Euro-International du Beton, Lausanne.

Skibbe E (2010). *A Comparison of Design Using Strut-and-Tie Modeling and Deep Beam Method for Transfer Girders in Building Structures*. MSc thesis, Kansas State University, Manhattan, Kansas.

Souza R, Kuchma D, Park JW and Bittencourt T (2009). Adaptable strut-and-tie model for design and verification of four-pile caps. *ACI Structural Journal* 106(2):142–150.

Tang CY and Tan KH (2004). Interactive mechanical model for shear strength of deep beams. *Journal of Structural Engineering* 130(10):1634–1544.

Timoshenko SP and Woinowsky-Kreiger S (1959). *Theory of Plate and Shells*. 2nd edition, McGraw-Hill, New York.

Tjhin TN and Kuchma DA (2007). Integrated analysis and design tool for the strut-and-tie method. *Engineering Structures* 29(11):3042–3052.

To NHT, Ingham JM and Sritharan S (2000). Cyclic strut-and-tie modeling of simple reinforced concrete structures. *Proceedings of 12WCEE* 1249:1–8.

Tuchscherer RG, Birrcher DB and Bayrak O (2011). Strut-and-tie model design provisions. *PCI Journal* 56(1):155–170.

Wang G-L and Meng S-P (2008). Modified strut-and-tie model for prestressed concrete deep beams. *Engineering Structures* 30(12):3489–3496.

Warner RF and Faulkes KA (1988). *Prestressed Concrete*. 2nd edition, Longman Cheshire, Melbourne.

Warner RF, Rangan BV and Hall AS (1989). *Reinforced Concrete*. 3rd edition, Longman Cheshire, Melbourne.

Warner RF, Rangan BV, Hall AS and Faulkes KA (1998). *Concrete Structures*. Longman, South Melbourne.

Wight JK and Parra-Montesinos GJ (2003). Strut-and-tie model for deep beam design – A practical exercise using Appendix A of the 2002 ACI Building Code. *Concrete International* May:63–70.

Wong YW and Loo YC (1985). Deflection analysis of reinforced and prestressed concrete beams under repeated loading. In: *Proceedings of the Concrete 85 Conference*, Brisbane, 23–25 October 1985, Institution of Engineers, Australia National Conference Publication no. 85/17, 138–142.

Yun YM and Lee WS (2005). Nonlinear strut-and-tie model analysis of pre-tensioned concrete deep beams. *Journal of Advances in Structural Engineering* 8(1):85–98.

Zhang N and Tan K-H (2007). Direct strut-and-tie model for single span and continuous deep beams. *Engineering Structures* 29:2987–3001.

INDEX

action effects, 3, 19, 35, 297
active earth pressure coefficient, 348
American Concrete Institute (ACI)
    ACI 318–1995 *Building Code Requirements for Structural Concrete*, 46, 114
    ACI 318–2002 *Building Code Requirements for Structural Concrete*, 387
    ACI 318–2005 *Building Code Requirements for Structural Concrete*, 386
    ACI 318–2008 *Building Code Requirements for Structural Concrete*, 386
    ACI 318–2014 *Building Code Requirements for Structural Concrete*, 63, 68, 100, 107, 112, 297, 299, 303
anchorage. *See* bond; end blocks
arbitrary beam sections. *See* beams:nonstandard sections
arbitrary column sections. *See* columns
Australian Standards (AS)
    AS 1012 *Concrete Test Methods*, 13–15, 18
    AS 1379–2007 *Specification and Supply of Concrete*, 13
    AS 1480–1982 *SAA Concrete Structures Code*, 3–4, 11, 26, 41, 46, 64, 66, 269
    AS 1481–1978 *SAA Prestressed Concrete Code*, 3
    AS 3600–1988 *Concrete Structures*, 3, 19, 127
    AS 3600–1994 *Concrete Structures*, 3, 103
    AS 3600–2001 *Concrete Structures*, 3, 24, 103, 159
    AS 3600–2009 *Concrete Structures*, 3–5, 9–10, 13, 15, 19, 22, 37, 39, 64, 94, 98, 103, 112, 127, 140, 157, 165, 172, 175, 180, 185, 193, 195, 210, 219, 257–8, 284, 296, 316, 377, 387–8, 390–1, 394, 416, 437, 458, 473–4
    AS 4678–2002 *Earth-Retaining Structures*, 348, 353
    AS 5100.5–2017 *Bridge Design Part 5 – Concrete*, 467
    AS/NZS 1170.0–2002 *Structural Design Actions – General Principles*, 5, 7
    AS/NZS 1170.1–2002 *Dead and Live Load Code*, 18, 211, 238
    AS/NZS 4671–2001 *Steel Reinforcing Materials*, 15
    AS/NZS 4672.1–2007 *Steel Prestressing Materials Part 1 – General Requirements*, 412, 475

backfills
    cohesionless, 353, 355
    definition of, 311
    granular, 353
    internal friction angle of, 353
    pressure exerted by, 311, 345–6
    properties of, 353–4
    slope angle of, 348
balanced failure
    of beams, 22, 46
    of columns, 260, 265
bar groups (areas of), 15–16
beams
    balanced failure of, 22–3, 46
    compression failure of, 23, 26
    definitions of analysis and design of, 19
    doubly reinforced, 19, 31, 36, 40, 44–5, 50–2, 54, 67, 70
    effective (reliable) moment capacity, 27
    flanged sections. *See* flanged beams
    free design of, 35
    nonstandard sections, 74
    over-reinforced, 23–4, 26, 29, 31, 51
    restricted design of, 35–6, 40, 69, 243, 331
    serviceability design of, 97, 102
    singly reinforced, 22, 28, 31, 35, 40, 47, 51, 69, 71
    tension failure of, 23, 25, 41, 80, 460, 463
    T-sections. *See* T-beams
    under-reinforced, 23–4, 28, 32
bearing stress, 473, 475, 477, 480
bond
    anchorage, 81, 171, 173
    deformed bars, 172
    failure, 20, 171, 474
    mechanism of resistance, 172
    plain bars, 172, 176, 178
    strength, 171–3, 177
    ultimate stress, 171

INDEX 497

box beams, 108, 475
box systems, 65
bundled bars, 9, 12, 178–9, 287
   development length of, 179
bursting stress, 473, 476–7

cantilever beams, 95–6, 104, 106, 120–1, 148, 151, 182, 451–2
capacity reduction factor
   for bending moment, 3, 20, 27
   for columns, 266, 279
   for shear, 129
   for torsion, 157
   for walls, 298
characteristic strength of concrete
   compressive, 4, 13, 22
   strength grades, 4, 10
   tensile, 14, 155
chemical bond, 172
coefficients
   active earth pressure, 348
   passive earth pressure, 348
   thermal expansion of concrete, 15
   thermal expansion of steel, 17
cogs, 173, 176
   development length of, 177
column strips. *See* slabs
columns
   arbitrary cross-sections, 269
   balanced failure of, 260, 264–5
   biaxial bending effects on, 285
   braced, 281–3
   capacity reduction factor for, 257–8
   centrally loaded, 256–7
   compression failure of, 260–1, 264, 267, 280
   in uniaxial bending, 256, 258
   interaction diagrams for, 256, 263, 266–7
   moment magnifiers for, 257, 282
   short, 281, 285
   tension failure of, 260–1, 264–5, 267, 280
   unbraced, 284
Comité Européen du Béton – Fédération Internationale de la Precontrainte (CEB-FIP), 63
compression failure
   of beams, 23, 26
   of columns, 260–1, 264, 267

concrete cover for reinforcement
   and exposure classifications, 4
   for fire resistance, 4, 7, 9, 35, 39, 81, 213
   for rigid formwork and intense compaction, 10
   for standard formwork and compaction, 10
   maximum, 9, 112
   minimum, 9, 134
continuous beams, 64, 81, 96, 102–3, 106
crack control
   beams, 9, 11, 94
   end blocks, 481
   slabs, 11, 192, 210, 222
   struts, 389
   walls, 304, 309
crack widths
   ACI Code on, 114
   average, 112, 114
   British Standard on, 112, 114
   Eurocode on, 112, 114
   formulas on, 114
   maximum, 112, 115
cracking moment, 14, 99, 109, 459, 463
creep, of concrete, 15, 104, 109, 193, 416
criteria for yielding of reinforcement
   in doubly reinforced beams, 40
   in singly reinforced beams, 31
   in tension, 46
critical shear perimeter, 227, 229, 326, 332, 336, 343, 485
critical stress state method
   analysis, 403, 414
   Case A prestressing, 425–7, 438, 450, 452
   Case B prestressing, 425–7, 429, 449
   criteria for, 437
   design, 403, 438, 440
   maximum external moments, 422, 425, 450, 453
   minimum external moments, 422, 425, 450, 453
   permissible stresses, 427, 458
   theory, 403, 407

damping, 94
decompression points, 261–2, 268
deflection
   accumulated, 108
   effect of creep and shrinkage, 40, 104
   effective span for, 96

deflection (cont.)
　formulas, 95
　limits, 95
　long-term, 94, 102, 104
　multiplier methods for, 104, 106
　of slabs, 193, 209, 236
　short-term, 98
　total, 94, 104, 108, 248
　under repeated loading, 107
density of concrete, 15, 18
design strips. *See* slabs
development length
　basic, 171, 474
　of bars in compression, 177
　of bundled bars, 178
　of cogs, 177
　of deformed bars, 173, 176
　of hooks, 177
　of lapped splices, 178
　of plain bars, 178
　refined, 173
diagonal cracks, 125
doubly reinforced beams. *See* beams
drop panels, 187, 216, 227, 248
durability design, 4, 7, 9, 14, 35, 81, 94, 437

effective flange width
　code recommendations for, 63
　definition of, 63
　formulas, 64–5, 116, 129
effective moment of inertia. *See* moment of inertia
effective span, 96
end blocks
　anchorage, 403, 473, 475–7
　post-tensioning systems, 475
　pretensioning systems, 473
　reinforcement, 473, 477
exposure classifications, 4, 8

fire resistance design, 4, 7, 9, 35, 39, 81, 94, 297, 437–8
flange-beams, 67, 69–70, 72–3
flanged beams, 61, 63, 100, 158, 224
flat plates
　definition of, 186–7
　multistorey, 236, 249
　reinforcement detailing for, 221
　with openings, 236
　with studs, 236
flat slabs
　definition of, 186–7
　with drop panels, 194
　without drop panels, 194, 248
footings
　asymmetrical, 312, 318–19, 325
　column, 311, 325, 332, 339, 344
　concentrically loaded, 317
　definition of, 311
　pad, 325, 328
　subjected to biaxial bending, 325
　symmetrical, 312, 318
　under eccentric loads, 317
　wall, 311, 313, 316, 318, 325, 328

grades of concrete, 13–14
grillage structures, 155, 187, 378

helices, 179, 287, 477
hooks, 173, 176, 182
　development length of, 177

idealised frames
　analysis of, 188, 216, 222–3, 226, 242
　definition of, 216, 222
　distribution of slab moments in, 225
　member rigidities of, 224
　thickness of, 223–4, 242
interaction diagrams (for columns), 255, 263, 266, 286, 289

lapped splices, 178–80
　development length of, 178
L-beams, 64, 70, 106, 112
limit state design philosophy, 3
load combinations, 5, 7, 100, 210, 224, 238, 241, 359, 361–2
load factors
　combination, 5
　dead, 315, 320
　earthquake combination, 6
　live, 6
　long-term, 6
　short-term, 7
loss of prestress, 416, 419, 428–9

middle strips. *See* slabs
minimum bending reinforcement
   for beams, 35
   for slabs, 192
   for walls, 303
minimum shear reinforcement, 132
minimum torsional reinforcement, 160, 230
mix design of concrete, 13
modulus of elasticity
   concrete, 14
   steel, 17
moment coefficients
   for one-way slabs, 191, 195
   for two-way slabs, 188, 202
moment magnifiers (for columns), 257, 283, 289, 301
moment of inertia
   Branson's formula for, 99
   effective, 98–100
   equivalent, 109
   for fully cracked sections, 99
   gross (of uncracked sections), 98, 101, 224, 407, 420
moment redistribution, 189
multibox system, 65
multispan two-way slabs
   bending moments in, 219
   definition of, 216
   layout, 207
   reinforcement detailing for, 221

neutral axis parameter, 23–4, 99, 463, 484
neutral axis position, 52, 66, 75, 99, 123, 260, 483

one-way slabs
   definition of, 185
   reinforcement detailing for, 192, 200
   reinforcement for, 192
   simplified analysis of, 190, 192
orthotropic slab, 187

passive earth pressure coefficient, 348
pile caps, 311, 339–40, 342–4
plastic centre, 257–8, 275
Poisson's ratio for concrete, 15

prestressed concrete
   analysis and design, 403, 431
   beams, 403, 407, 414, 421–2, 437, 443, 458, 460
   flat plates, 236
   fully prestressed, 403, 414, 431
   maximum permissible span, 403
   partially prestressed, 403, 409, 414, 437, 460
   precast girders, 413, 488
   principle, 406
prestressing
   forces, 414, 438–9, 451, 473
   jacks, 408, 416
   non-engineering examples of, 404
   post-tensioned, 407–8
   pretensioned, 407–8
   tendons, 411, 440, 473
punching shear
   basic strength, 227
   definition of, 226
   for column footings, 336
   in two-way slabs, 220
   of pile caps, 342
   strength of slabs, 227–8
   and unbalanced moment, 220

reinforcement
   bar areas, 15
   bursting, 389, 395, 398
   Class L, 15, 27, 190–1
   Class N, 15, 17, 27–9, 45, 71, 190, 203, 238
   for end blocks. *See* end blocks
   for shear. *See* shear
   for torsion. *See* torsion
   for walls. *See* walls
   ratios. *See* steel ratios
   spacing. *See* spacing of reinforcements
   splicing of, 178–80
   spread of, 31
retaining walls, xxxiv–xxxv, 311, 345–7, 353, 372

serviceability design, 4, 7, 65, 81, 94, 97, 100, 102, 111, 437
shear
   critical section for, 302, 328, 331
   failure modes, 125

shear (cont.)
  inclined reinforcement for, 127, 133, 160
  longitudinal, 14, 123, 138–9, 141, 157, 206
  mechanism of resistance, 127
  reinforcement, 125, 127–9, 132–3, 139, 160, 335, 371, 373, 387
  strength of beams, 130
  stress formula, 123
  transverse, 141, 157–60, 326, 340, 344
shear heads, 227–9, 233
shear planes, 138–41
shrinkage
  effects and reinforcement, 199, 209–10, 222, 304, 317, 324, 372
  of concrete, 15, 104, 107, 193, 416
singly reinforced beams. *See* beams
slabs
  column strip, 217, 219, 223, 225, 242, 244–5, 249
  design strips, 217–18
  middle strips, 217, 223, 225, 237, 242–3, 246–7
  minimum thickness for, 209
  serviceability design of, 193, 209
  See also flat plates; flat slabs; multispan two-way slabs; one-way slabs; two-way slabs
soil bearing capacity, 314, 325, 327, 357
spacing of reinforcements
  and crack control, 192
  maximum, 11, 210, 323
  minimum, 9, 192
spalling stresses, 473, 476, 480
spandrel beams
  definition of, 231
  design of, 228, 231–2
  minimum reinforcement for, 232
  reinforcement detailing for, 232
squash points, 262, 268
steel ratios
  for balanced beam sections, 24
  maximum for beams, 25
  maximum for columns, 287
  minimum for beams, 35
  minimum for columns, 287
  minimum for slabs, 204
stress blocks, 20–2, 26, 66, 69, 75, 139, 141, 164, 309, 458, 476

stress–strain curves
  for concrete, 14, 20, 22
  for steel, 17, 20
strut-and-tie modelling, 303, 377–9, 384–7, 392

T-beams
  criteria for, 66
  definition of, 61
  doubly reinforced, 70–1
  effective flange width formulas for, 63, 69
tension failure
  of beams, 23, 41, 80, 463
  of columns, 261, 264–5, 267, 280, 460
thermal expansion of concrete coefficient, 15
thermal expansion of steel coefficient, 17
torsion
  cracks, 155
  maximum allowed in beams, 158
  modulus, 158
  reinforcement, 155, 158, 160–1, 163, 230–1
torsion strips
  definition of, 229
  design of, 228–9
  minimum reinforcement for, 230
  reinforcement detailing for, 232
twisting moments, 155, 188, 208
two-way slabs
  bending moments in, 201, 204, 206, 219, 224
  distribution of loads in, 206
  edge conditions for, 202, 204
  moment coefficients, 202
  reinforcement detailing for, 206
  simplified analysis of, 201
  supported on four sides, 186, 188–9, 201, 203, 209

ultimate strain
  of concrete, 20
  of steel, 17, 20
ultimate strength
  definition of, 19
  theory, 19–20
ultimate stress blocks. *See* stress blocks
unit weights, 18, 315, 328, 354, 359–60

vibrations, 94, 437

walls
    axial strength of, 298
    capacity reduction factor for, 298
    column design method for, 297–8, 301
    definition of, 296
    eccentricity parameter, 301
    in one-way action, 298, 300
    in two-way action, 298, 301
    reinforcement, 299
    shear strength of, 302
    simplified method, 297–8, 305
    slenderness ratios for, 299, 301
    with openings, 299
web-beams, 67–70, 72, 74
welded mesh, 173, 179
wires, 81, 179, 434–6, 445, 458

yield criteria for reinforcement.
    See criteria for yielding of reinforcement
yield line analysis, 189
yield strength of steel, 15, 140, 230